MILLETS IN SUSTAINABLE AGRICULTURE

Genetic Improvement, Nutritional Quality, Cultivation, and Integrated Pest–Disease Management

MILLETS IN SUSTAINABLE AGRICULTURE

Genetic Improvement, Nutritional Quality, Cultivation, and Integrated Pest–Disease Management

Roshni Vijayan, PhD
A. John Joel, PhD
R. Ambika Rajendran, PhD
Adithya P. Balakrishnan, PhD
Editors

AAP | APPLE ACADEMIC PRESS

First edition published 2027

Apple Academic Press Inc.
1265 Goldenrod Circle, NE,
Palm Bay, FL 32905 USA

760 Laurentian Drive, Unit 19,
Burlington, ON L7N 0A4, Canada

CRC Press
2385 NW Executive Center Drive,
Suite 320, Boca Raton FL 33431

4 Park Square, Milton Park,
Abingdon, Oxon, OX14 4RN, UK

Library and Archives Canada Cataloguing in Publication

CIP data on file with Canada Library and Archives

Library of Congress Cataloging-in-Publication Data

CIP data on file with US Library of Congress

ISBN: 978-1-77964-232-5 (hbk)
ISBN: 978-1-77964-233-2 (ebk)

DOI: 10.1201/9781779642332

About the Editors

Roshni Vijayan, PhD
Assistant Professor, Division of Pulses, Regional Agricultural Research Station–Pattambi, Kerala Agricultural University, India.

Roshni Vijayan, PhD, is an Indian plant breeder currently working as an Assistant Professor at the Division of Pulses, Regional Agricultural Research Station–Pattambi, under Kerala Agricultural University. She is actively engaged in pulse breeding research in coordination with the All India Coordinated Research Project (AICRP) on Kharif pulses. She earned her undergraduate degree in Agricultural Sciences from Kerala Agricultural University. She holds a Master's degree in Plant Breeding and Genetics (Agriculture) from Tamil Nadu Agricultural University, India, and also earned a Master's degree in International Agriculture and Life Sciences with a specialization in Plant Breeding and Genetics from Cornell University. She later pursued her doctoral studies in Plant Breeding and Genetics (Agriculture) at the Kerala Agricultural University, India, with research focused on molecular profiling of fertility restorer genes in traditional and high yielding lines of rice for the development of CGMS red rice hybrids tailored to the agro-climatic conditions of Kerala. Dr. Roshni is an experienced pulse breeder combining conventional and molecular breeding expertise with computational biology to advance data driven crop improvement. In addition to her primary focus on pulses, her research and breeding experience also extends to maize and rice. She has authored several research articles and book chapters and has edited two books.

A. John Joel, PhD
Professor, Center for Plant Molecular Biology and Biotechnology, Tamil Nadu Agricultural University, India.

A. John Joel, PhD, is a dynamic Indian biotechnologist and currently serves as a Professor at the Center for Plant Molecular Biology and Biotechnology, Tamil Nadu Agricultural University, India. He has extensive experience in plant breeding and genetics, molecular biology and biotechnology, and plant genetic resources. Over the course of his nearly 35-year career, he has made distinguished contributions to the improvement of a wide range of crops, including rice, maize, pigeon pea, cassava, and others. He has mentored numerous doctoral students and is highly regarded for his expertise in quantitative genetics, which greatly benefits students pursuing plant breeding.

Dr. Joel has received several prestigious honors, including the International Development Research Center Fellowship (Canada), the Rockefeller Postdoctoral Fellowship (United States), and the Sri Ratan Tata Postdoctoral Fellowship (India). He has published approximately 110 journal articles, along with one book chapter, one review article, and five conference proceedings. His work is frequently cited and widely recognized in the scientific community.

R. Ambika Rajendran, PhD

Senior Scientist in plant breeding, Division of Genetics at the ICAR–Indian Agricultural Research Institute, Pusa Campus, New Delhi, India.

R. Ambika Rajendran, PhD, is a Senior Scientist in plant breeding in the Division of Genetics at the ICAR–Indian Agricultural Research Institute, Pusa Campus, New Delhi. She has been working for 18 years as a research scholar and scientist. Her current research focuses on abiotic stress tolerance and doubled haploidy in soybean. Her previous experience includes maize breeding for high-oil baby corn. Her significant research contributions include heterotic grouping and patterning of Quality Protein Maize using molecular markers, *in vitro* mutagenesis in rice, identification of genetic stocks with high oil content and resistance to post-*Fusarium* stalk rot in maize, and identification of tolerant sources for pre-germination anaerobic stress in soybean. Dr. Rajendran has edited four books: *Genetics and Breeding in Pulses, Handbook of Cereals, Soybean: History, Taxonomy, Genetics, Breeding, Disease and Biotechnology, and Phytohormones in Abiotic Stress*. She has also contributed approximately 11 book chapters and 29 research articles to the scientific community.

Adithya P. Balakrishnan, PhD

Adithya P. Balakrishnan, PhD, holds a Bachelor of Science in Agriculture from Mahatma Phule Krishi Vidyapeeth, Rahuri, and a Master's degree in Genetics and Plant Breeding from Sardarkrushinagar Dantiwada Agricultural University, Gujarat, where she was honored with the Chancellor's Gold Medal. She has published two research articles based on her MSc research, which focused on the genetic analysis of the medicinal plant Ashwagandha. As a DST-INSPIRE Fellow, she pursued her doctorate at the Division of Genetics, ICAR–Indian Agricultural Research Institute, Pusa Campus, New Delhi. Her doctoral research centers on genetic improvement for wheat grain quality, while her broader research interests encompass molecular breeding and quality enhancement.

Contents

Contributors

U. Amala
ICAR–National Bureau of Agricultural Insect Resources (ICAR–NBAIR), Bangalore, Karnataka, India

Padhiyil Amrutha
School of Agricultural Sciences, Karunya Institute of Technology and Sciences, Coimbatore, Tamil Nadu, India

N. Ananda
ICAR–AICRP on Linseed, Main Agricultural Research Station, University of Agricultural Sciences, Raichur, Karnataka, India

V. A. Anjali
University of Horticultural Sciences, Bagalkot, Karnataka, India

Charan Babu Ankela
ICAR–Central Research Institute for Dryland Agriculture, Santhoshnagar, Hyderabad, Telangana, India

N. Anuradha
Agricultural Research Station, Acharya N. G. Ranga Agricultural University, Vizianagaram, Andhra Pradesh, India

Adithya P. Balakrishnan
Division of Genetics, ICAR–Indian Agricultural Research Institute, Pusa, New Delhi, Delhi, India

Purabi Banerjee
ICAR–Central Research Institute for Dryland Agriculture, Hyderabad, Telangana, India

Sujata Bhat
Project Coordinating Unit, ICAR–AICRP on Small Millets, University of Agricultural Sciences, GKVK, Bangalore, Karnataka, India

Sai Bhavana
Project Coordinating Unit, ICAR–AICRP on Small Millets, University of Agricultural Sciences, GKVK, Bangalore, Karnataka, India

Senthilraja Chinnaiah
Texas A&M AgriLife Research, Amarillo, Texas, USA

Farooqkhan
Project Coordinating Unit, ICAR–AICRP on Small Millets, University of Agricultural Sciences, GKVK, Bangalore, Karnataka, India

Neethu Francis
School of Agricultural Sciences, Karunya Institute of Technology and Sciences, Karunya Nagar, Coimbatore, Tamil Nadu, India

Prabhu C. Ganiger
Project Coordinating Unit, ICAR–AICRP on Small Millets, University of Agricultural Sciences, GKVK, Bangalore, Karnataka, India

Ikkurti Gopinath
Division of Genetics, ICAR–Indian Agricultural Research Institute, Pusa, New Delhi, Delhi, India

K. A. Gopinath
ICAR–Central Research Institute for Dryland Agriculture, Hyderabad, Telangana, India

Mahalingam Govindaraj
International Crops Research Institute for the Semi-Arid Tropics, Patancheru, Hyderabad, Telangana, India

Karthika Gunasekaran
International Crops Research Institute for the Semi-Arid Tropics, Patancheru, Hyderabad, Telangana, India

Gopika P. Hassan
Department of Community Science, College of Agriculture, Kerala Agricultural University, Vellayani, Thiruvananthapuram, Kerala, India

Manjunath Hubballi
University of Horticultural Sciences, Bagalkot, Karnataka, India

S. M. Indhu
Department of Plant Biotechnology, Center of Plant Molecular Biology and Biotechnology, Tamil Nadu Agricultural University, Coimbatore, Tamil Nadu, India

Jayakumar Jaganathan
International Crops Research Institute for the Semi-Arid Tropics, Patancheru, Hyderabad, Telangana, India

R. Jaishreepriyanka
Department of Millets, Center for Plant Breeding and Genetics, Tamil Nadu Agricultural University, Coimbatore, Tamil Nadu, India

Atul Jayapal
Department of Agronomy, College of Agriculture, Kerala Agricultural University, Vellayani, Kerala, India

A. John Joel
Department of Bioinformatics, Center of Plant Molecular Biology and Biotechnology, Tamil Nadu Agricultural University, Coimbatore, Tamil Nadu, India

Lisha Jose Kappen
Department of Agronomy, College of Agriculture, Kerala Agricultural University, Vellayani, Kerala, India

T. V. Krishna
ICAR–AICRP on Kharif Pulses, Zonal Agricultural Research Station, University of Agricultural Sciences, GKVK, Bangalore, Karnataka, India

U. Krishnaja
Department of Community Science, College of Agriculture, Kerala Agricultural University, Vellayani, Thiruvananthapuram, Kerala, India

V. Krishnasree
Department of Community Science, College of Agriculture, Kerala Agricultural University, Padannakkad, Kerala, India

V. Visha Kumari
ICAR–Central Research Institute for Dryland Agriculture, Hyderabad, Telangana, India

Christopher R. Little
Kansas State University, Manhattan, Kansas, USA

Ancy G. Martin
Department of Agronomy, College of Agriculture, Kerala Agricultural University, Vellayani, Kerala, India

B. Mohanapriya
Department of Millets, Center for Plant Breeding and Genetics, Tamil Nadu Agricultural University, Coimbatore, Tamil Nadu, India

Vishnu Varthini Nachimuthu
International Crops Research Institute for the Semi-Arid Tropics, Patancheru, Hyderabad, Telangana, India

T. E. Nagaraja
Project Coordinating Unit, ICAR–AICRP on Small Millets, University of Agricultural Sciences, GKVK, Bangalore, Karnataka, India

R. Harischandra Naik
University of Horticultural Sciences, Bagalkot, Karnataka, India

C. Nandini
Zonal Agricultural and Horticultural Research Station, Babbur Farm, Hiriyur, KSNUAHS, Karnataka, India

C. Nandini
Regional Agricultural Research Station-Central Zone, Kerala Agricultural University, Pattambi, Palakkad, Kerala, India

M. S. Niji
College of Agriculture, Kerala Agricultural University, Vellayani, Thiruvananthapuram, Kerala, India

P. R. Nithya
Regional Agricultural Research Station, Kerala Agricultural University, Vellayani, Thiruvananthapuram, Kerala, India

K. B. Palanna
Project Coordinating Unit, ICAR–AICRP on Small Millets, University of Agricultural Sciences, GKVK, Bangalore, Karnataka, India

T. S. S. K. Patro
Agricultural Research Station, Acharya N. G. Ranga Agricultural University, Vizianagaram, Andhra Pradesh, India

T. Tharana Ponach
Project Coordinating Unit, ICAR–AICRP on Small Millets, University of Agricultural Sciences, GKVK, Bangalore, Karnataka, India

Govintharaj Ponnaiah
International Crops Research Institute for the Semi-Arid Tropics, Patancheru, Hyderabad, Telangana, India

A. V. S. Durga Prasad
Department of Plant Breeding and Genetics, Acharya N. G. Ranga Agricultural University, Vizianagaram, Andhra Pradesh, India

R. R. Rachana
ICAR–National Bureau of Agricultural Insect Resources (ICAR–NBAIR), Bangalore, Karnataka, India

Ambika Rajendran
Division of Genetics, ICAR–Indian Agricultural Research Institute, Pusa, New Delhi, Delhi, India

Y. Sandhya Rani
Agricultural Research Station, Acharya N. G. Ranga Agricultural University, Vizianagaram, Andhra Pradesh, India

T. Rathamani
Center for Plant Breeding and Genetics, Tamil Nadu Agricultural University, Coimbatore, Tamil Nadu, India

Swarna Ronanki
ICAR–Indian Institute of Millets Research, Hyderabad, Telangana, India

Govinda Rai Sarma
Division of Genetics, ICAR–Indian Agricultural Research Institute, Pusa, New Delhi, Delhi, India

P. Shanthi
Center for Plant Breeding and Genetics, Tamil Nadu Agricultural University, Coimbatore, Tamil Nadu, India

Sharifullah Sharifi
Afghanistan National Agricultural Sciences and Technology University (ANASTU), Kandahar, Afghanistan

P. Shruthi
Department of Community Science, College of Agriculture, Kerala Agricultural University, Vellayani, Thiruvananthapuram, Kerala, India

Ashok Singamsetti
Agricultural Research Station, Acharya N. G. Ranga Agricultural University, Vizianagaram, Andhra Pradesh, India, Institute of Agricultural Sciences, Banaras Hindu University, Varanasi, Uttar Pradesh, India

V. K. Singh
ICAR–Central Research Institute for Dryland Agriculture, Hyderabad, Telangana, India

J. Sowjanya
ICAR–Central Research Institute for Dryland Agriculture, Santhoshnagar, Hyderabad, Telangana, India

V. S. Susha
Department of Agronomy, College of Agriculture, Kerala Agricultural University, Vellayani, Kerala, India

U. Triveni
Agricultural Research Station, Acharya N. G. Ranga Agricultural University, Vizianagaram, Andhra Pradesh, India

M. Umadevi
Center for Plant Breeding and Genetics, Tamil Nadu Agricultural University, Coimbatore, Tamil Nadu, India

M. R. Umesh
ICAR–AICRP on Sunflower, Main Agricultural Research Station, University of Agricultural Sciences, Raichur, Karnataka, India

G. Venkatesh
ICAR–Central Research Institute for Dryland Agriculture, Hyderabad, Telangana, India

Roshni Vijayan
Assistant Professor, Division of Pulses, Regional Agricultural Research Station–Pattambi, Kerala Agricultural University, India.

P. Vinoth
Department of Plant Biotechnology, Center of Plant Molecular Biology and Biotechnology, Tamil Nadu Agricultural University, Coimbatore, Tamil Nadu, India

Sunaina Yadav
Division of Genetics, ICAR–Indian Agricultural Research Institute, Pusa, New Delhi, Delhi, India

Abbreviations

ASMIP	accelerated sorghum and millet improvement program
CEF	controlled environment facility
CMS	cytoplasmic-genetic male sterility
CP	crude protein
DM	downy mildew
EW	epicuticular waxes
FYM	farmyard manure
GABA	gamma-aminobutyric acid
GBS	genotyping by sequencing
GD	genetic distance
GFY	green forage yield
GS	genomic selection
GWAS	genome-wide association studies
HPLC	high-performance liquid chromatography
HPR	host plant resistance
ICAR	Indian Council of Agricultural Research
LT	lodging tolerance
MAS	marker-assisted selection
MTA	marker-trait association
NGS	next-generation sequencing
OPVs	open-pollinated varieties
PGR	plant genetic resource
PS	photoperiod-sensitive
QTL	quantitative trait loci
RNAi	RNA interference
SCA	specific combining ability
SDG	Sustainable Development Goal
SNP	single nucleotide polymorphism
TAA	Technicon Autoanalyzer
WGS	whole genome sequencing

Foreword

It is with great pleasure that I present this foreword for the book *Millets in Sustainable Agriculture: Genetic Improvement, Nutritional Quality, Cultivation, and Integrated Pest–Disease Management.* This comprehensive volume is a timely contribution to the global discourse on sustainable food production, resilience to climate change, and nutritional security. The increasing global population and the adverse effects of climate change on staple crop production necessitate a shift toward more resilient and sustainable crops, and millets have emerged as an outstanding choice in this regard.

Millets have historically been a staple in many traditional farming systems, particularly in arid and semi-arid regions, due to their remarkable adaptability to diverse climatic conditions. These crops can thrive in poor soils with minimal inputs, making them highly suitable for resource-limited farmers. Their ability to withstand harsh environmental conditions, including drought and heat stress, sets them apart as a crucial component in future food security strategies. Furthermore, millets offer significant health benefits, being rich in essential nutrients such as proteins, fiber, vitamins, and minerals. Their low glycemic index makes them particularly beneficial in addressing lifestyle diseases such as diabetes and obesity. As we navigate the challenges of food security in an era of climate uncertainty, millets offer a promising solution due to their drought tolerance, low input requirements, and exceptional nutritional profile.

This book serves as an essential resource, bringing together insights from leading experts on the genetic improvement, agronomic practices, nutritional benefits, and disease management strategies of these climate-resilient crops. The knowledge compiled within these pages will be invaluable for researchers, farmers, and policymakers alike, guiding efforts toward the broader adoption and integration of millets into mainstream agriculture. Additionally, the emphasis on disease management will help address one of the key challenges faced in millet production, ensuring higher productivity and quality.

Kerala Agricultural University has long been committed to advancing research and development in sustainable agriculture, particularly in the conservation and promotion of underutilized crops such as millets. Our institution has been actively engaged in millet research, focusing on breeding programs, sustainable agronomic practices, and value addition to enhance their commercial viability. The diverse expertise compiled in this book aligns with our vision to enhance agricultural resilience and improve the livelihoods of farmers, particularly smallholders, who stand to benefit the most from millet cultivation.

I commend the editors and contributors for their meticulous efforts in compiling this valuable reference, which will undoubtedly serve researchers, policymakers, and practitioners in their pursuit of a more sustainable and nutrition-secure future. I am confident that this book will inspire further research and practical innovations in the field of millet cultivation and utilization, ultimately contributing to a more resilient and diversified agricultural landscape.

—Dr. B. Ashok

Preface

Millets, belonging to the family *Poaceae*, are low-yielding yet nutrient-rich crops that hold significant agronomic and economic importance. Often referred to as "orphan crops," millets are ranked second to primary staple cereals. However, they have gained increasing attention for their roles in supporting human health, enhancing nutrition, promoting climate resilience, and offering agronomic advantages. Traditionally consumed as staple crops across various regions of the world, millets have re-emerged as a vital food source in the context of global challenges. Climate change and unprecedented disruptions such as the outbreak of Coronavirus Disease 2019 (COVID-19) have had severe impacts on agriculture, economies, and livelihoods. In this scenario, crops like millets offer a promising solution for addressing issues related to human health, nutritional security, and environmental sustainability. Rich in health-promoting phytochemicals such as polyphenols, lignans, phytosterols, phytoestrogens, and phytocyanins, millets function as antioxidants, immune modulators, and detoxifying agents. These compounds help protect against age-related degenerative diseases such as cardiovascular disorders, diabetes, cancer, and other lifestyle-related conditions. Millets are also an excellent source of essential nutrients, including vitamins, minerals, and essential fatty acids, and contribute to the prevention of both degenerative and nutritional deficiency diseases. Ranked sixth among the world's major cereal grains, millets are annual C_4 grasses, with foxtail millet serving as a model C_4 cereal. Beyond food, millets contribute to feed and fuel needs and show potential for biofuel production. As gluten-free grains, millets serve as a reliable dietary option for individuals with celiac disease. Moreover, they have been found to regulate the glycaemic index more effectively in individuals with impaired glucose tolerance.

Small-seeded millets are produced worldwide for use as feed, food, fuel, and forage. Millets serve as a nutritional source for both animals and humans. The most commonly consumed millets include sorghum, pearl millet, finger millet, foxtail millet, common millet, little millet, barnyard millet, and kodo millet. Except for pearl millet (*Pennisetum glaucum* (L.) R.Br.) and sorghum (*Sorghum bicolor* (L.) Moench), which are classified as major millets, the rest are referred to as small millets. Foxtail millet (*Setaria italica* (L.) P. Beauv.) is considered one of the earliest domesticated crops, and green foxtail millet (*Setaria viridis* (L.) P. Beauv.) is presumed to be its wild progenitor. As demand for millets grows, efforts by the United Nations and countries like India—such as the declaration of 2023 as the International Year of Millets—are creating increased business opportunities for entrepreneurs and enhancing the employability of farmers, especially women. There is an urgent need to strengthen the connection between all stakeholders in the value chain, from production to consumption, particularly linking smallholder dryland farmers with markets and consumers through reliable platforms. To achieve this, millets must be widely promoted among all farmers by ensuring quality-assured yields. Millet crop improvement programmes must be enhanced, and a diverse range of high-yielding cultivars should be made available to farmers. Agronomic practices should be updated using modern, eco-friendly techniques to ensure marketable yields. Similarly, pest and disease management practices, along with good and organic cultivation methods, need to be widely disseminated for easier adoption by farmers. It is equally important to raise public awareness about the health and nutritional benefits of millet consumption to create sustainable demand for millet-based products.

The book emphasizes current aspects of taxonomy and domestication of all millet species and aims to provide comprehensive information on genomics as well as conventional and modern breeding techniques in the section on genetic improvement. The nutritional and nutraceutical importance of each millet species, along with their fortification needs, is thoroughly presented in the section on nutritional quality enhancement. A detailed discussion on production, production challenges, and modern agronomic techniques for each millet is compiled in the section on improved cultivation practices. Finally, the current integrated pest and disease management practices are covered in a dedicated section. This compilation offers a holistic approach to the broad areas of millet research, cultivation, and consumption, which will ultimately help pave the way for a millet revolution.

We are grateful to all our contributors for sharing their knowledge and integrating their expertise into well-structured and informative chapters. We sincerely appreciate their suggestions and contributions. We also extend our thanks to the team at Apple Academic Press (CRC Press, Taylor & Francis Group) for their cooperation.

INTRODUCTION

Millets of the Semi-Arid Tropics: Key Crops for the New Green Revolution

ROSHNI VIJAYAN,[1] GOVINDA RAI SARMA,[2,*] AMBIKA RAJENDRAN,[2] A. JOHN JOEL,[3] ADITHYA P. BALAKRISHNAN,[2] M. UMADEVI,[4] P. SHANTHI,[4] and T. RATHAMANI[4]

[1]*Division of Pulses, Regional Agricultural Research Station–Pattambi, Kerala Agricultural University, India.*

[2]*Division of Genetics, ICAR–Indian Agricultural Research Institute, Pusa, New Delhi, Delhi, India*

[3]*Department of Bioinformatics, Center of Plant Molecular Biology and Biotechnology, Tamil Nadu Agricultural University, Coimbatore, Tamil Nadu, India*

[4]*Center for Plant Breeding and Genetics, Tamil Nadu Agricultural University, Coimbatore, Tamil Nadu, India*

*Corresponding author

MILLETS: THE MIRACLE CROPS

The global population is expected to rise drastically to 9.8 billion people by 2050. Cereals such as wheat, rice, and maize comprise the major crops used to meet the staple food requirements of the increasing population. However, these crops are highly resource-intensive and comparatively less nutritious in various ways (Dobermann et al., 2022; Fan et al., 2008; Welch et al., 2013). To satisfy the growing demands of the world's dietary and nutritional needs, the scientific community has a responsibility to address both the quantity and quality of food production. Simultaneously, the world faces the challenges of impending climate change, which calls for sustainable, resource-efficient crop production methods and environmentally friendly crops with a low carbon footprint. Millets are such resource-efficient, climate-resilient, and nutrient-dense crops that surpass staple crops in the long run (Kumari et al., 2024; Layek et al., 2023). Millets are, therefore, essential to the new green revolution that the world requires.

Millets include sorghum and pearl millet (major millets), and *tef*, *Job's tears*, finger, foxtail, barnyard, browntop, proso, kodo, little, and *fonio* millets (minor millets). These are low-resource, drought-tolerant crops suited to marginal and impoverished areas with sporadic and minimal rainfall (Karthika and Govintharaj, 2021). In the global context,

Millets in Sustainable Agriculture: Genetic Improvement, Nutritional Quality, Cultivation, and Integrated Pest–Disease Management.
Roshni Vijayan, A. John Joel, R. Ambika Rajendran, and Adithya P. Balakrishnan (Eds.)

millet farming has declined despite their high adaptability to various climate niches (Food and Agriculture Organization [FAO], 2022). Recently, people have become more aware of healthy diets due to the rising incidence of lifestyle-related diseases. Policymakers are also increasingly focused on climate-smart and adaptable food crops, prompted by frequent natural disasters and climatic uncertainties. To take advantage of the rising carbon dioxide (CO_2) levels in the atmosphere, millets—small-seeded grasses with C_4 photosynthetic pathways—are among the most promising choices. These grains are non-glutinous, high in fiber, nutritious, and easily digestible (Kumari et al., 2024).

According to Vetriventhan et al. (2020), millet crops are extremely nutritious and rich in essential minerals and antioxidant compounds. These grain cereals can be used in a variety of traditional medical practices due to their metabolic products, including anti-diabetic properties (Paschapur et al., 2021). Despite the many advantages of millets, their production and consumption are not increasing at the expected rate. These crops were marginalized from mainstream agriculture due to the expansion of various other cash and high-yielding food crops. For several millets, breeding efforts have been limited, leading to restricted cultivation beyond marginal areas. Consequently, the United Nations Organization (UNO) designated 2023 as the International Year of Millets to raise global awareness of these valuable crops.

PRODUCTION SCENARIO OF MILLET

Based on average data from 1961 to 2021, Asia and Africa are the two largest millet-growing regions (Figure 1). Despite the potential advantages of millets, the area under cultivation decreased dramatically during this period (FAO, 2022). Comparing ten-year averages (1961–1970 versus 2012–2021), the global decline in millet-growing area was 28%, with Asia experiencing a 59% decline (Figure 1; FAO, 2022), whereas Africa showed a 56% increase in overall cultivated area. In comparison to the first decade, the average production during the last ten years (2012–2021) increased by 5% globally and by 82% in Africa. Europe, by contrast, showed a 74% decline in millet production.

The Asian continent experienced a remarkable 111% gain in productivity, with only a 5% decline in production despite a significant reduction in cultivated land. This was followed by productivity increases in the Americas (38%), Europe (88%), and Asia (111%). The overall global increase in millet productivity was 48% (FAO, 2022). Statistics from 2021 show that Africa accounts for 60% of the total cultivated area, with Asia second at 37%. Nevertheless, Africa's share of global output is 40%, lagging behind Asia's 56%. This is due to Asia's higher productivity (1,457.6 kg/ha) compared to Europe (1,607.2 kg/ha) and Africa (651 kg/ha).

India stood as the largest producer of millet in 2021, accounting for 19% of global production, with 13.63 million hectares (Mha) of harvested land and an output of approximately 18 million tons (mt) (19% of total production) (FAO, 2022; Meena et al., 2021). Niger, Sudan, and Nigeria were the other main producers. Therefore, changes made to millet farming in India could have a significant impact on the global economy. Detailed information on millets will be covered in the following chapters; the current discussion is limited to providing a broad overview.

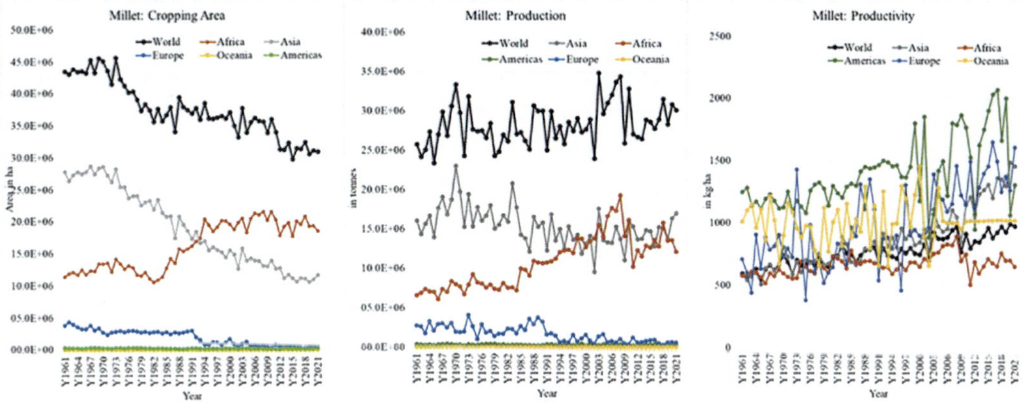

FIGURE 1 Global scenario of the millet cropping area, production, and productivity.

Source: Adapted from FAO, 2022

NUTRITIOUS AND NUTRACEUTICAL BENEFITS OF MILLETS

Given their well-documented nutritional and medical advantages, millets can be regarded as both nutritious and nutraceutical cereals. Due to the increasing health issues of obesity, diabetes, and malnutrition, biofortification of major cereals has been widely considered over the past ten years. Efforts are ongoing to reduce the complex effects of carbohydrates and enhance nutrient components like vitamins A and E, iron, zinc, and essential amino acids, and to impart nutraceutical properties through increased compounds like polyphenols (Hossain et al., 2022; Ibba et al., 2022; Senguttuvel et al., 2023). Millets are naturally fortified with all the vital qualities a modern diet should have to tackle the aforementioned issues. As millets are rich in several nutritional components, they are known as "nutri-cereals" and "smart food crops." They offer vitamins, antioxidants, proteins, minerals, low glycemic index, and are gluten-free, high in energy and dietary fiber. Millets are the healthiest and most beneficial food choice for the current sedentary lifestyle (Dwivedi et al., 2012; Kam et al., 2016; Vetriventhan et al., 2020). They possess 60–70% carbohydrates composed of non-starchy polysaccharides combined with significant amounts of bioactive substances like polyphenols, which improve fat absorption and increase the time taken for sugars to be released during digestion, hence lowering the glycemic index (Shivran, 2016). According to Goudar et al. (2023), millet flour has a glycemic index of 48–60, but food products made with wheat have an index of 75–110 (Shobana et al., 2007). Millets have vitamins and minerals such as riboflavin, niacin, and thiamine about 0.28–1.65 mg, 0.09–1.11 mg, and 0.15–0.60 mg per 100 g, respectively. The minerals include calcium (10–348 mg), potassium (113–408 mg), iron (2.2–17.6 mg), and zinc (32.6–60.7 mg) (values per 100 g), with many of these components being higher than those found in conventional cereals (Kumar et al., 2018).

Each millet has a unique nutritional advantage. In contrast to major cereals, finger millets are rich in calcium (>350 mg/100 g), followed by teff (159 mg/100 g). Additionally, millets such as proso, barnyard, foxtail, and Job's tears are rich in protein (>10%), while foxtail, little, and Job's tears are rich in fat (>4.0%) and are also good sources of oleic, linoleic, and palmitic acids, as well as tocopherols. Crude fiber (6.7–13.6%) is abundant in millets such as little, foxtail, barnyard, and fonio, while barnyard and little millets are rich in iron (9.3–18.6 mg per 100 g) (Dwivedi et al., 2012; Vetriventhan et al., 2020). Millets also supply vital amino acids like lysine in protein. The disadvantage of consuming millet is that it contains antinutritional, or "anti-nutrient," elements such as dietary fiber, polyphenols, tannins (0.61%), and phytates (0.48%) that block enzymes and can chelate metals (Chandrasekara and Shahidi, 2010; Paschapur et al., 2021).

Millet consumption has various dietary and digestive benefits due to its phenolics, mineral components, and dietary fibers. In addition, several health advantages of millet have been reported, such as antibacterial, anti-inflammatory, antihypertensive, antitumorigenic, antidiabetic, and antioxidant activities (Saleem et al., 2023). A diet of millet has a high content of magnesium, potassium, and plant lignin, which help lower blood pressure by having a vasodilator effect, thereby reducing the risk of cardiovascular disease. Additionally, the fiber content helps enhance HDL proteins and decrease LDL proteins to control cholesterol levels (Lee et al., 2010; Thayumanavan and Kumari, 1997). Similarly, a low glycemic index and easy-to-digest qualities help millet to prevent type II diabetes. Moreover, the increased magnesium content in millets boosts insulin levels and reduces blood sugar, as this element is a cofactor of insulin activity (Saleh et al., 2013). The dietary fibers and phenolics help prevent gastrointestinal tract issues such as bloating, constipation, and cramping. Millets like pearl millet food products help maintain an alkaline gut pH, unlike regular food products. This contributes to disease prevention and improved digestion (Chandrasekara and Shahidi, 2010; Muthamilarasan et al., 2016). The antioxidant properties of millets are attributed to their phenolic content. More than 50 phenolic compounds and their derivatives—namely, dehydrodiferulates, dehydrotriferulates, and phenolic acids—are abundantly present in millet grains and can be utilized as functional food ingredients and natural antioxidants (Devi et al., 2014; Mohamed et al., 2012). The phenolic compounds also possess antioxidant, metal-chelating, and metal-reducing properties. They are rich in curcumin, quercetin, ellagic acid, and catechins, which help the human body scavenge toxins and xenobiotics, thereby facilitating excretion and enzymatic neutralization (Tsao, 2010). Studies indicate that millet can significantly reduce the risk of esophageal, breast, and colon cancers (Shobana and Malleshi, 2007). The amount of dietary fiber consumed daily is linked to a lower risk of breast cancer when millet is included in the diet (Paschapur et al., 2021). The superior antioxidant and antitoxic chemicals found in millet may be responsible for its anticarcinogenic properties. Owing to their nutritive and nutraceutical qualities, millets are regarded as *Shree Anna* or nutri-cereals.

DIVERSITY IN MILLET CONSUMPTION

Millets are simple to incorporate into a regular diet. They can be eaten raw or cooked in recipes like *upma* and *kichadi*, especially millets like kodo, foxtail, and little; they can also be consumed directly as popped grains (sorghum, pearl, finger, and foxtail) and as germinated grains (finger, little, and kodo) (Malleshi et al., 1996; Paschapur et al., 2021). Consuming whole grains makes it easier to absorb the seed coat, which is high in phenolics, vitamins, minerals, and dietary fiber, offering more promising effects to humans than other cereals (Antony et al., 1996).

There is a wide variety of flour-based millet products, such as flatbreads like *kisra* and *injera* in Africa (Yetneberk et al., 2004); *roti* or *chapatis*, *idli*, and *dosa* in India (Murty and Kumar, 1995); *couscous* and millet semolina in Africa (Paschapur et al., 2021); and dumplings and/or steamed balls of finger millet and pearl millet, known as *mudde* in India and *dingwa* in Africa (Malleshi et al., 1996). Millets are also used to make non-alcoholic (*Oskikundu* and *Togwa*) and alcoholic (*opaque brews*, *Busa* or *Bouza*, *Tella* and *Katikalla*, *Chaang*, *Sulim*, *Burukutu*, *Dolo*, *talla*, or *Pito*) beverages (Bultosa and Taylor, 2004; Paschapur et al., 2021). Sufficient diversity and ease of preparation make millets simpler to modify for healthy eating. Fermentation of millets into beverages increases nutrient availability, decreases antinutritional factors, and improves protein and mineral bioavailability (Nkhata et al., 2018). Apart from starch, bioethanol, bio-coatings, and bioplastics, millets are also important in the industrial sector (Belton et al., 2006; Taylor et al., 2006).

CLIMATE RESILIENCE OF THE MILLET GROUP

Compared to other commercial crops and major cereals, millets are more resilient to harsh environments. It is abundantly clear from their distribution that millets thrive in semi-arid and arid areas that receive irregular and scanty rainfall of about 50 cm (Meena et al., 2021). The main crops in dry and dryland agricultural systems of the semi-arid tropics are pearl, sorghum, and finger millets. These crops have a greater ability to withstand adverse environments due to their unique morphological, physiological, biochemical, and molecular properties. Table 1 lists the characteristics of each millet that make it climate robust.

Millets have the following benefits that make them suitable for semi-arid environments:

1. **C4 Crops:** In the changing climate scenario, the C_4 photosynthetic system of millets gives them a fundamental advantage over traditional cereals. The RuBisCO units in C_4 plants perform well in warm and elevated CO_2 levels, ultimately providing higher water use efficiency and nitrogen use efficiency than C_3 crops (Shah et al., 2022). The C_4 system offers added benefits such as flexible biomass distribution and decreased hydraulic conductivity per unit leaf area, in addition to enhanced growth, development, and adaptability in warm environments (Sage and Zhou, 2011).

2. **Crop Maturity:** Millets take 85–100 days to reach maturity, whereas wheat and rice require 90–150 days, depending on the variety. Kodo millet varieties often mature in 100 to 140 days (Paschapur et al., 2021). Their short crop duration offers productive advantages to escape stress, facilitates repeated cropping, and enables contingency cropping (Maitra et al., 2022).

3. **Rhizosphere Microbe Interactions:** Each millet possesses a unique microbial association with its roots, affecting nutrient uptake, disease suppression, root and shoot growth, and tolerance-specific responses to abiotic stressors (Pudake et al., 2022). Since rhizoplane strains can precisely stimulate or repress certain genes in the host plant in response to external stimuli, these relationships are strain- and host-specific (Simmons et al., 2020; Wang et al., 2022).

4. **Physiological Advantages:** Morpho-physiological properties in millets—such as low plant stature, erect leaves with small leaf area, large radiation use efficiency, high leaf area index, thickened cell walls, and dense root systems—help keep them competitive under stress conditions.

According to several reports, millets have stress-induced responses that help them counteract the effects of stress. These responses include root elongation, increased ROS synthesis, antioxidant compounds, enzymes that scavenge ROS, osmolytes, and stress-related proteins (Ajithkumar and Panneerselvam, 2014; Lata et al., 2011a). The genetic constitution of each millet is linked to stress-specific tolerance. Millets are considered to be "climate-smart" crops due to their climate-hardy qualities. However, crop development programs are limited due to a lack of well-annotated genome sequences (Lata et al., 2011b; Paschapur et al., 2021; Satyavathi et al., 2019). Significant breeding programs are required to produce farmer-, consumer-, and industry-friendly millet varieties without sacrificing their climate-resilient features.

TABLE 1 Climate-Resilient Characteristics of Millet Crops

Crop	Duration (Days)	Climate resilient characteristics
Sorghum	100–110	Highly adaptable to heat, drought, very poor soils, and responsive to high-input management.
Pearl millet	80–95	
Finger millet	90–130	Adapted to a wide altitude range; moderately resistant to drought, heat, and humidity.
Foxtail millet	70–120	Adapted to high altitudes and low rainfall environments.
Proso millet	60–90	Short-duration crop adapted to high altitudes and low rainfall situations.
Little millet	70–110	Adapted to poor soils, low rainfall, and waterlogging to some extent.
Kodo millet	100–140	Very hardy crop with long duration; adapted to low rainfall and poor soils; responds well to improved agronomic practices.
Barnyard millet	45–60	Short-duration crop, well adapted to high altitudes and low rainfall conditions.

Source: Adapted from Paschapur et al. (2021).

BREEDING EFFORTS IN MILLETS

In light of crop improvement initiatives in other crops, millets are still in the early stages of improvement. Because of their increased popularity, larger grains, dual use, and higher yield, sorghum and pearl millet are the only crops included in most millet breeding programs. Due to factors like blossom size, pollination method, and sex differentiation, sorghum and pearl millet have hybrid cultivars as well. The inability of other millets to undergo hybridization is largely due to their autogamy and small flowers (Lydia et al., 2023; Vetriventhan et al., 2020). The primary breeding goals for millets are:

i. Resistance to lodging;
ii. Non-shattering nature;
iii. Bold seeds;
iv. Increased productivity;
v. Responsiveness to nutrients; and

vi. Resistance to economically important pests and diseases (Francis et al., 2022; Vetriventhan et al., 2020).

In addition to their innate resistance to drought and semi-arid climates, millet crops' genetic resources ought to be thoroughly investigated to increase their stress tolerance levels without compromising their exceptional agronomic features and yield capacity. To help breeders set breeding objectives, efforts have been focused on characterizing global germplasm over the past ten years (2011–2020) and providing validated core collections (Lata et al., 2011a; Nirmalakumari et al., 2010; Upadhyaya et al., 2011, 2014; Vetriventhan et al., 2020).

Genomic sequences for sorghum, Japanese barnyard and white fonio, pearl, foxtail, finger, proso, and tef are now available, which can accelerate the identification of QTLs and genes for traits such as quality, stress resistance, and agronomic performance (Vetriventhan et al., 2020; Wang et al., 2021; Zou et al., 2019). For stress tolerance and other features, many QTLs and genes in various millets are already accessible (Lydia Pramitha et al., 2023; Paschapur et al., 2021; Vetriventhan et al., 2020). To identify QTLs and genes, there is a need for scientific collaboration and funding directed toward omics research pipelines in millets. The genomes of the remaining millets—particularly kodo, tiny, and browntop millets—will likewise require significant scientific contributions.

CONCLUSION

Millets are a great option for resilient crops to tackle climate challenges due to their remarkable resistance to extreme environments. Millets are also well suited for conservation agriculture and zero-carbon output agriculture because of their efficient resource usage. Nevertheless, despite a significant rise in productivity, the area planted with millets is shrinking due to a lack of interventions from various stakeholders. These crops can be integrated into new cuisines owing to their adaptable qualities and diverse applications. Their nutritional, nutraceutical, and health benefits make them appealing to a wide range of contemporary consumers. To accelerate millet breeding and cultivation, the active engagement of policymakers and public–private stakeholders in millet research and development is essential. Millets have the potential to lead a new green revolution that is resilient to climate change, supporting the slogan "healthy food and a safe environment."

KEYWORDS

- **active engagement**
- **climate resilience**
- **millets**
- **miracle crops**
- **new green revolution**
- **policymakers**
- **stakeholders**

REFERENCES

Ajithkumar, I. P., & Panneerselvam, R. (2014). ROS scavenging system, osmotic maintenance, pigment and growth status of *Panicum sumatrense* Roth. under drought stress. *Cell Biochemistry and Biophysics, 68*(3), 587–595. https://doi.org/10.1007/s12013-013-9746-x (accessed on 7th November 2025).

Antony, U., Sripriya, G., & Chandra, T. S. (1996). Effect of fermentation on the primary nutrients in finger millet (*Eleusine coracana*). *Journal of Agricultural and Food Chemistry, 44*(9), 2616–2618. https://doi.org/10.1021/jf950787q (accessed on 7th November 2025).

Belton, P. S., Delgadillo, I., Halford, N. G., & Shewry, P. R. (2006). Kafirin structure and functionality. *Journal of Cereal Science, 44*(3), 272–286. https://doi.org/10.1016/j.jcs.2006.05.004 (accessed on 7th November 2025).

Bultosa, G., & Taylor, J. R. N. (2004). Paste and gel properties and *in vitro* digestibility of tef [*Eragrostis tef* (Zucc.) Trotter] starch. *Starke, 56*(1), 20–28. https://doi.org/10.1002/star.200200191 (accessed on 7th November 2025).

Chandrasekara, A., & Shahidi, F. (2010). Content of insoluble bound phenolics in millets and their contribution to antioxidant capacity. *Journal of Agricultural and Food Chemistry, 58*(11), 6706–6714. https://doi.org/10.1021/jf100868b (accessed on 7th November 2025).

Devi, P. B., Vijayabharathi, R., Sathyabama, S., Malleshi, N. G., & Priyadarisini, V. B. (2014). Health benefits of finger millet (*Eleusine coracana* L.) polyphenols and dietary fiber: A review. *Journal of Food Science and Technology, 51*(6), 1021–1040. https://doi.org/10.1007/s13197-011-0584-9 (accessed on 7th November 2025).

Dobermann, A., Bruulsema, T., Cakmak, I., Gerard, B., Majumdar, K., McLaughlin, M., Reidsma, P., Vanlauwe, B., Wollenberg, L., Zhang, F., & Zhang, X. (2022). Responsible plant nutrition: A new paradigm to support food system transformation. *Global Food Security, 33*(100636), 100636. https://doi.org/10.1016/j.gfs.2022.100636 (accessed on 7th November 2025).

Dwivedi, S. L., Upadhyaya, H. D., Senthilvel, S., Hash, C. T., Fukunaga, K., Diao, X., Santra, D., Baltensperge, D., & Prasad, M. (2012). Millets: Genetic and genomic resources. *Plant Breeding Reviews, 35*, 247–375.

Fan, M. S., Zhao, F. J., Fairweather-Tait, S. J., Poulton, P. R., Dunham, S. J., & McGrath, S. P. (2008). Evidence of decreasing mineral density in wheat grain over the last 160 years. *Journal of Trace Elements in Medicine and Biology, 22*(4), 315–324. https://doi.org/10.1016/j.jtemb.2008.07.002 (accessed on 7th November 2025).

Goudar, G., Manne, M., Sathisha, G. J., Sharma, P., Mokalla, T. R., Kumar, S. B., & Ziouzenkova, O. (2023). Phenolic, nutritional and molecular interaction study among different millet varieties. *Food Chemistry Advances, 2*(100150), 100150. https://doi.org/10.1016/j.focha.2022.100150 (accessed on 7th November 2025).

Hossain, F., Zunjare, R. U., Muthusamy, V., Bhat, J. S., Mehta, B. K., Sharma, D., Talukder, Z. A., Chhabra, R., Katral, A., Dutta, S., Chand, G., Bhatt, V., Mishra, S. J., Gain, N., Kasana, R., Ikkurti, G., & Duo, H. (2022). Biofortification of maize for nutritional security. In S. Kumar, H. K. Dikshit, G. P. Mishra, & A. Singh (Eds.), *Biofortification of Staple Crops* (pp. 147–174). Springer Singapore: Singapore.

Ibba, M. I., Gupta, O. P., Govindan, V., Johnson, A. A. T., Brinch-Pedersen, H., Nikolic, M., & Taleon, V. (2022). Editorial: Wheat biofortification to alleviate global malnutrition. *Frontiers in Nutrition, 9*. https://doi.org/10.3389/fnut.2022.1001443 (accessed on 7th November 2025).

Kam, J., Puranik, S., Yadav, R., Manwaring, H. R., Pierre, S., Srivastava, R. K., & Yadav, R. S. (2016). Dietary interventions for type 2 diabetes: How millet comes to help. *Frontiers in Plant Science, 7*. https://doi.org/10.3389/fpls.2016.01454 (accessed on 7th November 2025).

Karthika, G., & Govintharaj, P. (2022). Breeding climate-resilience crops for future agriculture. In A. Anjali, K. S. Arun, S. Chitra, & M. Maheswari (Eds.), *Climate Change and Crop Stress* (pp. 1–32). Elsevier.

Kumar, A., Tomer, V., Kaur, A., Kumar, V., & Gupta, K. (2018). Millets: A solution to agrarian and nutritional challenges. *Agriculture & Food Security, 7*(1). https://doi.org/10.1186/s40066-018-0183-3 (accessed on 7th November 2025).

Kumari, N., Prakash, A., Tiwari, P., Kumar, A., Ranjan, S., Ray, P., et al. (2024). Origin, diversity, floral biology, pollination, and genetics in foxtail millet. In S. Mishra, S. Kumar, & R. C. Srivastava (Eds.), *Genetic*

Improvement of Small Millets (pp. 351–364). Springer Nature: Singapore. https://doi.org/10.1007/978-981-99-7232-6_15 (accessed on 7th November 2025).

Kumari, S. K., & Thayumanavan, B. (1997). Comparative study of resistant starch from minor millets on intestinal responses, blood glucose, serum cholesterol and triglycerides in rats. *Journal of the Science of Food and Agriculture, 75*(3), 296–302. https://doi.org/10.1002/(sici)1097-0010(199711)75:3<296::aid-jsfa877>3.0.co;2-x (accessed on 7th November 2025).

Lata, C., Bhutty, S., Bahadur, R. P., Majee, M., & Prasad, M. (2011b). Association of an SNP in a novel DREB2-like gene SiDREB2 with stress tolerance in foxtail millet (*Setaria italica* (L.)). *Journal of Experimental Botany, 62*(10), 3387–3401. https://doi.org/10.1093/jxb/err016 (accessed on 7th November 2025).

Lata, C., Jha, S., Dixit, V., Sreenivasulu, N., & Prasad, M. (2011a). Differential antioxidative responses to dehydration-induced oxidative stress in core set of foxtail millet cultivars (*Setaria italica* (L.)). *Protoplasma, 248*(4), 817–828. https://doi.org/10.1007/s00709-010-0257-y (accessed on 7th November 2025).

Layek, J., Rangappa, K., Das, A., Ansari, M. A., Choudhary, S., Rajbonshi, N., Patra, S., Kumar, A., Mishra, V. K., Ravisankar, N., Kumar, S., Hazarika, S., Dutta, S. K., Babu, S., Tahasildar, M., & Shettigar, N. (2023). Evaluation of millets for physio-chemical and root morphological traits suitable for resilient farming and nutritional security in Eastern Himalayas. *Frontiers in Nutrition, 10.* https://doi.org/10.3389/fnut.2023.1198023 (accessed on 7th November 2025).

Lee, S. H., Chung, I. M., Cha, Y. S., & Park, Y. (2010). Millet consumption decreased serum concentration of triglyceride and C-reactive protein but not oxidative status in hyperlipidemic rats. *Nutrition Research, 30*(4), 290–296. https://doi.org/10.1016/j.nutres.2010.04.007 (accessed on 7th November 2025).

Lydia Pramitha, J., Ganesan, J., Francis, N., Rajasekharan, R., & Thinakaran, J. (2023). Revitalization of small millets for nutritional and food security by advanced genetics and genomics approaches. *Frontiers in Genetics, 13.* https://doi.org/10.3389/fgene.2022.1007552 (accessed on 7th November 2025).

Maitra, S., Praharaj, S., Hossain, A., Patro, T. S. S. K., Pramanick, B., Shankar, T., Pudake, R. N., Gitari, H. I., Palai, J. B., Sairam, M., Sagar, L., & Sahoo, U. (2022). Small millets: The next-generation smart crops in the modern era of climate change. In R. N. Pudake, A. U. Solanke, A. M. Sevanthi, & P. Rajendrakumar (Eds.), *Omics of Climate Resilient Small Millets* (pp. 1–25). Springer Nature Singapore: Singapore.

Malleshi, N. G., Hadimani, N. A., Chinnaswamy, R., & Klopfenstein, C. F. (1996). Physical and nutritional qualities of extruded weaning foods containing sorghum, pearl millet, or finger millet blended with mung beans and nonfat dried milk. *Plant Foods for Human Nutrition, 49*(3), 181–189. https://doi.org/10.1007/bf01093214 (accessed on 7th November 2025).

Meena, R. P., Joshi, D., & Bisht, J. K., & Kant, L. (2021). Global scenario of millets cultivation. In A. Kumar, M. K. Tripathi, D. Joshi, & V. Kumar (Eds.), *Millets and Millet Technology* (pp. 33–50). Springer Singapore: Singapore.

Mohamed, T. K., Issoufou, A., & Zhou, H. (2012). Antioxidant activity of fractionated foxtail millet protein hydrolysate. *International Food Research Journal, 19*(1), 207.

Murty, D. S., & Kumar, K. A. (1995). Traditional uses of sorghum and millets. In D. A. V. Dendy (Ed.), *Sorghum and Millets: Chemistry and Technology* (p. 221). American Association of Cereal Chemists: St Paul, MN.

Muthamilarasan, M., Dhaka, A., Yadav, R., & Prasad, M. (2016). Exploration of millet models for developing nutrient rich gramineous crops. *Plant Science, 242*, 89–97. https://doi.org/10.1016/j.plantsci.2015.08.023 (accessed on 7th November 2025).

N/A. (2012). *FAO–Statistical Year Book 2012: World Food and Agriculture.* Food & Agriculture Organization of the United Nations (FAO): Rome, Italy. https://www.fao.org/3/i2490e/i2490e00.htm (accessed on 7th November 2025).

N/A. (2022). *FAO–Statistical Year Book 2022. World Food and Agriculture.* Food & Agriculture Organization of the United Nations (FAO): Rome, Italy. https://www.fao.org/3/cc2211en/cc2211en.pdf (accessed on 7th November 2025).

Nirmalakumari, A., Salini, K., & Veerabadhiran, P. (2010). Morphological characterization and evaluation of little millet (*Panicum sumatrense* Roth. ex. Roem. and Schultz.) germplasm. *Electronic Journal of Plant Breeding, 1*(2), 148–155.

Nkhata, S. G., Ayua, E., Kamau, E. H., & Shingiro, J. B. (2018). Fermentation and germination improve nutritional value of cereals and legumes through activation of endogenous enzymes. *Food Science & Nutrition, 6*(8), 2446–2458. https://doi.org/10.1002/fsn3.846 (accessed on 7th November 2025).

Paschapur, A. U., Joshi, D., Mishra, K. K., Kant, L., Kumar, V., & Kumar, A. (2021). Millets for life: A brief introduction. In A. Kumar, M. K. Tripathi, D. Joshi, & V. Kumar (Eds.), *Millets and millet technology* (pp. 1–32). Springer Singapore: Singapore. https://doi.org/10.1007/978-981-16-0676-2_1 (accessed on 7th November 2025).

Pudake, R. N., Kumari, M., Sapkal, D. R., & Sharma, A. K. (2023). *Millet Rhizosphere*. Springer Nature: Cham, Switzerland.

Sage, R. F., & Zhu, X. G. (2011). Exploiting the engine of C4 photosynthesis. *Journal of Experimental Botany, 62*(9), 2989–3000. https://doi.org/10.1093/jxb/err179 (accessed on 7th November 2025).

Saleem, S., Mushtaq, N. U., Shah, W. H., Rasool, A., Hakeem, K. R., Seth, C. S., Tahir, I., & Rehman, R. U. (2023). Millets as smart future food with essential phytonutrients for promoting health. *Journal of Food Composition and Analysis, 124*(105669), 105669. https://doi.org/10.1016/j.jfca.2023.105669 (accessed on 7th November 2025).

Saleh, A. S. M., Zhang, Q., Chen, J., & Shen, Q. (2013). Millet grains: Nutritional quality, processing, and potential health benefits. *Comprehensive Reviews in Food Science and Food Safety, 12*(3), 281–295. https://doi.org/10.1111/1541-4337.12012 (accessed on 7th November 2025).

Satyavathi, C. T., Solanki, R. K., Kakani, R. K., Bharadwaj, C., Singhal, T., Padaria, J., Khandelwal, V., Srivastava, R., Tomar, R. S., & Iqubal, M. A. (2019). Genomics assisted breeding for abiotic stress tolerance in millets. In V. Rajpal, D. Sehgal, A. Kumar, & S. Raina (Eds.), *Genomics Assisted Breeding of Crops for Abiotic Stress Tolerance, Vol. II* (pp. 241–255). Springer International Publishing: Cham.

Senguttuvel, P., Padmavathi, S., Jasmine, N., Rao, S., C. N., N., Jaldhani, B., Beulah, G., Gobinath, R., Kumar, A., Sruthi, S., Shivani, R., R. M., S., & Govindaraj, M. (2023). Rice biofortification: Breeding and genomic approaches for genetic enhancement of grain zinc and iron contents. *Frontiers in Plant Science, 14*. https://doi.org/10.3389/fpls.2023.1138408 (accessed on 7th November 2025).

Shah, W. H., Saleem, S., Mushtaq, N. U., Rasool, A., Tahir, I., & Rehman, R. U. (2022). C4 and CAM plants with better resilience to environmental stresses. In A. Roychoudhury (Ed.), *Photosynthesis and Respiratory Cycles During Environmental Stress Response in Plants* (pp. 163–191). Apple Academic Press: Oakville, ON, Canada.

Shivran, A. C. (2016). Biofortification for nutrient-rich millets. In U. Singh, C. Praharaj, S. Singh, & N. Singh (Eds.), *Biofortification of Food Crops* (pp. 409–420). Springer India: New Delhi. https://doi.org/10.1007/978-81-322-2716-8_30 (accessed on 7th November 2025).

Shobana, S., & Malleshi, N. G. (2007). Preparation and functional properties of decorticated finger millet (*Eleusine coracana*). *Journal of Food Engineering, 79*(2), 529–538. https://doi.org/10.1016/j.jfoodeng.2006.01.076 (accessed on 7th November 2025).

Shobana, S., Usha Kumari, S. R., Malleshi, N. G., & Ali, S. Z. (2007). Glycemic response of rice, wheat and finger millet based diabetic food formulations in normoglycemic subjects. *International Journal of Food Sciences and Nutrition, 58*(5), 363–372. https://doi.org/10.1080/09637480701252229 (accessed on 7th November 2025).

Simmons, T., Styer, A. B., Pierroz, G., Gonçalves, A. P., Pasricha, R., Hazra, A. B., Bubner, P., & Coleman-Derr, D. (2020). Drought drives spatial variation in the millet root microbiome. *Frontiers in Plant Science, 11*. https://doi.org/10.3389/fpls.2020.00599 (accessed on 7th November 2025).

Taylor, J. R. N., Schober, T. J., & Bean, S. R. (2006). Novel food and non-food uses for sorghum and millets. *Journal of Cereal Science, 44*(3), 252–271. https://doi.org/10.1016/j.jcs.2006.06.009 (accessed on 7th November 2025).

Tsao, R. (2010). Chemistry and biochemistry of dietary polyphenols. *Nutrients, 2*(12), 1231–1246. https://doi.org/10.3390/nu2121231 (accessed on 7th November 2025).

Upadhyaya, H. D., Dwivedi, S. L., Singh, S. K., Singh, S., Vetriventhan, M., & Sharma, S. (2014). Forming core collections in barnyard, kodo, and little millets using morpho agronomic descriptors. *Crop Science, 54*(6), 2673–2682. https://doi.org/10.2135/cropsci2014.03.0221 (accessed on 7th November 2025).

Upadhyaya, H. D., Ravishankar, C. R., Narasimhudu, Y., Sarma, N. D. R. K., Singh, S. K., Varshney, S. K., Reddy, V. G., Singh, S., Parzies, H. K., Dwivedi, S. L., Nadaf, H. L., Sahrawat, K. L., & Gowda, C. L. L. (2011). Identification of trait-specific germplasm and developing a mini core collection for efficient use of foxtail millet genetic resources in crop improvement. *Field Crops Research, 124*(3), 459–467. https://doi.org/10.1016/j.fcr.2011.08.004 (accessed on 7th November 2025).

Vetriventhan, M., Azevedo, V. C. R., Upadhyaya, H. D., Nirmalakumari, A., Kane-Potaka, J., Anitha, S., Ceasar, S. A., Muthamilarasan, M., Bhat, B. V., Hariprasanna, K., Bellundagi, A., Cheruku, D., Backiyalakshmi, C., Santra, D., Vanniarajan, C., & Tonapi, V. A. (2020). Genetic and genomic resources, and breeding for accelerating improvement of small millets: Current status and future interventions. *Nucleus (Calcutta), 63*(3), 217–239. https://doi.org/10.1007/s13237-020-00322-3 (accessed on 7th November 2025).

Wang, X., Chen, S., Ma, X., Yssel, A. E. J., Chaluvadi, S. R., Johnson, M. S., Gangashetty, P., Hamidou, F., Sanogo, M. D., Zwaenepoel, A., Wallace, J., Van de Peer, Y., Bennetzen, J. L., & Van Deynze, A. (2021). Genome sequence and genetic diversity analysis of an under-domesticated orphan crop, white fonio (*Digitaria exilis*). *Gigascience, 10*(3). https://doi.org/10.1093/gigascience/giab013 (accessed on 7th November 2025).

Wang, Y., Wang, X., Sun, S., Jin, C., Su, J., Wei, J., Luo, X., Wen, J., Wei, T., Sahu, S. K., Zou, H., Chen, H., Mu, Z., Zhang, G., Liu, X., Xu, X., Gram, L., Yang, H., Wang, E., & Liu, H. (2022). GWAS, MWAS and mGWAS provide insights into precision agriculture based on genotype-dependent microbial effects in foxtail millet. *Nature Communications, 13*(1). https://doi.org/10.1038/s41467-022-33238-4 (accessed on 7th November 2025).

Welch, R. M., Graham, R., & Cakmak, I. (2013). Linking agricultural production practices to improving human nutrition and health. http://www.fao.org/3/a-as574e.pdf (accessed on 7th November 2025).

Yetneberk, S., de Kock, H. L., Rooney, L. W., & Taylor, J. R. N. (2004). Effects of sorghum cultivar on injera quality. *Cereal Chemistry, 81*(3), 314–321. https://doi.org/10.1094/cchem.2004.81.3.314 (accessed on 7th November 2025).

Zou, C., Li, L., Miki, D., Li, D., Tang, Q., Xiao, L., Rajput, S., Deng, P., Peng, L., Jia, W., Huang, R., Zhang, M., Sun, Y., Hu, J., Fu, X., Schnable, P. S., Chang, Y., Li, F., Zhang, H., Feng, B., Zhu, X., Liu, R., Schnable, J. C., Zhu, J. K., & Zhang, H. (2019). The genome of broomcorn millet. *Nature Communications, 10*(1). https://doi.org/10.1038/s41467-019-08409-5 (accessed on 7th November 2025).

PART I

Genetic Improvement

CHAPTER 1

Genetic Improvement in Sorghum

GOVINTHARAJ PONNAIAH,[1,*] KARTHIKA GUNASEKARAN,[1] VISHNU VARTHINI NACHIMUTHU,[1] ADITHYA P. BALAKRISHNAN,[2] and JAYAKUMAR JAGANATHAN[1]

[1]*International Crops Research Institute for the Semi-Arid Tropics, Patancheru, Hyderabad, Telangana, India*

[2]*Division of Genetics, ICAR–Indian Agricultural Research Institute, Pusa, New Delhi, Delhi, India*

*Corresponding author

1.1 INTRODUCTION

Sorghum (*Sorghum bicolor* (L.) (Moench)) is the most important food crop grown widely in the areas of Africa and Asia. Currently, this crop is grown in nearly 100 countries in the world (Hao et al., 2021; Upadhyaya et al., 2019). Being a C4 species with the ability to produce higher dry matter productivity, greater nitrogen and water use efficiency, and also the inherent ability to adapt to hot and dry agroecologies where other food crops could not perform well (Griebel et al., 2019; Hao et al., 2021), this crop is considered a nutritious cereal with 68% carbohydrates, 10% proteins, and micronutrients, as well as the presence of various chemical compounds having nutraceutical properties (Visarada and Aruna, 2019). The daily consumption of sorghum in sorghum-growing areas in India is high at 75 kg/year, and it meets >50% of the grain iron and zinc needs of low-income communities living in the country (Ashok Kumar et al., 2011; Ashok Kumar et al., 2013; Rao et al., 2006). Considering all these traits, this crop is designated as a climate-resilient crop to tackle current and future climatic changes (Hao et al., 2021; Ndlovu et al., 2021).

The average annual sorghum harvested area worldwide is about 40.9 million hectares over the last two decades (2001–2020). This crop accounts for almost 80% of the world's total sorghum harvest, which comes from just 10 countries, sharing 7.3 million ha from India, 6.8 million ha from Sudan, 6.1 million ha from Nigeria, 3.1 million ha from Niger, 2.5 million ha from the United States, 1.8 million ha from Mexico, 1.7 million ha from Burkina Faso, 1.6 million ha from Ethiopia, 1.2 million ha from Mali, and 0.7 million ha from Brazil (Khalifa and Eltahir, 2023). However, the average sorghum production per year is about 58.7 million t over the last two decades (2001 to 2020). Globally, just 10 countries cumulatively contributed almost 78.6% (46.1 million t) of the world's total

Millets in Sustainable Agriculture: Genetic Improvement, Nutritional Quality, Cultivation, and Integrated Pest–Disease Management. Roshni Vijayan, A. John Joel, R. Ambika Rajendran, and Adithya P. Balakrishnan (Eds.)

DOI: 10.1201/9781779642332-2

sorghum production. These countries are the United States, Nigeria, India, Mexico, Sudan, Ethiopia, Argentina, China, Brazil, and Australia, contributing about 10.1, 7.3, 6.2, 5.9, 4.2, 3.4, 2.8, 2.6, 1.9, and 1.8 million t, respectively, to global sorghum production (Khalifa and Eltahir, 2023).

Unlike other countries, sorghum is widely grown in two different seasons: the kharif (rainy) season (June/July, September/October) and the rabi (post-rainy) season (October, December/January) under limited irrigation and residual soil moisture conditions in India. As well as in other countries, sorghum is majorly cultivated for food and fodder usage in the drier regions of India (Maharashtra, Andhra Pradesh, Karnataka, and Tamil Nadu) and other regions receiving low rainfall (400–500 mm) per year (Pray and Nagarajan, 2009). Cultivar options significantly differ for these two seasons (Rana et al., 1997). The total cropped area is about 62% in kharif and 38% in rabi season during 1965–1966. However, these cropping areas shifted to 37% in kharif and 63% in rabi, respectively, during 2014–2015 (Charyulu et al., 2016). The lower productivity during the rabi season was attributed to the lack of high-yielding cultivars, higher seed rates, little or no use of chemical fertilizers, narrow spacing, and cultivation under residual moisture conditions (Sanjana and Patil, 2015). However, the grain quality is better in rabi than in the kharif season, which further supports that rabi crops are specifically used for consumption and kharif crops are mainly utilized for industrial and poultry purposes. However, cultivated areas of sorghum drastically declined from 18.24 Mha in 1966–1971 to 4.89 Mha in 2016–2020 based on a five-year average in India (Figure 1.1). Also, the total production declined from 9.38 m t to 4.40 m t, whereas the productivity has increased from 514 kg/ha to 902 kg/ha (Figure 1.2) during the same periods, which might be due to the adoption of improved production technologies by the farmers.

The public sector sorghum breeding programs started in the early 20s in India with the involvement of more state agricultural research centers (Tonapi et al., 2011). The Accelerated Sorghum and Millet Improvement Program (ASMIP) was established by the Indian Council of Agricultural Research (ICAR) in 1962 with the goal of hybrid breeding. Exactly seven years later, the All India Coordinated Research Project on Sorghum (AICRP-Sorghum) established 18 centers in 13 state agricultural universities (SAUs) covering 11 sorghum-growing states, with focused research activities on sorghum grain and forage improvement. Later, more centers were added to the AICRP system in the subsequent five-year plans. In 1987, the Hyderabad regional centre was renamed the National Research Center for Sorghum (NRCS), and the All India Coordinated Sorghum Improvement Project (AICSIP) was integrated with it later. Further, NRC-Sorghum was elevated to the Directorate of Sorghum Research (DSR) and the Indian Institute of Millets Research (IIMR) in 2009 and 2014, respectively, by ICAR to strengthen the basic and strategic research to enhance sorghum productivity, product utilization, and increased profitability. Currently, breeding research work on sorghum in India is mainly carried out at IIMR with AICRP on sorghum research centers and the International Crops Research Institute for Semi-arid Tropics (ICRISAT). The coordinated research activities by these research centers have led to the identification and release of 35 hybrids and 30 varieties of sorghum having different uses (Aruna and Cheruku, 2019). Over the past 30 years, breeding efforts have improved grain yield, fodder and forage, nutrient-dense grains, and bioenergy production traits.

This book chapter presents an overview of the methods and techniques used in sorghum genetic improvement. We also look at the possibility of accelerating genetic advancements even further to fulfill the increased demand for sorghum and increase the profitability of its cultivation.

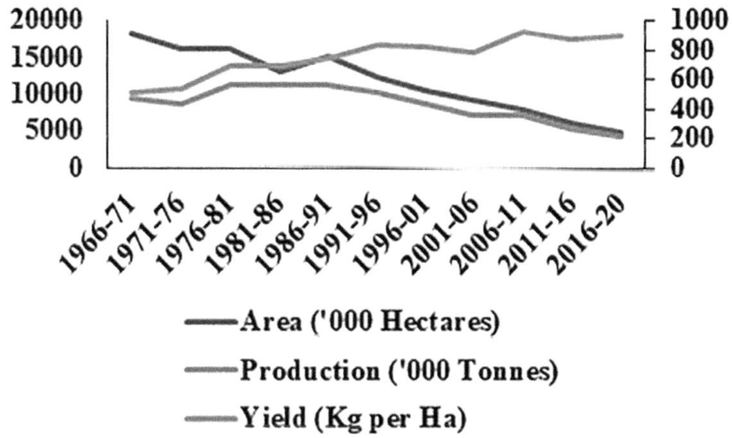

FIGURE 1.1 Sorghum area, production, and productivity in India since 1966 (based on 5-year average).

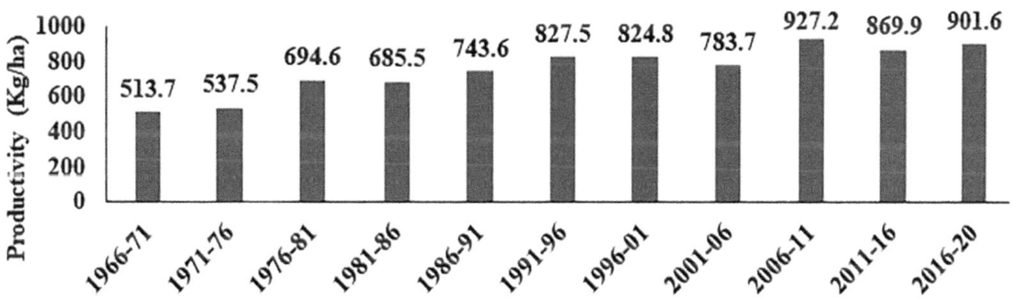

FIGURE 1.2 Trends in productivity (Kg/ha) over the years (based on a 5-year average).

1.2 GENETIC RESOURCES

Sorghum genetic resources are conserved at many research institutes across the globe. Around 235,711 sorghum germplasm collections are available worldwide (FAO, 2009), particularly in the United States, Ethiopia, Sudan, South Africa, India, and China (Rosenow and Dahlberg, 2000). For example, 45,192 accessions in the United States were from Ethiopia, Sudan, Yemen, Mali, India, and the United States (http://www.ars-grin.gov/cgibin/npgs/html/tax_stat.pl). Similarly, ICRISAT has a collection of 37,949 accessions from 92 countries (nearly 16% of the global total), and it represents the major diversity centers of sorghum (Upadhyaya et al., 2014). Cultivated sorghums are grouped into five basic races; among them, durra accounted for 21.21%, caudatum 20.12%, guinea 12.89%, bicolor 4.59%, and kafir 3.49% of the ICRISAT collection (Upadhyaya et al., 2014). Also, the Indian National Gene Bank holds around 19,844 sorghum accessions

(Radhamani et al., 2011). The development of core and mini-core collections (10% and 1%, respectively, of the entire collection) is the best strategy for germplasm utilization, and it represents the total genetic diversity. For instance, at ICRISAT, Patancheru, India, nearly 2247 germplasm accessions have been developed and are being maintained (Grenier et al., 2001). A tremendous amount of variability in the sorghum core and mini-core collection was maintained at the ICRISAT gene bank, which serves as a reservoir for traits of interest (Upadhyaya et al., 2016) (Figure 1.3).

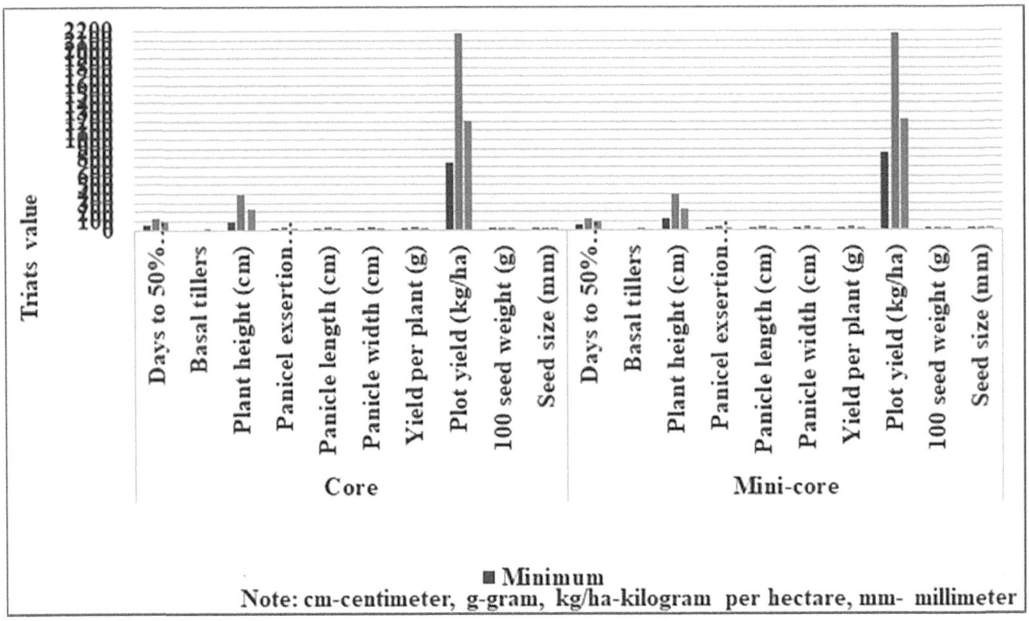

FIGURE 1.3 Traits variation for sorghum collections stored at ICRISAT gene bank.

Source: Modified from Upadhyaya et al. (2009).

1.3 MALE STERILITY

Sorghum hybrids are superior to pure lines and populations in terms of yield and other agronomically important traits. The discovery of a cytoplasmic-nuclear male sterile system in sorghum helped to produce plenty of commercial hybrid seeds on a large scale using a three-line breeding system (A, B, and R) for the cultivation of hybrids (Stephens and Holland, 1954). Nearly eight different types of genetic male sterility have been identified in the form of recessive alleles (homozygous condition) at a number of loci and reported in countries like India and the USA (Table 1.1). Among all, only ms3 and ms7 have been extensively used in population improvement because they are stable performers (Reddy and Stenhouse, 1994), although the al (antherless) allele is also useful (House, 1985). Using these genetic male sterility alleles, population improvement can be successfully deployed in sorghum to derive new hybrid parents/superior varieties (Reddy et al., 2008). Half of the sorghum hybrid parents at ICRISAT were developed through population improvement breeding methods.

TABLE 1.1 Male Sterility Genes, Gene Symbols, and Mechanism of Genetic Male Sterility in Sorghum

Character/Mechanism	Sterility Gene/ Gene Symbol	Maintained in/Source	References
Anthers without pollen	*ms1*	N316	Ayyangar and Ponnaiya (1937); Stephens and Quinby (1945)
Anthers without pollen	*ms2*	N317	Stephens (1937); Stephens and Quinby (1945)
Anthers without pollen	*ms3*	N318	Webster (1965)
Empty pollen cells	*ms4*	Not available	Ayyangar (1942)
Aborted pollen	*ms5*	Not available	Barabas (1962)
Micro-anthers without pollen	*ms6*	Not available	Barabas (1962)
Empty pollen cells	*ms7*	N319	Andrews and Webster (1971)
Anther less stamens	*msal*	N320	Karper and Stephens (1936)
Do not produce pollen grains	*ms8*	mutant from an inbred BTx623	Xin et al. (2017)
No pollen grains	*ms9*	M4 mutant line, M2P0110 (ARS178)	Chen et al. (2019)

Source: Modified from Rooney (2000).

The cytoplasmic-genetic male sterility (CMS) is caused by incompatible interaction between nuclear and cytoplasmic genes. Stephens and Holland first reported the CMS system in sorghum in 1954, which is the backbone for hybrid product development. The A_1 male sterile system has been controlled by one to two major genes and several modifier genes. Also, the A_1 male sterility was caused by the single gene *msc1* (Maunder and Pickett, 1959) and the *msc2* gene. The different types of CMS systems (Table 1.2) have been reported in sorghum (Schertz, 1994) (Table 1.2).

1.4 GENETIC IMPROVEMENT IN GRAIN, FORAGE AND SWEET SORGHUM

Breeding cultivars during 1962–1969 in sorghum breeding was aimed at developing heterotic hybrids by involving male sterile lines. It converted dwarf lines, and the resultant hybrids CSH1, CSH2, and CSH3 derive from temperate × temperate and temperate × tropical crosses (Rao and Ranan, 1982). Of these, CSH1 produced the highest yield, wider adaptability, and was well suited for rainfed areas. However, continuous cultivation for several years led to susceptibility to leaf disease, grain mold, and reduced fodder quantity. A brief overview of crop improvement in grain, forage, and sweet sorghum is provided below.

1.4.1 GRAIN SORGHUM

The genetic improvement of grain sorghum focuses mainly on grain yield, color, appearance, quality, and adaptation. The integration of genes that induce reduced plant height, improved grain/straw ratio, and heightened nutrient responsiveness has been shown to result in a remarkable increase in sorghum grain yield under rainfed conditions (Aruna

TABLE 1.2 Sources of Cytoplasmic Male Sterility Systems in Sorghum

Classified Based on Pollen Development and Anther Morphology	Cytoplasm	Identity	Race/Hybrid Race	Origin
Small anthers but without fertile pollen which degenerates during microsporogenesis	A_1	Milo	*durra*	USA
		IS 6771C	*guinea-caudatum*	India
		IS 2266C	*durra*	Sudan
		IS 6705C	*guinea*	Burkino Faso
		IS 7502C	*guinea*	Nigeria
		IS 3579C	*caudatum*	Sudan
		IS 8232C	*(kafir-caudatum)-caudatum*	India
		IS 1116C	*guinea*	India
		IS 7007C	*guinea*	Sudan
	A_2	IS 1262C	*guinea*	Nigeria
		IS 12662C		USA
		TAM 428		USA
		IS 2573C	*caudatum*	Sudan
		IS 2816C	*caudatum*	Zimbabwe
	A_5	IS 7506C	*bicolor*	Nigeria
	A_6	IS 1056C	*durra*	India
		IS 2801C	*durra*	Zimbabwe
		IS 3063C	*durra*	Ethiopia
Large non-dehiscent anthers that may contain some viable pollen	A_3	IS 1112C	*durra-(durra-bicolor)*	India
		IS 12565C	*caudatum*	Sudan
		IS 6882C		USA
	A_4	IS 7920C	*guinea*	Nigeria
	9E	IS 7218		Nigeria
		IS 112603C	*guinea*	Nigeria
		9E		Ghana

Source: Modified from Schertz (1994).

and Cherku, 2019; House et al., 2000). The first hybrid developed in sorghum, CSH 1, was rainy-season sorghum. The utilization of exotic germplasm has improved the grain yield of rainy-season sorghum, but it has not had a positive impact on its quality parameters compared to indigenous sources. When it comes to genetic variability, rainy-season sorghum holds much variability, represented by caudatum, bicolor, and their intermediary races, but post-rainy-season sorghum holds much less variability, represented majorly by durra races. Post-rainy-season sorghum suffers from low productivity induced by post-flowering drought stress, difficulty developing heterotic groups, etc (Aruna and Audilakshmi, 2008; Patil et al., 2013). These issues should be addressed by expanding its genetic base. Grain color is also targeted while working on grain sorghum as it affects overall product quality and consumer acceptance. At least 10 gene pairs that influence the color of sorghum grains have been identified so far (Valencia et al., 2009).

1.4.2 FORAGE SORGHUM

The success of any crop breeding program largely depends on the presence of genetic variability in the germplasms and breeding materials. Several studies have been conducted and identified the traits linked/used for forage improvement in sorghum. Traits identified for forage quantity are early vigor, plant height, number of leaves per plant, larger leaf length and broader leaf width, stem thickness, biomass weight per unit area, delayed or no flowering in the growing season (higher accumulation of biomass managed through photosensitivity), and rapid regrowth (ratoonability). In contrast, traits identified for forage quality include a higher leaf-to-stem ratio, increased digestibility, higher protein, absence of wax/bloom, lower lignin, and lower HCN (hydrocyanic acid or prussic acid) released during the degradation of dhurrin upon cutting/injury to the sorghum plants. Furthermore, tan color is also responsible for a broad spectrum of foliar diseases. Traits like leaf-to-stem ratio, stem girth, and total soluble solids were found to have high heritability coupled with genetic advancement, indicating opportunities for developing new forage cultivars with better forage quality (Vedansh et al., 2010).

The yield-contributing traits like plant height, number of tillers, leaf length, and leaf breadth were predominantly controlled by additive gene effects. In contrast, both additive and dominance effects were significant for the number of leaves and fodder yield (Pahuja et al., 2003). Furthermore, traits controlled by additive effects are improved through simple pedigree selection, while traits controlled by both additive and dominance effects are improved through S_2 and reciprocal recurrent selection (Pahuja et al., 2003). Traits like protein content and *in vitro* dry matter digestibility (IVDMD) were predominantly controlled by additive genetic variance in addition to non-additive genetic variance, suggesting that reciprocal recurrent selection could be effective in improving fodder quality traits in sorghum (Ahluwalia and Rao, 1980). Khatri et al. reported that additive (d) and dominance (h) gene effects were responsible for leaf breadth. In contrast, non-additive gene effects were responsible for leaf length and number of leaves (2001a). Also, the duplicate gene interaction was responsible for leaf number and leaf length in some crosses (Khatri et al., 2001a). The additive and dominance effects, as well as high narrow-sense heritability, were observed for ratoonability or regrowth traits in sorghum (Khatri et al., 2001b). Dhurrin (HCN) content was predominantly controlled by non-additive gene effects, suggesting that low dhurrin cultivars could be developed through recurrent selection, population improvement methods, and heterosis breeding in sorghum (Pandey and Shrotria, 2009).

Low dhurrin (HCN) transgenic sorghum was developed through antisense technology (Pandey, 2010). The same study also identified stable T_3 plants that exhibited lower HCN compared to control genotypes, and these transgenes can be introgressed into elite forage genotypes by the backcross method to breed lower or no HCN in sorghum (Pandey, 2010). Tannins present in plants are tightly associated with reduced digestibility and animal intake as they reduce protein digestibility by rumen microflora (Barnes and Gustine, 1973). Tannin content varies from one plant part to another, and this is continuously synthesized in younger leaves and during grain-filling stages (Gourley and Lusk, 1978). Whole-plant tannin levels are positively associated with fibrous constituents and negatively associated with forage quality traits such as crude protein (CP) and IVDMD (Montgomery et al., 1988).

The epicuticular waxes (EW) (bloom or glaucousness) are primarily accumulated in the whitish powder on the sorghum sheaths, leaves, and culms (Peterson et al., 1982). Bloomless (absence of wax) types of sorghum have 22% higher IVDMD in comparison to bloom (presence of heavy wax) types. Therefore, forage quality can be improved simply by selecting bloomless types (Cummins and Dobson, 1972; Cummins and Sudweeks, 1976). The glossy (seedlings with shiny and light yellow-green leaf surfaces) trait is responsible for reduced wax (sparse-bloom) or bloomless surfaces on leaves (Traore et al., 1989), which is most important where sorghum is grown for grazing purposes.

The leaf-to-stem (LS) ratio is another important trait that has a direct effect on protein content and digestibility. For example, a higher LS ratio increases IVDMD and CP (Hoveland and Monson, 1980). Compared to other fodder crops, the leaf sheath of forage sorghum is consistently higher in NDF, ADF, and cellulose and lower in IVDMD than the leaf blade or stem (Cherney et al., 1991).

The photoperiod-sensitive (PS) line produces higher biomass because it stays in vegetative production for longer in the growing season rather than diverting to reproduction. The PS sorghum forage produces higher biomass tonnage under limited water conditions (Bean et al., 2002). Further, they reported that PS forage sorghum could produce 2.5 wet tons of forage, which is three times higher than maize silage produced per acre inch of water. It has been reported that plants in nine F_2 populations derived from sorghum inbred photoperiod insensitive (PI) × inbred photoperiod insensitive (PI) segregated for photoperiod response, and eight of the nine F_2 populations fit a 9:7 PS/PI ratio, indicating that the interaction of two independent gene loci was interacting in complementary dominant epistasis (Rooney and Aydin, 1999). In the USA, the best-performing photoperiod and/or thermosensitive genotype, Tentka, was identified for its high dry matter (40 t/ha) potential in a single late-season harvest (Venuto and Kindiger, 2008).

The utility of sorghum as a multi-cut forage crop is majorly hindered by the presence of cyanogenic glucoside, dhurrin [(S)-p-hydroxymandelonitrile-β-D-glucopyranoside] in sorghum, sorghum-Sudan grass, and Sudan grass. The digestive enzymes present in grazing cattle hydrolyze the compound into hydrocyanic acid (HCN), and the presence of 0.5 g of dhurrin is enough to kill the animal. It has also been reported that HCN increased under drought stress conditions (Teutsch, 2002). Inheritance studies of HCN potential (HCN-p) revealed that a single major gene pair was responsible, and HCN-p levels in F1s were intermediate, indicating that neither high nor low HCN-p was completely dominant (Gorz et al., 1986a, b), quantitatively inherited with additive genetic effects (Lamb et al., 1987), and multiple loci affecting the HCN-p trait (Kalton, 1988). Elder and Dennan reported that forage breeding lines/cultivars with less than 750 ppm of HCN on a fresh biomass basis have no detrimental effect on ruminants (1966).

Increased forage quality was observed in bloomless than bloom types of sorghum cultivars (Cummins and Dobson, 1972; Cummins and Sudweeks, 1976). Sorghum cultivar seedlings with light yellow-green and shiny leaf surfaces (glossy) are associated with a lack of wax deposits on leaf surfaces, which ultimately improves forage quality (Traore et al., 1989).

The availability of diverse parental lines, diverse cytoplasmic male sterility sources, good general combiners, and good knowledge of the genetics of traits would allow breeders to exploit heterosis in their breeding programs. To date, the forage sorghum hybrid market is predominantly dominated by private seed sectors/companies. To identify promising forage-type hybrid parents, a series of studies have been conducted on the combining ability of forage-type parents for biomass traits by several Indian researchers (Bhatt and Baskheti, 2011; Parmar et al., 2004; Parmar and Tikka, 2005; Sanghi, 1982; Singh and Shrotria, 2008; Singh et al., 2010; Yadav and Pahuja, 2007) through various breeding designs. With these studies, a few good general combiners identified for biomass yield were Pusa Chari-6 and Pusa Chari-1 (Sanghi, 1982), 2077A, 36A, and M31-2A, C10-2, Vidisha, Rio, PC6, IS685, IS3287, and IS4776 (Pathak and Sanghi, 1992), TNAU M 1A (Manickam and Das, 1995), 18-3A and Ruchira (Sumalini et al., 2005), 2077A and 111A (Iyanar and Khan, 2005), ICS 13A, ICS 10286A, S 241 (Yadav and Pahuja, 2007), Pant Chari-5 and UTMC-532 (Singh and Shrotria, 2008), HB 94004 A, SPA2 94012, 993100A, UTMC 532, Pant Chari-6, and PMC 03 1212 R (Singh et al., 2010), 2219A, ICSA 363, M-20, Pant Chari-5, and UPMC-504 (Bhatt and Baskheti, 2011), and for multicut traits were ICS 4A, ICS 79A, and IS 3289 (Mohan et al., 2007a,b).

In India, forage sorghum cultivars are predominantly bred for single-cut types other than interspecific hybrid sorghum-Sudan grass. The varieties "Pusa Chari" (selection from IS 6090, notified in 1974) for single cut and "Meethi Sudan" (SSG 59-3, notified in 1978) for multi-cut were the first improved cultivars released in India. Mahyco seed company released the first forage sorghum hybrid MFSH 3 available to Indian markets in 1990. Thereafter, the first public sector forage sorghum hybrid PCH 106 was developed and released in 1997. Most recently, the public sector released hybrid CSH 24MF, which is popular among growers because of its high biomass potential with improved quality and good seed yield. Also, the latest release of the variety CSV 33MF has yielded more than that of the best hybrids available in the market (Bhat, 2019).

1.4.3 GENETIC IMPROVEMENTS OF SWEET SORGHUM

Sorghum with juicy and sweet stalks is generally referred to as "sweet sorghum." This type of sorghum is generally cultivated for syrup or forage production, while kafirs and milos sorghum are cultivated for grain usage. Sweet sorghum plants have accumulated higher amounts of soluble sugars (10 to 15%) in the sweet stalks' sap or juice. Due to its inherent drought-tolerant nature, this crop is called 'the camel among crops' or 'the sugarcane of the desert' (Sanderson et al., 1992).

Sweet sorghum is characterized by having broad leaves and a round or elliptical head full of grains at maturity, higher tonnage production under irrigation, dormant during extremely dry spells, and resuming its growth when there is enough moisture (Gnansounou et al., 2005). The sweet sorghum grain, stalk juice, and bagasse (fibrous residue leftover after juice extraction) are utilized to produce human food, animal feed, ethanol, and power generation. Mostly, the sugars accumulate more in its stems, and these can be extracted following the same protocols used for sugarcane (Mascia et al., 2010). The stalk juice (Brix or stalk sucrose percentage) contains glucose, fructose, and sucrose, which are suitable for direct fermentation to ethanol (Sipos et al., 2009) and/or other biobased chemicals production (Ou et al., 2016; Wang et al., 2015). Bagasse can be used for power generation and the development of byproducts, and it contains higher nutrient contents that offer feed to livestock (Ashok Kumar et al., 2010; Blummel et al., 2009). In addition, the cellulose and hemicellulosic polysaccharides found in bagasse can be utilized to produce second-generation biofuels.

Sweet sorghum contains higher soluble and insoluble carbohydrates and bioactive compounds, which help improve human health (Antar et al., 2021; Vanamala et al., 2018). It can produce a higher stem biomass yield (60 t/ha) and seed yield of 2250–6000 kg/ha when grown in non-cultivable lands (Tian et al., 2009). It also performs better under low input conditions than maize (Maw et al., 2019; Xie and Xu, 2019) and is tolerant of several abiotic stresses, including salinity, drought, and alkalinity. Generally, the sorghum growing period is about 4.5 months, which is four times less than sugarcane (12–16 months), and the required water for sorghum is about 8000 m^3/two crops annually, which is again much lower than that required for sugarcane, about 36000 m^3/crop. It is also estimated that the cost of cultivation for sorghum is less than for sugarcane (Reddy et al., 2007). Under favorable conditions, sweet sorghum can yield up to 13.2 t ha^{-1} of total sugars, or 7682 L of ethanol ha^{-1} (Murray et al., 2009).

The inheritance pattern found for a sweet stalk of sorghum was governed by a single recessive gene with additive and dominant effects (Guiying et al., 2000; Ritter et al., 2007). A study by Audilakshmi et al. found that the stalk sugar of sorghum was predominantly controlled by dominant genes, and the traits (stalk yield, plant height, and juice yield) associated with high ethanol production were controlled by overdominance (2010). The stalk sugar content of progenies derived from sweet × non-sweet lines indicated that the trait is controlled by a quantitative nature (Saballos, 2008). The sweet stalk-linked traits like plant height, stem girth, TSS, millable stalk yield, and extractable juice were predominantly governed by non-additive gene action, indicating traits improved through heterosis breeding (Sankarapandian et al., 1994). Further, additive and non-additive gene effects might affect the inheritance of stalk biomass, Brix, and stalk weight in sugar stalks, although non-additive genes dominated the control (Zhou et al., 2005). A study by Saballos et al. (2008) found that enzymatic saccharification following the pretreatment of high-biomass *bmr* mutants produced more fermentable sugars than their counterparts. Also, double *bmr* mutants showed higher biomass conversion than single *bmr* mutants (Dien et al., 2009). Regarding stalk sugar storage loss and juice quality, the researchers need to be careful about self-fermentation due to strong invertase activity. This led to a loss of both quantity and quality of the stalks after harvesting. For example, there was a loss of fermentable sugars of up to 20% and 50% after three days and one week of harvest, respectively (Ratnavathi et al., 2011).

The sweet sorghum research work was first initiated in India by ICRISAT with the availability of global sorghum collections. The research program mainly focused on the development of hybrid parental lines having high biomass traits with specific adaptation to rainy and rabi seasons due to higher genotype × environment interactions (G × E). Through their systematic development of parental lines for sweetness, the Brix content varied from 10 to 15% and 8 to 13% in rainy and post-rainy seasons, respectively, in the A/B lines. Meanwhile, the R lines exhibited 12 to 24% and 9 to 19% of Brix during rainy and post-rainy seasons, respectively (Srinivasarao et al., 2009). The National Agricultural Research

System (NARS) partners in India are the Directorate of Sorghum Research (DSR), Hyderabad, and the All India Coordinated Sorghum Improvement Project (AICSIP) centers that carried out sweet sorghum research. With this research and development, many cultivars are available in the markets. With continued research, they have contributed to the release of several parental lines having higher sweet content.

Numerous mapping studies have been conducted to identify the gene/QTL linked to bioenergy-related traits in sorghum (de Souza et al., 2021; Kajiya-Kanegae et al., 2020). A study by de Souza et al. reported 33 QTLs for bioenergy traits in 272 RIL plants derived from Brandes × Wray (2021). Also, Jin et al. have identified 43 QTLs for biomass traits, and several of them were stable across years (2021). Most recently, eight major and two minor QTLs were identified for bioenergy traits in the F_3 population derived from a cross between sweet sorghum (cv. "Erdurmus") and grain sorghum (cv. "Ogretmenoglu") (Guden et al., 2023).

1.5 BIOTIC STRESS RESISTANCE

Sorghum crop productivity is primarily affected by biotic stresses, including diseases, insects, and parasitic plants. Nearly 150 insect pests have been reported in sorghum (Reddy and Davies, 1979). However, only 50 were reported as causing significant economic damages globally or regionally/locally (Thakur et al., 2007). The genetics and resistance mechanisms to diseases, insect pests, and weeds are discussed individually hereunder.

1.5.1 DISEASE RESISTANCE

The major diseases affecting sorghum production include grain mold (caused by a complex of several fungal species), downy mildew (*Peronosclerospora sorghi* [Wetson and Uppal (Shaw)]), anthracnose [*Colletotrichum graminicola* (Ces.) GW Wilson], rust (*Puccinia purpurea* Cooke), leaf blight [*Exserohilum turcicum* (Pass.) Leonard and EG Suggs], charcoal rot/stalk rot [*Macrophomina phaseolina* (Tassi) Goid], and several virus diseases.

1.5.1.1 GRAIN MOLD

Grain mold is the most destructive biotic stress in rainy sorghum. The availability of standardized greenhouse and field screening protocols led to the identification of resistance sources for grain mold (Thakur et al., 2006). Through these screening protocols, red pericarp, grain hardness, and pigmented testa were responsible for grain mold resistance (Thakur et al., 2006). Grain mold resistance in grain sorghum was mostly found in colored grain sorghum germplasms with and without tannin content and also in a few white grain sorghum germplasms (Bandyopadhyay et al., 1988ab). The most preferred sorghum for food use in India is white-grain sorghum, while colored-grain sorghum is used by the rest of the world. The inheritance pattern for grain mold disease indicated that disease resistance is controlled by dominance. Thereby, any one parent has enough disease resistance to breed a hybrid cultivar (Kumar et al., 2008). It has been observed that grain mold resistance is complex in nature as it is controlled by major and minor genes, additive and epistatic effects, and strong G × E interactions (Audilakshmi et al., 2000, 2005; Stenhouse et al., 1996). Five QTLs (phenotypic variance explained between 10% and 23%) were identified to have linkages with grain mold resistance in sorghum (Klein et al., 2001). In addition, two significant SNP loci were found to be linked with grain mold resistance in an association-mapping panel consisting of 242 mini-core sorghum genotypes (Upadhyaya et al., 2013b). Most recently, a significant SNP was associated with grain mold resistance in the sorghum association panel (Ahn et al., 2023).

1.5.1.2 ANTHRACNOSE

The most frequently reported foliar disease in sorghum is anthracnose, which infects all aerial tissue parts of the plant and reduces the yield by up to 50% (Thakur and Mathur, 2000). Inheritance studies on sorghum anthracnose disease revealed that anthracnose resistance is governed by multiple genes with different modes of action (Well, 1989), a single gene with multiple alleles (Tenkouano and Miller, 1993), two to three linked loci with dominant effects (Cuevas et al., 2014), a single recessive locus in SC326-6 (Boora et al., 1998), and a single dominant locus in BTx378 and SC784-5 lines of sorghum (Erpelding, 2007). It has also been reported that progenies obtained between two resistant inbred lines have shown segregation for anthracnose resistance, indicating inbred parents differ for resistance loci (Mehta et al., 2005).

Eight SNPs were found to be linked with anthracnose resistance across all environments, and two out of eight SNPs were validated (Upadhyaya et al., 2013a) and found co-located with the reported two major QTL (QAnt3 and QAnt2) by Mohan et al. (2010). The sorghum anthracnose resistance gene was mapped on LG 5 (Ramasamy et al., 2009). Also, two anthracnose resistance genes (Sobic.009G049500 and Sobic.009G049800—nucleotide-binding site-leucine-rich repeat (NBS-LRR)) were also located on LG 9 (Biruma et al., 2012). Most recently, nine significant SNPs were found to have linkages with anthracnose resistance in the Senegalese sorghum population (Ahn et al., 2023). Additionally, two anthracnose resistance genes (ARG4 and ARG5) were identified in the sorghum parental lines SAP135 and P9830, which encoded canonical nucleotide-binding leucine-rich repeat (NLR) receptors (Habte et al., 2023).

1.5.1.3 DOWNY MILDEW (DM)

Sorghum DM is reported worldwide, and it occurs epidemically, which can cause severe yield reduction (Perumal et al., 2008; Sharma et al., 2010). Inheritance studies revealed that the number of genes controlling DM resistance varied from 1 to 6, and in most cases, it has been inherited in oligogenic manners (Rana et al., 1982; Sifuentes and Frederiksen, 1988). The sorghum DM inheritance pattern is qualitative with dominant to partially dominant loci (Craig and Schertz, 1985; Reddy et al., 1992). The number of genes and gene actions mainly depended on the genotypes studied and the pathotypes involved. Six DM resistance accessions, IS 28747, IS 31714, IS 23992, IS 27697, IS 28449, and IS 30400, were identified from the 242 mini-core accessions of germplasms (Sharma et al., 2010).

1.5.1.4 ERGOT

Ergot disease of sorghum is one of the most destructive biotic stresses that affect both seed quantity and seed quality of seed parents, especially male sterile parents. The ergot disease infects ovaries and exudes sticky sweet honeydew in the panicles, with yield losses of up to 80% reported in India and Zimbabwe (Bandyopadhyay et al., 1998). Furthermore, feeding ergot-infected (Sclerotia-vegetative fungal bodies) sorghum grain to cattle can be harmful (Blaney et al., 2000). Ergot infestation was higher and heavily influenced by low temperatures, wind, cloudiness, and rain during flowering time, which reduced pollen viability (PV) and pollen quantity (PQ) (Bandyopadhyay et al., 1998; Brooking, 1976; Wang et al., 2000), affecting pollination and fertilization, thereby increasing the risk of ergot severity. Three different ergot species have been reported in sorghum globally. The species *Claviceps africana* is the most common pathogen observed throughout the Americas, Australia, Asia, and Africa, whereas *Claviceps sorghicola* is found only in Japan, and *Claviceps sorghi* is restricted to Asia (Bandyopadhyay et al., 1998; Pažoutová et al., 2000; Tooley et al., 2000).

Efforts are being made to develop resistant hybrid parental lines and cultivars. Resistance to ergot diseases has been identified through pollen-mediated (effective pollination and fertilization before infection occurs) (Bandyopadhyay et al., 1998; McLaren, 1997; Watkins and Littlefield, 1976) and non-pollen-mediated (reduced stigma exposure time to inoculum before pollination, faster stigma drying after pollination, and smaller stigmas) (Dahlberg et al., 2001; Reed et al., 2002) mechanisms. Additionally, a study conducted by Parh et al. found that ergot resistance in sorghum is controlled by several minor loci, and the pollen traits PV and PQ had a moderate negative genetic association with the percentage of ergot infection (PCERGOT) (2006). The same study also concluded that the higher levels of genetic diversity and broad sense of heritability suggest that ergot resistance will probably respond well to conventional selection. Parh et al. identified QTL for both pollen and non-pollen-mediated mechanisms for ergot resistance (2008). Four major QTL linked to PCERGOT (SBI-01-11.8% near sPb-8261, SBI-06-14.1% near sPb-1543, SBI-08-11% near AGG + CAG6, and SBI-09-19.5% near Sb4-32) and one major QTL each for PV (SBI-07-12.5% near sPb-5594) and PQ (SBI-06-19.9% near AAG + CTT6) in sorghum.

1.5.1.5 CHARCOAL ROT

Charcoal rot is the most destructive stalk and root rot disease in sorghum grown under residual soil moisture conditions during post-rainy seasons (Rana et al., 1982; Seetharama et al., 1987), leading to severe grain yield losses as well as reduced stover quantity and quality. Studies on resistance to charcoal rot have been reported to be under non-additive (Indira et al., 1983), recessive genes (Rosenow and Frederiksen, 1982), quantitative with additive and additive × dominance and dominance × dominance interactions (Rana et al., 1982; Rao and Shinde, 1985). One inbred parent conferring resistance is sufficient to confer resistance in the hybrid (Deshpande et al., 2011; Indira et al., 1983; Rana et al., 1982), and it is controlled by a multiple-locus complex (Bramel-Cox et al., 1988) regulated by both dominant and recessive epistatic interactions between two gene loci, with a third locus having a modifying effect (Tenkouano et al., 1993). Multi-year studies identified nine stable QTLs for three agro-morphological traits (number of internodes crossed by the rot, length of infection, and percentage lodging) and two biochemical traits (lignin and total phenols) in a RIL population IS22380 × E36-1 (Patil et al., 2012; Reddy et al., 2008). The major QTL linked with the number of internodes crossed and length of infection was also reported on LG 2.

1.5.1.6 RUST

Sorghum rust is widely distributed in almost all sorghum-growing areas of the world, with yield losses of up to 65% (Bandyopadhyay, 2000). Susceptibility to sorghum rust was dominant over resistance, and additive effects were predominant (Dabholkar et al., 1980). Contrary to this, rust resistance was shown to be dominant (Pu) in sweet sorghum crosses between the resistant cultivar MN 960 and the susceptible cultivar Planter (Coleman and Dean, 1961). Studies have also found that rust resistance in sorghum exhibits polygenic inheritance (Mohan et al., 2010; Upadhyaya et al., 2013b; Wang et al., 2014). A study by Tao et al. reported four major QTLs tightly associated with rust resistance on SBI-01, SBI-02, SBI-03, and SBI-08 in a sorghum RIL population (1998). Upadhyaya et al. found five significant SNPs linked to rust resistance, two of which showed homology to the wheat rust resistance gene (Lr1) and to the maize rust resistance gene (Rp1-D) (2013b).

1.5.2 INSECT-PESTS

At least 150 insects have been identified as pests of sorghum worldwide, of which aphid [*Melanaphis saccari* (Zehntner)], head bug (*Calocoris angustatus* Leth), [*Schizaphis graminum* (Rondani)], midge (*Contarinia sorghicola* Coq.), stem borer [*Chilo partellus* (Swinhoe)] and [*Sesamia inferens* (Walker)], and shoot fly [*Atherigona soccata (*Rondani)] are the major pests of sorghum.

1.5.2.1 APHIDS

Sorghum aphids are the most important insect pest in India (Balikai, 1997). Both field and greenhouse studies involving high (PI257595) and moderately (129-3A) resistant parents and susceptible (RTx430) parents suggested that aphid resistance is governed by a monogenic dominant gene (Hsieh and Pi, 1982; Pi and Hsieh, 1982; Tan et al., 1985). Hsieh reported that both dominant and additive gene actions are required for aphid resistance, and the additive gene action is responsible for the resistance expression (1988). Studies have concluded that the inheritance of sorghum aphid resistance is controlled by complementary gene action (Chang and Fang, 1984) and non-additive gene action (Deshpande et al., 2011). Also, the latter study reported that at least one parent possessing a resistant allele in a hybrid combination could be sufficient to confer aphid resistance (Deshpande et al., 2011). Recently, a study has identified two significant SNPs for field-based aphid visual ratings on SBI-08 and SBI-10 for aphid count and plant damage, as well as three linked SNPs for reflectance-based aphid resistance traits on SBI-02, SBI-03, and SBI-05

(Punnuri et al., 2022). Further, these studies have identified candidate genes encoding leucine-rich repeats (LRR), Avr proteins, lipoxygenases (LOXs), WRKY transcription factors, flavonoid biosynthesis genes, calmodulin (CAM) dependent protein kinase, and 12-oxo-phytodienoic acid reductase on near SNPs that had significant associations with different sorghum aphid-linked resistance traits.

1.5.2.2 HEAD BUG

Sorghum head bug is one of the most important insect pests of grain sorghum. Resistance to sorghum head bug was recessive, and no maternal effects were observed (Ratnadass et al., 1995). Inheritance studies on resistance to sorghum head bugs revealed that both additive and non-additive gene effects are important for resistance, and resistance was due to dominance to partial dominance type of gene action (Sharma et al., 2000). Ratnadass et al. reported that additive gene action is the sole factor responsible for head bug resistance, and high heritability is also important (2022). Genotypes IS2761, IS6984, IS9692, IS17610, IS17618, and IS17645 showed the non-preference type of host plant resistance (HPR) at both anthesis and milky stages (Sharma and Lopez, 1990).

1.5.2.3 GREEN BUG

Sorghum green bug is a major insect pest of grain sorghum in India. This sap-sucking aphid removes amounts of phloem sap and injects phytotoxins into the plants. The major symptom caused by green bug insect pests is a dark red spot that appears at the feeding site, which is surrounded by pale-yellow discoloration. The green bug is also a carrier of many viral pathogens (Harvey et al., 1996). Several biotypes (C, E, I, and K) of green bugs were found infecting sorghum, of which Biotype I is the most prominent and dangerous one (Punnuri et al., 2013). Inheritance studies on sorghum green bugs reported that resistance was governed by incomplete dominant genes (Boozaya-Angoon, 1983; Teetes, 1975). A study has found that a combination of genes from different sources provides increased resistance to green bug (Dixon et al., 1990). Among green bug biotypes, the resistance to biotype I revealed that it was incompletely dominant and controlled by two genes, which may rely on complementary gene action (Tuinstra et al., 2001).

Numerous studies have identified significant QTLs and defense signaling compounds linked to aphids. Three significant loci identified in the SBI05, SBI06, and SBI07 regions showed resistance to green bug biotype I (Katsar et al., 2002). Nine significant QTLs associated with both resistance and tolerance to biotypes I and K of green bugs have been identified, with individual QTL phenotypic variance explained by 5.6–38.4% (Agrama et al., 2002). Nagaraj et al. utilized chlorophyll loss as an important indicator of green bug damage and identified three significant QTLs on SBI-01 and SBI-04 for biotype I that showed resistance and tolerance (2005). Wu and Huang detected one major QTL on SBI-09 for resistance to biotype I of green bug (2008). Using the choice and no-choice assays, the study identified that the nested association mapping panel (NAM) parental line SC265 showed a combination of antibiotic and antibiotic-mediated resistance against green bugs, whereas the Segaolane (susceptible) expressed the least resistance to green bugs (Grover et al., 2019).

1.5.2.4 MIDGE

Sorghum midge is the most dangerous pest of grain sorghum globally (Harris, 1976), particularly in Asia, Australia, Africa, the Americas, and Mediterranean Europe. The most important morphological trait, "glume size," is tightly associated with midge resistance. The size of the glume (shorter and tighter glumes) makes oviposition difficult for sorghum midge. Therefore, it is the most preferred trait for midge resistance breeding. The rapid grain development and higher tannin content in sorghum are associated with midge resistance (Sharma et al., 1994). Singh et al. found a high degree of resistance in nine working groups, viz., Caudatum, Durra, Caudatum/Nigricans-Zerazera, Caudatum/Kafir (Hegari), Nigricans, Caffrorum-Darso, Durra-Nigricans, Durra-Nandyal, and Caudatum-Bicolor (2011). Further, the authors also identified the morphological traits like short and compact glumes that are closed during anthesis and confer resistance to sorghum midge. The presence of higher tannin in sorghum grain is linked to resistance, specifically against the larva. The inheritance of sorghum midge resistance showed a quantitative nature. The open and closed glume traits are expressed as dominance and recessiveness, respectively.

Studies on sorghum midge resistance revealed that it is controlled by polygenic gene action with some cytoplasmic effects (Agrawal et al., 1988), recessive genes at two or more loci (Boozaya-Angoon et al., 1984; Hanna et al., 1989), and more than one gene in TAM2566 (Johnson, 1974). Also, good general combiners (GCA) IS15107, DJ6514, TAM2566, and ICSV197 were identified to have resistance to sorghum midge (Ratnadass et al., 2002; Sharma et al., 2004). Further, both GCA and specific combining ability (SCA) are important for sorghum midge resistance (Agrawal et al., 1988; Patil and Thombre, 1985). It is also noted that both the hybrid parents must possess resistance to achieve midge-resistant hybrids (Sharma et al., 2004).

For sorghum midge resistance, QTLs were found for HPR mechanisms like antixenosis and antibiosis in an ICSV745 × 90562 RIL population (Tao et al., 2003). The identification of DNA markers for the antixenosis and antibiosis midge resistance mechanisms would be very helpful in identifying new midge resistance sources and in gene pyramiding these mechanisms to achieve long-lasting resistance through marker-assisted selection (MAS).

1.5.2.5 STEM BORER

Stem borer is the major insect pest that causes significant yield reduction in sorghum. The inheritance studies on resistance to stem borer revealed that it is controlled by polygenic and partially dominant factors (Pathak, 1985; Pathak and Olela, 1983; Rana et al., 1984). Resistance to stem borer symptoms like leaf feeding, dead heart, and stem tunneling are inherited differently (Singh et al., 1983). Both additive and non-additive gene actions are important for stem borer resistance, but additive gene action is most important for dead heart formation and leaf-feeding symptoms (Pathak, 1990). The inheritance patterns of morphological traits like early panicle initiation (Taneja and Woodhead, 1989) and ligular hairs (Woodhead and Taneja, 1987), and biochemical traits like low sugar, high amino acids, total phenols, high tannins, neutral detergent fibers, acid detergent fibers (Khurana and Verma 1982, 1983), and high silica content (Narwal, 1974) associated with stem borer resistance are not clear. Most recently, a study by Kumar et al. reported that all tested sorghum genotypes had lower total protein, starch, and total sugar, and higher antioxidant enzymes [ascorbate oxidase (AO), catalase (CAT), ascorbate peroxidase (APX), phenyl ammonia lyase (PAL), tyrosine ammonia lyase (TAL)] and non-enzymatic antioxidants [total phenols, total antioxidants, and ferric ion reducing antioxidant power (FRAP)] at seedlings compared to the susceptible genotype (2023).

1.5.2.6 SHOOT FLY

The sorghum shoot fly is the most dangerous insect pest in sorghum cultivated areas worldwide. Generally, the shoot fly incidence is higher in late- and early-sown rainy and post-rainy seasons, respectively (Jotwani and Srivastava, 1970). Several morphological traits have been identified, including seedling glossiness (Maiti et al., 1984), trichome density (Maiti and Bidinger, 1979), seedling vigor, leaf and sheath pigmentations, epicuticular wax, and other biochemical factors linked with shoot fly resistance (Tarumoto, 2005). Breeding for resistance to the sorghum shoot fly is a complex trait in nature (Dhillon, 2004). Studies have concluded that the major resistance mechanism to the shoot fly is non-preference (Young, 1972) and the cumulative effects of non-preference and antibiosis (Rana et al., 1981).

Sorghum shoot fly resistance is quantitatively (Agrawal and Abraham, 1985) and polygenically inherited (Halalli et al., 1983). Studies have also concluded that both additive and non-additive gene actions are responsible for shoot fly resistance (Aruna et al., 2011b; Mohammed et al., 2016a, b), with additive gene action being more predominant (Biradar and Borikar, 1985; Dhillon et al., 2006; Nimbalkar and Bapat, 1987). However, Agrawal et al. and Saikiran et al. reported that non-additive gene action was more predominant in controlling sorghum shoot flies (1988; 2023). Partial dominance was also reported to be responsible for controlling the shoot fly (Borikar and Chopde, 1980).

Molecular markers like SSR have been identified to have linkages with shoot fly resistance in sorghum (Apotikar et al., 2011; Aruna et al., 2011a; Kiranmayee et al., 2016; Satish et al., 2009, 2012). 29 significant QTLs for morphological traits like seedling vigor, seedling glossiness, and trichomes have been identified (Satish et al., 2009). Most interestingly, the IS18551 entry contributed resistant loci to most of the detected QTLs for shoot fly resistance. Furthermore, Satish et al. identified the QTLs for leaf glossiness on LG 5 and trichome density, oviposition, and deadhearts on LG 10 (2012). The earlier reported QTLs for shoot fly resistance for different traits on SBI-01, SBI-07, and SBI-10 were introgressed into elite parental lines (Gorthy et al., 2023).

1.5.2.7 FALL ARMYWORM (FAW)

FAW is the most devastating chewing insect pest reported in sorghum recently and can cause severe yield reduction. The FAW directly affects yield loss through FAW larvae feeding on developing or mature seeds and also indirectly causes yield reduction through defoliation, thereby reducing photosynthetic activity and/or causing the loss of seedlings. The application of insecticides could be an immediate solution to control FAW, but it disturbs the environment. HPR is the best option to control FAW because it is a natural plant defense mechanism. The plant can induce an array of secondary chemical compounds produced in different pathways that show a defense response to insect pests. Among several pathways, the phenylpropanoid pathway produces several molecules such as monolignols, coumarins, phenolic glycosides, flavonoids, benzoic acids, phenolamides, and stilbenes, which show a range of important functions, particularly in plant-environment interactions (Fraser and Chapple, 2011). Recently, gene expression and metabolomic studies on the nested association mapping population (NAM) panel founder lines suggested that 10 days after FAW infestation, FAW feeding suppressed the gene expression involved in the monolignol pathway and their associated phenolic intermediates (Grover et al., 2022). Furthermore, following 10 days of FAW feeding, the SC1345 genotype showed increased amounts of flavonoid molecules, indicating a diversion of precursors from lignin biosynthesis to the flavonoid pathway (Grover et al., 2022). Additionally, the presence of flavonoids in sorghum showed significant mortality and reduced the body weights of FAW (Chatterjee et al., 2023). Recently, Shinde et al. concluded that FAW oral secretions (OS)/regurgitant induced the genes in flavonoid biosynthesis and total flavonoids, thereby increasing the defense mechanism against FAW larvae (2024). The same study also identified that FAW saliva suppressed the expression of genes in the flavonoid pathways, thereby enhancing FAW caterpillar growth, glucose oxidase (GOX), independent of the FAW salivary component.

1.5.3 WEED RESISTANCE

Striga is a parasitic weed (commonly known as "witchweed") present widely in sub-Saharan Africa. One or more species of Striga together infect and destroy the whole crop, resulting in a yield loss of up to 100% and affecting over 60% of farmland in those areas (Ejeta, 2007). Studies have identified Striga resistance mechanisms such as low germination stimulant production, germination inhibition, and low historical initiation activity observed in wild sorghum (Rich et al., 2004). Three wild sorghum accessions (WSA 1, WSE 1, and WSA 2) were identified to have higher resistance levels than the striga-resistant check N13 (Mbuvi et al., 2017). A study by Gobena et al. identified the gene regulating Striga resistance in sorghum (2017). The presence of mutant alleles at the LGS1 (Low Germination Stimulant 1) locus drastically reduces the *Striga* germination stimulant activity.

Numerous studies have been conducted to map the QTLs for Striga resistance in sorghum (Haussmann et al., 2004). These were further introgressed into parental line improvement through MAS (Deshpande et al., 2013). The Striga resistance QTL was originally identified from the N13 parent, which has been subsequently utilized and introgressed into Wad Ahmad and Tabat (cultivars) in the Sudanese sorghum breeding program (Ali et al., 2016), introgressed into Ochuti in the Kenya sorghum breeding program (Ngugi et al., 2015), and introgressed into Danyana and Samsorgh 39 cultivars in the Nigerian sorghum breeding program (Danquah et al., 2019) with the help of molecular markers.

1.6 ABIOTIC STRESS RESISTANCE

1.6.1 DROUGHT STRESS

Sorghum is the major food crop for people living in the arid and semi-arid regions of the world where other crops like rice and maize fail to grow well. These regions are characterized by low rainfall and resource-poor agronomic and management practices. However, studies were conducted to assess the yield loss under different drought durations. For example, drought stress at 16 and 28 days during the vegetative stage of the sorghum crop decreased the grain yield by 16% and 36%, respectively (Inuyama et al., 1976). In another study, early booting and early grain filling stages of sorghum for 28 days under drought stress resulted in a 27% and 12% reduction in grain yield, respectively (Eck and Musick, 1979). On the other hand, at the start of the booting stage, prolonged drought stress lasting 35 and 42 days reduced grain yield by 43% and 54%, respectively. Numerous studies have been conducted to understand the drought resistance mechanisms as it is grown in water-limited areas in those regions. This crop responds to drought stress differently depending on the occurrence of stress during the pre-flowering and post-flowering stages. Of these two critical stages, post-flowering stress is the major constraint for sorghum productivity. Studies have identified agro-morpho-physiological traits like decreased leaf growth, deep root systems, high leaf epicuticular wax, osmotic adjustment, and quiescence (Dugas et al., 2011). Numerous studies have identified that stay-green (delayed senescence) is an important trait linked with drought tolerance as it sustains nitrogen during the grain-filling stage under water stress conditions (Borrell and Hammer, 2000; Sanchez et al., 2002). A study by Mace et al. identified traits like nodal roots, deep root traits, and metaxylem vessels per plant with higher water extraction efficiency associated with drought tolerance (2012). Also, limiting transpiration rate (TR) under high vapor pressure deficit (VPD) tends to increase transpiration efficiency (TE, biomass produced per unit of water utilized), which is an emergent consequence of the relationship between stomatal conductance and VPD. With the pre-flowering water savings from the TE, it could enhance water availability during the grain filling stage, particularly at the end-of-season drought stress (Vadez et al., 2013a ; Zaman-Allah et al., 2011). Additionally, the increased post-flowering water availability to the plant leads to the expression of stay-green (Kholová et al., 2014; Vadez et al., 2011), which is strongly associated with increased grain yield under drought stress (Borrell et al., 2014; Sinclair et al., 2005). Recently, the expression of limited TR under high VPD studied on three different cereals—maize, sorghum, and pearl millet—revealed that pearl millet showed better adaptation than maize, followed by sorghum (Karthika et al., 2018). Furthermore, limited TR under high VPD and progressive soil drying conditions experiments on parental lines of an existing recombinant inbred lines (RILs) of sorghum mapping population revealed that donor parents had lower transpiration than recurrent parents during midday/high VPD conditions (Karthika et al., 2019). The same study also found significant genotypic variation for leaf area normalized with received radiation. A recent review study by Chadalavada et al. suggested that deploying high-throughput genotyping and phenotyping technologies, along with a good understanding of genotype × environment (G×E) interactions, gives greater scope to improve grain quality under drought stress or water-limited conditions in the future decades (2022). Most recently, a comparative study by Rotundo et al. between historical maize and sorghum cultivars' performances under water-limited conditions revealed that the 1970 and older cultivars of sorghum outperformed the maize cultivars (2024). Further, it has been concluded that modern maize cultivars are superior to sorghum cultivars under water-stressed conditions. Efforts were undertaken to identify genomic regions linked to pre- and post-flowering drought stress with several identified donor parents, such as B 35, SC 56, and QL 41 (Sabadin et al., 2012). Researchers at ICRISAT chose six candidate QTLs from the published reports and introgressed them into genetically diverse and tropically adapted parental lines of Asia, Africa, and Latin America (Hash et al., 2003). Additionally, several of the identified stay-green QTLs have been validated in diverse genetic backgrounds of breeding materials (Harris et al., 2007; Kassahun et al., 2010; Vadez et al., 2011). The parents of pre-flowering drought-tolerant (Tx7000 from *Durra*) and post-flowering drought-tolerant (BTx642 from *Kafir*) were re-sequenced and aligned to the sorghum reference genome BTx623 (from *Caduatum)* and identified 1.2 million SNPs and 120,969 indels that distinguished the Tx7000 sequence from BTx623, and 1.6 million SNPs and 152,836 unique indels from the BTx642 sequence that clearly distinguished it from BTx623 (Evans et al., 2013). A study by Johnson et al. assessed the gene expression between stay-green (B35) and senescent (R16) parents (2015). It revealed that delta1-pyrroline-5-carboxylate synthase 2 (P5CS2) was found to be higher in the stay-green line compared to its counterpart, which also exhibited a significant correlation with higher proline levels.

1.6.2 SALINITY STRESS

Soil salinity is a critical abiotic stress that negatively affects both economic and biological yield. It is a major problem in arid and semi-arid tropical regions (Koca et al., 2007), where sorghum is an important staple food crop prone to lower productivity. Sorghum is moderately tolerant to soil salinity stress (Maas et al., 1986) and shows salinity tolerance up to 6.8 dS m^{-1} (Roy et al., 2018). However, further improvement is required to fulfill the demands coming from challenging environments. The germination (Tabatabaei and Anagholi, 2012) and early seedlings (Kausar et al., 2012) are the two most sensitive stages in sorghum to salinity stress, which reduces the grain yield significantly (Almodares and Sharif, 2005) and modifies the plant physiological and biochemical processes (Netondo et al., 2004a, b). and modifies plant physiological and biochemical processes (Netondo et al., 2004a, b). It has been reported that salinity-sensitive sorghum genotypes exhibit increased Na^+ levels and reduced K^+, Ca^{2+}, and K^+/Na^+ ratios in the leaves and roots under saline conditions (Lacerda et al., 2005; Omari and Nhiri, 2015; Zhang et al., 2020). Efforts have been made to improve salinity tolerance through conventional breeding programs (Ahmar et al., 2020; Shahbaz and Ashraf, 2013) or enhance salinity tolerance through the exogenous application of various agents (Desoky et al., 2018; Mofokeng et al., 2017; Nimir et al., 2017; Wang et al., 2014) in sorghum. Also, few studies have reported that gene transformation is one approach to enhance salinity resistance under saline conditions (Maheswari et al., 2010; Reddy et al., 2015; Wang et al., 2014). Limited studies have been found on QTL mapping for salinity tolerance traits in the germination and seedling stages of sorghum (Tang et al., 2015; Wang et al., 2017). Recently, Wang et al. found six major effect QTLs, qTB6, qSFW9, qJW9, qBrix2, qBrix10, and qSTI-Brix9 linked to the salinity tolerance traits in sorghum (2020).

1.6.3 COLD TOLERANCE

Among the various abiotic stresses, cold stress plays a key role in limiting agricultural productivity globally (Pearce and Fuller, 2001). Around 64% of the land worldwide has an average minimum temperature below 0°C, and 48% of the land areas have an average minimum temperature below −10°C, thus making cold tolerance a priority trait in climate-resilient crops (Deane, 1994). The sorghum crop is popular among smallholder farmers because it thrives under heat- and drought-stressed conditions; cold stress restricts the geographic expansion of crop plants and affects plant growth and developmental stages from germination to flowering (Doggett, 1995; Lyons, 1973; Peacock, 1982). The minimum temperature of 16°C is required for normal crop growth and development in sorghum (Paul, 1990). The germination and seedling establishment stages are highly vulnerable to low temperatures and cause a significant grain yield reduction in sorghum. Cold stress causes reduced physiological traits, including root hydraulic conductivity, leaf water potential, and loss of turgor pressure. Several earlier studies have found that cold stress reduced the germination rate, plant height, leaf number, and plant weight when temperatures are below 8 to 10°C in sorghum (Lyons, 1973; Major et al., 1982; Peacock, 1982; Taylor et al., 1972). Marla et al. identified multiple QTLs associated with cold tolerance that overlap with agronomically important genes, including those linked to tannin content (*Tan1* and *Tan2*) and plant height (*Dw1* and *Dw3*) (Hilley et al., 2016; Multani et al., 2003; Wu et al., 2019). Further, a study by Casto et al. (2021) reported cold-tolerance-linked markers located on chromosomes 2, 3, and 6 in sorghum. A study conducted with 48 sorghum lines and hybrids grown under early, mid-, and standard planting dates assessed cold-tolerance-linked traits at the seedling, early vegetative, and maturity stages and found that a delayed panicle initiation period and reduced panicle size under early and mid-planting dates resulted in reduced yield (Emendack et al., 2021). Most recently, a study identified two and three major QTLs in the CT19 × TX430 (C_1) and ICSV700 × M81E (C_2) RIL mapping populations, respectively, that were linked to cold tolerance at the seedling stage in sorghum (La Borde et al., 2023).

1.6.4 HIGH TEMPERATURE

Sorghum is primarily cultivated in arid and semi-arid regions of the world where the current mean air temperature is higher than required for sorghum growth and development, and a further increase in temperature will significantly reduce sorghum yield (Prasad et al., 2006). Both the air temperature and the frequency of warm nights and days are expected to rise by 0.2°C every 10 years in the future. According to the climate model, the mean grain yield of sorghum will decrease by 8–9% for every 1°C mean temperature increase in the sorghum growing season (Lobell and Field, 2007).

Sorghum genotypes varied in pollen germination and seed set at short-term high temperatures in a controlled environment facility (CEF) with two maximum temperatures (high 38°C and optimal 32°C), revealing that pollen germination and seed set traits were highly correlated (Nguyen et al., 2013). The genotypic differences were found for seed set percentage for both threshold temperature and increased temperature above threshold (Singh et al., 2015). Singh et al. studied the sorghum genotypes under optimum (31 vs 21°C) and higher (38 vs 21°C) day vs night temperatures for 5-day periods starting just before flag leaf full expansion and ending 20 days later (2016). They found that higher temperature before anthesis is the most critical period as it reduces pollen germination, seed set, and grain yield. Further, Prasad et al. reported that early morning flowering and canopy temperature depression were tightly linked to higher temperatures in sorghum (2019). Limited studies have been conducted and found that cultivars with stay-green traits seem to have heat tolerance in sorghum (Khizzah et al., 1993). They assessed reproductive high-temperature-stress tolerance in four sorghum inbred parental lines, BTx3197, RTx7000, RTx430, and B35, and their direct and reciprocal F_1 and F_2 populations and found that the inbreds were more tolerant than the F_1s. Further, they reported that two loci were responsible for the expression of heat tolerance and complete dominance at both gene pairs, but one gene, when dominant, is epistatic to the other. The sorghum breeding programs at ICRISAT identified that breeding lines ICSB 55, ICSB 84, ICSB 603, ICSR 14001, ICSR 8, ICSR 21, ICSV 162, and ICSV 376 had 100% seed set in both rainy and post-rainy seasons, indicating these lines possess heat tolerance (Ashok Kumar et al., 2015).

1.6.5 FLOODING (WATERLOGGING) TOLERANCE

Under the scenario of climate change, the frequency of heavy rains has increased, making waterlogging one of the most severe abiotic stresses that put agricultural crops under selection pressure (Gebrehiwot, 2018). This crop is affected by intermittent or long-term waterlogging due to excessive irrigation, heavy rain, and storms (Tari et al., 2013). Plants that are growing in flood conditions experience hypoxia (lack of oxygen) or anoxia (absence of oxygen). The most significant physiological function in crop plants that is highly vulnerable to waterlogging is photosynthesis (Zhang et al., 2019a). Stomatal closure is the first reaction of plants expressed when they are exposed to waterlogging, leading to a restrained decrease in gas exchange, which limits photosynthesis (Zhang et al., 2019b). The flood-tolerant sorghum genotype SSG-59-3 showed higher levels of ethanol and alcohol dehydrogenase and lactate dehydrogenase activities than the sensitive genotype S-308 in the roots (Jain et al., 2010). Promkhambut et al. found that tolerant sorghum genotypes showed aerenchyma formation from the roots to the stalk base in flooded soils (2011). Most recently, a waterlogging experiment was conducted at the five-leaf stage (T1), flowering stage (T2), grain-filling stage (T3), and no waterlogging conditions (control) and reported that sorghum genotypes responded differently to waterlogging timings, with maximum negative effects observed at T1, followed by T2 and T3 stages (Zhang et al., 2023). The tissue-specific transcripts of aquaporin genes (*PIP2-6, PIP2-7, TIP2-2, TIP4-4,* and *TIP5-1*) were expressed under waterlogging stress conditions (Kadam et al., 2017).

1.6.6 ALUMINUM TOLERANCE

Aluminum (Al) toxicity is a major abiotic stress in the arid and semi-arid tropical regions around the globe, where acidic soil is prevalent. Al mainly affects root growth, thereby limiting the absorption of water and nutrients, leading to reduced grain quality under acidic conditions (Čiamporová, 2002). The monogenic dominant gene was responsible for Al tolerance in sorghum (Furlani and Bastos, 1990). Studies have also concluded that Al tolerance is controlled in either an additive or epistatic manner with the major Al gene in sorghum (Magalhaes et al., 2013). The Al tolerance locus, AltSB, located on the terminal region of sorghum chromosome 3, is driven by the activity of the Al-induced and Al-activated root citrate transporter, SbMATE, in sorghum (Magalhaes et al., 2007). The root citrate transporter SbMATE found in the guinea sorghum line SC283 has shown enhanced grain yield under acid and Al toxic soil (Carvalho et al., 2016).

1.6.7 LODGING TOLERANCE (LT)

Sorghum is mainly cultivated by smallholder farmers in areas with limited water availability and most often lodges due to post-flowering drought, which leads to reduced grain yield and quality. Lodging incidence and severity mainly depend on the developmental stages of the crop plant and environmental conditions. The LT mechanism in sorghum has been explored less. However, it is believed that lodging occurs during the grain-filling stage under water stress, which leads to grain loss globally, particularly where grains are harvested mechanically (Johnson et al., 1997). Due to its complex nature, limited studies have been conducted to understand the genetics of LT in sorghum (Kebede et al., 2001; Murray et al., 2008; Srinivasa Reddy et al., 2008). These studies have identified seven QTLs linked to LT in sorghum; however, no common QTL was detected across all these studies. Recently, a genome-wide association study (GWAS) was conducted to dissect the genetic architecture of lodging on 2308 unique grain sorghum hybrids grown across 17 Australian sorghum trials evaluated over three years and detected 213 QTLs, most of which exhibited significant associations with plant height and leaf senescence (71% and 72%, respectively), while 16 lodging tolerance QTLs had no association with either plant height or leaf senescence (Wang et al., 2020).

1.7 NUTRIENTS USE EFFICIENCY

Due to the current climatic changes, world agriculture faces a severe threat in terms of feeding a healthy diet to over 3 billion people, and it is expected to reach 9 billion by 2050 (FAO, 2013). Sorghum is an important food crop cultivated widely in arid and semi-arid tropical regions worldwide. These regions are characterized by intermittent rain, severe drought, limited water availability, and infertile and low-input soils. It is well known that sorghum performs well under these conditions. Improving and adapting sorghum cultivars under low N and P conditions are crucial for agricultural productivity, food security, and farmers' livelihoods in sorghum-growing regions. Under nutrient-deficient soil conditions, crop plants evolved with two basic adaptation strategies: higher nutrient acquisition efficiency from soils and nutrient-use efficiency. Sorghum productivity improved through enhancing resource-use efficiency, yield, and quality, and protecting environmental quality that would affect a large number of people around the world. This crop is well adapted and provides optimal yield under marginal soils with no supplies of nutrients (Qi et al., 2016). It was reported that higher grain yields were achieved under intensive agricultural systems with an adequate supply of nitrogen in sorghum (Bollam et al., 2021; Mahama et al., 2014). Extensive studies have been conducted on nitrogen use efficiency in sorghum (Bollam et al., 2021; Mahama et al., 2014; Melaku et al., 2017). The grain yield of sorghum is majorly determined by both grain number and weight. The formation of grain number is mostly influenced during the early phase of the plant developmental cycle (panicle initiation, gametogenesis, and anthesis), whereas grain weight is influenced by post-flowering (Maman et al., 2004). Worland et al. reported that nitrogen application time could influence grain size because of increased nitrogen supply during the grain filling stage, which further supports an increase in grain yield (2017). It has been shown that maintaining grain number after a post-flowering water deficit was more crucial than maintaining individual grain weight (Adotey et al., 2021). It was noted that leaf sheaths and leaf blades are the predominant N sinks in sorghum, and high levels of nitrogen application (10 mM KNO_3) at 55 days after sowing led to four times higher nitrate in the leaf sheaths than in the control (Worland et al., 2017). Multiple genes have been identified and reported in sorghum (mainly nitrate/peptide transporter 1 (*NRT1/PTR*) genes contributing to N uptake), which could be utilized in breeding programs for further enhancement of nitrogen use efficiency (Massel et al., 2016). The low phosphorus soil is the key sorghum productivity barrier in West and Central Africa (WCA) (Buerkert et al., 2001). Phosphorus is the second most limiting nutrient after nitrogen affecting crop productivity. Several studies have been conducted to assess both adaptation strategies in West African sorghum (Haussmann et al., 2012; Leiser et al., 2012, 2014). A study by Leiser et al. assessed grain yields of 70 West African diverse sorghum varieties under −P (no P fertilization) and +P conditions at two locations over five years in Mali (2012). The grain yield was found to have high estimates of genetic variance and broad-sense heritability's in both the +P ($h^2 = 0.92$) and −P ($h^2 = 0.93$) environments (Leiser et al., 2012). The substantial genetic correlation (rG = 0.89) between grain yield performance under −P and +P conditions indicates that West African sorghum cultivars are generally well-adapted to low-P settings. Significant genotypic variation was observed for both P uptake and P use efficiency traits among all genotypes and within landraces (Leiser et al., 2014). The P uptake traits showed a larger variation compared to P use efficiency traits among studied entries (Leiser et al., 2014). Furthermore, landraces showed higher P uptake and grain P concentration, whereas earlier bred genotypes exhibited a higher P use efficiency. Also, photoperiod sensitivity is highly linked with higher P uptake (Leiser et al., 2014). The same study used *AltSB*-aluminum tolerance-linked specific markers and found strong associations of *SbMATE* with grain yield production under low-P conditions in multiple environments in West Africa. They concluded that *SbMATE* confers P use efficiency (PUE) in addition to Al tolerance (Drouillon and Merckx, 2003) and by enhancing root development in Al-tolerant genotypes, thereby increasing P uptake (Magalhaes et al., 2018). Leiser et al. found that no single P parameter (P-utilization or P-uptake) linked trait was appropriate as a single selection criterion for increased crop productivity under limited P conditions (2015).

1.8 GRAIN NUTRIENT TRAITS

Sorghum is an important staple cereal crop for millions of people living in developing countries around the globe. According to United Nations Food and Agriculture Organization estimates, 780 million of the 792.5 million malnourished people worldwide reside in developing nations (McGuire, 2015). Apart from this, almost two billion people worldwide suffer from "hidden hunger or micronutrient malnutrition," caused by insufficient daily dietary energy intake but a nutrient-poor diet (Hedge, 2016; Muthayya et al., 2013), even though increased food crop production (Gould, 2017).

However, to meet the daily requirements, the population's diet, which primarily consists of cereals like rice, wheat, cassava, and maize, is deficient in several minerals, including iron, zinc, vitamin A, calcium, manganese, copper, iodine, and selenium. Efforts have been made to address micronutrient malnutrition through food fortification, dietary supplementation, and diversification. The low-cost solution to address this issue sustainably is staple food crops biofortification. Sorghum is the fifth most important cereal staple crop grown in more than 30 countries in the arid and semi-arid regions of Africa and Asia. Along with other millets, sorghum is also considered "nutri-cereals" by providing the cheapest sources of energy, protein content, iron, and zinc, and it contributes over 50% of the grain iron and zinc requirement in low-income groups in India (Ashok Kumar et al., 2011, 2013; Parthasarathy Rao et al., 2006).

Enormous genetic variability exists in the ICRISAT sorghum breeding programs for grain Fe and Zn contents; the elite parental lines generally vary from 20 to 40 ppm for grain Fe, while grain Zn concentration ranges from 20 to 30 ppm (Ashok Kumar et al., 2013). Several studies have shown that extensive genetic variability available for grain Fe ranged from 6.3 to 168.1 ppm, and grain Zn varied from 3.6 to 91.3 ppm (Ashok Kumar et al., 2013; Hariprasanna et al., 2014; Kayode et al., 2006; Kumar et al., 2010; Ng'uni et al., 2012; Reddy et al., 2005; Reddy et al., 2010) in sorghum.

The grain Fe concentration is predominantly controlled by additive and non-additive gene action, whereas grain Zn is controlled by additive gene action (Ashok Kumar et al., 2013). There were highly significant positive associations observed between grain Fe and Zn concentrations ($r^2 = 0.79$; $P < 0.01$), and these two grain nutrients had no negative association with grain yield (Ashok Kumar et al., 2013; Phuke et al., 2017; Reddy et al., 2005). Also, the protein content was positively associated with zinc concentration, while copper showed significant positive associations with grain yield, protein, and all other micronutrients except zinc. Additionally, grain Fe concentration is positively associated with manganese and calcium contents.

1.9 HETEROTIC POOLS

The success of any crop breeding program depends on the availability of genotypic variation in the germplasm/breeding pools. Genetic similarity (or genetic distance (GD)) estimates among available germplasms and/or elite genotypes are helpful for the classification of these germplasms and genotypes into heterotic groups/pools. Heterotic groups can be formed or established based on pedigree information, agronomical and morphological traits, geographical origin, and molecular marker information (Melchinger, 1999). Efforts have been made to identify heterotic groups in sorghum. For example, Menz et al. reported that studied germplasms were grouped into heterotic pools based on the working groups of sorghum like Zera-Zera, Kafir, Kafir-Milo, Durra, and Feterita, with the help of molecular markers (2004). Studies have also found a positive association between the GD between parental lines and grain yield heterosis in sorghum (Gabriel, 2005; Jordan et al., 2003). In Ethiopian sorghum breeding programs, six heterotic groups were identified from 184 sorghum genotypes based on geographical origin, agro-ecological adaptation, and whole genome sequence-based SNP genetic differentiation (Mindaye et al., 2015). Like many hybrid crops, the B- and R- parental lines in sorghum are considered two different heterotic pools, but these two pools may not genetically diverge into two clear-cut heterotic groups in sorghum (Menz et al., 2004; Mindaye et al., 2015). However, recently, Silva et al. reported 160 sorghum lines (B-lines and R-lines) genotyped with 29,649 GBS (Genotyping-by-Sequencing) identified SNPs (Single Nucleotide Polymorphism) and found four heterotic groups based on the maintainer and restorer response with pedigree information (2020). Further, a total of 435 inbred lines of sorghum from the Texas A&M program were genotyped by GBS, and these lines were divided into three groups, namely B-lines, R-lines, and forage lines using K-means clustering based on genetic similarity. They also noted that these groups were not highly diverged from one another (Crozier et al., 2020). More recently, two heterotic groups were found in rabi sorghum based on their maintainer and restorer responses (Verma et al., 2022).

1.10 GENOMIC SELECTION

The development and advancement of molecular markers and genotyping technologies help in improving genetic gain rapidly in crop breeding programs. The application of MAS in breeding programs improved the efficiency of selection, especially for traits controlled by major genes (Lande and Thompson, 1990). The agronomically important traits of crops are controlled by many minor genes (Riedelsheimer et al., 2012), where MAS is ineffective. To overcome this limitation, genomic selection (GS) uses genome-wide markers distributed across all linkage groups to predict breeding value (BV). The GS method accelerates genetic gain in crop breeding by reducing the cost per breeding cycle, reducing the breeding cycle time (Hickey et al., 2014), and reducing labor costs (Bhat et al., 2016) in comparison to traditional breeding.

Limited studies have been conducted on GS in sorghum and found it effective (de Oliveira et al., 2018; dos Santos et al., 2020; Fernandes et al., 2018; Habyarimana et al., 2020; Hunt et al., 2018; Sapkota et al., 2020; Velazco et al., 2019; Yu et al., 2016). All these studies have focused mainly on model training on training populations like testcross hybrids, natural populations, and mixed populations. However, GS is less explored in most sorghum breeding due to limited model testing and optimization. Over the years, many GS statistical models have been developed and applied in sorghum breeding, and the rrBLUP model was found to be promising (Dos Santos et al., 2020; Yu et al., 2016). Due to variable production environments, the inclusion of G × E (genotype by environment) interaction effects in genomic prediction models improves selection efficiency and hybrid deployment (Hunt et al., 2018). Furthermore, Habyarimana et al. reported that multi-trait GS performs better than single-trait GS (2020). Recently, Fonseca et al. applied classical GCA-SCA–based (BRR) and kernel-based Bayesian GBLUP models (GB) in grain sorghum and observed improved prediction ability of these models and the mean prediction accuracies (0.31–0.54) for grain yield in one parent out strategy (2021). They also found that the GB model had higher accuracy when

G × E interaction effects were included in the model than that of the BRR model (Fonseca et al., 2021). Most recently, higher prediction accuracy was observed in single cross hybrids of grain sorghum when both parents (T2) data were included in the prediction models. In contrast, the inclusion of female parent-only (T1F) in the prediction models showed a drastic reduction (12–90%) in genomic prediction accuracy (Sapkota et al., 2023). Additionally, a study by Maulana et al. reported that increasing the training population size showed higher prediction accuracy for all the studied traits of 204 hybrids (2023). They also found that the prediction accuracies for thousand kernel weight (TKW) and grain yield (GY) were 0.03 and 0.58 with the partial (additive) model, while in the whole (additive and dominance) model, the prediction accuracies varied from 0.06 to 0.67 for TKW and GY, respectively.

1.11 ACCELERATED CROP IMPROVEMENT BREEDING TECHNIQUES

1.11.1 SPEED BREEDING

In general, breeding cultivars through traditional breeding programs requires parental selection with the breeder's traits of interest, followed by crossing to develop F_1 and superior progenies advanced through selection to release the cultivars that fulfill market demands (Shimelis and Laing, 2012), and it normally takes 8 to 12 years to develop and release a cultivar. The speed breeding method mainly relies on adjustments of photoperiod, temperature, light intensity, soil moisture and nutrition, and high-density planting, thereby inducing early flowering. For instance, Yang et al. showed that early flowering was induced when sensitive genotypes (BTx642/Tx7000 RIL population and parental lines) were exposed to either LD conditions (14/10 hr light/dark) or SD conditions (10/14 hr light/dark) at 30/23°C day/night temperature in sorghum (2014). Also, temperatures above the critical period (33°C) led to a reduction in pollen viability and caused an increase in male sterility in sorghum (Singh et al., 2015). It has also been reported that sorghum genotypes grown in well-watered conditions flowered nine days earlier than the genotypes grown in drought-stressed conditions (Munamava and Riddoch, 2001). Additionally, plant development and grain yield were unaffected by plant density of four to eight plants m^{-2} (Jones and Johnson, 1991). Another study reported that flowering days reduced from 59 to 50 days when increased plant density (increased from 16 to 38 plants m^{-2}) was maintained in sorghum (Villar et al., 1989). The availability and use of glasshouses and plant growth chambers with controlled temperature, photoperiod, pot size, irrigation management, and *in vitro* embryo rescue procedures enable the acceleration of sorghum growth cycles. With this procedure, sorghum can be grown for up to six generations in a year with the help of management techniques such as tiny pots, limited watering, continuous lighting, and embryo rescue (dan Anisiyah, 2020; Ghanim et al., 2014).

A study was also conducted to shorten the breeding cycle of sorghum. It showed that both crossing and selfing in uni-culm plants and the embryo rescue method in sorghum were used and shortened the breeding cycle from 17 to 11 weeks, which could help breeders develop breeding materials much faster (Rizal et al., 2014). More recently, Ghanim et al. manipulated the temperature, watering, photoperiod, pot size, and the use of immature embryo rescue to speed up the growth cycle in sorghum (2023). They reported that the time required for seed to seed or embryos was reduced to 2 to 3 months under optimized protocols in the limited rooting areas and with the embryo rescue technique from 5 to 6 months under natural field conditions, and this process can be repeated four to six times annually (Ghanim et al., 2014).

1.11.2 APOMIXIS

The apomixis method is a breeding strategy of interest that can fix and propagate genetic heterozygosity and rapidly increase hybrid vigor. It also reduces the time required to generate new cultivars and lowers seed production costs. Studies have shown that apomixis is a complex phenomenon that depends on a combination of one genotype: parthenogenesis of the egg cell, autonomous endospermogenesis (or formation of pseudogamous endosperm), and the ability to form unreduced embryo sacs (either diplosporous, arising from disturbances in meiosis, or aposporous, developing from somatic cells located near the macrospore mother cell or meiotic embryo sac) (Barcaccia and Albertini, 2020; Fiaz et al., 2020). The progenies derived from apomictic lines × sexual lines exhibited reduced apomictic seed formation and differed in their ability to transmit the characteristic of facultative apomixis (Tang et al., 1980).

Several diverse breeding lines/accessions have been most often used to study the apomixis mechanism in sorghum. The most frequently used breeding line to express any level of apomixis is R473 (Murthy, 1993). There are many reports available on the identification of aposporous or diplosporous embryo sacs (ESs) (Carman et al., 2011; El'konin et al., 1995, 2012; Hanna et al., 1970; Ping et al., 2004; Rao et al., 1978; Wu et al., 1994) and the autonomous development of embryo and endosperm by cytological methods (El'konin et al., 1995, 2012; Ping et al., 2004; Rao et al., 1978; Wu et al., 1994). The line AS-1a utilized in the apomixis study revealed that apomictic properties were expressed under hot and drought conditions, whereas they did not express in moderate and wet conditions (El'konin et al., 2012). Recently, the presence of apomixis functional properties (apospory, parthenogenesis, and autonomous endospermogenesis) was reported in the grain sorghum line AS-3 (Belyaeva et al., 2021).

1.11.3 DOUBLE HAPLOID

Doubled haploid (DH) technology has shown great promise for hastening the development of complete homozygous lines in a shorter time compared to traditional plant breeding methods. Recently, Hussain and Franks screened 4000 sorghum germplasms collected worldwide and identified two haploid inducer lines, SMHI01 and SMHI02D, which were shown to have a frequency of haploid generation of about 1–2% (2019).

1.12 PRECISION PHENOTYPING METHODS

Phenotyping a large number of germplasms/genotypes has been difficult through conventional breeding, requiring extensive time- and resource-consuming measurements (Qiu et al., 2018). However, advancing high-throughput phenotyping (HTP) technologies in recent times will greatly benefit and help achieve food production to feed 10 billion people (Ray et al., 2013; Tester and Langridge, 2010). The HTP is a non-destructive and non-invasive approach to evaluating complex traits associated with plant growth and development and targeted traits in agriculturally important crops. Over the past 10 years, studies have been conducted to identify the traits linked to drought adaptation and found that the ability of plants to limit transpiration rates per unit green leaf area (TR) under high evaporative demand (Kholová et al., 2010; Sinclair et al., 2005; Zaman-Allah et al., 2011). The availability of the LeasyScan platform allows for the simultaneous measurement of transpiration rates, leaf area, and environmental conditions (VPD) without harvesting the plants (Vadez et al., 2015). Recent updates on the LeasyScan platform, such as continuous monitoring of water use along with 3D scanning of the crop canopy, would help identify key traits for drought adaptation (Vadez et al., 2017).

The protein content is phenotyped using the Technicon autoanalyzer (TAA) method (Jambunathan et al., 1983). The β-carotene content present in the yellow-colored endosperm lines did not exceed 1.1 ppm and is phenotyped using high-performance liquid chromatography (HPLC). The grain Fe and Zn concentrations are accurately estimated using Atomic Absorption Spectrometry and inductively coupled plasma-optical emission spectrometry (ICP-OES) (Houk, 1986). However, in recent times, the X-ray fluorescence spectrometer (XRF) method has emerged as the most robust method for estimating grain Fe and Zn concentrations in sorghum (Ashok Kumar et al., 2013b; Ashok et al., 2015). Additionally, with recent advancements in the histochemical method, NanoSIMS showed that $^{56}Fe+$ and $^{64}Zn+$ are present in the sub-aleurone layer and starchy endosperm. Furthermore, the presence of $^{31}P^{16}O^{+}$ in the same locations indicates that Fe and Zn are present as phytates (Gaddameedi et al., 2022). Sorghum forage quality traits were measured using NIRS (near-infrared spectroscopy) without grinding the samples (Blümmel et al., 2009).

1.13 SYNERGIZING CULTIVAR IMPROVEMENT AND PRODUCTION SYSTEMS

The Indian grain sorghum is widely grown in *kharif* (rainy), *rabi* (post-rainy), and summer (limited irrigations) in several states of India (Kumar et al., 2010). In recent times, sorghum *kharif* hybrids have largely been cultivated in the nontraditional areas in Guntur and East Godavari districts of Andhra Pradesh, where farmers have been reporting high grain yields of 5–7 Mg ha^{-1} (Mishra et al., 2011). It is also grown in rice fallows under zero tillage in the December months of every year, and it matures during dry April, facilitating quality grain harvest. The sorghum breeding program in India is mainly targeted at grain yield, biotic and abiotic stress tolerance, and grain quality traits across longer (kharif cultivars) and shorter (rabi cultivars) photoperiod seasons to achieve higher productivity. To date, only one cultivar, CSH 13, has shown improved crop productivity across all three seasons (Kumar et al., 2009).

A multi-locational trial on crop management practices like conventional, reduced, and minimum tillage in sorghum crops revealed that the cost of cultivation and energy requirements were reduced under reduced and minimum tillage treatments compared to conventional tillage (Mishra et al., 2014). Crop establishment studies conducted across multi-locational trials at the All-India level showed that early-sown crops in the kharif season achieved a 10% greater yield. In contrast, in the rabi season, early and delayed sowing could reduce the grain yield by 9% and 11%, respectively (Ravi et al., 2009). The use of both organic and inorganic sources of nitrogen management trials was conducted multi-locationally in India, revealing that location-specific recommendations had great promise in improving crop productivity. For instance, 25% of farmyard manure (FYM) helped boost productivity in Akola, Maharashtra, whereas 25% of vermicompost (VER) combined with the 75% recommended N rate helped increase yield at Indore, Madhya Pradesh. The row spacing also greatly influences the crop productivity. Under low-yielding environments, conventional row spacing (75 cm) seems to be the best option compared to narrow row spacing (25 cm). Narrow spacing has several advantages, including canopy closure, improved weed control, and reduced evaporation (Sanabria et al., 1995). Clump planting (e.g., planting in a group of three plants) compared to uniformly spaced plants exhibited comparable or superior yield. A study by Zougmore et al. found that sorghum-cowpea intercropping doubled the grain yield compared to the monocrop (2000). Simulation studies have concluded that maize production is more sensitive to temperatures and rainfall compared to other cereals like sorghum (Lobell et al., 2008), and it has been proposed that sorghum could substitute maize in some regions of the world in the future (Lobell et al., 2008).

1.14 CONCLUSION

Sorghum is an important, versatile dryland crop for resource-poor farmers. Given the growing consciousness among urban consumers regarding the nutritional significance of superfoods, such as sorghum, there appears to be a promising future for sorghum as a nutrient-dense food. The recent emphasis on increasing grain productivity and quality, fodder (green and dry), and industrial bioenergy conversion usage will enable breeders to diversify genotypes, and end uses will help sorghum face the competition of other cereals under better soils and rainfall. Applying genomic technologies, readily available, flexible bioinformatics databases and services that can both capture and provide information across research platforms allow for the smooth transfer of data with partners, open databases, the acquisition of high-throughput bioinformatics instruments, and facilities for data analysis and storage. Being a cheaper source of nutritious food and fodder, this crop can be easily incentivized to grow and consume in both growing and non-growing cropping areas.

KEYWORDS

- **accelerated breeding**
- **apomixis method**
- **genomic selection**
- **molecular markers**
- **precision phenotyping**
- **soil salinity**
- **sorghum midge**
- **stem borer**

REFERENCES

Afolayan, G., Aladele, S. E., Deshpande, S. P., Oduoye, O. T., Nwosu, D. J., Michael, C., Blay, E. T., & Danquah, E. Y. (2019). Marker assisted foreground selection for identification of striga resistant backcross lines in *Sorghum bicolor*. *Covenant Journal of Physical and Life Sciences*, *7*(1), 29–36.

Ahluwalia, M., & Rao, V. M. J. (1980). Fodder quality in sorghum. *Indian Journal of Genetics and Plant Breeding*, *40*(3), 554–557.

Ahmar, S., Gill, R. A., Jung, K., Faheem, A., Qasim, M. U., Mubeen, M., & Zhou, W. (2020). Conventional and molecular techniques from simple breeding to speed breeding in crop plants: Recent advances and future outlook. *International Journal of Molecular Sciences*, *21*, 2590.

Ahn, E., Prom, L. K., & Magill, C. (2023). Multi-trait genome-wide association studies of *Sorghum bicolor* regarding resistance to anthracnose, downy mildew, grain mold and head smut. *Pathogens*, *12*, 779.

Ali, R., Hash, C. T., Damris, O., Elhussein, A., & Mohamed, A. H. (2016). Introgression of striga resistance into popular Sudanese sorghum varieties using marker assisted selection. *World Journal of Biotechnology*, *1*(1), 48–55.

Almodares, A., & Sharif, M. E. (2005). Effect of water quality on yield of sugar beet and sweet sorghum. *Journal of Environmental Biology*, *26*(3), 487–493.

Antar, M., Lyu, D., Nazari, M., Shah, A., Zhou, X., & Smith, D. (2021). Biomass for a sustainable bioeconomy: An overview of world biomass production and utilization. *Renewable and Sustainable Energy Reviews*, *139*, 110691.

Aruna, C., & Audilakshmi, S. (2008). A strategy to identify potential germplasm for improving yield attributes using diversity analysis in sorghum. *Plant Genetic Resources*, *6*(3), 187–194. https://doi.org/10.1017/s1479262108994144 (accessed on 7th November 2025).

Aruna, C., & Cheruku, D. (2019). Genetic improvement of grain sorghum. In *Breeding Sorghum for Diverse End Uses* (pp. 157–173). Elsevier. https://doi.org/10.1016/B978-0-08-101879-8.00010-3 (accessed on 7th November 2025).

Ashok Kumar, A., Reddy, B. V. S., & Grando, S. (2015). Global millets improvement and its relevance to India and developing world. In *Millets: Promotion for Food, Feed, Fodder, Nutritional and Environment Security. Proceedings of Global Consultation on Millets Promotion for Health & Nutritional Security* (pp. 154–172). Society for Millets Research, Indian Council of Agricultural Research Indian Institute of Millets Research: Hyderabad, India.

Ashok Kumar, A., Reddy, B. V. S., Ramaiah, B., Sahrawat, K. L., & Pfeiffer, W. H. (2013b). Gene effects and heterosis for grain iron and zinc concentration in sorghum (*Sorghum bicolor* (L.) Moench). *Field Crops Research*, *146*, 86–95.

Ashok Kumar, A., Reddy, B. V. S., Sharma, H. C., Hash, C. T., Srinivasa, R. P., Ramaiah, B., & Reddy, P. S. (2011). Recent advances in sorghum genetic enhancement research at ICRISAT. *American Journal of Plant Sciences*, *2*, 589–600.

Ashok Kumar, A., Sharma, H. S., Sharma, R., Blummel, M., Reddy, P. S., & Reddy, B. V. S. (2013). Phenotyping in sorghum (*Sorghum bicolor* (L.) Moench). In Panguluri, S. K., & Ashok Kumar, A. (Eds.), *Phenotyping for Plant Breeding* (pp. 73–109). Springer, New York.

Ashok, K. A., Anuradha, K., Ramaiah, B., Frederick, R., Virk, P., Peiffer, W., & Stefania, G. (2015). Recent advances in sorghum biofortification research. *Plant Breeding Reviews*, *39*, 89–124.

Bandyopadhyay, R., Frederickson, D. E., McLaren, N. W., Odvody, G. N., & Ryley, M. J. (1998). Ergot: A new disease threat to sorghum in the Americas and Australia. *Plant Disease*, *82*, 356–367.

Barcaccia, G., & Albertini, E. (2013). Apomixis in plant reproduction: A novel perspective on an old dilemma. *Plant Reproduction*, *26*(3), 159–179.

Barnes, R. F., & Gustine, D. L. (1973). Allelochemistry and forage crops. In Metches (Ed.), *Anti-Quality Components of Forages* (p. 1). Crop Science Society of America, Madison.

Bean, B., et al. (2002). 2001 Texas Panhandle Irrigated Sorghum Silage Trial. Accessed at: http://amarillo.tamu.edu/amarillo-center-programs/agronomy/forage-sorghum/ (accessed on 7th November 2025).

Belyaeva, E. V., Elkonin, L. A., Vladimirova, A. A., & Panin, V. M. (2021). Manifestation of apomictic potentials in the line AS-3 of *Sorghum bicolor* (L.) Moench. *Planta*, *254*, 1–14.

Bhat, B. V. (2019). Breeding forage sorghum. In C. Aruna, K. B. R. S. Visarada, B. V. Bhat, & V. A. Tonapi (Eds.), *Breeding Sorghum for Diverse End Uses* (pp. 175–191). Woodhead Publishing.

Bhat, J. A., Ali, S., Salgotra, R. K., Mir, Z. A., Dutta, S., Jadon, V., Tyagi, A., Mushtaq, M., Jain, N., Singh, P. K., Singh, G. P., & Prabhu, K. V. (2016). Genomic selection in the era of next generation sequencing for complex traits in plant breeding. *Frontiers in Genetics*, *7*, 221.

Blaney, B., McKenzie, R., Walters, J., Taylor, L., Bewg, W., Ryley, M. J., & Maryam, R. (2000). Sorghum ergot (*Claviceps africana*) associated with agalactia and feed refusal in pigs and dairy cattle. *Australian Veterinary Journal*, *78*, 102–107.

Blümmel, M., Samad, M., Singh, O. P., & Amede, T. (2009). Opportunities and limitations of food–feed crops for livestock feeding and implications for livestock–water productivity. *The Rangeland Journal*, *31*, 207–212.

Bollam, S., Romana, K. K., Rayaprolu, L., Vemula, A., Das, R. R., Rathore, A., Gandham, P., Chander, G., Deshpande, S. P., & Gupta, R. (2021). Nitrogen use efficiency in sorghum: Exploring native variability for traits under variable N-regimes. *Frontiers in Plant Science*, *12*, 643192.

Brooking, I. R. (1976). Male sterility in *Sorghum bicolor* (L.) Moench induced by low night temperature: I. Timing of the stage of sensitivity. *Australian Journal of Plant Physiology*, *3*, 589–596.

Buerkert, A., Bationo, A., & Piepho, H. P. (2001). Efficient phosphorus application strategies for increased crop production in sub-Saharan West Africa. *Field Crops Research*, *72*, 1–15.

Carman, J., Jamison, M., Elliott, E., Dwivedi, K. K., & Naumova, T. N. (2011). Apospory appears to accelerate onset of meiosis and sexual embryo sac formation in sorghum ovules. *BMC Plant Biology*, *11*(9).

Carvalho, G., Schaffert, R. E., Malosetti, M., Viana, J. H. M., Menezes, C. B., Silva, L. A., Guimaraes, C. T., Coelho, A. M., Kochian, L. V., van Eeuwijk, F. A., & Magalhaes, J. V. (2016). Back to acid soil fields: The citrate transporter SbMATE is a major asset for sustainable grain yield for sorghum cultivated on acid soils. *G3: Genes, Genomes, Genetics*, *6*, 475–484.

Casto, A. L., & Murphy, K. M., & Gehan, M. A. (2021). Coping with cold: Sorghum cold stress from germination to maturity. *Crop Science*, *61*(6), 3894–3907.

Chadalavada, K., Guna, K., Kumari, B. R., & Kumar, T. S. (2022). Drought stress in sorghum: impact on grain quality. In *Climate Change and Crop Stress* (pp. 113–134). Academic Press.

Chatterjee, D., Lesko, T., Peiffer, M., Elango, D., Beuzelin, J., Felton, G. W., & Chopra, S. (2023). Sorghum and maize flavonoids are detrimental to growth and survival of fall armyworm *Spodoptera frugiperda*. *Journal of Pest Science*, *96*(4), 1551–1567.

Cherney, J. H., Patterson, J. A., Cherney, J. H., & Axtell, J. D. (1991). Fiber and soluble phenolic monomer composition of morphological components of sorghum stover. *Journal of the Science of Food and Agriculture*, *54*(4), 645–649.

Čiamporová, M. (2002). Morphological and structural responses of plant roots to aluminum at organ, tissue, and cellular levels. *Biologic Plantarum*, *45*, 161–171.

Craig, J., & Schertz, K. F. (1985). Inheritance of resistance in sorghum to three pathotypes of *Peronosclerospora sorghi*. *Phytopathology*, *75*(9), 1077–1078.

Crozier, D., Hoffmann, L., Jr., Klein, P. E., Klein, R. R., & Rooney, W. L. (2020). Predicting heterosis in grain sorghum hybrids using sequence-based genetic similarity estimates. *Journal of Crop Improvement*, *34*(5), 600–617.

Cummins, D. G., & Dobson, J. W. (1972). Digestibility of bloom and bloomless sorghum leaves as determined by a modified *in vitro* technique. *Agronomy Journal*, *64*, 682–683.

Cummins, D. G., & Sudweeks, E. M. (1976). *In vivo* performance of bloom and bloomless sorghum forage. *Agronomy Journal*, *68*, 735–737. Elder, W. C., & Dennan, C. E. (1966). Sudangrass and Sudangrass Hybrid Research. *Oklahoma Agriculture Experimental Station Processed Series*, p. 543.

Dahlberg, J. A., Bandyopadhyay, R., Rooney, W. L., Odvody, G. N., & Madera-Torres, P. (2001). Evaluation of sorghum germplasm used in US breeding programmers for sources of sugary disease resistance. *Plant Pathology*, *50*, 681–689.

dan Anisiyah, I. W. M. (2020). Accelerated purification of sorghum mutant line by using rapid cycling methods. In *Journal of Physics: Conference Series*, *1436*(1), 012028. IOP Publishing.

de Oliveira, A. A., Pastina, M. M., de Souza, V. F., da Costa Parrella, R. A., Noda, R. W., Simeone, M. L. F., Schaffert, R. E., de Magalhães, J. V., Damasceno, C. M. B., & Margarido, G. R. A. (2018). Genomic prediction applied to high-biomass sorghum for bioenergy production. *Molecular Breeding*, *38*, 49.

de Souza, V. F., da Silva Pereira, G., Pastina, M. M., da Costa Parrella, R. A., Ferreira Simeone, M. L., de Almeida Barros, B., et al. (2021). QTL mapping for bioenergy traits in sweet sorghum recombinant inbred lines. *G3: Genes, Genomes, Genetics, 11*, jkab314. https://doi.org/10.1093/g3journal/jkab314 (accessed on 7th November 2025).

Deane, C. (1994). Whole plant and tissue culture studies on frost tolerance in cauliflower [PhD thesis]. *The National University of Ireland.*

Deshpande, S. P., Mohamed, A., & Thomas Hash Jr, C. (2013). Molecular breeding for *Striga* resistance in sorghum. *Translational Genomics for Crop Breeding: Biotic Stress*, *1*, 77–93.

Desoky, E. M., Merwad, A. M., & Elrys, A. S. (2017). Response of pea plants to natural bio-stimulants under soil salinity stress. *American Journal of Plant Physiology*, *12*, 28–37.

Doggett, H. (1995). *Sorghum*. Wiley.

dos Santos, J. P. R., Fernandes, S. B., McCoy, S., Lozano, R., Brown, P. J., Leakey, A. D. B., Buckler, E. S., Garcia, A. A. F., & Gore, M. A. (2020). Novel Bayesian networks for genomic prediction of developmental traits in biomass sorghum. *G3: Genes, Genomes, Genetics*, *10*, 769–781.

Drouillon, M., & Merckx, R. (2003). The role of citric acid as a phosphorus mobilization mechanism in highly P-fixing soils. *Gayana Botanica*, *60*, 55–62.

Elder, W. C., & Dennan, C. E. (1966). Sudangrass and Sudangrass Hybrid Research. *Oklahoma Agriculture Experimental Station Processed Series*, p. 543.

El'konin, L. A., Belyaeva, E. V., & Fadeeva, I. Y. (2012). Expression of the apomictic potential and selection for apomixis in sorghum line AS-1a. *Russian Journal of Genetics*, *48*(1), 32–40.

El'konin, L. A., Enaleeva, N. K., Belyaeva, E. V., & Ishin, A. G. (1995). Partially fertile line with apospory obtained from tissue culture of male sterile plant of sorghum. *Annals of Botany*, *76*(4), 359–364.

Emendack, Y., Sanchez, J., Hayes, C., Nesbitt, M., Laza, H., & Burke, J. (2021). Seed-to-seed early-season cold resiliency in sorghum. *Scientific Reports*, *11*(1), 7801.

Fernandes, S. B., Dias, K. O. G., Ferreira, D. F., & Brown, P. J. (2018). Efficiency of multi-trait, indirect, and trait-assisted genomic selection for improvement of biomass sorghum. *Theoretical and Applied Genetics*, *131*, 747–755.

Fiaz, S., Wang, X., Younas, A., Alharthi, B., Riaz, A., & Ali, H. (2020). Apomixis and strategies to induce apomixis to preserve hybrid vigor for multiple generations. *GM Crops & Food*, *12*(1), 57–70.

Fonseca, J. M., Klein, P. E., Crossa, J., Pacheco, A., Perez-Rodriguez, P., Ramasamy, P., Klein, R., & Rooney, W. L. (2021). Assessing combining abilities, genomic data, and genotype × environment interactions to predict hybrid grain sorghum performance. *The Plant Genome*, *14*(3), e20127.

Fraser, C. M., & Chapple, C. (2011). The phenylpropanoid pathway in *Arabidopsis*. *The Arabidopsis Book*, *9*, e0152.

Furlani, P., & Bastos, C. (1990). Genetic control of aluminum tolerance in sorghum. In *Genetic Aspects of Plant Mineral Nutrition*. Springer.

Gabriel, K. (2005). A study of heterotic relationships in sorghum [PhD thesis]. *Texas A & M University*.

Gaddameedi, A., Sheraz, S., Kumar, A., Li, K., Pellny, T., Gupta, R., Wan, Y., Moore, K. L., & Shewry, P. R. (2022). The location of iron and zinc in grain of conventional and biofortified lines of sorghum. *Journal of Cereal Science*, *107*, 103531.

Gebrehiwot, K. A. (2018). A review on waterlogging, salinization, and drainage in Ethiopian irrigated agriculture. *Sustainable Water Resources Management*, *4*(1), 55–62.

Ghanim, A. M. A., Ali, A. B., Sen, A., Ingelbrecht, I., & Sivasankar, S. (2024). Rapid cycling and generation advancement for accelerated mutation breeding in sorghum. In Ghanim, A. M. A., Sivasankar, S., & Rich, P. J. (Eds.), *Mutation Breeding and Efficiency Enhancing Technologies for Resistance to Striga in Cereals*. Springer, Berlin, Heidelberg.

Ghanim, A. M. A., Bado, S., Sen, A., Ahmed, M. S., Tassoob-Shirazi, F., & Amos, E. K. (2014). Techniques in accelerating plant mutation breeding in wheat and sorghum. *VISCEA*, 19–20 Feb, Vienna, Austria, p. 30.

Gnansounou, E., Dauriat, A., & Wyman, C. E. (2005). Refining sweet sorghum to ethanol and sugar: Economic trade-offs in the context of North China. *Bioresource Technology*, *96*, 985–1002.

Gorthy, S., Gaddameedi, A., Jagannathan, J., Are, A. K., Deshpande, S. P., Govindaraj, M., & Kasanaboina, K. (2023). Comparison of shoot fly resistance QTLs in sorghum introgression lines using SNP genotyping. *Indian Journal of Entomology*, *85*, 40–44.

Gorz, H. J., Haskins, F. A., & Vogel, K. P. (1986a). Inheritance of dhurrin content in mature sorghum leaves. *Crop Science*, *26*, 65–67.

Gorz, H. J., Haskins, F. A., & Vogel, K. P. (1986b). Registration of NP23 and NP24 Sudan grass germplasms. *Crop Science*, *27*, 212–213.

Gould, J. (2017). Nutrition: A world of insecurity. *Nature Outlook*, *544*, S7.

Gourley, L. M., & Lusk, J. W. (1978). Genetic parameters related to sorghum silage quality. *Journal of Dairy Science*, *61*, 1821–1827.

Grenier, C., Bramel-Cox, P. J., & Harmon, P. (2001). Core collection of sorghum: I. Stratification based on eco-geographical data. *Crop Science*, *41*(1), 234–240.

Grover, S., Shinde, S., Puri, H., Palmer, N., Sarath, G., Sattler, S. E., & Louis, J. (2022). Dynamic regulation of phenylpropanoid pathway metabolites in modulating sorghum defense against fall armyworm. *Frontiers in Plant Science*, *13*, 4858.

Grover, S., Wojahn, B., Varsani, S., Sattler, S. E., & Louis, J. (2019). Resistance to greenbugs in the sorghum nested association mapping population. *Arthropod-Plant Interactions*, *13*, 261–269.

Guden, B., Yol, E., Erdurmus, C., Lucas, S. J., & Uzun, B. (2023). Construction of a high-density genetic linkage map and QTL mapping for bioenergy-related traits in sweet sorghum (*Sorghum bicolor* (L.) Moench). *Frontiers in Plant Science, 14*, 1081931. https://www.frontiersin.org/articles/10.3389/fpls.2023.1081931/full (accessed on 7th November 2025).

Habte, N., Girma, G., Xu, X., Liao, C. J., Adeyanju, A., Hailemariam, S., Lee, S., Okoye, P., Ejeta, G., & Mengiste, T. (2023). Haplotypes at the sorghum ARG4 and ARG5 NLR loci confer resistance to anthracnose. *The Plant Journal*. https://doi.org/10.1111/tpj.16594 (accessed on 7th November 2025).

Habyarimana, E., Lopez-Cruz, M., & Baloch, F. S. (2020). Genomic selection for optimum index with dry biomass yield, dry mass fraction of fresh material, and plant height in biomass sorghum. *Genes (Basel)*, *11*, 61–76.

Hanna, W. W., Schertz, K. F., & Bashaw, E. C. (1970). Apospory in *Sorghum bicolor* (L.) Moench. *Science*, *170*(3955), 338–339.

Hariprasanna, K., Agte, V., Elangovan, M., & Patil, J. V. (2014). Genetic variability for grain iron and zinc content in cultivars, breeding lines and selected germplasm accessions of sorghum (*Sorghum bicolor* (L.) Moench). *Indian Journal of Genetics*, *7*, 42–49.

Haussmann, B. I. G., Hess, D. E., Omanya, G. O., Folkertsma, R. T., Reddy, B. V. S., Kayentao, M., Welz, H. G., & Geiger, H. H. (2004). Genomic regions influencing resistance to the parasitic weed *Striga hermonthica* in two recombinant inbred populations of sorghum. *Theoretical and Applied Genetics*, *109*, 1005–1016.

Haussmann, B. I. G., Rattunde, H. F., Weltzien-Rattunde, E., Traoré, P. S. C., vom Brocke, K., & Parzies, H. K. (2012). Breeding strategies for adaptation of pearl millet and sorghum to climate variability and change in West Africa. *Journal of Agronomy and Crop Science*, *198*, 327–339.

Hickey, J. M., Dreisigacker, S., Crossa, J., Hearne, S., Babu, R., Prasanna, B. M., Grodon, M., Zambelli, A., Windhaussen, V. S., Mathews, K., & Gorjan, G. (2014). Evaluation of genomic selection training population designs and genotyping strategies in plant breeding programs using simulation. *Crop Science*, *54*, 1476–1488.

Hilley, J., Truong, S., Olson, S., Morishige, D., & Mullet, J. (2016). Identification of Dw1, a regulator of sorghum stem internode length. *PLOS ONE*, *11*(3), e0151271.

Hodge, J. (2016). Hidden hunger: Approaches to tackling micronutrient deficiencies. In Gillespie, S., Hodge, J., Yosef, S., & Pandya-Lorch, R. (Eds.), *Nourishing Millions: Stories of Change in Nutrition* (pp. 35–43). International Food Policy Research Institute (IFPRI), Washington.

House, L. R., Murty, D. S., Gomez, M., Verma, B. N., & Sun, Y. (2000). *The Development of Agricultural Industries in Several African and Asian Countries.*

Hoveland, C. S., & Monson, W. G. (1980). Genetic and environmental effects on forage quality. In *Crop Quality, Storage and Utilization* (pp. 139–168). American Society of Agronomy and Crop Science Society of America, Madison, WI, USA.
Hunt, C. H., van Eeuwijk, F. A., Mace, E. S., Hayes, B. J., & Jordan, D. R. (2018). Development of genomic prediction in sorghum. *Crop Science*, *58*, 690.
Hussain, T., & Franks, C. (2019). Discovery of sorghum haploid induction system. *Methods in Molecular Biology*, *1931*, 49–59.
Jain, V., Singla, N. K., Jain, S., & Gupta, K. (2010). Activities of enzymes of fermentation pathways in the leaves and roots of contrasting cultivars of sorghum (*Sorghum bicolor* L.) during flooding. *Physiology and Molecular Biology of Plants*, *16*(3), 241–247.
Jambunathan, R., Rao, N. S., & Gurtu, S. (1983). Rapid methods for estimating protein and lysine in sorghum (*Sorghum bicolor* (L.) Moench). *Cereal Chemistry*, *60*(3), 192–194.
Jin, P., Wang, L., Zhao, W., Zheng, J., Wang, Y. H., Liu, Y., et al. (2021). Construction of high-density genetic map and QTL mapping in sorghum × Sudan grass. *Euphotic*, *217*, 162. https://doi.org/10.1007/s10681-021-02895-9 (accessed on 7th November 2025).
Jones, O. R., & Johnson, G. L. (1991). Row width and plant density effects on Texas high plains sorghum. *Journal of Production Agriculture*, *4*(4), 613–621.
Jordan, D. R., Tao, Y., Godwin, I. D., & Henzell, R. G. (2003). Prediction of hybrid performance in grain sorghum using RFLP markers. *Theoretical and Applied Genetics*, *106*, 559–567.
Kadam, S., Abril, A., Dhanapal, A. P., Koester, R. P., Vermerris, W., Jose, S., & Fritschi, F. B. (2017). Characterization and regulation of aquaporin genes of sorghum (*Sorghum bicolor* (L.) Moench) in response to waterlogging stress. *Frontiers in Plant Science*, *8*, 862.
Kalton, R. R. (1988). Overview of the forage sorghums. In *Proceedings of American Seed Trade Association Annual Corn Sorghum Research Conference* (Vol. 43, pp. 1–12).
Karthika, G., Choudhary, S., Kholova, J., & Vadez, V. (2018). A comparative analysis of transpiration response to atmospheric increasing vapor pressure deficit conditions in cereal crops. *Electronic Journal of Plant Breeding*, *9*(4), 1461–1468.
Karthika, G., Kholova, J., Alimagham, S., Ganesan, M., Chadalavada, K., Kumari, R., & Vadez, V. (2019). Measurement of transpiration restriction under high vapor pressure deficit for sorghum mapping population parents. *Plant Physiology Reports*, *24*, 74–85.
Kausar, A., Ashraf, M. Y., Ali, I., Niaz, M., & Abbass, Q. (2012). Evaluation of sorghum varieties/lines for salt tolerance using physiological indices as screening tool. *Pakistan Journal of Botany*, *44*, 47–52.
Kayode, A. P. P., Linnemann, A. R., & Hounhouigan, J. D. (2006). Genetic and environmental impact on iron, zinc, and phytate in food sorghum grown in Benin. *Journal of Agricultural and Food Chemistry*, *54*, 256–262.
Khalifa, M., & Eltahir, E. A. B. (2023). Assessment of global sorghum production, tolerance, and climate risk. *Frontiers in Sustainable Food Systems*, *7*, 1184373. https://doi.org/10.3389/fsufs.2023.1184373 (accessed on 7th November 2025).
Khatri, R. S., Lodhi, G. P., Grewal, R. P. S., & Pahuja, S. K. (2001b). Inheritance of regrowth in forage sorghum (*Sorghum bicolor* (L.) Moench). *Forage Research*, *27*(2), 147–148.
Khatri, R. S., Pahuja, S. K., Avatar, R., & Phogat, D. S. (2001a). Inheritance of leafiness in forage sorghum (*Sorghum bicolor* (L.) Moench). *Forage Research*, *27*(1), 35–38.
Khizzah, B. W., Miller, F. R., & Newton, R. J. (1993). Inheritance and heritability of heat tolerance in several sorghum cultivars during the reproductive phase. *African Crop Science Journal*, *1*(Suppl. 2), 81–85.
Kholová, J., Hash, C. T., Kumar, L. K., Yadav, R. S., Kocŏvá, M., & Vadez, V. (2010). Terminal drought tolerant pearl millet (*Pennisetum glaucum* (L.) R. Br.) have high leaf ABA and limit transpiration at high vapor pressure deficit. *Journal of Experimental Botany*, *61*, 1431–1440.
Koca, H., Bor, M., Ozdemir, F., & Turkan, I. (2007). The effect of salt stress on lipid peroxidation, antioxidative enzymes and proline content of sesame cultivars. *Environmental and Experimental Botany*, *60*, 344–351.
Kumar, A. A., Reddy, B. V. S., Sahrawat, K. L., & Ramaiah, B. (2010). Combating micronutrient malnutrition: Identification of commercial sorghum cultivars with high grain iron and zinc. *Journal of SAT Agricultural Research*, *8*.

Kumar, H., Dhillon, M. K., & Bhoi, T. K. (2023). Stress-induced defense in sorghum in response to attack by the spotted stem borer, *Chilo partellus* (Swinhoe). *Phytoparasitica, 51*(1), 49–61.

Kumar, R., Nandan, R., Kumar, V., Prasad, S., & Singh, D. (2009). Response of summer mungbean cultivars to sowing time and seed rate. *Indian Journal of Agricultural Sciences, 79*(4), 309–312.

Kumar, S. R., Ramanjaneyulu, A. V., & Krishna, A. (2010). A decadal analysis of improved sorghum (*Sorghum bicolor*) cultivar response to fertilizer application in rainy season and a hypothetical grain production model. *Indian Journal of Agricultural Sciences, 80*, 786–790.

La Borde, N., Rajewski, J., & Dweikat, I. (2023). Novel QTL for chilling tolerance at germination and early seedling stages in sorghum. *Frontiers in Genetics, 14*, 1129460.

Lamb, J. F. S., Haskins, F. A., Gorz, H. J., & Vogel, K. P. (1987). Inheritance of seedling hydrocyanic acid potential and seed weight in sorghum-Sudan grass crosses. *Crop Science, 27*, 522–525.

Lande, R., & Thompson, R. (1990). Efficiency of marker-assisted selection in the improvement of quantitative traits. *Genetics, 124*(3), 743–756.

Leiser, W. L., Rattunde, H. F. W., Piepho, H. P., Weltzien, E., Diallo, A., Melchinger, A. E., Parzies, H. K., & Haussmann, B. I. (2012). Selection strategy for sorghum targeting phosphorus-limited environments in West Africa: Analysis of multi-environment experiments. *Crop Science, 52*(6), 2517–2527.

Leiser, W. L., Rattunde, H. F. W., Piepho, H. P., Weltzien, E., Diallo, A., Toure, A., & Haussmann, B. I. G. (2015). Phosphorus efficiency and tolerance traits for selection of sorghum for performance in phosphorus-limited environments. *Crop Science, 55*, 1152–1162.

Leiser, W. L., Rattunde, H. F. W., Weltzien, E., & Haussmann, B. I. (2014). Phosphorus uptake and use efficiency of diverse West and Central African sorghum genotypes under field conditions in Mali. *Plant and Soil, 377*, 383–394.

Lobell, D. B., & Field, C. B. (2007). Global scale climate-crop yield relationships and the impacts of recent warming. *Environmental Research Letters, 2*, 014002.

Lobell, D. B., Burke, M. B., Tebaldi, C., Mastrandea, M. D., Falcon, W. P., & Naylor, R. L. (2008). Prioritizing climate change adaptation needs for food security in 2030. *Science, 319*, 607–610.

Lyons, J. M. (1973). Chilling injury in plants. *Annual Review of Plant Physiology, 24*(1), 445–466.

Maas, E. V., Poss, J. A., & Hoffman, G. J. (1986). Salinity sensitivity of sorghum at three growth stages. *Irrigation Science, 7*, 1–11.

Magalhaes, J. V., Liu, J., Guimarães, C. T., Lana, U. G. P., Alves, V. M. C., Wang, Y. H., Schaffert, R. E., Hoekenga, O. W., Piñeros, M. A., Shaff, J. E., Klein, P. E., Carneiro, N. P., Coelho, C. M., Trick, H. N., & Kochian, L. V. (2007). A gene in the multidrug and toxic compound extrusion (MATE) family confers aluminum tolerance in sorghum. *Nature Genetics, 39*, 1156–1161.

Magalhaes, J. V., Maron, L. G., Piñeros, M. A., Guimarães, C. T., & Kochian, L. V. (2013). Aluminum tolerance in sorghum and maize. In *Translational Genomics for Crop Breeding: Abiotic Stress, Yield and Quality* (Vol. 2, pp. 83–98).

Magalhaes, J. V., Piñeros, M. A., Maciel, L. S., & Kochian, L. V. (2018). Emerging pleiotropic mechanisms underlying aluminum resistance and phosphorus acquisition on acidic soils. *Frontiers in Plant Science, 9*, 1–12.

Mahama, G. Y., Vara Prasad, P. V., Mengel, D. B., & Tesso, T. T. (2014). Influence of nitrogen fertilizer on growth and yield of grain sorghum hybrids and inbred lines. *Agronomy Journal, 106*, 1623–1630.

Maheswari, M., Varalaxmi, Y., Vijayalakshmi, A., Yadav, S. K., Sharmila, P., Venkateswarlu, B., Vanaja, M., & Saradhi, P. P. (2010). Metabolic engineering using *mtlD* gene enhances tolerance to water deficit and salinity in sorghum. *Biologic Plantarum, 54*, 647–652.

Major, D. J., Hamman, W. M., & Rood, S. B. (1982). Effects of short-duration chilling temperature exposure on growth and development of sorghum. *Field Crops Research, 5*, 129–136.

Marla, S. R., Shiva, S., Welti, R., Liu, S., Burke, J. J., & Morris, G. P. (2017). Comparative transcriptome and lipidome analyses reveal molecular chilling responses in chilling-tolerant sorghums. *The Plant Genome, 10*(3), 1–16.

Mascia, P. N., Portereiko, M., Sorrells, M., & Flavell, R. B. (2010). Designing plants to meet feedstock needs. In P. N. Mascia, J. Scheffran, & J. M. Widholm (Eds.), *Plant Biotechnology for Sustainable Production of Energy and Co-Products* (pp. 57–84). Springer.

Maulana, F., Perumal, R., Serba, D. D., & Tesso, T. (2023). Genomic prediction of hybrid performance in grain sorghum (*Sorghum bicolor* L.). *Frontiers in Plant Science*, *14*, 1139896.

Maw, M. J. W., Houx, J. H., III, & Fritschi, F. B. (2019). Nitrogen content and use efficiency of sweet sorghum grown in the lower Midwest. *Agronomy Journal*, *111*(6), 2920–2928.

McGuire, S. (2015). FAO, IFAD, and WFP. The state of food insecurity in the world 2015: Meeting the 2015 international hunger targets: Taking stock of uneven progress. Rome: FAO. *Advances in Nutrition*, *6*(5), 623–624.

McLaren, N. W. (1997). Changes in pollen viability and concomitant increase in the incidence of sorghum ergot with flowering date and implications in selection for escape resistance. *Phytopathology*, *145*, 261–265.

Melaku, N. D., Bayu, W., Ziadat, F., & Strohmeier, S. (2017). Effect of nitrogen fertilizer rate and timing on sorghum productivity in Ethiopian highland Vertisols. *Archives of Agronomy and Soil Science*, *64*, 480–491.

Melchinger, A. E. (1999). Genetic diversity and heterosis. In *The Genetics and Exploitation of Heterosis in Crops* (pp. 99–118). American Society of Agronomy, Madison, WI.

Menz, M. A., Klein, R. R., Unruh, N. C., Rooney, W. L., Klein, P. E., & Mullet, J. E. (2004). Genetic diversity of public inbreds of sorghum determined by mapped AFLP and SSR markers. *Crop Science*, *44*, 1236–1244.

Mindaye, T. T., Mace, E. S., Godwin, I. D., & Jordan, D. R. (2015). Genetic differentiation analysis for the identification of complementary parental pools for sorghum hybrid breeding in Ethiopia. *Theoretical and Applied Genetics*, *128*, 1765–1775.

Mishra, J. S., Subbarayudu, B., Chapke, R. R., & Seetharama, N. (2011). Evaluation of sorghum (*Sorghum bicolor*) cultivars in rice (*Oryza sativa*) fallows under zero tillage. *Indian Journal of Agricultural Sciences*, *81*, 277–279.

Mishra, J. S., Thakur, N. S., Pushpendra, S., Kubsad, V. S., Kalpana, R., Alse, U. N., & Nemade, S. M. (2014). Tillage and integrated nutrient management in rainy season grain sorghum (*Sorghum bicolor*). *Indian Journal of Agronomy*, *59*, 619–623.

Mofokeng, A. M., Shimelis, H., & Laing, M. (2017). Breeding strategies to improve sorghum quality. *Australian Journal of Crop Science*, *11*, 142–148.

Montgomery, C. R., Nelson, B. D., Joost, R., & Mason, L. F. (1988). Tannin concentration and quality changes in sorghum as affected by maturity and sorghum type. *Crop Science*, *26*, 372–375.

Munamava, M., & Riddoch, I. (2001). Response of three sorghum (*Sorghum bicolor* L. Moench) varieties to soil moisture stress at different developmental stages. *South African Journal of Plant and Soil*, *18*(2), 75–79.

Murray, S. C., Rooney, W. L., Hamblin, M. T., Mitchell, S. E., & Kresovich, S. (2009). Sweet sorghum genetic diversity and association mapping for Brix and height. *The Plant Genome*, *2*, 48–62.

Murty, U. R. (1993). Appraisal on the present status of research on apomixis in sorghum. *Current Science*, *64*(5), 315–318.

Muthayya, A., Rah, J. H., Sugimoto, J. D., Roos, F. F., Kraemer, K., & Black, R. E. (2013). The global hidden hunger indices and maps: An advocacy tool for action. *PLoS One*, *8*(6), e67860.

N/A. (2009). Commission on Genetic Resources for Food and Agriculture. *Draft Second Report on the World's Plant Genetic Resources for Food and Agriculture (Final Version)*. Rome, 330 p. Available from: http://www.fao.org/3/a-k6276e.pdf (accessed on 7th November 2025).

Netondo, G. W., Onyango, J. C., & Beck, E. (2004a). Sorghum and salinity: I. Response of growth, water relations, and ion accumulation to NaCl salinity. *Crop Science*, *44*, 797–805.

Netondo, G. W., Onyango, J. C., & Beck, E. (2004b). Sorghum and salinity: II. Gas exchange and chlorophyll fluorescence of sorghum under salt stress. *Crop Science*, *44*, 806–811.

Ng'uni, D., Geleta, M., Hofvander, P., Fatih, M., & Bryngelsson, T. (2012). Comparative genetic diversity and nutritional quality variation among some important Southern African sorghum accessions (*Sorghum bicolor* (L.) Moench). *Australian Journal of Crop Science*, *6*, 56.

Ngugi, K., Ngugi, A. J., Osama, S., & Mugoya, C. (2015). Combating *Striga* weed in sorghum by transferring resistance quantitative trait loci through molecular marker assisted introgression. *Journal of Plant Breeding and Genetics*, *3*(3), 67–76.

Nguyen, T. C., Singh, V., van Oosterom, E. J., Chapman, S. C., Jordan, D. R., & Hammer, G. L. (2013). Genetic variability in high temperature effects on seed-set in sorghum. *Functional Plant Biology*, *40*, 439–448.

Nimir, N. E. A., Zhou, G., Guo, W., Ma, B., Lu, S., & Wang, Y. (2017). Effect of foliar application of GA3, kinetin, and salicylic acid on ions content, membrane permeability, and photosynthesis under salt stress of sweet sorghum (*Sorghum bicolor* (L.) Moench). *Canadian Journal of Plant Science*, *97*, 525–535.

Pahuja, S. K., Yadav, R., & Grewal, R. P. S. (2003). Genetics of fodder yield and its components in multi-cut × single cut forage sorghum crosses. *Forage Research*, *29*(3), 139–141.

Pandey, A. K. (2010). *Developing Low-HCN Producing Forage Sorghum (Sorghum bicolor [L.]) Using Anti-Sense Approach* (Ph.D. thesis). *Jawaharlal Nehru Technological University*, Hyderabad, India.

Pandey, S., & Shrotria, P. K. (2009). Genetic parameters for hydrocyanic acid content in forage sorghum (*Sorghum bicolor* (L.) Moench). *Forage Research*, *35*(1), 17–19.

Parh, D. K., Jordan, D. R., Aitken, E. A. B., Gogel, B. J., McIntyre, C. L., & Godwin, I. D. (2006). Genetic components of variance and the role of pollen traits in sorghum ergot resistance. *Crop Science*, *46*, 2387–2395.

Parh, D. K., Jordan, D. R., Aitken, E. A. B., Mace, E. S., Jun-Ai, P., McIntyre, C. L., & Godwin, I. D. (2008). QTL analysis of ergot resistance in sorghum. *Theoretical and Applied Genetics*, *117*, 369–382.

Parthasarathy Rao, P., Birthal, B. S., Reddy Belum, V. S., Rai, K. N., & Ramesh, S. (2006). Diagnostics of sorghum and pearl millet grains-based nutrition in India. *International Sorghum and Millets Newsletter*, *47*, 93–96.

Patil, J. V., Rakshit, S., & Khot, K. B. (2013). Genetics of post-flowering drought tolerance traits in post-rainy sorghum (*Sorghum bicolor* (L.) Moench). *Indian Journal of Genetics and Plant Breeding, 73*(1), 44. https://doi.org/10.5958/j.0019-5200.73.1.006 (accessed on 7th November 2025).

Pažoutová, S., Olšovská, J., Linka, M., Kolínská, R., & Flieger, M. (2000). Chemoraces and habitat specialization of *Claviceps purpurea* populations. *Applied and Environmental Microbiology*, *66*, 5419–5425.

Peacock, J. M. (1982). Response and tolerance of sorghum to temperature stress. In *International Crops Research Institute for the Semi-Arid Tropics. Sorghum in the Eighties: Proceedings of the International Symposium on Sorghum* (pp. 143–158). ICRISAT.

Pearce, R. S., & Fuller, M. P. (2001). Freezing of barley studied by infrared video thermography. *Plant Physiology*, *125*, 227–240.

Perumal, R., Nimmakayala, P., Erattaimuthu, S. R., No, E. G., Reddy, U. K., Prom, L. K., Odvody, G. N., Luster, D. G., & Magill, C. W. (2008). Simple sequence repeats markers useful for sorghum downy mildew (*Peronosclerospora sorghi*) and related species. *BMC Genetics*, *9*, 77.

Peterson, G. C., Suksayretrup, K., & Weibel, D. E. (1982). Inheritance of some bloomless and sparse-bloom mutants in sorghum. *Crop Science*, *22*, 63–67.

Phuke, R. M., Anuradha, K., Radhika, K., Jabeen, F., Anuradha, G., Ramesh, T., Hariprasanna, K., Mehtre, S. P., Deshpande, S. P., Anil, G., Das, R. R., Rathore, A., Hash, T., Reddy, B. V. S., & Kumar, A. A. (2017). Genetic variability, genotype × environment interaction, correlation and GGE biplot analysis for grain iron and zinc concentration and other agronomic traits in RIL population of sorghum (*Sorghum bicolor* (L.) Moench). *Frontiers in Plant Science*, *8*, 712.

Prasad, P. V. V., Boote, K. J., & Allen, L. H. (2006). Adverse high temperature effects on pollen viability, seed set, seed yield and harvest index of grain-sorghum (*Sorghum bicolor* (L.) Moench) are more severe at elevated carbon dioxide due to high tissue temperature. *Agricultural and Forest Meteorology*, *139*, 237–251.

Prasad, P. V. V., Djanaguiraman, M., Jagadish, S. V. K., & Ciampitti, I. A. (2019). Drought and high temperature stress and traits associated with tolerance. In I. A. Ciampitti & P. V. V. Prasad (Eds.), *Sorghum: A State of the Art and Future Perspectives* (Vol. 58, pp. 245–265). ASA, CSSA, SSSA: Madison, WI, USA.

Promkhambut, A., Polthanee, A., Akkasaeng, C., & Younger, A. (2011). A flood-free period combined with early planting is required to sustain yield of pre-rice sweet sorghum (*Sorghum bicolor* L. Moench). *Acta Agriculture Scandinavica, Section B*, *61*, 345–355.

Punnuri, S. M., Ayele, A. G., Harris-Shultz, K. R., Knoll, J. E., Coffin, A. W., Tadesse, H. K., Armstrong, J. S., Wiggins, T. K., Li, H., & Wallace, J. G. (2022). Genome-wide association mapping of resistance to the sorghum aphid in *Sorghum bicolor*. *Genomics*, *114*(4), 110408.

Qi, G., Li, N., Sun, X. S., Wang, D., Ciampitti, I., & Prasad, V. (2016). Overview of sorghum industrial utilization. In I. Ciampitti & V. Prasad (Eds.), *Sorghum: A State of the Art and Future Perspectives* (pp. 463–476). American Society of Agronomy and Crop Science Society of America, Inc., Madison, WI.

Qiu, R., Wei, S., Zhang, M., & Li, H. (2018). Sensors for measuring plant phenotyping: A review. *International Journal of Agricultural and Biological Engineering, 11*(2), 1–17.

Radhamani, J., Pandey, A., Srinivasan, K., & Tyagi, V. (2011). Conserving millet genetic resources in India. *Proceedings of the Indian National Science Academy, 77*(4), 295–304.

Rakshit, S., Gomashe, S. S., Ganapathy, K. N., Elangovan, M., Ratnavathi, C. V., Seetharama, N., & Patil, J. V. (2012). Morphological and molecular diversity reveal wide variability among sorghum Maldandi landraces from India. *Journal of Plant Biochemistry and Biotechnology, 21*(2), 145–156. https://doi.org/10.1007/s13562-011-0079-9 (accessed on 7th November 2025).

Rana, B., Anahosur, K., Rao, V., Parameshwarappa, R., & Rao, N. (1982). Inheritance of field resistance to sorghum charcoal rot and selection for multiple disease resistance. *Indian Journal of Genetics and Plant Breeding, 42*, 302–310.

Rao, N. G. P., Narayana, L. L., & Reddy, V. R. (1978). Apomixis and its utilization in grain sorghum. I. Embryology of two apomictic parents. *Caryologia, 31*(4), 427–433.

Ravi, K. S., Graeme, L. H., Ian, B., Peter, H., & Greg, M. (2009). Characterizing environmental effects on phenology and canopy development of diverse sorghum genotypes. *Field Crops Research, 111*, 157–165.

Ray, D. K., Mueller, N. D., West, P. C., & Foley, J. A. (2013). Yield trends are insufficient to double global crop production by 2050. *PLoS One, 8*(6), e66428.

Reddy, B. V. S., Kumar, A. A., & Ramesh, S. (2007). Sweet sorghum: a water saving bio-energy crop. In *International Conference on Linkages between Energy and Water Management for Agriculture in Developing Countries*, 29–30 Jan 2007, ICRISAT, Patancheru, India.

Reddy, B. V. S., Ramesh, S., & Longvah, T. (2005). Prospects of breeding for micronutrients and carotene dense sorghums. *International Sorghum and Millets Newsletter, 46*, 10–14.

Reddy, B. V., Mughogho, L. K., Narayana, Y. D., Nicodemus, K. D., & Stenhouse, J. W. (1992). Inheritance pattern of downy mildew resistance in advanced generations of sorghum. *Annals of Applied Biology, 121*, 249–255.

Reddy, K. S., & Davies, J. (1979). Pests of sorghum and pearl millet, and their parasites and predators, recorded at ICRISAT Center, India up to August 1979.

Reddy, P. S., Jogeswar, G., Rasineni, G. K., Maheswari, M., Reddy, A. R., Varshney, R. K., & Kishor, P. B. K. (2015). Proline over-accumulation alleviates salt stress and protects photosynthetic and antioxidant enzyme activities in transgenic sorghum (*Sorghum bicolor* (L.) Moench). *Plant Physiology and Biochemistry, 94*, 104–113.

Reddy, P. S., Reddy, B. V. S., Ashok Kumar, A., Ramesh, S., Sahrawat, K. L., & Rao, P. V. (2010). Association of grain Fe and Zn contents with agronomic traits in sorghum. *Indian Journal of Plant Genetic Resources, 23*, 280–284.

Reed, J. D., Tuinstra, M. R., McLaren, N. W., Kofoid, K. D., Ochanda, N. W., & Claflin, L. E. (2002). Analysis of combining ability for ergot resistance in grain sorghum. *Crop Science, 42*, 1818–1823.

Riedelsheimer, C., Lisec, J., Czedik-Eysenberg, A., Sulpice, R., Flis, A., Grieder, C., Altmann, T., Stitt, M., Willmitzer, L., & Melchinger, A. E. (2012). Genome-wide association mapping of leaf metabolic profiles for dissecting complex traits in maize. *Proceedings of the National Academy of Sciences, 109*(23), 8872–8877.

Rizal, G., Karki, S., Alcasid, M., Montecillo, F., Acebron, K., Larazo, N., Garcia, R., Slamet-Loedin, I. H., & Quick, W. P. (2014). Shortening the breeding cycle of sorghum, a model crop for research. *Crop Science, 54*(2), 520–529.

Rooney, W. (2000). Genetics and cytogenetics. In *Sorghum: Origin, History, Technology, and Production* (Vol. 1).

Rooney, W. L., & Aydin, S. (1999). The genetic control of a photoperiod sensitive response in sorghum. *Crop Science, 39*, 397–400.

Rosenow, D. T., & Dahlberg, J. A. (2000). Collection, conversion and utilization of sorghum. In C. W. Smith & R. A. Frederiksen (Eds.), *Sorghum: Origin, History, Technology and Production* (pp. 309–328). Wiley.

Rotundo, J. L., Salinas, A., Gomara, N., Borras, L., & Messina, C. (2024). Maize outyielding sorghum under drought conditions helps explain land use changes in the US. *Field Crops Research, 308*, 109298.

Roy, R. C., Sagar, A., Jannat-E-Tajkia, A. M., & Zakir Hossain, A. K. M. (2018). Effect of salt stress on growth of sorghum germplasms at vegetative stage. *Journal of Bangladesh Agricultural University, 16*, 67–72.

Saikiran, V., Shivani, D., Ramesh, S., Maheswaramma, S., Sujatha, K., Sravanthi, K., Yamini, K. N., Varaprasad, B. V. V., & Kumar, C. S. (2023). Genetic inheritance of component traits associated with shoot fly resistance in sorghum (*Sorghum bicolor* (L.) Moench). *Cereal Research Communications, 51*(3), 567–575.

Sanabria, J., Stone, J. F., & Weeks, D. L. (1995). Stomatal response to high evaporative demand in irrigated grain sorghum in narrow and wide row spacing. *Agronomy Journal, 87*, 1010–1017.

Sanderson, M. A., Jones, R. M., & Ward, J., & Wolfe, R. (1992). Silage sorghum performance trial at Stephenville. *Forage Research in Texas*, Report PR-5018. Texas Agricultural Experimental Station, Stephenville, USA.

Sapkota, S., Boatwright, J. L., Jordan, K., Boyles, R., & Kresovich, S. (2020). Multi-trait regressor stacking increased genomic prediction accuracy of sorghum grain composition. *Agronomy, 10*(9), 1221.

Sapkota, S., Boatwright, J. L., Kumar, N., Myers, M., Cox, A., Ackerman, A., Caughman, W., Brenton, Z. W., Boyles, R. E., & Kresovich, S. (2023). Genomic prediction of hybrid performance for agronomic traits in sorghum. *G3: Genes, Genomes, Genetics, 13*(4), jkac311.

Schertz, K. (1994). Male-sterility in sorghum: its characteristics and importance. In *Use of Molecular Markers in Sorghum and Pearl Millet Breeding for Developing Countries: Proceedings of an ODA Plant Sciences Research Programmer Conference on Genetic Improvement* (pp. 35–37).

Shahbaz, M., & Ashraf, M. (2013). Improving salinity tolerance in cereals. *Critical Reviews in Plant Sciences, 32*, 237–249.

Sharma, R., Rao, V. P., Upadhyaya, H. D., Reddy, V. G., & Thakur, R. P. (2010). Resistance to grain mold and downy mildew in a mini-core collection of sorghum germplasm. *Plant Disease, 94*, 439–444.

Shimelis, H., Gwata, E. T., & Laing, M. D. (2019). Crop improvement for agricultural transformation in Southern Africa. In R. A. Sikora, E. R. Terry, P. L. G. Vlek, & J. Chitja (Eds.), *Transforming Agriculture in Southern Africa* (1st ed., pp. 97–103). Routledge.

Shinde, S., Kundu, P., & Louis, J. (2024). Beyond bites: Differential role of fall armyworm oral secretions and saliva in modulating sorghum defenses. *Molecular Plant-Microbe Interactions*. https://doi.org/10.1094/MPMI-12-23-0213-FI (accessed on 7th November 2025).

Sifuentes, J., & Frederiksen, R. A. (1988). Inheritance of resistance to pathotypes 1, 2, and 3 of *Peronosclerospora sorghi* in sorghum. *Plant Disease, 72*, 332–333.

Silva, K. J. D., Pastina, M. M., Guimarães, C. T., Magalhães, J. V., Pimentel, L. D., Schaffert, R. E., Pinto, M. D. O., Souza, V. F. D., Bernardino, K. D. C., Silva, M. J. D., Borém, A., & Menezes, C. B. D. (2020). Genetic diversity and heterotic grouping of sorghum lines using SNP markers. *Scientia Agricola, 78*, e20200039.

Sinclair, T. R., Hammer, G. L., & van Oosterom, E. J. (2005). Potential yield and water-use efficiency benefits in sorghum from limited maximum transpiration rate. *Functional Plant Biology, 32*, 945–952.

Singh, B. U. (1987). Varietal resistance in sorghum to midge, *Contarinia sorghicola* Coquillett (Diptera: Cecidomyiidae). *Insect Science and Its Application, 8*(2), 129–144. https://doi.org/10.1017/S174275840000713X (accessed on 7th November 2025).

Singh, V., Nguyen, C. T., van Oosterom, E. J., Chapman, S. C., Jordan, D. R., & Hammer, G. L. (2015). Sorghum genotypes differ in high temperature responses for seed set. *Field Crops Research, 171*, 32–40.

Singh, V., Nguyen, C. T., Yang, Z., Chapman, S. C., van Oosterom, E. J., & Hammer, G. L. (2016). Genotypic differences in effects of short episodes of high-temperature stress during reproductive development in sorghum. *Crop Science, 56*(4), 1561–1572.

Tabatabaei, S. A., & Anagholi, A. (2012). Effects of salinity on some characteristics of forage sorghum genotypes at germination stage. *International Journal of Agriculture and Crop Sciences, 4*, 979–983.

Tang, C. C., Gao, J. M., Han, Y., Pei, Z. Y., & Suo, S. J. (2015). Identification of QTLs associated with complex salt-alkaline tolerance at the seedling stage in sorghum. *Acta Agriculture Boreali-Sinica, 30*, 42–47.

Tang, C. Y., Schertz, K. F., & Bashaw, E. C. (1980). Apomixis in sorghum lines and their F1 progenies. *Botanical Gazette, 141*(3), 294–299.

Tari, I., Laskay, G., Takács, Z., & Poór, P. (2013). Response of sorghum to abiotic stresses: A review. *Journal of Agronomy and Crop Science, 199*(4), 264–274.

Taylor, A. O., Jepsen, N. M., & Christeller, J. T. (1972). Plants under climatic stress: III. Low temperature, high light effects on photosynthetic products. *Plant Physiology, 49*(5), 798–802.

Tester, M., & Langridge, P. (2010). Breeding technologies to increase crop production in a changing world. *Science, 327*(5967), 818–822.

Teutsch, C. (2002). *Warm-Season Annual Grasses for Summer Forage*. Virginia Polytechnic and State University Cooperative Extension, Blacksburg.

Thakur, R., Reddy, B. V., & Mathur, K. (2007). *Screening Techniques for Sorghum Diseases* (Information Bulletin No. 76). International Crops Research Institute for the Semi-Arid Tropics, Patancheru, India.

Tian, Y. S., Zhao, L. X., Meng, H. B., Sun, L. Y., & Yan, J. Y. (2009). Estimation of unused land potential for biofuels development in the People's Republic of China. *Applied Energy*, *86*(Suppl.), S77–S85.

Tooley, P. W., O'Neill, N. R., Goley, E. D., & Carras, M. M. (2000). Assessment of diversity in *Claviceps africana* and other *Claviceps* species by RAM and AFLP analyses. *Phytopathology*, *90*, 1126–1130.

Traore, M., Sullivan, C. Y., Rosowski, J. R., & Lee, K. W. (1989). Comparative leaf surface morphology and the glossy characteristic of sorghum, maize and pearl millet. *Annals of Botany*, *64*, 447–450.

Upadhyaya, H. D., Pundir, R. P. S., Dwivedi, S. L., Gowda, C. L. L., Reddy, V. G., & Singh, S. (2009). Developing a mini core collection of sorghum for diversified utilization of germplasm. *Crop Science*, *49*, 1769–1780.

Upadhyaya, H. D., Sharma, S., Dwivedi, S. L., & Singh, S. K. (2014). Sorghum genetic resources: Conservation and diversity assessment for enhanced utilization in sorghum improvement. In Y. H. Wang, H. D. Upadhyaya, & C. Kole (Eds.), *Genetics, Genomics and Breeding of Sorghum* (pp. 28–55). CRC Press, Taylor & Francis Group.

Upadhyaya, H. D., Vetriventhan, M., & Deshpande, S. (2016). Sorghum germplasm resources characterization and trait mapping. In *The Sorghum Genome* (pp. 77–94). Springer.

Vadez, V., Kholová, J., Hummel, G., Zhokhavets, U., Gupta, S. K., & Hash, C. T. (2015). LeasyScan: A novel concept combining 3D imaging and lysimetry for high-throughput phenotyping of traits controlling plant water budget. *Journal of Experimental Botany*, *66*(18), 5581–5593.

Vadez, V., Kholova, J., Srikanth, M., Rekha, B., Tharanya, M., Sivasakthi, K., Alimagham, M., Karthika, G., & Keerthi, C. (2017). LeasyScan: 3D scanning of crop canopy plus seamless monitoring of water use to harness the genetics of key traits for drought adaptation. *Interdrought, 112*.

Valencia, C., & Rooney, W. L. (2009). Genetic control of sorghum grain color.

Vanamala, J. K. P., Massey, A. R., Pinnamaneni, S. R., Reddivari, L., & Reardon, K. F. (2018). Grain and sweet sorghum (*Sorghum bicolor* L. Moench) serve as a novel source of bioactive compounds for human health. *Critical Reviews in Food Science and Nutrition*, *58*(17), 2867–2881.

Vedansh, S. K., Kerkhi, S. A., Singh, A., Kumar, M., & Kumar, A. (2010). Variability, heritability and genetic advance for forage yield and quality traits in forage sorghum (*Sorghum bicolor* L. Moench). *Progressive Agriculture*, *10*(2), 400–401.

Velazco, J. G., Jordan, D. R., Mace, E. S., Hunt, C. H., Malosetti, M., & van Eeuwijk, F. A. (2019). Genomic prediction of grain yield and drought-adaptation capacity in sorghum is enhanced by multi-trait analysis. *Frontiers in Plant Science*, *10*, 1–12.

Venuto, B., & Kindiger, B. (2008). Forage and biomass feedstock production from hybrid forage sorghum and sorghum–Sudan grass hybrids. *Grassland Science*, *54*, 189–196.

Verma, L. K., Biradar, B. D., & Meena, A. K. (2022). Heterotic grouping in rabi sorghum (*Sorghum bicolor* (L.) Moench) through diallel analysis. *The Pharma Innovation Journal*, *11*(2), 1955–1957.

Villar, J. L., Maranville, J. W., & Gardner, J. C. (1989). High density sorghum production for late planting in the Central Great Plains. *Journal of Production Agriculture*, *2*(4), 333–338.

Wang, E., Meinke, H., & Ryley, M. (2000). Event frequency and severity of sorghum ergot in Australia. *Australian Journal of Agricultural Research*, *51*, 457–466.

Wang, H. L., Zhang, H. W., Liu, B., Yang, Y. B., Qin, L., Chen, E. Y., & Guan, Y. A. (2017). QTL mapping for traits related to salt tolerance at seedling stage of sorghum under low salt stress. *Molecular Plant Breeding*, *15*, 604–610.

Wang, H., Wang, R., Liu, B., Yang, Y., Qin, L., Chen, E., Zhang, H., & Guan, Y. (2020). QTL analysis of salt tolerance in *Sorghum bicolor* during whole-plant growth stages. *Plant Breeding*, *139*(3), 455–465.

Wang, T. T., Ren, Z. J., Liu, Z. Q., Feng, X., Guo, R. Q., Li, B. G., Li, L. G., & Jing, H. C. (2014). *SbHKT1;4*, a member of the high-affinity potassium transporter gene family from *Sorghum bicolor*, functions to maintain optimal Na^+/K^+ balance under Na^+ stress. *Journal of Integrative Plant Biology*, *56*, 315–332.

Wang, X., Mace, E., Tao, Y., Cruickshank, A., Hunt, C., Hammer, G., & Jordan, D. (2020). Large-scale genome-wide association study reveals that drought-induced lodging in grain sorghum is associated with plant height and traits linked to carbon remobilization. *Theoretical and Applied Genetics*, *133*, 3201–3215.

Watkins, J. E., & Littlefield, L. J. (1976). Relationship of anthesis in Waldron wheat to infection by *Claviceps purpurea*. *Transactions of the British Mycological Society*, *66*, 362–363.

Wu, S. B., Shang, Y. J., Han, X. M., Wang, J., Niu, T., Zhang, F., Wei, Y., Meng, C., Ximei, Y., & Zheng, J. (1994). Embryological study on apomixis in a sorghum line SSA-1. *Acta Botanica Sinica*, *36*(11), 833–837.

Wu, Y., Guo, T., Mu, Q. I., Wang, J., Li, X., Wu, Y., Tian, B., Wang, M., Li, L., Bai, G., Perumal, R., Trick, H. N., Bean, S. R., Dweikat, I. M., Tuinstra, M. R., Morris, G., Tesso, T. T., & Yu, J. (2019). Allelochemicals targeted to balance competing selections in African agroecosystems. *Nature Plants*, *5*(12), 1229–1236.

Xie, Q., & Xu, Z. H. (2019). Sustainable agriculture: from sweet sorghum planting and ensiling to ruminant feeding. *Molecular Plant*, *12*(5), 603–606.

Yang, S., Weers, B. D., Morishige, D. T., & Mullet, J. E. (2014). CONSTANS is a photoperiod regulated activator of flowering in sorghum. *BMC Plant Biology*, *14*, 148.

Yu, X. Q., Li, X. R., Guo, T. T., Zhu, C. S., Wu, Y. Y., Mitchell, S. E., Roozeboom, K. L., Wang, D. H., Wang, M. L., Pederson, G. A., Tesso, T. T., Schnable, P. S., Bernardo, R., & Yu, J. M. (2016). Genomic prediction contributing to a promising global strategy to turbocharge gene banks. *Nature Plants*, *2*, 16150.

Zaman-Allah, M., Jenkinson, D. M., & Vadez, V. (2011). Chickpea genotypes contrasting for seed yield under terminal drought stress in the field differ for traits related to the control of water use. *Functional Plant Biology*, *38*, 270–281.

Zhang, R. D., Zhou, Y. F., Yue, Z. X., Chen, X. F., & Huang, R. D. (2019a). Changes in photosynthesis, chloroplast ultrastructure, and antioxidant metabolism in leaves of sorghum under waterlogging stress. *Photosynthetic*, *57*(4), 1076–1083.

Zhang, R., Yue, Z., Chen, X., Huang, R., & Zhou, Y. (2023). Effects of waterlogging at different growth stages on the photosynthetic characteristics and grain yield of sorghum (*Sorghum bicolor* L.). *Scientific Reports*, *13*(1), 7212.

Zhang, R., Zhou, Y., Yue, Z., Chen, X., & Xing, Y. (2019b). The leaf-air temperature difference reflects the variation in water status and photosynthesis of sorghum under waterlogged conditions. *PLoS ONE*, *14*(7), e0219209.

Zougmore, R., Kambou, F. N., Ouattara, K., & Guillobez, S. (2000). Sorghum–cowpea intercropping: an effective technique against runoff and soil erosion in the Sahel (Saria, Burkina Faso). *Arid Soil Research and Rehabilitation*, *14*, 329–342.

CHAPTER 2

Genetic Improvement in Pearl Millet

MAHALINGAM GOVINDARAJ,* GOVINTHARAJ PONNAIAH, and KARTHIKA GUNASEKARAN

International Crops Research Institute for the Semi-Arid Tropics, Patancheru, Hyderabad, Telangana, India

*Corresponding author

2.1 INTRODUCTION

Pearl millet [*Pennisetum glaucum* (L.) R. Br.] is a highly cross-pollinated crop having a diploid genome (2n=14) with an annual growth habit, considered a warm-season cereal, and belongs to the grass family Poaceae. This is an important food and fodder crop widely grown by marginal and smallholder farmers in the dry-land regions of Africa and Asia. Being a C4 crop, it has a higher dry matter production capacity due to photosynthetic efficiency and water use efficiency, with lesser input requirements, tolerance to low soil fertility, salinity, and drought; being short-duration and having significant tolerance to pests and diseases, hence pearl millet is considered a "climate-resilient crop for the future." Also, it has the inherent ability to withstand harsh climatic conditions where other crops fail to achieve economic yield (Baltensperger, 2002).

Worldwide, more than two billion people suffer from micronutrient deficiency (Narayan et al., 2019; Nguyen et al., 2021). Micronutrients like iron (Fe), zinc (Zn), vitamin A, and iodine (I) deficiencies are the most common micronutrients responsible for malnutrition, and they are more prevalent in developing countries, with the most affected people being women and children. This crop is rich in nutrient contents like carbohydrates, proteins, iron, calcium, zinc, fats, fibers, resistant starch, vitamins, and antioxidants. It has received special attention from health-conscious consumers since it has several health benefits, including fighting against cardiovascular disease, diabetes, celiac disease, high blood pressure, and thyroid disorders (Ambati and Sucharitha, 2019). Recently, a study emphasized the importance of millet consumption in combating cancer diseases (Gupta et al., 2023). Realizing the resilient nature and nutritional values of the crop, it is continuously expanding in other regions of the world, particularly the Americas, Canada, Brazil, and some parts of the European Union.

Besides grain consumption and dual purpose, this crop is also grown exclusively for forage use. The countries currently cultivating this crop for forage/feed use include the United States (Sheahan, 2014), Canada (Brunette et al., 2014), Australia (Hanna, 1996),

Millets in Sustainable Agriculture: Genetic Improvement, Nutritional Quality, Cultivation, and Integrated Pest–Disease Management. Roshni Vijayan, A. John Joel, R. Ambika Rajendran, and Adithya P. Balakrishnan (Eds.)

DOI: 10.1201/9781779642332-3

and as a mulch crop in Brazil (Dias-Martins et al., 2018). Owing to its multiple-tillering potential and quick and faster ratooning ability, this crop offers multi-cutting behavior, which allows the livestock-rearing community to get enough green/dry forage/fodder supply throughout the year.

Over the past 60 years, India has made significant improvements toward increasing productivity through the development and release of high-yielding cultivars and better agronomic management practices for them (Jukanti et al., 2016). The achievements of the Indian pearl millet breeding program are often referred to as the greatest success stories in Indian agriculture (Yadav et al., 2019). This book chapter presents an overview of the methods and techniques followed in pearl millet genetic improvement. We also look at the possibility of accelerating genetic advancements even further to fulfill the increased demand for pearl millet and increase the profitability of its cultivation.

2.2 PEARL MILLET PRODUCTION

Pearl millet is cultivated over 30 million ha globally, with Africa and Asia accounting for over 18 million ha (60% of cultivation areas) and over 10 million ha (35% of cultivation areas), respectively (Raheem et al., 2021). It is the staple food for 90 million people living in the Sahelian region of Africa and northwestern India (Srivastava et al., 2020b). It is the fifth most important cereal grown in the world (Khairwal, 2008), the fourth most important cereal food crop after major cereals like rice, wheat, and maize in India (Project Coordinators Review (AICMIP), 2021–2022), and the second most important cereal grown for dual purposes (grain and fodder/stover) after sorghum. The pearl millet cultivated area, production, and productivity over the years are shown in Figure 2.1. This crop is mainly cultivated for grain use on 7.41 million ha, with an average productivity of 1391 kg/ha, and advanced estimates for 2021–22 showed 9.41 million tons of production in India (Projects Coordinator Review (AICMIP), 2021–2022). The intensive breeding efforts in pearl millet breeding programs have led to increased productivity of 1275 kg/ha during the last 3 years (2017–2020) compared to the productivity of 362 kg/ha achieved during 1966–1969 (Figure 2.2). The pearl millet cultivated area, production, and productivity over the years are shown in Figure 2.1. In India, the pearl millet growing states Rajasthan (45%), Uttar Pradesh (19%), Gujarat (9%), Haryana (9%), Maharashtra (6%), and Tamil Nadu (2%) contributed to 90% of total production. This crop is largely grown as a kharif (rainy) season crop (June–September) in India. Recently, it has expanded as a summer season crop (February–May) in the states of Gujarat, Rajasthan, and Uttar Pradesh, and is also grown as a rabi (post-rainy) season crop (November–February) on a small scale in the states of Maharashtra and Gujarat. Additionally, this crop will have huge demand in the future due to urbanization, environmental degradation, rapidly increasing human populations, global warming, and livestock populations in Africa and Asia (Rai et al., 2008). Therefore, productivity needs to be increased more quickly in these challenging agroecologies.

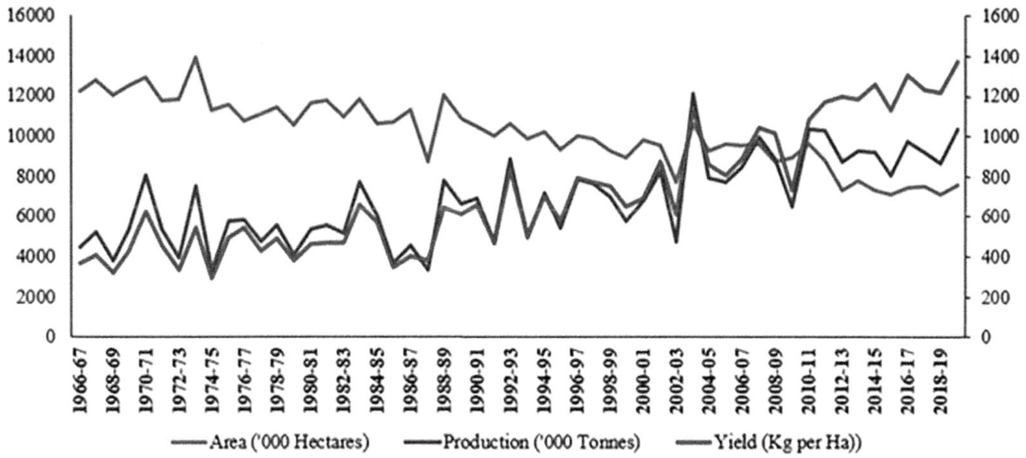

FIGURE 2.1 Pearl millet area, production, and productivity in India since 1966.

Further, this crop is also largely grown in the eastern/southern African (ESA) countries, including Sudan, Ethiopia, and Tanzania, and in the western/central African (WCA) countries, including Nigeria, Niger, Chad, Mali, and Senegal. Of these, WCA is the largest pearl millet producing region in the African continent (Jukanti et al., 2016). It is grown on 18.50 million hectares across 28 countries in Africa, with a yield of 11.36 million tons, spanning 30% of the continent in a variety of agroecologies (FAO, 2019).

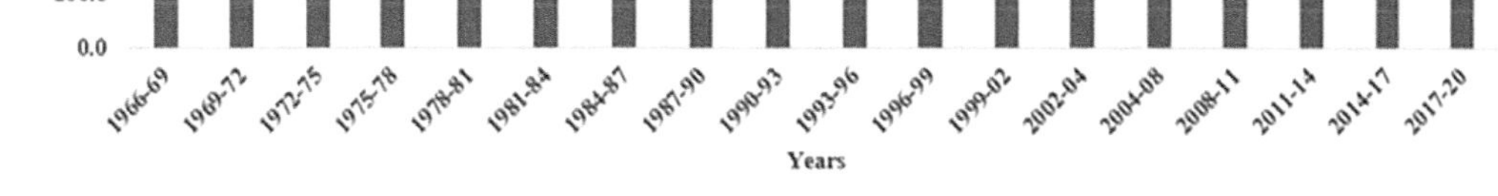

FIGURE 2.2 Trends in pearl millet productivity (Kg/ha) over the years (based on 3-year average).

2.3 TRAITS PRIORITIZATION: GENETIC IMPROVEMENT

Pearl millet crop is often cultivated in diverse agro-dry ecologies, which has led to the prioritization of farmers' traits research interests to improve the productivity of various locations/regions. Hence, the pearl millet growing areas have been divided into three zones (mega-environments), viz., A1, A, and B, based on the rainfall pattern, local adaptation, geographical location, and other environmental conditions (Figure 2.3). The areas in northern India that receive annual rainfall of >400 mm are considered an A zone; the areas in northwestern India receiving <400 mm of annual rainfall are considered an A1 zone; and the areas in peninsular India receiving >400 mm of annual rainfall are considered a B zone. Currently, A and A1 zones account for 75% of pearl millet cultivation, whereas the B zone accounts for 25%. Depending on the mega-environment needs, Indian pearl millet breeding programs have developed their product profile portfolio to achieve yield gains. A study by Yadav and Rai (2013) found that high grain yield, disease resistance, and maturity (75–85 days) traits are focused on their mega-environment needs. However, few traits were found common across mega-environments, including good exertion, good seed set, compact panicles, and lodging resistance.

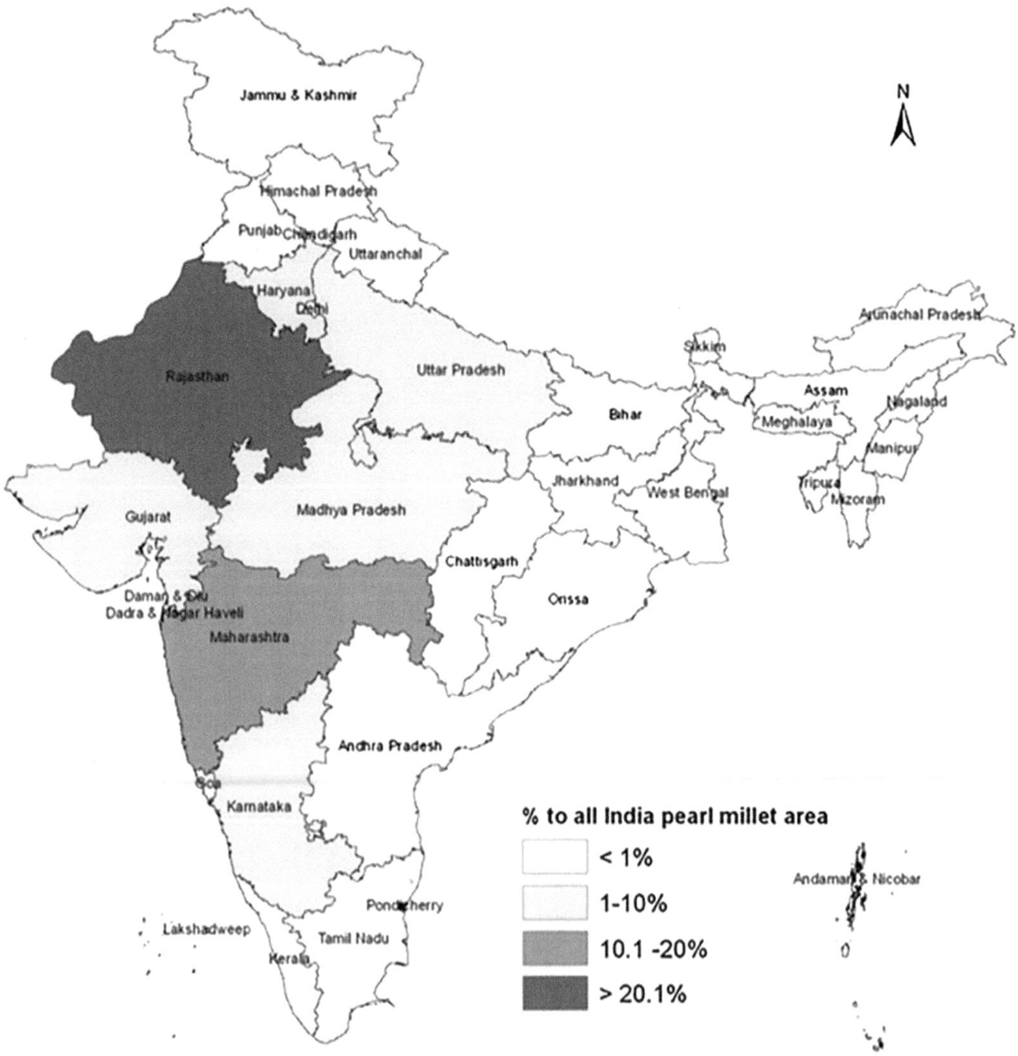

FIGURE 2.3 Distribution of pearl millet in different states and its growing proportion (%) in India.

Source: Reprinted from All India Coordinated Research Project on Pearl Millet. https://aicpmip.res.in/aboutus.html

2.4 HISTORIC BREEDING ACTIVITIES: CULTIVAR DEVELOPMENT

2.4.1 GERMPLASMS COLLECTIONS: FOUNDATION FOR GENETIC GAINS

The global pearl millet germplasm collections consist of about 56,000 accessions conserved in 70 different genebanks across 46 countries (FAO, 2010). Of these, landraces alone represent 88% of the germplasm conserved in genebanks globally; the remaining 12% includes wild relatives, improved varieties, and breeding materials (Mathur, 2012). The largest collection of pearl millet germplasm (22,288 accessions) is conserved at the International Crops Research Institute for the Semi-Arid Tropics (ICRISAT) genebank, followed by the Institute of Research for Development (IRD) in France (about 3968 accessions from 16 countries) and the Canadian Genetic Resources Programme (CGRP) in Saskatoon, Canada (about 3821 accessions, including 3390 cultivated pearl millet). The USDA Griffin, Georgia Agricultural Research Station maintains about 1314 accessions (Mathur, 2012). About 7268 and 193 accessions of pearl millet were conserved at the NBPGR in New Delhi and in Pakistan, respectively. Around 27 countries reported having pearl millet germplasm collections in their genebanks on the African continent (Mathur, 2012). Apart from this, countries like Australia (253), Brazil (52), China (102), Germany (54), the Russian Federation (406), and Lebanon (110) are conserving pearl millet germplasm. Large phenotypic diversity in the pearl millet core collection is stored at the ICRISAT genebank for several traits (Table 2.1) and is being utilized effectively and efficiently in their breeding programs (Upadhyaya et al., 2009).

TABLE 2.1 Traits Variation for Pearl Millet Collection Stored at ICRISAT Gene Bank

Traits	Range	Mean
Time to 50% flowering-R (days)	33–159	72.7
Time to 50% flowering-PR (days)	32–138	71.7
Plant height (cm)-R	30–490	248.5
Plant height (cm)-PR	25–425	161
Total tillers (No.)	1–35	2.7
Productive tillers (No.)	1–19	2.1
Panicle exertion (cm)	–45 to 29	3.5
Panicle length (cm)-R	5–135	28.9
Panicle length (cm)-PR	4–125	25.8
Panicle width (mm)-R	8–58	24
Panicle width (mm)-PR	8–61	22.8
1000-Seed weight (g)	1.5–21.3	8.5

R- rainy season; PR- post-rainy season

Source: Reprinted with permission from Upadhyaya et al. (2009). Copyright 2009 © Crop Science Society of America

Earlier, in the 1930s, the breeding activities in pearl millet mainly focused on the utilization and improvement of local and exotic germplasms through mass selection and progeny testing. With mass selection, many varieties were developed and cultivated in many states of India. For example, C1, C2, Co.2, and Co.3 in Tamil Nadu; AKP1, AKP2, and AKP3 in Andhra Pradesh; T5, T55, A1/3, S350, and S530 in Punjab; LM 38-39, Bajra 1, Bajra 2, S-14, S 28-25-2, Puri, and Sadas 11 in Maharashtra; Baroda 5, B117, B119, and 14D in Madhya Pradesh; RSJ, RSK, Bari N207, and S14 in Rajasthan; and Bijapur 11-11-7-14-6, Golagiri 1-8-5-5, and Sindagi 3-16-13-9-4 in Karnataka (Ahluwalia, 1964; Krishnaswamy, 1962; Rachie and Majmudar, 1980). Also, germplasms imported from Africa during the 1960s were improved for phenological uniformity in a "Ghana" landrace at the Indian Agricultural Research Institute (IARI), New Delhi, which led to the development of "Improved Ghana" (Joshi et al., 1961) and was further improved through mass selection, leading to the release of Pusa Moti (Ahluwalia, 1964).

Being a cross-pollinated crop, the natural way of maintaining the genetic architecture is through natural open pollination, and it is maintained as open-pollinated varieties (OPVs). Selfing in cross-pollinated crops leads to inbreeding depression and a reduction in plant vigor and yield. Hence, a variety refers to a heterozygous and heterogeneous randomly mating population that can interbreed among OPVs well operated by Hardy-Weinberg's equilibrium. Variety (OPVs) improvement aims to increase the superiority of alleles responsible for higher yield, adaptation to low soil nutrients, drought tolerance, disease, and pest resistance in pearl millet. In the part of population improvement programs, the OPVs development started in the 1970s at ICRISAT, Patancheru. Various recurrent selection breeding methods (mass selection, gridded-mass selection, S_1 and S_2 progeny selection, half-sib selection, full-sib selection, and recurrent restricted phenotypic selection) have been used, and several trait-specific composites (like dwarf composite, early composite, high-tillering composite, medium composite, late composite, bold-seeded composite, smut-resistant composite, and high head volume composite) were developed (Gill, 1991; Govil et al., 1986; Rai and Anand Kumar, 1994; Rattunde and Witcombe, 1993; Singh et al., 1988; Witcombe, 1999; Zaveri et al., 1989). Furthermore, these trait-specific composites were improved at different cycles through recurrent selection, and OPVs like WC-C75, ICMV 221, ICTP 8203, HC 4, Raj 171, JBV 2, JBV 3, HC 20, RCB 2, CZP 9802, and Pusa 266 were developed (Khairwal et al., 2009).

Pearl millet is highly cross-pollinated and exhibits significant levels of heterosis for both grain and stover yields. Efforts were made in the early 1950s to exploit higher heterosis in hybrid breeding programs with the advantage of the protogyny nature of flowering. The chance hybrids were produced by mixing parental populations having the same flowering time and allowing them to cross-pollinate, and the resultant hybrids produced 40% true seeds (Gill, 1991). The chance hybrids had a 10 to 15% yield advantage over local landraces but couldn't become popular because of a narrow genetic base, limited superiority over OPVs, and lack of quality seed production protocols.

The discovery and availability of the novel cytoplasmic-nuclear male sterility (CMS) line Tift 23A in 1958 at Tifton, Georgia (Burton, 1965), which was characterized by semi-dwarf stature, high tillering, synchronous flowering, and good combining ability, were made readily available to Indian pearl millet breeding programs (Athwal, 1961). Commercial hybrids were developed using the Tift 23A line and the availability of good fertility restorers in Indian germplasm, which had a huge impact on pearl millet productivity (Gill, 1991). However, the intensive and regular cultivation of a limited number of hybrids (HB 1, HB

2, HB 3, HB 4, and HB 5 released between 1965 and 1969) derived from the Tift 23A line for several years led to a downy mildew (DM) outbreak in the 1970s (Safeeulla, 1976). Also, hybrids (BJ 104 and BK 560) developed involving the cytoplasmic male sterile line 514A bred at IARI, New Delhi, and restorer lines became popular in pearl millet growing areas during the periods of 1977 to 1984, but they were also affected by DM (Govila et al., 1996). However, the limited hybrids grown year after year in larger areas led to DM epidemics (Singh, 1995). Breeding efforts were intensified on the diversification of cytoplasm in the seed parental lines, particularly after the 1980s. Along with the A1 CMS source (Burton, 1965), six additional CMS sources, namely A2, A3 (Athwal, 1961, 1966), Av (Marchais and Pernes, 1985), A4 (Hanna, 1989), Aegp (Rai and Hash, 1993), and A5 (Rai, 1995), were identified (Table 2.2). Based on the extensive characterization of these CMS sources, it was found that A4 and A5 were promising, whereas A2 and A3 sources exhibited instability (Rai et al., 1996). The maintainer frequency was highest in A5 cytoplasm, followed by A4 and A1 cytoplasms, whereas the restorer frequency was highest in A1 cytoplasm, followed by A4 and A5 cytoplasms. Studies also concluded that A5 cytoplasm had no negative effects on grain yield, whereas A4 cytoplasm exhibited a 5% reduction in grain yield compared with A1 cytoplasm and showed a high genotype × environment interaction.

TABLE 2.2 CMS Systems in Pearl Millet

CMS system	Source	References
A_1	Narrow-based population	Burton (1958)
A_2	Narrow-based population	Athwal (1961)
A_3	Narrow-based population	Athwal (1966)
A_v	Wild germplasm	Marchais and Pernes (1985)
A_4	Wild germplasm	Hanna (1989)
A_{egp}	Broad-based genepool	Rai and Hash (1993)
A_5	Broad-based genepool	Rai (1995)

2.4.2 SEED PARENTS (A- & B-LINES) AND RESTORER PARENTS (R-LINES)

Breeding of seed parents requires male-sterile lines (A-lines) that must have complete and stable male sterility, while maintainer lines (B-lines) have been bred for profuse pollen production across the seasons and sites. The most important factors to consider in breeding male-sterile lines (A-lines) are their high grain yield potential, both as lines *per se* and in hybrids (i.e., combining ability). However, achieving a higher grain yield potential requires a combination of an agronomically elite genetic background and farmers' preferred traits. Traits like good exertion, seed set, compact panicle, and lodging resistance are common for hybrid parents across all regions, while maturity, tillering ability, plant height, seed color, and seed size have regional preferences (Yadav et al., 2012).

In seed parent breeding, the dwarf plant height (d2) has been the most prevalent plant type (Rai and Hanna, 1990; Rai and Rao, 1991). It possesses many desirable features, including the ability to breed hybrids with variable heights, reduced risk of lodging, and rapid detection and efficient roguing of off-types and pollen shedders in A-lines. Pearl millet breeding programs at ICRISAT and AICPMIP centers in India are developing new plant types with high productivity potential, such as A-lines with long (30–80 cm compared to a normal length of 10–20 cm), thick (40–50 mm diameter compared to a normal thickness of 20–30 mm) panicles, and large seed size (17–20 g of 1000-seed mass compared to a normal seed mass of 9–12 g) (Yadav et al., 2012).

The most basic requirement for good restorer lines is to have a higher pollen load (pollen grains) that should remain viable at higher air temperatures (42 to 44°C) and produce highly fertile hybrids. The restorer parents (R-lines) should be taller, measuring 150 to 180 cm compared to the seed parents. They should also possess significant levels of tolerance to diseases and lodging (Yadav et al., 2012).

2.5 BIOTIC STRESS RESISTANCE

Like in other crops, pearl millet has also been affected by many diseases and insect pests. However, DM (*Sclerospora graminicola* (Sacc.) J. Schroet), rust (*Puccinia substriata* var. indica Ramachar and Cumm (Syn*: Puccinia substriata* Ell. and Barth. var. penicillariae; *Puccinia penniseti* Zimm), smut (*Moesiziomyces penicillariae* Bref. Venky (Syn. *Tolyposporium penicillariae* Bref.)), ergot (*Claviceps fusiformis* Lov. (Syn. *Claviceps microcephala* (Wallr.) Tul.)), and blast (*Pyricularia grisea* (Cooke) Sacc.) are considered economically important diseases that cause severe yield reduction. Being a resource-poor farmers' crop, the biotic stresses like diseases and insect pests can be controlled by manipulating host plant resistance (HPR) mechanisms without any additional cost to farmers. A good understanding of disease biology and epidemiology, the development of effective greenhouse and field screening techniques, and the availability of resistance sources are integral parts of HPR.

2.5.1 DOWNY MILDEW

Downy mildew (DM) is the major biotic stress that causes significant yield reduction from 20% up to 90%, depending upon the locations, seasons, and specific cultivars (ICAR-AICRP on Pearl Millet (aicpmip.res.in)). Higher relative humidity (>95%) and moderate temperatures (20 to 22°C) favor disease occurrence. Generally, DM symptoms appear on the second leaf and onwards, with the whitish powdery pathogen growth observed on the abaxial leaf surface. Green ear symptoms appear at the panicle emergence stage. The DM resistance gene is controlled by a dominant gene, partially dominant (Deswal and Govila, 1994; Singh and Talukdar, 1998), and/or a complex trait (Joshi and Ugale, 2002). The mapping studies have identified Quantitative Trait Loci (QTL) for DM resistance and both major and minor genes responsible for disease resistance (Jones, 1995). Later, even a single major gene provided higher levels of resistance in hybrid parental lines (Hash et al., 2006).

2.5.2 BLAST

The pearl millet blast disease was once a minor disease but has now become the most destructive in crop-growing areas (Rai et al., 2012). The blast pathogen was more prevalent when pearl millet was exposed to high relative humidity (>90%) and moderate temperatures (25 to 30°C). Continuous monitoring of pathogen virulence and screening of germplasms and breeding materials under well-established greenhouse and field screening methods allows for the rapid identification of stable resistant germplasms or breeding lines, thereby being used to breed blast-resistant hybrid parent lines and hybrids of pearl millet (Sharma et al., 2013; Sharma et al., 2019; Yella Goud et al., 2016). Understanding the inheritance pattern of disease resistance would guide breeders to utilize the gene(s) effectively in their breeding programs. Numerous studies have been conducted in pearl millet to understand the inheritance of blast disease resistance and concluded that blast disease resistance is governed by partially dominant or dominant genes (Gupta et al., 2012; Singh et al., 2018; Wilson et al., 1989). Hanna and Wells (1989) identified three independent blast dominant genes from the wild species *Pennisetum glaucum* subsp. *monodii.* Blast resistance genes in pearl millet inbreds to Indian isolates of *Magnaporthe grisea* are controlled by a single dominant gene (Gupta et al., 2012b; Singh et al., 2018b), making them easier to use in introgression breeding. The pearl millet mini-core germplasm accessions were screened against five blast pathotype isolates under greenhouse conditions, and three accessions, IP 7846, IP 11036, and IP 21187, exhibited higher levels of resistance to four isolates (Upadhyaya and Ortiz, 2001). Thakur et al. (2009) identified breeding lines such as 863B, ICMB 01333, ICMB 01777, ICMB 02111, ICMB 03999, ICMB 93222, ICMB 97222, ICMR 06222, ICMR 06444, and ICMR 07555 as having higher levels of resistance. A study was conducted to identify molecular markers for blast resistance and reported one RAPD marker (OP-D11_{700}, 5.6 cM) was found to be linked to blast resistance (Morgan et al., 1998).

2.5.3 RUST

Rust is the most important disease affecting both grain yield and forage quality traits in pearl millet. Rust disease shows symptoms in all growth stages. Nearly 51% of digestible dry matter yield loss is caused by the rust pathogen (Monson et al., 1986). The cooler night temperature (<20°C) and dew formation are the two major factors contributing to rust development. In peninsular India, pearl millet is mainly grown in the post-rainy season, where temperatures go below the required level, leading to the development of rust disease. Additionally, pearl millet rust disease has been reported in summer-growing ecologies in central and peninsular India where seed production is carried out. In most cases, rust resistance is governed by single dominant genes in pearl millet lines (Andrews et al., 1985b; Pannu et al., 1996; Ramamoorthi et al., 1995). Seed (ICMB 96222) and restorer (ICMR 0699, ICMP 451-P8, and ICMP 451-P6) parental lines were found to be resistant to rust disease in pearl millet (Sharma et al., 2009). Three molecular markers, one RFLP marker (Xpsm306, 10.0 cM) from ICMP 83506, and two RAPD markers (OP-$K19_{350}$, 8.8 cM and OP-08_{350}, 19.6 cM) from Tift 89D_2 parental lines, are linked to rust-resistant loci in pearl millet (Morgan et al., 1998).

2.5.4 SMUT

Smut is the most dangerous yield-reducing disease in pearl millet growing areas worldwide. The pearl millet florets infected by the smut disease produce smut sori instead of grains.

The occurrence of smut disease was higher when pearl millet was exposed to higher relative humidity (80 to 95%) and an optimal temperature of 25 to 35°C during anthesis time. Earlier studies have concluded that smut disease resistance is controlled by a dominant gene (Thakur and Chahal, 1987) and is also quantitatively inherited (additive and non-additive gene effects) (Chavan et al., 1988; Phookan, 1987), and thus could be utilized for population improvement. The genes responsible for trichomelessness offer some level of resistance to smut disease (Wilson, 1995).

2.5.5 *ERGOT*

Ergot is a floral disease commonly observed in areas where pearl millet is popularly grown, causing significant reductions in both grain quantity and quality. The incidence of ergot disease was severe when pearl millet was exposed to high relative humidity (80 to 85%) and medium temperature ranges from 25 to 30°C. Ergot-infected pearl millet panicles ooze out creamy to pink mucilaginous droplets known as "honeydew," which is the first sign of ergot disease. The severity of ergot disease was found to be greater in hybrids derived from CMS sources (Natarajan et al., 1974), and the incidence of susceptibility was higher in hybrids than in OPVs (Thakur et al., 1983b), leading to speculation about the correlation between CMS source and ergot susceptibility incidence. The ergot-infected grains produce neurotoxic alkaloids that pose a threat to human and cattle health. The inheritance of ergot disease resistance in pearl millet is controlled by recessive polygenic genes with significant cytoplasmic × nuclear interaction, implying that resistance genes must be present in both seed (male-sterile lines) and restorer (pollen) parents to breed ergot-resistant hybrids (Rai and Thakur, 1995; Thakur et al., 1989b; Thakur and Rai, 2002).

2.5.6 *INSECT-PESTS*

Nearly 300 insect pests have been reported to have caused yield losses in pearl millet (Nwanze and Harris, 1992). However, only a few insect pests, like stem borers (*Chilo partellus* in India and *Coniesta ignefusalis* in Western Africa), shoot fly (*Atherigona approximata*), bugs (*Blissus leucopterus*), grasshoppers (*Diabolocatantops axillaris*), desert locust (*Schistocerca gregaria*), and head miner (*Heliocheilus albipunctella*), have been causing significant damage to pearl millet crops. Recently, the fall armyworm (FAW), *Spodoptera frugiperda* (J.E. Smith; Noctuidae, Lepidoptera), has been reported as an invasive polyphagous pest that attacks 353 plant species, including pearl millet. However, it is mostly a pest of grasses, with maize being the most preferred host plant. Among millets, sorghum (60.41%) is the most preferred host, followed by pearl millet (41.4%), based on whorl damages (ICAR-IIMR, 2019). The fall armyworm is a voracious pest of maize and pearl millet crops native to the Americas (Walton and Lunginbill, 1916).

A good understanding of insect pest bioecology and pest management practices is essential for effective and economical control. Due to climate shocks, the distribution range and frequency of insect outbreaks may vary in different agro-climatic conditions. Insect pests in any crop are controlled by integrated approaches, mostly cultural and chemical methods (Sharma and Youm, 1999). Considering environmental safety, it is important to understand pest behavior in relation to semiochemicals (natural chemical compounds like synthetic sex attractants and plant-derived kairomones), non-toxic electromagnetic radiation (light), and combinations of both to control insect pests. Due to the continuous evolution of new insect pests under climate change, insect pest resistance should be considered a target trait in breeding programs. For FAW, HPR mechanisms like extreme non-preference or antibiosis were previously reported in pearl millet inbreds Tifton no. 240 (Leuck, 1970). Recently, researchers from WA pearl millet breeding programs screened a set of 40 genotypes against head miner and identified genotypes (like Gamoji, ICMP 177001, ICMP 177002, ICMV 177003, ICMV IS 90311, LCIC9702, Souna 3, ICMV IS 94206, and PE08043) that exhibited the antibiosis HPR mechanism to head miner with appreciable grain yield and agronomic performance compared to the susceptible open-pollinated check variety ICMV IS 92222. Additionally, some genotypes, including Faringuero, ICMV 167005, ICMV IS 99001, Sadore local, SOSAT-C88, and ICMP 177004, showed tolerance to head miner and good per se performance (Mohammed et al., 2020).

2.6 ABIOTIC STRESS RESISTANCE

2.6.1 DROUGHT TOLERANCE

Pearl millet is the most drought-tolerant crop grown by smallholder farmers living in the arid and semi-arid regions of Africa and Asia. These areas are generally characterized by highly variable environments and longer dry seasons, where other cereals fail to perform better. Drought stress can cause significant grain yield reductions of up to 60% in pearl millet (Bidinger et al., 1987). Numerous studies have been conducted to identify the critical growth periods and assess the effects of drought stress on pearl millet. Many critical periods have been identified under drought stress: seedling emergence and early seedling stages cause severe yield reductions due to poor plant stand (Carberry et al., 1985; Soman et al., 1987), the vegetative stage has little effect on productivity traits (Bidinger et al., 1987; Mahalakshmi et al., 1987) but has shown an increased number of panicles per plant, which directly reflects the compensation mechanism for damaged main panicles (van Oosterom et al., 2003, 2006), and at flowering and grain filling stages, heavy yield losses occur due to reduced grain number and grain size (Fussel et al., 1991; Mahalakshmi et al., 1987).

Breeding efforts have been undertaken to identify traits linked to drought tolerance in pearl millet. Morphological traits are easily identifiable and incorporated by breeders in their breeding programs. Traits like early vigor (Manga and Yadav, 1995), low tillering, and long panicles (Bidinger et al., 1987; Yadav, 1994), early flowering (Bidinger et al., 1987; Fussel et al., 1991), stay-green (Jordan et al., 2012), and panicle harvest index (PNHI) (Bidinger and Mahalakshmi, 1993; Bidinger and Witcombe, 1989) are important. The trait early vigor (EV) is defined as "the rapid development of leaf area in the early developmental stage of the plant cycle" (López-Castañeda et al., 1996). The EV in crop plants shows increased biomass production at the early phase of the plant life cycle, and rapid canopy closure leads to reduced evaporation from the soil, thereby increasing water availability to plants (Condon et al., 2004). Additionally, reduced canopy closure decreases solar radiation on the soil and increases the crop's ability to compete with weeds

(Bertholsson, 2005; Lemerle et al., 2001). Low tillering, longer panicles, and the number of grains per panicle traits are routinely used by pearl millet breeders as selection criteria. For example, Bidinger et al. (1987) selected these traits in his breeding and observed higher grain yield per panicle. Early flowering time (EFT) is an important trait for pearl millet production, especially where this crop experiences terminal drought, as this trait reduces the risk of dehydration at flowering and grain filling stages. EFT is a drought escape (DE) mechanism rather than drought tolerance (DT) (Bidinger et al., 1987). The EFT has been well exploited in pearl millet breeding by utilizing Iniadi-type landraces originating from Western Africa (Andrews and Bramel-Cox, 1993; Andrews and Kumar, 1996), and cultivars derived from Iniadi-type landraces have been widely adopted in drought-prone regions in India and Africa. The stay-green (SG) or delay of senescence is characterized by a crop's ability to maintain greenness for a longer time at crop maturity during water stress. Studies have reported that stay-green cultivars possess drought tolerance at crop maturity, having higher leaf nitrogen and more basal stem sugars compared to non-stay-green cultivars in several cereal crops like wheat (Christopher et al., 2008; Lopes and Reynolds, 2012), rice (Hoang and Kobata, 2009; Woonho et al., 2018), sorghum (Rosenow et al., 1983; Victor et al., 1989), and maize (Ceppi et al., 1987; Russell et al., 1986). However, a study by Awala and Wilson (2005) found that stay-green genotypes had more retention of chlorophyll than non-stay-green genotypes in pearl millet. Debieu et al. (2018) evaluated 188 pearl millet inbred parents under drought conditions and identified four stay-green linked markers on chromosome 6.

The panicle harvest index (PNHI) or panicle threshing percentage is another trait used to improve/develop cultivars with drought tolerance by the way plants exploit the assimilation and translocation process to set and fill more grains under water-stressed conditions. The PNHI showed higher variation for grain yield among studied genotypes in cereal crops (Bidinger and Mahalakshmi, 1993; Bidinger and Witcombe, 1989). With this selection criteria, ICRISAT scientists from India developed an open-pollinated variety (OPV) (ICMV 221) from an early maturing large-seeded composite. Germplasms/landraces responsible for drought tolerance are provided in Table 2.3. Researchers have also identified physiological traits like transpiration rate (Kholová et al., 2010 a, b), osmotic potential (Henson, 1982), and canopy temperature (Singh and Kanemasu, 1983) linked to drought tolerance in pearl millet. Four drought-tolerant QTLs were identified and linked to enhanced transpiration in pearl millet. Of these, one major QTL was mapped on LG 6, and the QTLs responsible for leaf weight and biomass traits were mapped on LG 7 (Aparna et al., 2015).

TABLE 2.3 Pearl Millet Germplasms or Cultivars for Abiotic Stress Tolerance

Pearl millet	Drought	CZMS 44A (landrace 3072); IP 8210; landraces 220, 184, 235, 238	Kusaka et al. (2005)
	Salinity	93613, KAT/PM-2, Kitui, Kitui local, 93612; 10876, 10878, 18406, 18570; IP 3757, 3732; Birjand pearl millet; IP 6112; IP 3616, 6104, 6112; ZZ ecotype	Nadaf et al. (2010)

2.6.2 SALINITY TOLERANCE

Pearl millet is mainly cultivated for food and fodder purposes in dry-land conditions where salt-affected soils are often increasing due to flooding and water-logging conditions. A total of 20 to 33% of cultivable irrigation land was affected by soil salinity worldwide, and it is expected to increase up to 50% by 2050 (Karthika and Govintharaj, 2022; Machado and Serralheiro, 2017; Soda et al., 2016). The pearl millet crop is well-suited for these salt-affected areas (Ali et al., 2006). However, very few studies have identified salinity tolerance-linked traits in pearl millet. For example, Krishnamurthy et al. (2007) determined the critical growth stages: i) seed germination and ii) seedling establishment, which are highly sensitive to soil salinity stress. The same study also found that K^+/Na^+ and Ca^{2+}/Na^+ ratios were positively associated with whole-plant salinity tolerance, but not as closely as the Na^+ content, which makes it easier for researchers to breed high crop productivity traits with enhanced levels of salinity tolerance. Additionally, the trait of shoot Na^+ concentration could be considered when breeding for salinity tolerance at the vegetative stage. In another study, Yakubu et al. (2010) found a significant reduction in plant growth traits and nutrient content with increasing soil salinization and identified the tolerant cultivar "Maiwa," which shows increased tolerance levels compared to the control cultivar. Recently, a total of 33 pearl millet genotypes (including landraces and inbred lines) were screened using an *in vitro* screening method at the seedling stages for salinity tolerance, revealing that salt-sensitive genotypes showed reduced total biomass and shoot growth compared to the salt-tolerant genotypes in pearl millet (Jha et al., 2022). Germplasms identified for salinity tolerance are provided in Table 2.3.

2.6.3 HEAT TOLERANCE

Current climate shocks, including heat stress, pose a significant threat to agricultural production. Heat stress is another important stress that pearl millet encounters, which reduces crop productivity by affecting the germination, seedlings, and reproductive stages of the crop. The optimum threshold temperatures required for seed germination (10 to 34°C), seedling (35°C), vegetative (26°C), and reproduction (22 to 26°C) stages in pearl millet are noted (Ashraf and Hafeez, 2004; Ong, 1983a,b). The pearl millet growing areas in the northwestern arid zone of India and western and southern Africa often experience maximum air temperatures of about 43°C, and soil surface temperatures may reach up to 60 to 62°C during germination in these regions (Yadav and Rai, 2013; Yadav et al., 2021). Studies have reported that the rate of germination and final germination percentage decreased when pearl millet seeds were exposed to higher temperatures (45°C) for a period of 8 days (Ashraf and Hafeez, 2004). Additionally, in pearl millet, the seedlings are highly vulnerable to increased temperatures for the first 10 days from sowing to seedling establishment Peacock et al., 1993; Soman et al., 1981). To assess seedling heat tolerance, a study identified seedling survival under high soil surface temperature (Peacock et al., 1993), greater seedling emergence under artificial screening (Soman and Peacock, 1985), and membrane thermostability (Howarth et al., 1997) methods/protocols developed and found effective. In the recent past, pearl millet has become increasingly important and a preferred crop for farmers in northwestern India (Yadav and Rai, 2013) due to its prolific and productive nature under drier and hotter regions where other staple cereals fail to grow. The higher air temperatures (>42°C) in these regions coincide with the flowering period of pearl millet, which causes sterile spikelets, leading to significant yield reduction (Djanaguiraman et al., 2018; Gupta et al., 2015b). With the availability of standardized heat stress screening protocols in these ecologies, the identified boot-leaf stage is highly sensitive to heat compared to the panicle-emergence stage in pearl millet. Furthermore, female reproductive (stigma) organs are more heat-sensitive than male reproductive (pollen) organs, as evaluated in a larger group of germplasms, improved populations, and parental lines of hybrids (Gupta et al., 2015b). The flowering stage heat tolerance trait in pearl millet is controlled by dominant genes, indicating that one heat-tolerant parent is sufficient to develop heat-tolerant hybrids. Several germplasm accessions (IP 19799, IP 19877, and IP 19743), hybrid parental lines (ICMB 92777, ICMB 05666, ICMB 00333, ICMB 01888, ICMB 02333, and ICMB 03555), and improved populations (ICMV 82132, MC 94, ICTP 8202, and MC-Bulk) have been identified as having higher seed set under higher temperatures (Yadav et al., 2012).

2.6.4 LODGING TOLERANCE

Lodging tolerance (LT) is a quantitative/complex genetic trait, highly influenced by various morpholo-biophysical and physiological traits and difficult to measure in breeding materials which underlie finding alternative ways to improve LT use of indirect selection criteria. Pearl millet hybrids are mainly cultivated in kharif (rainy) season and/ as irrigated crop during the summer season where it often exposed to intermittent heavy rains and high wind velocity at crop maturity leading to complete crop losses. Recently, based on multi-seasons and multi-years investigation, Gupta et al. (2020) found culm strength is an important trait for LT in pearl millet.

2.6.5 COLD TOLERANCE

Millets, belongs to Poaceae family, were adapted and grown in varied environments and showed different levels of tolerance to different types of stresses. For instance, they are same family, but pearl millet and finger millet showed sensitive and tolerant responses to cold stress, respectively. Pearl millet was domesticated in semi-arid areas and is cold-sensitive in nature thereby limiting agricultural production in higher altitude growing areas. Recently, Meng et al. (2021) screened the pearl millet, foxtail millet, proso millet, and switchgrass to cold stress at three leaf seedling stage and identified cold-responsive genes conserved in pearl millet. Also, the genes linked to forage quality traits in pearl millet were responsible for cold tolerance (Govintharaj et al., 2021).

2.6.6 WATER-LOGGING TOLERANCE

The larger amount of water stagnates in lowlands due to the presence of an undulating topography in many areas of semi-arid regions worldwide (Van Staveren and Stoop, 1985).

Millets are commonly grown in these areas for grain and fodder purposes. Among millets, pearl millet is their choice in the semi-arid regions that are generally known for being too dry and hot, and low in soil fertility. Limited studies have been conducted to identify water-logging tolerance traits in pearl millet. For instance, Zegada-Lizarazu and Iijima (2005) compared the dep water uptake efficiency and water use efficiency (WUE) in pearl millet and other small millets. The study concluded that pearl millet is resilient to drought tolerance in comparison to barnyard millet as explained by higher WUE rather than water uptake efficiency in lowlands.

2.7 GRAIN NUTRITIONAL TRAITS

The grain pearl millet is a highly nutritious energy-rich crop with a balanced amino acid (protein: 9 to 13%), higher micronutrients (Fe: 35 to 85 ppm; Zn: 33 to 50 ppm), crude fibres [including SDF (soluble detergent fiber), IDF (insoluble detergent fiber), and resistant starch], fat with better digestibility and higher antioxidant in comparison to other cereals (Gupta et al., 2023). Due to its climate resilience nature and higher grain nutritional properties, this crop is considered as "nutricereals" by the Government of India (Gazette of India, No. 133 dated 13 April 2018). Further, Government of India has declared the year 2018 as "National Year of Millets (NYoM)" by promoting millets cultivation in a view of its nutritional potential and diversify food diet among small-holder and marginal farmers. Also, the FAO (Food and Agriculture Organization) committee on agriculture (COAG) has declared the year 2023 as "International Year of Millets (IYoM)" by aiming at creating awareness among farmers and consumers about its nutrition, to diversify food systems under climate change and ensure food security worldwide.

Pearl millet shows enormous variability for grain Fe and Zn contents and has shown huge potential for further improvement through plant breeding (Govindaraj et al., 2013; Kanatti et al., 2014; Rai et al., 2014). The germplasms and biparental mapping population, which are originally derived from the Iniadi landrace, showed higher levels of grain Fe and Zn (Govindaraj et al., 2016 Velu et al., 2008). More than 120 commercial pearl millet hybrids were screened for grain Fe and Zn contents and showed 46–56 ppm for grain Fe and 37–44 ppm for grain Zn (Rai et al., 2016). By realizing the importance of grain nutrients, the Indian national testing (Indian Council of Agricultural Research (ICAR)-All India Coordinated Pearl Millet Improvement Project (AICPMIP)) and cultivar release proposal set the baseline (42 ppm for grain Fe and 32 ppm for grain Zn) for the release of any cultivar, which has also encouraged breeders to mainstream Fe and Zn in pearl millet (AICPMIP, 2018).

Pearl millet grain micronutrients Fe and Zn are predominantly controlled by additive gene action, indicating that hybrid parents could have higher values to improve these grain micronutrients further (Govindaraj et al., 2013; Kanatti et al., 2014). This was further supported by a highly significant positive association between the per se of the hybrids and mid-parental values for grain Fe and Zn micronutrients, being largely governed by additive genetic control. Lower or smaller G × E interaction for grain Fe and Zn contents in pearl millet (Govindaraj et al., 2016; Kanatti et al., 2014) translated into higher narrow-sense heritability, indicating that progeny selection in pedigree breeding is effective in developing lines/cultivars with an enhanced level of grain Fe and Zn contents. The predominance of additive genetic effects implies that recurrent selection is effective for the improvement of grain Fe and Zn contents within breeding populations and OPVs (Govindaraj et al., 2019).

It has also been noted that grain Fe and Zn contents were significantly and positively correlated with each other in pearl millet (Govindaraj et al., 2013, 2016, 2020; Kanatti et al., 2014; Pujar et al., 2020; Rai et al., 2014). Further, pearl millet seed size was also highly associated with these two micronutrients (Govindaraj et al., 2016; Gupta et al., 2009; Kanatti et al., 2014). Pearl millet grain color has no effect on grain Fe and Zn contents (Govindaraj et al., 2018). These studies would help breeders to breed cultivars having high grain Fe and Zn contents in a larger seed size (bold seed) background with no influence of grain color in their breeding programs. Pujar et al. (2020) identified 18 SNPs for grain Fe, 43 for grain Zn, and 17 for grain protein content, which can be utilized in future pearl millet biofortification programs.

2.8 FORAGE QUANTITY AND QUALITY TRAITS

This crop is high tillering in nature and has a rapid ratooning ability, which offers multiple-time harvesting, thereby supplying green/dry forages/fodder year-round. In most developing countries, this crop is mainly cultivated for dual purposes and exclusively for forage during the summer season, while it is also importantly cultivated for the feed industry in developed countries. As a forage crop, it is widely cultivated for livestock grazing, green forage chopping (cut and carry system), silage, and hay (Newman et al., 2010), and at any crop growth stage, it can be fed to animals without causing any adverse effects (Arya et al., 2013).

The non-availability of sufficient feed and fodder has been a major hindrance to achieving increased livestock production in most countries in the drier regions. Considering the current fodder scarcity, there will be an 18.4% green fodder and 13.2% dry fodder deficit in the year 2050 in India (IGFRI, 2013). To fulfill this fodder deficit, green forage supply alone should grow at 1.7% annually (IGFRI, 2013). Little breeding efforts are being made to increase forage/fodder yield by utilizing new germplasms, breeding materials, OPVs, and hybrid cultivars of pearl millet under a single cut system (Bidinger et al., 2010; Blümmel and Rai, 2003; Blümmel et al., 2007; Gupta et al., 2015; Gupta et al., 2018; Hash et al., 2006). However, dairy farming communities are trying to get more forage yield from a single crop; in such cases, a multi-cut forage system might be the best option, where the first cut of forage takes place earlier (45 to 50 days) and more often with 3 to 4 cuts/season, every 30 days interval. Very few studies have been conducted, estimating the forage yield and quality traits of pearl millet under a multi-cut system (Govintharaj et al., 2018, 2021; Ponnaiah et al., 2019; Rai et al., 2012). A study by Govintharaj et al. (2017) noted that green forage yield (GFY) and dry forage yield (DFY), as well as total green forage yield (TGFY), had moderate heritability and high GA, indicating these traits can be improved through selection in later generations. Also, the genotypic coefficient of variation (GCV) was low compared to the phenotypic coefficient of variation (PCV) for the leaf to stem ratio, suggesting that selection is ineffective for this trait. Large phenotypic variability for forage quantity traits (GFY and DFY: 15.0 to 29.0 t/ha and 3.0 to 6.0 t/ha at the first cut, and 12.0 to 42.0 t/ha and 5.0 to 9.0 t/ha at the second cut, respectively), and forage quality traits (stover nitrogen, metabolizable energy (ME), and *in vitro* organic matter digestibility (IVOMD): 1.84 to 2.34%, 7.42 to 7.76 MJ/kg, and 54.0 to 56.0% at the first cut, and 1.77 to 2.00%, 6.95 to 7.68 MJ/kg, and 51.0 to 55.0% at the second cut, respectively) have been reported in pearl millet (Govintharaj et al., 2018). The low positive significant association observed between genetic distance (GD between parents of hybrid) based on SSRs and forage traits heterosis value indicated that SSR-based GD could be used to predict the heterosis for forage quantity traits (GFY and DFY) and forage quality trait CP in pearl millet (Ponnaiah et al., 2019). Traits like total dry forage yield (TDFY), CP, and IVOMD had lower barker predictability ratio (PR $\leq$ 0.80), indicating the importance of non-additive gene action; hence, they can be improved through heterosis breeding (Govintharaj et al., 2021). The same study also found that additive gene action plays a crucial role for TGFY, which can be improved through simple selection. Higher heterosis (Better parent heterosis, BPH) was observed for TDFY in both single cross and top cross hybrids in pearl millet. The mean BPH ranged from −11 to 7% and −13 to 11% for IVOMD in single and top cross hybrids, respectively, under multi-cut pearl millet (Govintharaj et al., 2021). In a cultivars comparison study, the three-way top-cross hybrids had higher mean TDFY than that of top-cross hybrids, followed by OPVs. Also, the forage quality traits of three-way top-cross hybrids were higher or at par with other types of cultivars (Gupta et al., 2022).

The brown midrib (bmr) trait is linked to lower lignin and increased digestibility. To date, three bmr mutants Pbmr from Purdue University (Cherney et al., 1988), SDML 89107 from Zimbabwe (Gupta et al., 1993), and a spontaneous mutant from Tifton (Degenhart et al., 1991) have been reported earlier in pearl millet. Recently, bmr trait having pearl millet inbreds were developed from two populations, namely ICMV 155 bmr and WRajPop bmr. The inheritance study revealed that the bmr trait is controlled by the same single recessive gene in both populations and had no effect of a modifier gene. The same study identified four inbreds that stably expressed the bmr trait (three in ICMV 155 and one in the WRajPop background) in a high biomass background; this can be used as donor parents in forage pearl millet breeding programs in view of improving forage quantity (Gupta and Govintharaj, 2023).

The recent study has identified one to nine significant single nucleotide polymorphisms (SNPs) for CP and IVOMD identified through genome-wide association studies (GWAS) in pearl millet, which can be used for trait improvement after validation of these markers.

2.9 HYBRID REPLACEMENT

To achieve higher genetic gains, it is important to breed high-yielding hybrids and replace these high-yielding hybrids with newer hybrids having the highest yielding potential. To achieve this, the ICAR has formulated the All India Coordinated Research Project (AICRP) on Pearl Millet, which has a standardized multi-location and multi-year testing evaluation system for the identification of the best hybrids over the recently released hybrids (AICPMIP, 2023). This evaluation system particularly involves close monitoring of the yield performances, nutrient-dense traits (grain Fe and Zn), and disease incidences of hybrids in the areas where they are produced.

Generally, from initial hybrid cross-development to commercial release, hybrids (products) used to take 8 to 12 years. The factors that determine the product are demand, competition, branding (promotional strategies), internal management, environmental, political, socio-economic, and other external factors. The life cycle of each product passes through different phases such as introduction, growth, maturity, saturation, decline, and replacement. For sustained seed company growth, at least one product must be present at every phase of the product lifecycle, from introduction to decline.

The lifespans of commercially released hybrids/products have shortened over the years, and this trend continues at a faster rate, thus causing earlier replacement of older products with newer ones. Competition among seed companies, low sales volume of products, and breakdown of resistance genes play important roles in hybrid/product replacement. As a result of competition, new hybrids/products with higher yields, nutrient density, and disease resistance have been introduced often, leading to an acceleration of genetic gain.

2.10 GENETIC DIVERSITY AND GENOMICS ASSISTED BREEDING

Classical breeding, along with the application of advanced/high-throughput genomic tools, might help to identify the desirable alleles from the breeding materials (Varshney et al., 2005). The trait introgression into agronomically superior or desirable genetic backgrounds with the use of marker-assisted selection (MAS) has great promise for crop trait improvement by reducing extensive phenotypic selection and evaluation. The molecular markers were developed by DFID-JIC-ICRISAT and utilized for pearl millet crop improvement in 1991 (Gale et al., 2005). This further led to the use of different marker systems including RFLPs (Bhattacharjee et al., 2002), RAPDs (Chowdari et al., 1998), AFLPs (Vom Brocke et al., 2003), SSCP-SNPs (Bertin et al., 2005), SSRs (Gupta et al., 2015; Mariac et al., 2006; Nepolean et al., 2012; Oumar et al., 2008; Stich et al., 2010), and most recently GBS-SNPs (Hu et al., 2015; Ponnaiah et al., 2019) used to assess the genetic diversity among pearl millet germplasms, hybrid parents, and populations. With the existence of larger variability in a cross-pollinated crop like pearl millet, genetic analysis is a straightforward approach for trait dissection, which includes genome mapping and gene tagging of agronomically important traits. Many breeders' interest traits, including disease resistance (downy mildew, blast disease, and rust), drought tolerance, stover yield, and grain nutrients, were mapped in the recent past using different types of molecular markers in bi-parental mapping populations (Gulia et al., 2007; Morgan et al., 1998; Yadav et al., 2011).

A set of 250 PMiGAP (pearl millet inbred germplasm association panel) germplasms collected from 23 countries representing three continents was used to capture wider genetic variations for various traits (Sehgal et al., 2015). Also, Multi-parent advanced generation intercross (MAGIC) in pearl millet would attract assessing larger genetic variation in the germplasm compared to GWAS in naturally occurring haplotypes (Rakshit et al., 2012). The pearl millet genome was sequenced and provides resources for improving agronomically important traits (Varshney et al., 2017), which has helped the international pearl millet community (Jarquin et al., 2020; Srivastava et al., 2020a). With the help of currently available SNPs discovered from high-throughput genotyping platforms, the already reported QTLs can be remapped (Srivastava et al., 2020a,b). The sequencing-based mapping technique helps to identify superior haplotypes for different traits to establish haplotype-based breeding (Sinha et al., 2020). Furthermore, the availability of high-quality reference genome, the largest whole genome resequencing data, affordable/cost-effective genotyping methods, and phenotyping platforms with high precision have now made it possible to map genes linked to breeding-related traits in a fast manner (Bohra et al., 2020).

2.10.1 HETEROTIC GROUPING/PATTERNS OF HYBRID PARENTAL LINES

Hybrid cultivars are important for enhancing the crop productivity of both self- and cross-fertilizing crops. Heterotic grouping or patterns of hybrid parental lines is the key element in breeding programs to increase the magnitude of heterosis over a longer time (Melchinger and Gumber, 1998). At ICRISAT, Patancheru, the pearl millet breeding program regularly utilizes diverse breeding materials/germplasms to develop trait-specific seed (B-lines) and restorer (R-lines) parents (Rai et al., 2006). These hybrid parents were routinely assessed by molecular markers and identified that two different groups exist in the breeding programs: one for seed parents and one for pollinator parents (Gupta et al., 2015a; Nepolean et al., 2012; Ponnaiah et al., 2019; Singh et al., 2018a).

Efforts are being made to develop heterotic pools in pearl millet. Pucher et al. (2016) studied combining ability patterns among 100 population hybrids (derived from Senegal × Niger or Senegal × Nigeria) developed from 20 West African (WA) pearl millet landraces and identified that no clear-cut heterotic groups are available in the WA pearl millet breeding programs, which might be due to genetic admixture caused by protogyny and wind- and insect-borne pollen outcrossing. Ramya et al. (2018) identified seven heterotic groups from 99 hybrids that were developed from selected lines of 9 of the 180 B-lines and 11 of the 182 R-lines based on SSR GD. Furthermore, two heterotic pools, each for seed parents and restorer parents, were identified from 136 hybrids involving 17 (9 B-lines and 8 R-lines) representative parents selected from 147 (73 B-lines and 74 R-lines) hybrid parents in pearl millet based on SSR GD. Also, Sattler et al. (2019) identified good combining ability patterns among hybrids derived from 17 pearl millet OPVs originated from Niger and Senegal and identified some OPVs with high combining ability recommended to develop new divergent heterotic pools for long-panicled pearl millet hybrids. Recently, based on heterotic performance and combining ability, two heterotic groups, each for B-lines and R-lines, were identified from 320 hybrids generated using 37 (15 B-lines and 22 R-lines) representative parents from 580 (260 B-lines and 320 R-lines) parental lines from 6 diverse pearl millet breeding programs in India (Gupta et al., 2020). The pearl millet breeding program at ICRISAT, Patancheru, India, has seed parental (B-lines) and pollinator parental (R-lines) groups that behave as two clear-cut broad heterotic pools, as B × R hybrids produced greater heterosis than hybrids derived from B × B or R × R (Gupta et al., 2020; Ramya et al., 2018; Singh et al., 2018a). In contrast, in the West African pearl millet breeding program, the eastern WA and western WA populations have higher per se performance and GCA effects and thus could be used to develop heterotic pools (Sattler and Haussmann, 2020). The population hybrids generated from Senegal and Mali germplasms/populations or Burkina Faso germplasms/populations and Niger produced higher yields across all environments (Dutta et al., 2021).

Combining ability studies in pearl millet revealed that the association between per se performances and GCA (general combining ability) effects was either positive or negative. To date, breeding lines have been selected or advanced based on per se performance through trait-specific pedigree breeding programs in India. Lines with higher per se grain yield are more likely to have high general combiners than any other yield group. Recently, there has been a shift toward assessing the combining ability of lines at early stages. Considering the seed production cost in hybrid breeding, the high-yielding pearl millet hybrids are generally produced from high GCA inbred parental lines and should also have higher grain yield *per se.*

The specific combining ability (SCA) and GCA variances in the B- and R-gene pools of Indian pearl millet breeding are not well understood. A recent study revealed that high SCA variance was two times higher than GCA variance, indicating the predominance of SCA variance in hybrid parents of pearl millet, which is entirely different from well-established maize heterotic pools in the United States and European breeding programs where SCA variance is lower than GCA variance (Gupta et al., 2020). Considering this scenario, it is of utmost importance to investigate the contributions of SCA and GCA variances to heterosis in the established pearl millet B- and R-line heterotic pools. In WA pearl millet breeding, high GCA and SCA variances were observed, particularly in hybrids derived from different geographical origins with no availability of heterotic groups. By reducing the geographic barriers of studied cultivars from Sahelian Africa to WA and then to Niger and Senegal, it became feasible to provide specific recommendations for forming initial heterotic groups (Dutta, 2019; Pucher et al., 2016; Sattler et al., 2019).

2.10.2 GENOMIC SELECTION (GS)

Genomic selection (GS) refers to the selection of candidate genotypes based on genomic estimated breeding values (GEBV) of individuals obtained from high-density genome-wide markers. It offers quicker selection of desired genotypes, thereby fast-tracking the breeding cycle by reducing cost and time. In pearl millet, genomic prediction accuracy varied among traits: days to flowering (0.87 to 0.89), plant height (0.72 to 0.73), thousand grain weight (0.73 to 0.74), and grain yield (0.48 to 0.51) when data were utilized from 320 hybrids (Liang et al., 2018). Furthermore, hybrid alone and hybrid and/or inbred prediction schemes performed equally for traits with little to no heterosis. On the other hand, the inclusion of inbred phenotypic data in the selection schemes showed reduced genomic prediction accuracy for traits with significant mid-parent heterosis (Liang et al., 2018). Recently, the study concluded that the inclusion of GCA and SCA components in the GS schemes could increase the efficiency for the rapid identification of superior/promising cultivars (Jarquin et al., 2020). They also observed improvements in genomic prediction accuracy in untested or target hybrids when including inbred phenotypic information in the calibration datasets (Jarquin et al., 2020).

2.10.3 PRECISION PHENOTYPING

Recently, several precision phenotyping methods have been used to achieve higher genetic gain in breeding programs. High-throughput automation systems like LeasyScan have been used to evaluate larger breeding materials/germplasms for drought tolerance (Vadez et al., 2015). Precision phenotyping tools like near-infrared spectroscopy (NIRS) for forage quality traits and X-ray fluorescence (XRF) for micronutrient traits like grain micronutrients Fe and Zn have also been utilized.

2.11 NUTRIENT USE EFFICIENCY

Macronutrients like nitrogen (N) and phosphorus (P) are important for plant growth and development. In the agricultural sector, these two macronutrients hold the largest share of fertilizer production and consumption. Despite their versatility, N and P suffer significant losses from volatilization, leaching, and fixation. The surplus amount of nitrogen left in the soil pollutes water and air, which is highly dangerous to most living things (Russo et al., 2017). Therefore, it is important to breed cultivars with improved nutrient absorption and utilization efficiency to help mitigate these problems (Good et al., 2004). Nitrogen and phosphorus-containing fertilizers have net efficiencies of about 30–35% and 18–20%, respectively (Umar et al., 2020). With the current and future climate shocks, there will be a huge demand for food production to feed billions of people, with limited/reduced external inputs, and to increase the availability and improve the uptake of nutrients present in the soil (Cordell et al., 2009). The effective utilization of nitrogen is a vital factor for sustainable agricultural production. Pearl millet is widely grown in drier, hotter, low-nutrient soils, and rain-fed areas of arid and semi-arid regions of the world. Few studies have been conducted on nutrient use efficiency and have informed the possibility of breeding cultivars with nutrient-use efficiency in pearl millet (Alagarswamy and Bidinger, 1987; Gemenet et al., 2015; Pujarula et al., 2021). These studies were conducted on diverse landraces and promising hybrids of pearl millet and identified significant genotypic differences for nitrogen use efficiency and biomass production (Alagarswamy and Bidinger, 1987). The same study also found that the West African genotype (Souna B) had 32% higher N use efficiency (NUE) than the Indian genotype (BJ 104) when they received similar N uptake. Furthermore, a study conducted by Pujarula et al. (2021) on a set of 380 diverse pearl millet lines under three different N levels (0%-N0, 50%-N50, 100%-N100 of recommended N, that is, 100 kg ha^{-1}) identified nitrogen-insensitive genotypes under low nitrogen fertilization.

Pearl millet crop production is hampered by low fertile soils, particularly low phosphorus soils in West Africa (WA). Studies have been conducted that identified morpho-physiological traits linked with low phosphorus-tolerant genotypes in pearl millet (Bationo et al., 1993; Buerkert et al., 2001; Faye et al., 2006). Furthermore, significant genotypic differences were observed for root traits and response to phosphorus fertilizer in WA pearl millet (Brück et al., 2003; Faye et al., 2006). All these studies have used a limited number of genotypes. Recently, Gemenet et al. (2015) conducted research on 180 West African pearl millet inbred lines under low- and high-phosphorus (LP & HP, respectively) soils in field and controlled conditions, revealing the existence of significant genotypic variation for both P uptake (PU) and utilization (PUE) in both seedling and adult plant stages. The same study concluded that PUE increased under low-phosphorus (LP) conditions, whereas total PU was of utmost importance for economic yield compared to PUE under low-P (LP) conditions.

2.12 ACCELERATED MODERN CROP BREEDING TECHNOLOGIES

Development and release of cultivars require a minimum of 10 to 12 years through conventional breeding procedures. In the majority of the cultivar development programs, larger resources, space, and time are invested in selection and generation advancement, as well as multi-environment evaluation procedures. The duration of time required in these processes slows down product development and commercialization. The genetic gains of breeding programs are mainly determined by the inherent nature of the crop breeding cycles completed in a year under conventional breeding. These crop breeding cycles sometimes vary in some agro-ecologies due to weather extremes such as erratic and poor rainfall distribution, extreme temperatures, and day length, which allows crops to complete one breeding cycle per year. For example, pearl millet can only be cultivated once a year in North India, whereas in Western, Central, and Peninsular India, this crop is cultivated twice per year. With the help of modern technologies like double haploid and speed breeding, the time duration of crop breeding cycles can be reduced.

Efforts have been made on *in vitro* production of haploid plants in pearl millet. Anther culture studies were conducted on seven pearl millet genotypes tested with 12 different culture media. The highest frequency of induced androgenic embryos, about 13.7%, was obtained from the 841-P3 genotype. Furthermore, 4% maltose included as an additional carbon source resulted in a greater number of multicellular microspores among the responsive genotypes (Sastry and Mallikarjuna, 2014).

Apomixis has also been exploited in pearl millet breeding programs for fixing heterosis or hybrid vigor. Hanna and Dujardin (1982) showed the possibility of transferring apomixis genes from *Panicum orientale* to pearl millet through inter-specific crosses. The resulting interspecific hybrid with 2n = 25 comprised seven large *Panicum glaucum* and 18 small *Panicum orientale* chromosomes. Three selected hybrids were examined and found to be types of apomictic lines: 1) highly apomictic (<5% sexual embryo sacs), 2) obligate apomict (progeny from 100% maternal parent), and 3) facultative apomict (progeny obtained through normal meiosis and/or fertilization of the egg cell). The interspecific hybrids derived from this study showed an immune response to rust and leaf spot disease. The interspecific hybrids were backcrossed with pearl millet pollen, resulting in approximately 25% of plants with 2n = 32 chromosomes (14 *Panicum glaucum* and 18 *Panicum orientale*), indicating that gene exchange/transfer is possible between these species.

The current development of "RapidGen," new environmentally controlled facilities, will drastically reduce the 6–7-year window. RapidGen has the potential to enable four crop breeding cycles annually when combined with the whole set of breeding acceleration procedures (https://www.icrisat.org/first-public-research-facility-to-put-agriculture-on-fast-forward-launched-at-icrisat/). We are entering a new era of pearl millet speed breeding with these new facilities in place, where genetic improvements should continue to accelerate.

2.13 SYNERGIZING CULTIVAR IMPROVEMENT AND PRODUCTION SYSTEMS

Sustained crop productivity is achieved by improving both productivity and suitable production systems simultaneously. Improving crop management practices and seeking advanced farming concepts like precision agriculture helps to minimize inputs such as irrigation and fertigation, while maximizing outcomes like productivity and quality of the cultivar through technological innovations (such as automation, sensing, spectra, and aerial phenotyping). Using appropriate agrotechniques can readily increase grain yield by about 20–25% in pearl millet at the farm level, according to on-farm demonstrations of better cultivars and production systems (Yadava et al., 2012a). The key to maximizing the yield potential of improved pearl millet cultivars would be intensive management practices, which include increased planting density, irrigation timing, and the recommended dose of mineral fertilizers in better-endowed ecologies. Likewise, spaced planting, intercropping with legumes, and nutrient microdosing are crucial for improving the productivity of pearl millet cultivars in drought-prone ecologies. Recently, deep neural networks (DNNs), a type of machine learning technique, have been used to identify DM incidence, and the model accuracy was 95.00% (Coulibaly et al., 2019; Shetty et al., 2016).

2.14 CONCLUSION

Pearl millet is a dual-use (food and fodder) crop grown by smallholder farmers across the globe. In recent years, this crop has garnered particular interest from farmers, consumers, and policymakers as a "crop of choice" due to its inherent potential to combat climate shocks, hidden hunger, and several health benefits. Pearl millet productivity has increased significantly through breeding activities in the past, with the availability of methods and techniques. The increased and better use of hybrid technologies, application of modern tools, multi-institutional and inter-institutional partnerships, and efficient crop management practices would play a significant role in attaining higher genetic gains for grain yield, forage yield, grain micronutrient and forage quality traits, disease and insect-pest resistance, and also climate-resilient traits.

KEYWORDS

- **Pearl millet breeding**
- **Abiotic stress tolerance**
- **Disease resistance**
- **Biofortification**
- **Genomics-assisted breeding**
- **Drought tolerance**
- **Hybrid breeding**
- **Nutritional quality**
- **Molecular markers**

REFERENCES

Ahluwalia, M. (1964). A worthy successor to 'Improved Ghana' Bajra Pusa Moti. *Indian Farm, 14*, 4–5.

Alagarswamy, G., & Bidinger, F. R. (1985). Genotypic variation in biomass production and nitrogen use efficiency in pearl millet (*Pennisetum americanum* (L.) Leeke). In W. H. Gabelman & B. C. Loughman (Eds.), *Genetic Aspects of Plant Mineral Nutrition: Proceedings of the Second International Symposium on Genetic Aspects of Plant Mineral Nutrition, University of Wisconsin, Madison, June 16–20, 1985* (pp. 281–286). Springer Netherlands.

Ali, G. M., Murtaza, N., Collins, J. C., & McNeilly, T. (2006). Study of salt tolerance parameters in pearl millet *Pennisetum americanum. Journal of Central European Agriculture, 7*, 365–376.

Ambati, K., & Sucharitha, K. V. (2019). Millets–review on nutritional profiles and health benefits. *International Journal of Recent Scientific Research, 10*, 33943–33948.

Andrews, D. J., & Anandkumar, K. (1996). Use of the West African pearl millet landrace Inlaid in cultivar development. *Plant Genetic Resources Newsletter, 105*, 15–22.

Andrews, D. J., & Bramel-Cox, P. J. (1993). Breeding cultivars for sustainable crop production in low-input dryland agriculture in the tropics. In D. A. Buxton et al. (Eds.), *Proceedings International Crop Science I* (pp. 211–223). Crop Science Society of America Inc.

Aparna, K., Nepolean, T., Srivastava, R. K., Kholová, J., Rajaram, V., Kumar, S., & Vadez, V. (2015). Quantitative trait loci associated with constitutive traits control water use in pearl millet (*Pennisetum glaucum* (L.) R. Br.). *Plant Biology, 17*(5), 1073–1084.

Ashraf, M., & Hafeez, M. (2004). Thermotolerance of pearl millet and maize at early growth stages: growth and nutrition relations. *Biologic Plantarum, 48*, 81–86.

Athwal, D. S. (1961). Recent developments in the breeding and improvement of bajra (pearl millet) in the Punjab. *Madras Agricultural Journal, 48*, 18–19.

Athwal, D. S. (1966). Current plant breeding research with special reference to *Pennisetum. Indian Journal of Genetics and Plant Breeding, 26A*, 73–85.

Awala, S. K., & Wilson, J. P. (2005). Expression and segregation of stay-green in pearl millet. *International Sorghum and Millets Newsletter, 46*, 87–100.

Bationo, A., Christianson, B. C., & Klaij, M. C. (1993). The effect of crop residue and fertilizer use on pearl millet yields in Niger. *Fertilizer Research, 34*, 251–258.

Bertholsson, N. O. (2005). Early vigor and allelopathy–two useful traits for enhanced barley and wheat competitiveness against weeds. *Weed Research, 45*, 94–102.

Bertin, I., Zhu, J. H., & Gale, M. D. (2005). SSCP-SNP in pearl millet–a new marker system for comparative genetics. *Theoretical and Applied Genetics, 110*, 1467–1472. https://doi.org/10.1007/s00122-005-1981-0 (accessed on 7th November 2025).

Bhattacharjee, R., Bramel, J., Hash, T., Kolesnikova-Allen, A., & Khairwal, S. (2002). Assessment of genetic diversity within and between pearl millet landraces. *Theoretical and Applied Genetics, 105*, 666–673. https://doi.org/10.1007/s00122-002-0917-1 (accessed on 7th November 2025).

Bidinger, F. R., & Mahalakshmi, V. (1993). Selection for drought tolerance. In *Cereals Program, ICRISAT 1993. Annual Report* (pp. 57–59). ICRISAT, Patancheru, India.

Bidinger, F. R., & Witcombe, J. R. (1989). Evaluation of specific drought avoidance traits as selection criteria for improvement of drought resistance. In F. G. W. Baker (Ed.), *Drought Resistance in Cereals* (pp. 151–164). CAB International.

Bidinger, F. R., Mahalakshmi, V., & Rao, G. D. P. (1987). Assessment of drought resistance in pearl millet (*Pennisetum americanum* (L.) Leeke). Estimation of genotype response to stress. *Australian Journal of Agricultural Research, 38*(1), 49–59.

Bohra, A., Jha, U. C., Godwin, I. D., & Varshney, R. K. (2020). Genomic interventions for sustainable agriculture. *Plant Biotechnology Journal, 18*, 12. https://onlinelibrary.wiley.com/doi/full/10.1111/pbi.13285 (accessed on 7th November 2025).

Brück, H., Sattelmacher, B., & Payne, W. A. (2003). Varietal differences in shoot and rooting parameters of pearl millet on sandy soils in Niger. *Plant and Soil, 251*, 175–185.

Buerkert, A., Bationo, A., & Piepho, H. P. (2001). Efficient phosphorus application strategies for increased crop production in sub-Saharan West Africa. *Field Crops Research, 72*, 1–15.

Burton, G. W. (1965). Pearl millet Tift 23A release. *Crops and Soils, 17*, 19.

Carberry, P. S., Cambell, L. C., & Bidinger, F. R. (1985). The growth and development of pearl millet as affected by plant population. *Field Crops Research, 11*, 20–193.

Ceppi, D., Sala, M., Gentinetta, E., Verderio, A., & Motto, M. (1987). Genotype-dependent leaf senescence in maize: inheritance and effects of pollination-prevention. *Plant Physiology, 85*, 720–725.

Christopher, J. T., Manschadi, A. M., Hammer, G. L., & Borrell, A. K. (2008). Developmental and physiological traits associated with high yield and stay-green phenotype in wheat. *Australian Journal of Agricultural Research, 59*, 354–364.

Condon, A. G., Richards, R. A., Rebetzke, G. J., & Farquhar, G. D. (2004). Breeding for high water-use efficiency. *Journal of Experimental Botany, 55*, 407.

Cordell, D., Drangert, J. O., & White, S. (2009). The story of phosphorus: global food security and food for thought. *Global Environmental Change, 19*(2), 292–305.

Coulibaly, S., Kamsu-Foguem, B., Kamissoko, D., & Traore, D. (2019). Deep neural networks with transfer learning in millet crop images. *Computers in Industry, 108*, 115–120.

Debieu, M., Sine, B., Passot, S., Grondin, A., Akata, E., Gangashetty, P., & Laplaze, L. (2018). Response to early drought stress and identification of QTLs controlling biomass production under drought in pearl millet. *PLoS ONE, 13*(10), e0201635. https://doi.org/10.1371/journal.pone.0201635 (accessed on 7th November 2025).

Deswal, D. P., & Govila, O. P. (1994). Genetics of resistance to downy mildew (*Sclerospora graminicola*) in pearl millet (*Pennisetum glaucum*). *Indian Journal of Agricultural Sciences, 64*, 661–663.

Dutta, S., Sattler, F. T., Pucher, A., Drabo, I., Issaka, A., Sy, O., Sanogo, M. D., Angarawai, I. I., & Haussmann, B. I. G. (2021). Heterosis and combining abilities in a diverse seven-parent pearl millet population diallel tested in West Africa. *Euphotic, 215*. https://doi.org/10.1007/s10681-021-02831-x (accessed on 7th November 2025).

Faye, I., Diouf, O., Guissé, A., Sène, M., & Diallo, N. (2006). Characterizing root responses to low phosphorus in pearl millet (*Pennisetum glaucum* (L.) R. Br.). *Agronomy Journal, 98*, 1187–1194.

Fussell, L. K., Bidinger, F. R., & Bieler, P. (1991). Crop physiology and breeding for drought tolerance: research and development. *Field Crops Research, 27*, 183–199.

Gale, M. D., Devos, K. M., Zhu, J. H., Allouis, S., Couchman, M. S., Liu, H., Pittaway, T. S., Qi, X. Q., Kolesnikova-Allen, M., & Hash, C. T. (2005). New molecular marker technologies for pearl millet improvement. *Journal of SAT Agricultural Research, 1*, 1–7.

Gemenet, D. C., Hash, C. T., Sanogo, M. D., Sy, O., Zangre, R. G., Leiser, W. L., & Haussmann, B. I. G. (2015). Phosphorus uptake and utilization efficiency in West African pearl millet inbred lines. *Field Crops Research, 171*, 54–66. https://doi.org/10.1016/j.fcr.2014.11.001 (accessed on 7th November 2025).

Gill, K. S. (1991). *Pearl Millet and Its Improvement*. Indian Council of Agricultural Research, New Delhi. 456p.

Good, A. G., Shrawat, A. K., & Muench, D. G. (2004). Can less yield more? Is reducing nutrient input into the environment compatible with maintaining crop production? *Trends in Plant Science, 9*(12), 597–605.

Govil, J. N., Pokhriyal, S. C., & Murty, B. R. (1986). Effect of recurrent mass selection on grain yield in pearl millet. *Crop Improvement, 13*, 67–70.

Govila, O. P., Rai, K. N., Chopra, K. R., Andrews, D. J., & Stegmeier, W. D. (1996). Breeding pearl millet hybrids for developing countries: Indian experience. In *Proceedings of International Conference on Genetic Improvement of Sorghum and Pearl Millet* (pp. 22–27).

Govindaraj, M., Kanatti, A., Rai, K. N., & Satyanarayana, T. (2019). Does soil micronutrient variability in test locations influence performance of biofortified pearl millet in India? *Better Crops South Asia, 11*, 31–35. http://oar.icrisat.org/11285 (accessed on 7th November 2025).

Govindaraj, M., Rai, K. N., & Shanmugasundaram, P. (2016). Intra-population genetic variance for grain iron and zinc contents and agronomic traits in pearl millet. *The Crop Journal, 4*, 48–54. https://doi.org/10.1016/j.cj.2015.11.002 (accessed on 7th November 2025).

Govindaraj, M., Rai, K. N., Shanmugasundaram, P., Dwivedi, S. L., Sahrawat, K. L., Muthaiah, A. R., & Rao, A. S. (2013). Combining ability and heterosis for grain iron and zinc densities in pearl millet. *Crop Science, 53*, 505–517. https://doi.org/10.2135/cropsci2012.08.0477 (accessed on 7th November 2025).

Govindaraj, M., Rao, A. S., Shivade, H., & Rai, K. N. (2018). Effect of grain color on iron and zinc density in pearl millet. *Indian Journal of Genetics and Plant Breeding, 78*, 247–251. https://doi.org/10.5958/0975-6906.2018.00032.9 (accessed on 7th November 2025).

Govindaraj, M., Virk, P. S., Kannati, A., Cherian, B., Rai, K. N., Anderson, M. S., & Pfeiffer, W. H. (2020). Biofortified pearl millet cultivars offer potential solution to tackle malnutrition in India. In M. S. Kang

(Ed.), *Quantitative Genetics, Genomics and Plant Breeding* (pp. 385–396). CAB International. https://doi.org/10.1079/9781789240214.0385 (accessed on 7th November 2025).

Gulia, S. K., Hash, C. T., Thakur, R. P., Breese, W. A., & Sangwan, R. S. (2007). Mapping new QTLs for downy mildew [*Sclerospora graminicola* (Sacc.) J. Schroet.] resistance in pearl millet (*Pennisetum glaucum* (L.) R. Br.). In D. S. Singh, V. S. Tomar, R. K. Behl, S. D. Upadhyaya, & M. S. Bhale (Eds.), *Crop Production in Stress Environments–Genetic and Management Options* (p. 456). Agrobios International.

Gupta, M., Asfaha, D., & Ponnaiah, G. (2023). Millets: A nutritional powerhouse with anti-cancer potential. *Cureus, 15*(10), e47769. https://doi.org/10.7759/cureus.47769 (accessed on 7th November 2025).

Gupta, S. K., Nepolean, T. V., Sankar, S. M., Rathore, A., Das, R. R., Rai, K. N., & Hash, C. T. (2015a). Patterns of molecular diversity in current and previously developed hybrid parents of pearl millet (*Pennisetum glaucum* (L.) R. Br.). *American Journal of Plant Sciences, 6*, 1697–1712.

Gupta, S. K., Rathore, A., Yadav, O. P., Rai, K. N., Khairwal, I. S., Rajpurohit, B. S., & Das, R. R. (2013). Identifying mega-environments and essential test locations for pearl millet cultivar selection in India. *Crop Science, 53*, 2444–2453.

Gupta, S. K., Shaikh, C. G., Ponnaiah, G., Kalleshappa, C. C., Kholova, J., Jayalekha, A. K., & Verma, Y. S. (2020). Culm strength: key trait to identify lodging tolerant genotypes in pearl millet. In *16th ROME International Conference on Advances in Agricultural, Biological and Environmental Sciences (RAABE-20)*, Feb. 3–5, 2020, Rome, Italy.

Hanna, W. W. (1989). Characteristics and stability of a new cytoplasmic-nuclear male sterile source in pearl millet. *Crop Science, 29*, 1457–1459.

Hash, C. T., Thakur, R. P., Rao, V. P., & Bhaskar, R. A. G. (2006). Evidence for enhanced resistance to diverse isolates of pearl millet downy mildew through gene pyramiding. *International Sorghum and Millets Newsletter, 47*, 134–138.

Henson, I. E. (1982). Osmotic adjustment to water stress in pearl millet (*Pennisetum americanum* (L.) Leeke) in a controlled environment. *Journal of Experimental Botany, 33*(132), 78–87.

Hoang, T. B., & Kobata, T. (2009). Stay-green in rice (*Oryza sativa* L.) of drought-prone areas in desiccated soils. *Plant Production Science, 12*, 397–408.

Hu, Z., Mbacké, B., Perumal, R., Guèye, M. C., Sy, O., Bouchet, S., Varaprasad, P. V., & Morris, G. P. (2015). Population genomics of pearl millet (*Pennisetum glaucum* (L.) R. Br.): comparative analysis of global accessions and Senegalese landraces. *BMC Genomics, 16*, 1048. https://doi.org/10.1186/s12864-015-2255-0 (accessed on 7th November 2025).

Jarquin, D., Howard, R., Liang, Z., Gupta, S. K., Schnable, J. C., & Crossa, J. (2020). Enhancing hybrid prediction in pearl millet using genomic and/or multi-environment phenotypic information of inbreds. *Frontiers in Genetics, 10*, 1294.

Jha, S., Singh, J., Chouhan, C., Singh, O., & Srivastava, R. K. (2022). Evaluation of multiple salinity tolerance indices for screening and comparative biochemical and molecular analysis of pearl millet (*Pennisetum glaucum* (L.) R. Br.) genotypes. *Journal of Plant Growth Regulation, 41*(4), 1820–1834.

Jones, E. S., Liu, C. J., Gale, M. D., Hash, C. T., & Witcombe, J. R. (1995). Mapping quantitative trait loci for downy mildew resistance in pearl millet. *Theoretical and Applied Genetics, 91*, 448–456.

Jordan, D. R., Hunt, C. H., Cruickshank, A. W., Borrell, A. K., & Henzell, R. G. (2012). The relationship between the stay-green trait and grain yield in elite sorghum hybrids grown in a range of environments. *Crop Science, 52*, 1153–1161.

Joshi, A. B., Ahluwalia, M., & Shankar, K. (1961). 'Improved Ghana' is a better bajra. *Indian Farm, 11*, 12.

Joshi, V. J., & Ugale, S. D. (2002). Involvement of higher order interactions addressing complex polygenetically controlled inheritance of downy mildew [*Sclerospora graminicola* (Sacc.) Schrot] resistance in pearl millet (*Pennisetum glaucum* (L.) R. Br.). *Euphotic, 127*, 149–161.

Kanatti, A., Rai, K. N., Radhika, K., Govindaraj, M., Sahrawat, K. L., & Rao, A. S. (2014). Grain iron and zinc density in pearl millet: combining ability, heterosis and association with grain yield and grain size. *Springer Plus, 3*, 1–12. https://doi.org/10.1186/2193-1801-3-763 (accessed on 7th November 2025).

Karthika, G., & Govintharaj, P. (2022). Breeding climate-resilience crops for future agriculture. In *Climate Change and Crop Stress* (pp. 1–32). Academic Press.

Khairwal, I. S., Rai, K. N., Yadav, O. P., Rajpurohit, B. S., & Negi, S. (2009). *Pearl Millet Cultivars: Seeds of Choice*. All India Coordinated Pearl Millet Improvement Project, Mandor, pp. 1–67.

Kholová, J., Hash, C. T., Kakkera, A., Kočová, M., & Vadez, V. (2010a). Constitutive water-conserving mechanisms are correlated with the terminal drought tolerance of pearl millet (*Pennisetum glaucum* (L.) R. Br.). *Journal of Experimental Botany, 61*(2), 369–377.

Kholová, J., Hash, C. T., Lavakumar, P., Yadav, R. S., Kočová, M., & Vadez, V. (2010b). Terminal drought-tolerant pearl millet (*Pennisetum glaucum* (L.) R. Br.) have high leaf ABA and limit transpiration at high vapor pressure deficit. *Journal of Experimental Botany, 61*, 1431–1441.

Krishnamurthy, L., Serraj, R., Rai, K. N., Hash, C. T., & Dakheel, A. J. (2007). Identification of pearl millet (*Pennisetum glaucum* (L.) R. Br.) lines tolerant to soil salinity. *Euphytica, 158*, 179–188.

Krishnaswamy, N. (1962). *Bajra: Pennisetum typhoides S & H*. Indian Council of Agricultural Research, New Delhi, p. 93.

Lemerle, D., Gill, G. S., Murphy, C. E., Walker, S. R., Cousens, R. D., Mokhtari, S., Peltzer, S. J., Coleman, R., & Luckett, D. J. (2001). Genetic improvement and agronomy for enhanced wheat competitiveness with weeds. *Australian Journal of Agricultural Research, 52*, 527–548. https://doi.org/10.1071/AR00056 (accessed on 7th November 2025).

Leuck, B. D. (1970). The role of resistance in pearl millet in control of the fall armyworm. *Journal of Economic Entomology, 63*, 1679–1681.

Liang, Z., Gupta, S. K., Yeh, C. T., Zhang, Y., Ngu, D. W., Kumar, R., Patil, H. T., Mungra, K. D., Yadav, D. V., Rathore, A., Srivastava, R. K., Gupta, R., Yang, J., Varshney, R. K., Schnable, P. S., & Schnable, J. C. (2018). Phenotypic data from inbred parents can improve genomic prediction in pearl millet hybrids. *G3: Genes, Genomes, Genetics, 8*(7), 2513–2522.

Lopes, M. S., & Reynolds, M. P. (2012). Stay-green in spring wheat can be determined by spectral reflectance measurements (normalized difference vegetation index) independently from phenology. *Journal of Experimental Botany, 63*, 3789–3798.

López-Castañeda, C., Richards, R. A., Farquhar, G. D., & Williamson, R. E. (1996). Seed and seedling characteristics contributing to variation in early vigor among temperate cereals. *Crop Science, 36*, 1257–1266.

Machado, R. M. A., & Serralheiro, R. P. (2017). Soil salinity: effect on vegetable crop growth. Management practices to prevent and mitigate soil salinization. *Horticulture, 3*, 30.

Mahalakshmi, V., Bidinger, F., & Raju, D. (1987). Effect of timing of water deficit on pearl millet (*Pennisetum americanum*). *Field Crops Research, 15*, 327–339. https://doi.org/10.1016/0378-4290(87)90020-7 (accessed on 7th November 2025).

Manga, V. K., & Yadav, O. P. (1995). Effect of seed size on development traits and ability to tolerate drought in pearl millet. *Journal of Arid Environments, 29*, 169–172.

Mariac, C., Luong, V., Kapran, I., Mamadou, A., Sagnard, F., Deu, M., Chantereau, J., Gerard, B., Ndjeunga, J., Bezançon, G., Pham, J. L., & Vigouroux, Y. (2006). Diversity of wild and cultivated pearl millet accessions (*Pennisetum glaucum* [L.] R. Br.) in Niger assessed by microsatellite markers. *Theoretical and Applied Genetics, 114*, 49–58.

Mathur, P. N. (2012). *Global Strategy for the ex-situ Conservation of Pearl Millet and its Wild Relatives* (pp. 1–89). Global Crop Diversity Trust, Rome, Italy.

Melchinger, A. E., & Gumber, R. K. (1998). Overview of heterosis and heterotic groups in agronomic crops. In K. R. Lamkey & J. E. Staub (Eds.), *Concepts and Breeding of Heterosis in Crop Plants* (pp. 29–44). CSSA. https://doi.org/10.2135/cssaspecpub25.c3 (accessed on 7th November 2025).

Mohammed, R., Gangashetty, P. I., & Karimoune, L. B. N. M. (2020). Genetic variation and diversity of pearl millet (*Pennisetum glaucum* (L.)) genotypes assessed for millet head miner, *Heliocheilus albipunctella* resistance, in West Africa. *Euphotic, 216*(10), 158.

Morgan, R. N., Wilson, J. P., Hanna, W. W., & Ozias-Akin, P. (1998). Molecular markers for rust and *Pyricularia* leaf spot disease resistance in pearl millet. *Theoretical and Applied Genetics, 96*, 413–420.

N/A. (2010). FAO (Food and Agriculture Organization of the United Nations). http://www.fao.org (accessed on 7th November 2025).

N/A. (2018). *Proceedings of the 53rd Annual Group Meeting of ICAR–All India Coordinated Research Project on Pearl Millet (AICPMIP)*. Jodhpur: Agriculture University. http://www.aicpmip.res.in/pw2018.pdf (accessed on 7th November 2025).

N/A. (2019). *Annual Report–2019*. ICAR-Indian Institute of Millets Research, Rajendranagar, Hyderabad, p. 152.

N/A. (2023). *Proceedings of Centre Wise Presentations & Pearl Millet Technical Program Planning Meeting 2023–24 of ICAR-All India Coordinated Research Project on Pearl Millet*. http://aicpmip.res.in/pw2023.pdf (accessed on 7th November 2025).

Narayan, J., John, D., & Ramadas, N. (2019). Malnutrition in India: status and government initiatives. *Journal of Public Health Policy, 40*, 126–141.

Nepolean, T., Gupta, S. K., Dwivedi, S. L., Bhattacharjee, R., Rai, K. N., & Hash, C. T. (2012). Genetic diversity in maintainer and restorer lines of pearl millet. *Crop Science, 52*, 2555.

Nguyen, P. H., Scott, S., Headey, D., Singh, N., Tran, L. M., Menon, P., & Ruel, M. T. (2021). The double burden of malnutrition in India: trends and inequalities (2006–2016). *PLOS ONE, 16*(2), e0247856.

Nwanze, K. F., & Harris, K. M. (1992). Insect pests of pearl millet in West Africa. *Review of Agricultural Entomology, 80*, 1132–1185.

Ong, C. K. (1983a). Response to temperature in a stand of pearl millet (*Pennisetum typhoides* S&H): I. Vegetative development. *Journal of Experimental Botany, 34*, 322–336.

Ong, C. K. (1983b). Response to temperature in a stand of pearl millet (*Pennisetum typhoides* S&H): II. Reproductive development. *Journal of Experimental Botany, 34*, 337–348.

Oumar, I., Mariac, C., Pham, J. L., & Vigouroux, Y. (2008). Phylogeny and origin of pearl millet (*Pennisetum glaucum* [L.] R. Br.) as revealed by microsatellite loci. *Theoretical and Applied Genetics, 117*, 489–497.

Peacock, J. M., Soman, P., Jayachandran, R., Rani, A. U., Howarth, C. J., & Thomas, A. (1993). Effects of high soil surface temperature on seedling survival in pearl millet. *Experimental Agriculture, 29*, 215–225.

Ponnaiah, G., Gupta, S. K., Blümmel, M., Marappa, M., Pichaikannu, S., Das, R., & Rathore, A. (2019). Utilization of molecular marker-based genetic diversity patterns in hybrid parents to develop better forage quality multi-cut hybrids in pearl millet. *Agriculture, 9*, 97. https://doi.org/10.3390/agriculture9050097 (accessed on 7th November 2025).

Pucher, A., Sy, O., Sanogo, M. D., Angarawai, I. I., Zangre, R., Ouedraogo, M., Boureima, S., Hash, C. T., & Haussmann, B. I. G. (2016). Combining ability patterns among West African pearl millet landraces and prospects for pearl millet hybrid breeding. *Field Crops Research, 195*, 9–20.

Pujar, M., Govindaraj, M., Gangaprasad, S., Kanatti, A., & Shivade, H. (2020). Genetic variation and diversity for grain iron, zinc, protein and agronomic traits in advanced breeding lines of pearl millet (*Pennisetum glaucum* (L.) R. Br.) for biofortification breeding. *Genetic Resources and Crop Evolution, 67*, 2009–2022.

Pujarula, V., Pusuluri, M., Bollan, S., Das, R. R., Ratnala, R., Adapala, G., Thuraga, V., Rathore, A., Srivastava, R. K., & Gupta, R. (2021). Genetic variation for nitrogen uses efficiency traits in global diversity panel and parents of mapping populations in pearl millet. *Frontiers in Plant Science, 12*, 625915.

Rachie, K. O., & Majmudar, J. V. (1980). *Pearl Millet* (p. 307). The Pennsylvania State University Press.

Rai, K. N. (1995). A new cytoplasmic-nuclear male sterility system in pearl millet. *Plant Breeding, 114*, 445–447.

Rai, K. N., & Anandkumar, K. (1994). Pearl millet improvement at ICRISAT—an update. *International Sorghum and Millets Newsletter, 35*, 1–29.

Rai, K. N., & Hanna, W. W. (1990). Morphological characteristics of tall and dwarf pearl millet isolines. *Crop Science, 30*, 23–25.

Rai, K. N., & Rao, A. S. (1991). Effect of d2 dwarfing gene on grain yield and yield components in pearl millet near-isogenic lines. *Euphotic, 52*, 25–31.

Rai, K. N., Gupta, S. K., Sharma, R., Govindaraj, M., Rao, A. S., Shivade, H., & Bonamigo, L. A. (2014). Pearl millet breeding lines developed at ICRISAT: a reservoir of variability and useful source of non-target traits. *SAT eJournal, 1*, 1–13. http://oar.icrisat.org/id/eprint/8629 (accessed on 7th November 2025).

Rai, K. N., Virk, D. S., Harinarayana, G., & Rao, A. S. (1996). Stability of male sterile sources and fertility restoration of their hybrids in pearl millet. *Plant Breeding, 115*, 494–500.

Rattunde, H. F., & Witcombe, J. R. (1993). Recurrent selection for increased grain yield and resistance to downy mildew. *Plant Breeding, 110*, 63–72.

Rosenow, D. T., Quisenberry, J. E., Wendt, C. W., & Clark, L. E. (1983). Drought tolerant sorghum and cotton germplasm. *Agricultural Water Management, 7*, 207–222.

Russell, W. A. (1986). Contribution of breeding to maize improvement in the United States, 1920s–1980s. *Iowa State Journal of Research, 61*, 5–34.

Russo, T. A., Tully, K., Palm, C., & Neill, C. (2017). Leaching losses from Kenyan maize cropland receiving different rates of nitrogen fertilizer. *Nutrient Cycling in Agroecosystems, 108*, 195–209. https://doi.org/10.1007/s10705-017-9852-z (accessed on 7th November 2025).

Sastry, P. S., & Mallikarjuna, N. (2014). Induction of androgenesis in pearl millet. *Universal Journal of Agricultural Research, 2*(6), 216–223.

Sharma, R., Thakur, R. P., Rai, K. N., Gupta, S. K., Rao, V. P., Rao, A. S., & Kumar, S. (2009). Identification of rust resistance in hybrid parents and advanced breeding lines of pearl millet. *Journal of SAT Agricultural Research, 7*, 1–4.

Shetty, H. S., Raj, S. N., Kini, K. R., Bishnoi, H. R., Sharma, R., Rajpurohit, B. S., & Yadav, O. P. (2016). *Downy Mildew of Pearl Millet and Its Management*. Indian Council of Agricultural Research, Mandor, Jodhpur, 342304, p. 55.

Singh, P., & Kanemasu, E. T. (1983). Leaf and canopy temperatures of pearl millet genotypes under irrigated and non-irrigated conditions. *Agronomy Journal, 75*, 497–501.

Singh, P., Rai, K. N., Witcombe, J. R., & Andrews, D. J. (1988). Population breeding methods in pearl millet improvement (*Pennisetum americanum*). *L'Agronomie Tropicale, 43*, 185–193.

Singh, S. D. (1995). Downy mildew of pearl millet. *Plant Disease, 79*, 545–550.

Singh, S. D., & Talukdar, B. S. (1998). Inheritance of complete resistance to pearl millet downy mildew. *Plant Disease, 82*, 791–793.

Singh, S., Gupta, S. K., Thudi, M., Das, R. R., Vemula, A., Garg, V., Varshney, R. K., Rathore, A., Pahuja, S. K., & Yadav, D. V. (2018a). Genetic diversity patterns and heterosis prediction based on SSRs and SNPs in hybrid parents of pearl millet. *Crop Science, 58*, 2379–2390.

Singh, S., Sharma, R., Pushpavathi, B., Gupta, S. K., Durgarani, C. V., & Raj, C. (2018). Inheritance and allelic relationship among gene(s) for blast resistance in pearl millet (*Pennisetum glaucum* (L.) R. Br.). *Plant Breeding, 137*(4), 573–584.

Soda, N., Sharan, A., Gupta, B. K., Singla-Pareek, S. L., & Pareek, A. (2016). Evidence for nuclear interaction of a cytoskeleton protein (OsIFL) with metallothionein and its role in salinity stress tolerance. *Scientific Reports, 6*, 34762.

Soman, P., Bidinger, F. R., Peacock, J. M., & Walker, T. S. (1981). Seedling establishment – a preliminary survey taken up in Aurepally during kharif. *ICRISAT Internal Report*. ICRISAT, Patancheru, India.

Soman, P., Jayachandran, R., & Bidinger, F. R. (1987). Uneven variation in plant-to-plant spacing in pearl millet. *Agronomy Journal, 79*, 891–895.

Srivastava, R. K., Bollam, S., Pujarula, V., Pusuluri, M., Singh, R. B., Potupureddi, G., & Gupta, R. (2020b). Exploitation of heterosis in pearl millet: a review. *Plants, 9*, 807.

Srivastava, R. K., Singh, R. B., Srikanth, B., Satyavathi, C. T., Yadav, R., & Gupta, R. (2020a). Genome-wide association studies (GWAS) and genomic selection (GS) in pearl millet: advances and prospects. *Frontiers in Genetics, 10*, 1389.

Stich, B., Haussmann, B. I., Pasam, R., Bhosale, S., Hash, C. T., Melchinger, A. E., & Parzies, H. K. (2010). Patterns of molecular and phenotypic diversity in pearl millet (*Pennisetum glaucum* (L.) R. Br.) from West and Central Africa and their relation to geographical and environmental parameters. *BMC Plant Biology, 10*, 216.

Thakur, R. P., Sharma, R., Rai, K. N., Gupta, S. K., & Rao, V. P. (2009). Screening techniques and resistance sources for foliar blast in pearl millet. *Journal of SAT Agricultural Research, 7*, 1–5.

Umar, W., Muhammad, A. A., Muhammad, Z. R., Hamaad, R. A., Zia, U. R. F., Ahsan, S., Umar, R., Adnan, M., & Muhammad, N. (2020). Nitrogen and phosphorus use efficiency in agroecosystems. In S. Kumar, R. S. Meena, & M. K. Jhariya (Eds.), *Resources Use Efficiency in Agriculture* (pp. 213–257). Springer, Singapore.

Upadhyaya, H. D., & Ortiz, R. (2001). A mini core subset for capturing diversity and promoting utilization of chickpea genetic resources in crop improvement. *Theoretical and Applied Genetics, 102*(8), 1292–1298.

Upadhyaya, H. D., Gowda, C. L. L., Reddy, K. N., & Singh, S. (2009). Augmenting the pearl millet core collection for enhancing germplasm utilization in crop improvement. *Crop Science, 49*(2), 573–580.

van Oosterom, E. J., Bidinger, F. R., & Weltzien, R. E. (2003). A yield architecture framework to explain adaptation of pearl millet to environmental stress. *Field Crops Research, 80*, 33–56.

van Oosterom, E. J., Weltzien, E., Yadav, O. P., & Bidinger, F. R. (2006). Grain yield components of pearl millet under optimum conditions can be used to identify germplasm with adaptation to arid zones. *Field Crops Research, 96*, 407–421.

Van Staveren, J., & Stoop, W. A. (1985). Adaptation to topo sequence land types in West Africa of different sorghum genotypes in comparison with local cultivars of sorghum, millet, and maize. *Field Crops Research, 11*, 13–35.

Varshney, R. K., Graner, A., & Sorrells, M. E. (2005). Genic microsatellite markers in plants: features and applications. *Trends in Biotechnology, 23*(1), 48–55.

Velu, G., Rai, K. N., Sahrawat, K. L., & Sumalini, K. (2008). Variability for grain iron and zinc contents in pearl millet hybrids. *Journal of SAT Agricultural Research, 6*, 1–4. http://oar.icrisat.org/5043/1/SATeJou6_1-4_2008.pdf (accessed on 7th November 2025).

Victor, D. M., Cralle, H. T., & Miller, F. R. (1989). Partitioning of 14C photosynthate and biomass in relation to senescence characteristics of sorghum. *Crop Science, 29*, 1049–1053.

vom Brocke, K., Christinck, A., Weltzien, E. R., Presterl, T., & Geiger, H. H. (2003). Farmers' seed systems and management practices determine pearl millet genetic diversity patterns in semi-arid regions of India. *Crop Science, 43*, 1680–1689.

Walton, W. R., & Luginbill, P. (1916). *The Fall Army Worm, or 'Grass Worm' and Its Control. Farmers Bulletin No. 752*, USDA, pp. 1–20.

Witcombe, J. R. (1999). Population improvement. In I. S. Khairwal, K. N. Rai, D. J. Andrews, & G. Harinarayana (Eds.), *Pearl Millet Breeding* (pp. 213–256). Oxford and IBH.

Woonho, Y., Shingu, K., Jeong-Hwa, P., Sukjin, K., Jong-Seo, C., & Sunggi, H. (2018). Association of grain filling duration and leaf activity with the grain yield in field-grown temperate japonica rice. *Korean Journal of Crop Science, 63*, 120–130.

Yadav, O. P., & Rai, K. N. (2013). Genetic improvement of pearl millet in India. *Agricultural Research, 2*, 275–292.

Yadav, O. P., Gupta, S. K., Govindaraj, M., Sharma, R., Varshney, R. K., Srivastava, R. K., Rathore, A., & Mahala, R. S. (2021). Genetic gains in pearl millet in India: insights into historic breeding strategies and future perspective. *Frontiers in Plant Science, 12*, 396.

Yadav, O. P., Manga, V. K., & Saxena, M. B. L. (1994). Ontogenetic approach to grain production in pearl millet (*Pennisetum glaucum*) based on path-coefficient analysis. *Indian Journal of Agricultural Sciences, 64*, 233–236.

Yadav, O. P., Rai, K. N., Rajpurohit, B. S., Hash, C. T., Mahala, R. S., Gupta, S. K., Shetty, H. S., Bishnoi, H. R., Rathore, M. S., Kumar, A., Sehgal, S., & Raghvani, K. L. (2012). *Twenty-Five Years of Pearl Millet Improvement in India* (pp. 1–130). ICAR.

Yadav, O. P., Rajpurohit, B. S., Kherwa, G. R., & Kumar, A. (2012a). Prospects of enhancing pearl millet (*Pennisetum glaucum*) productivity under drought environments of north-western India through hybrids. *Indian Journal of Genetics and Plant Breeding, 72*, 25–30.

Yadav, R. S., Hash, C. T., Bidinger, F., Cavan, G., & Howarth, C. J. (2002). Quantitative trait loci associated with traits determining grain and stover yield in pearl millet under terminal drought-stress conditions. *Theoretical and Applied Genetics, 104*, 67–83. https://doi.org/10.1007/s001220200008 (accessed on 7th November 2025).

Yadav, R. S., Hash, C., Bidinger, F., Devos, K., & Howarth, C. J. (2004). Genomic regions associated with grain yield and aspects of post-flowering drought tolerance in pearl millet across stress environments and tester background. *Euphotic, 136*, 265–277. https://doi.org/10.1023/B:EUPH.0000032711.34599.3a (accessed on 7th November 2025).

Yadav, R. S., Sehgal, D., & Vadez, V. (2011). Using genetic mapping and genomics approaches in understanding and improving drought tolerance in pearl millet. *Journal of Experimental Botany, 62*, 397–408. https://doi.org/10.1093/jxb/erq265 (accessed on 7th November 2025).

Yakubu, H., Ngala, A. L., & Dugje, I. Y. (2010). Screening of millet (*Pennisetum glaucum* L.) varieties for salt tolerance in semi-arid soil of Northern Nigeria. *World Journal of Agricultural Sciences, 6*, 374–380.

Zaveri, P. P., Phul, P. S., & Dhillon, B. S. (1989). Observed and predicted genetic gains from single and multi-trait selections using three mating designs in pearl millet. *Plant Breeding, 103*, 270–277.

Zegada-Lizarazu, W., & Iijima, M. (2005). Deep root water uptake ability and water use efficiency of pearl millet in comparison to other millet species. *Plant Production Science, 8*(4), 454–460.

CHAPTER 3

Genetic Improvement in Finger Millet

ASHOK SINGAMSETTI,[1,2] N. ANURADHA,[1,*] T. S. S. K. PATRO,[1] Y. SANDHYA RANI,[1] and U. TRIVENI[1]

[1]*Agricultural Research Station, Acharya N. G. Ranga Agricultural University, Vizianagaram, Andhra Pradesh, India*

[2]*Institute of Agricultural Sciences, Banaras Hindu University, Varanasi, Uttar Pradesh, India*

*Corresponding author

3.1 INTRODUCTION

Finger millet (*Eleusine coracana* (L.) Gaertn) is an herbaceous plant widely cultivated in the arid and semi-arid areas of Africa and Asia as a millet crop. It is an allotetraploid and highly self-pollinated crop with a chromosome number of 36 (2n=4X=36). Finger millet belongs to the Poaceae family and Chloridoideae subfamily under the tribe Chlorideae. It stands as the fourth most important millet crop worldwide, followed by sorghum, pearl millet, and foxtail millet (Gupta et al., 2012). It is realized that finger millets hold great promise for food security and nutrition amid the ever-increasing world population and the present climate change scenario. In production, it is in sixth position in India after wheat, rice, corn, sorghum, and pearl millet. According to final estimates of 2021–22 by the Directorate of Economics and Statistics (DES), Government of India, India is one of the largest producers of finger millet, with an area of 1.01 million hectares, a production of 1.67 million tons, and a productivity of 1747 kg of yield per hectare, respectively. Although the precise global area and production figures were unclear (Sood et al., 2019), some reports on the production of finger millet indicated that about 4.6 million tons were being produced every year globally (Sakamma et al., 2018).

Many reports have proven that finger millet is the richest source of minerals like iron, zinc, calcium, and manganese. It has a high nutritional value and is also rich in bioactive components, such as flavonoids (5.54 mg/g CE), phytic acid (0.48%), polyphenol (10.2 mg/g TAE), and dietary fiber (15–20%). Furthermore, Chandra et al. (2016) state that it is an adequate supply source of several important vitamins and amino acids, and it is also often regarded as the greatest source of lysine (Chandra et al., 2013). Polyphenols are well known for their antioxidant and anti-diabetic roles. Thus, finger millet is becoming an attractive diet for patients suffering from lifestyle diseases like diabetes (Singh et al., 2022).

Millets in Sustainable Agriculture: Genetic Improvement, Nutritional Quality, Cultivation, and Integrated Pest–Disease Management. Roshni Vijayan, A. John Joel, R. Ambika Rajendran, and Adithya P. Balakrishnan (Eds.)

DOI: 10.1201/9781779642332-4

Being an orphan crop for many decades, it has indeed gained great importance as a potential nutraceutical cereal because of its high dietary value and low glycemic index (Hatakeyama et al., 2018).

3.2 ORIGIN AND DIVERSITY

Although the exact center of domestication is not clear, many studies suggest that the area between western Uganda and the Ethiopian land of Eastern Africa is the primary center of origin for finger millet (de Wet, 1995). Archaeological and historical evidence revealed that farming communities in Eastern Africa started cultivating this millet around 5000 years ago (Klichowska, 1984). Approximately 3000 years ago, the crop was introduced to India, which became its secondary center of diversity in the continental region. The studies have also proven that cultivated finger millet, that is, Asiatic lowland race (*Eleusine coracana* subsp. *coracana*), was derived from a wild type with large-grained species, that is, African highland race (*E. coracana* subsp. *africana*) (Padulosi et al., 2009), through selection and domestication. The cultivated finger millet species is an allotetraploid (*E. coracana*; $2n = 4x = 36$), and it shows a morphology similar to that of *Eleusine indica* ($2n = 18$) and *Eleusine africana* ($2n = 36$). Evidence from chloroplast DNA sequence analysis using restriction fragment patterns, as well as morphological (Hilu and de Wet, 1976), molecular (Dida, 1998; Hilu, 1988), and cytological (Hiremath and Salimath, 1992) studies, revealed that one of the genomes (AA) from *E. indica* was transferred to the cultivated *E. coracana* and served as the maternal genome donor to *E. africana*. More recently, Zhang et al. (2019) reported that relationships among species of the genus *Eleusine* and their evolutionary history were elucidated through transcriptome analysis.

3.3 GERMPLASM RESOURCES

More than 28,000 finger millet germplasm are available in various organizations worldwide. Among these, the National Bureau of Plant Genetic Resources (NBPGR), India, maintains the largest collection of 10,507 germplasm accessions of finger millet under medium- and long-term conservation with good coverage and no gaps. The International Crop Research Institute for the Semi-Arid Tropics (ICRISAT), India, has 5957 germplasm accessions. The All India Small Millets Improvement Programme (AISMIP), Bangalore; ICRISAT; and NBPGR together account for three-quarters of the world's finger millet collections in India. Most of these germplasm collections in India are indigenous in nature, while very few are of exotic origin. Almost 60% of Indian collections are being maintained by AISMIP in Bangalore as medium-term storage that can be used as working collections. Finger millet germplasm conservation can be used to identify desired qualities, including early flowering and maturity, high yield, biotic and abiotic stress tolerance, and other secondary characteristics for crop development on a regular basis through germination testing and data documentation. The creation of core and mini-core sets, as well as the release of a crop germplasm library, improved the effective use of germplasm. Other organizations, like the Kenya Agricultural Research Institute (KARI), Kenya, with 2875 accessions; the Institute of Biodiversity Conservation (IBC), Ethiopia, with 2156 accessions; the

USDA Agricultural Research Service (USDA-ARS), United States, with 1452 accessions; and the Serere Agricultural and Animal Production Research Institute (SAARI), Uganda, have maintained a reasonable collection of finger millet germplasm (Dwivedi et al., 2012; Goron and Raizada, 2015; Gupta et al., 2017; Saha et al., 2016). A global-level plant genetic resources database, GENESYS (www.genesyspgr.org), developed by Biodiversity International with financial support from the Trust and the International Treaty on Plant Genetic Resources for Food and Agriculture (ITPGRFA), contains information on nearly 2.5 million germplasm accessions, including 7516 accessions under the genus *Eleusine*. Apart from GENESYS, a few other databases, such as the World Information and Early Warning System by the Food and Agricultural Organization (FAO-WIEWS; http://apps3.fao.org/wiews/wiews.jsp?i_l=EN), the System-wide Information Network for Genetic Resources (SINGER; http://singer.cgiar.org/), the Global Biodiversity Information Facility (GBIF; http://eurisco.ecpgr.org/), and the European Search Catalogue for Plant Genetic Resources (PGR) (EURISCO; http://eurisco.ecpgr.org/), hold information pertaining to finger millet gene pools, which assists breeders in understanding the accession origin and distribution patterns. *See* Table 3.1.

TABLE 3.1 Details of Databases Holding Information on Finger Millet Global Collection

Sl. No.	Database	Contact Details	Developed by
1.	GENESYS	www.genesyspgr.org	Genesis, ICRISAT
2.	FAO-WIEWS (FAO-World Information and Early Warning System)	http://apps3.fao.org/wiews/wiews.jsp?i_l=EN	Plant Genetic Resources for Food and Agriculture (PGRFA), by FAO
3.	SINGER (System-wide Information Network for Genetic Resources)	http://singer.cgiar.org/	An information exchange network of the Future Harvest Centers of the Consultative Group on International Agricultural Research (CGIAR) and associated partners
4.	EURISCO (European Plant Genetic Resources Search Catalogue)	http://eurisco.ecpgr.org/	Biodiversity International, Rome, Italy (from 2003 to 2014). Institute of Plant Genetics and Crop Plant Research (IPK), Gatersleben, Germany (from 2014)
5.	GBIF (Global Biodiversity Information Facility)	https://www.gbif.org/species/search	Coordinated through its Secretariat in Copenhagen, the GBIF network of participating countries and organizations

3.4 MORPHOLOGY AND FLORAL BIOLOGY

Finger millet is an annual crop that takes 90 to 150 days to mature and grows to a height of around 50 to 150 cm. The stem is straight and erect, compressed laterally with or without pigment at the nodes, and glabrous. The leaves have parallel veining, are grass-like, and can produce a large number of nodal branches and tillers. Leaf blades mostly have ciliated borders and are straight with an acute tip (Rachie and Peters, 1977). The panicle is made up of a group of spikes, often referred to as "fingers," that range in number from 5 to 20 and are arranged like the feet of a bird. Each spike contains about 70 spikelets with florets, which are perfect flowers except for the terminal ones, which may be infertile sometimes. The florets are covered with two large barren leaves, each being packed with a pair of scales referred to as 'palea' (Figure 3.1). The florets appear in the axil of lower flowering glumes known as the lemma. Two little scaly lodicules are located close to the base of the ovary (Dodake and Dhonukshe, 1998). Three stamens are present with anther lengths of 0.5–0.8 mm. The gynoecium bears a bi-capillary and uni-locular superior ovary with two styles with plumose stigma (Seetharam et al., 2003). The androecium is almost covered with the stigma, which ensures self-pollination. The filaments are shorter than the anthers, whereas the stigma with feathery branches is 0.83 mm in length (Dodake and Dhonukshe, 1998). Botanically, the fruit is a caryopsis (seed), which is smaller in size (1.2–1.8 mm in diameter) with an oval or round to oblong shape, smooth texture, and the color can range from yellow to reddish-brown, black, or purple. However, white-grained finger millet varieties have also been developed recently by various research institutes in India to enhance consumer acceptance by improving value-added products.

FIGURE 3.1 (A) Inflorescence; (B) Fully opened spikelets and floret (1–6); Parts of flower including (1) Outer glumes; (2) Anther; (3) Lemma; (4) Palea (5) Ovary and (6) stigma.

Finger millet is mostly self-pollinated (with less than 1% cross-pollination mediated by wind) due to a close association of elongated styles and filaments with large-sized anthers (Dodake and Dhonukshe, 1998). The anthesis pattern of finger millet is significantly influenced by external weather conditions. At high humidity and low temperature, both stigma and anthers are observed hanging out from the closed flowers, whereas only empty dehisced anthers are seen hanging out at low humidity and high temperature. The spikelet opens from top to bottom within a spike, while the floret opens from bottom to top within each spikelet. It is observed that only one floret in a spikelet opens in a day, which takes 5–7 days to complete flowering. A few studies confirm that the best suitable time for anthesis in finger millet is from 1:00 to 5:00 AM (Gowda, 1997). The anther dehiscence occurs longitudinally just prior to the floret opening, and the stigma and anthers attain the same height inside the floret, which leads to anthers pollinating their own stigma (Seetharam et al., 2003). Hot water treatment is widely used by finger millet breeders to alleviate the drawbacks of manual anther removal. This procedure involves dipping the female panicles for five minutes in hot water (48°C to 52°C). Similar to the contact method of pollination, the female panicles are then tied with male panicles after being air-dried. The GMS method is the source of the genotype "INFM 95001." In the varietal background of GPU-28, a partial GMS line (PS 1-IC0598201; INGR14015) is found; nevertheless, because of the variable levels of sterility/fertility in different geographical locations and genetic backgrounds, its practical utility is limited (Nagaraja et al., 2023).

3.5 THE SCOPE OF GENETIC RESEARCH ON FINGER MILLET

In the last two decades, interest in research pertaining to finger millet has increased significantly because of its potential as a nutri-cereal and its climate resilience. Although it is considered an "orphan crop" or "underutilized crop," it is a crop of great antiquity with historical, cultural, and nutritional importance, especially in Asia and Africa (Sood et al., 2019). It is also known as a hardy crop that shows resilience with great adaptability and capability towards any adverse environmental conditions. The prime focus of advancements in finger millet research is bio-fortification and resistance to biotic and abiotic stresses. Being an orphan crop for many years, researchers have more scope to reveal the diversification among germplasm, genetic architecture, mapping, and QTL identification. The rapid advances in finger millet research enlighten the way to elucidate various omics technologies such as genomics, proteomics, and transcriptomics. The deployment of high-throughput technologies like DNA sequencing, whole genome sequencing, gene annotation, and editing methods results in the genetic improvement of finger millet. The milestones in finger millet research, along with the area of work, are given in Figure 3.2.

3.6 CROP IMPROVEMENT STRATEGIES

Finger millet is highly self-pollinated, with a negligible amount of cross-pollination (1%) mediated by wind (Jansen and Ong 1996; Purseglove 1972). This mode of pollination and the small size of flowers are the major constraints to performing hand emasculation as easily as in other cereals. Later in the 1960s, new crosses were made successfully through hot water emasculation (after 3 to 4 days of inflorescence emergence at 52°C for five minutes), combined with contact pollination. The hot-water method of emasculation was more efficient (52–80%) compared with the hand emasculation method (16–40%) in finger millet (Krishna et al., 2019). The maximum percentage of the released varieties is from germplasm selections in India as well as in Africa (Dida, 2007). In finger millet, mass and pure-line selection techniques are extensively used to improve inter-varietal traits such as disease and stress resistance, early flowering and maturity, and grain yield (Harinarayana, 1986).

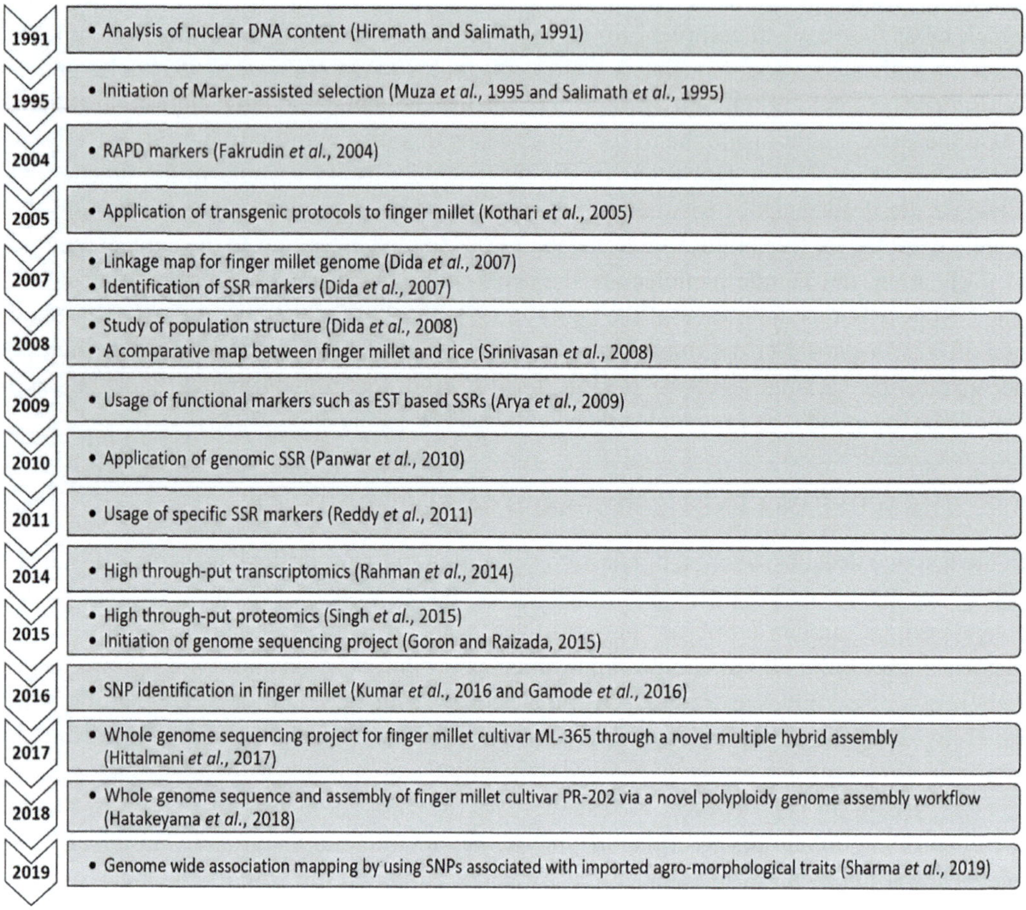

FIGURE 3.2 Major key events in finger millet research.

Mass selection has been extensively used in finger millet varietal improvement for the purification of the landrace and multiplication of varieties bred by pure line or pedigree breeding.

However, several attempts have been made to deploy common breeding methods practiced in other cereals, such as recombination breeding to generate stable and economically improved hybrids, breeding nurseries, and inbred lines. Many of these attempts have been unsuccessful due to the practical difficulties in making biparental crosses, performing emasculation, and manual hybridization in finger millet. Apart from hot water treatment to induce male sterility, genetic male sterility has also been reported in finger millet. This source of male sterility is governed by the *ms1* allele; it was developed through chemical mutagen treatment (1.5% aqueous solution of EMS at 25°C for 6 hours), and this genetic male-sterile line was designated as "INFM 95001" (Gupta et al., 1997). This induced mutant with genetic male sterility developed by the ICRISAT (International Crops Research Institute for the Semi-Arid Tropics) paved a path for finger millet breeders, as it is evident for another efficient breeding tool for enhancing grain yield and disease resistance through creating new recombinants and segregating populations. Later, a partial GMS line (PS 1-IC0598201; INGR14015) in the GPU-28 varietal background was identified, but the practical utility of this line is limited due to varying levels of fertility or sterility across locations and genetic backgrounds (Gowda et al., 2014). There is an immediate need for extensive studies on male sterile systems, mechanisms like protandry, and the usage of gametocides (Oduori and Kanyenji et al., 2007). Until now, there is no evidence of cytoplasmic male-sterility systems in finger millet. Heterosis or hybrid seed development is not commercially exploited in finger millet as in the case of other cereals such as sorghum and pearl millet (Ganapathy, 2017). Mutation breeding plays a significant role in the crop improvement of finger millet for the development of early maturing and photoperiod-insensitive cultivars, the generation of polygenic variability, and the development of complete/partial male-sterile lines. A wide range of polymorphic markers would be required for finger millet breeding programs due to the presence of narrow variation within the cultivated species.

3.6.1 MARKER-ASSISTED BREEDING IN FINGER MILLET

When it comes to identifying and enhancing a specific characteristic in terns, molecular markers are essential for yielding better crops. Marker-assisted research in finger millet crops was first reported by Muza et al. and Sailamath et al. (1995; 2005). Molecular markers, specifically randomly amplified polymorphic DNA (RAPD), were identified for finger millet by Fakruddin et al. (2004), followed by simple sequence repeats (SSR) markers by Dida et al. (2002). Gupta et al. (2010) studied three genotypes of finger millet with different seed coat colors by evaluating morphological traits along with a few physiological and biochemical characteristics using 10 RAPD and 10 ISSR (Inter-simple sequence repeats) markers. A variety of molecular marker breeding techniques, which are widely employed in crop breeding improvement, were made possible by the development of DNA-based markers (Babu et al., 2007). According to Babu et al. (2014b), functional SSR markers for blast resistance were used to estimate the genetic diversity among the finger millet germplasm collection across the globe. A combination of RAPD (10), ISSR (9), and SSR (36) markers was used to assess molecular diversity among 42 genotypes of finger millet procured from various geographical regions of southern India (Rajendran et al., 2016). For the first time, an SNP-based diversity study was performed by Kumar et al. (2016), who identified 23,000 SNPs from 113 different finger millet genotypes through the genotyping by sequencing (GBS) method. Gimode et al. (2016) generated 23,285 SNPs through next-generation sequencing (NGS) of two genotypes of cultivated finger millet and also found that 80 out of 92 selected SNP markers for further validation for assessing the diversity of cultivated and wild finger millet genotypes were polymorphic. However, RAPDs have been the only choice by default for molecular diversity studies in finger millet for many years; recent developments, like the identification of SSR markers by the above studies, have accelerated genomic studies in the crop. Although there is a lack of literature on SNP studies in finger millet breeding, the recently released WGS would aid in the development of SNP-based diversity analysis in finger millet accessions. The microsatellite markers and association mapping-based identification of QTLs associated with nutritional traits and stress-tolerant traits are helpful in bio-fortification techniques for ameliorating nutritional deficiencies and stress tolerance towards various adverse conditions. The list of identified QTLs for important agronomic and stress-tolerant traits in finger millet is given in Table 3.2. A total of 545 primer pairs were developed from 1956 ESTs of finger millet, and the microsatellite types, their distribution, and frequency were found through conducting *in silico* mining or comparative genomic assembly analysis of finger millet with rice (Babu et al., 2014a). The nucleotide-binding site-leucine-rich repeat (NBS-LRR) region of rice and finger millet exhibited significant synteny in the expressed sequence tags (EST) sequences. These sequences were also found in nearly identical positions on the rice chromosome map and included the distinctive R-genes with kinase-2 and kinase-3a motifs at the NBS-LRR region (Babu et al., 2014a).

TABLE 3.2 List of Identified QTLs in Finger Millet Related to Agro-Morphological and Nutrient Traits

Sl. No.	QTL Identified	Associated Trait or Character	Germplasm and Markers Used for the Study	Reference
1.	9 QTLs	Calcium content	113 genotypes of finger millet using 33 SSR markers anchored	Kumar et al. (2015)
2.	4 QTLs	Agro-morphological traits such as number of productive tillers, days to 50% anthesis, width of flag leaf blade and plant height	190 finger millet genotypes using 46 genomic SSR markers	Babu et al. (2014a)
3.	UGEP81, UGEP24, FMBLEST32, and RM262	Plant height, Finger blast, Flag leaf blade length, Leaf blast	104 SSR markers	Babu et al. (2014c)
4.	OM5 and FM8	Tryptophan content	190 genotypes of finger millet using 120 SSR markers	Babu et al. (2014b)
5.	FMO2EST1	Protein content		
6.	7 QTLs	Agronomic traits. Including number of productive tillers, seed yield and blast resistance	128 genotypes with 87 genomic SSR markers	Ramakrishnan et al. (2016c)
7.	qLRDW.1, qLRDW.2, qHSDW.1 and qHRL.1	Root and shoot dry weight and length of the root	Association mapping of finger millet under P-deficient and P-sufficient conditions	Ramakrishnan et al. (2017)

The markers such as RFLP, AFLP, EST, and SSR were used to create a genetic map of the cultivated tetraploid finger millet (*Eleusine coracana subsp. coracana*) genome ($2n = 4x = 36$) in an intersubspecific F_2 population through a cross between *E. coracana* subsp. *coracana cv.* Okhale-1 and its progenitor, *E. coracana* subsp. *africana*. It consists of 22 linkage groups in nine homology groups, along with four small linkage groups of unknown origin (Dida et al., 2006). This indicates that the map spans all 18 chromosomes and is approximately 721 cM on the A genome and 787 cM on the B genome. The same team of finger millet workers laid a way to use marker-assisted selection by developing the first set of 82 SSR markers as small-insert genomic libraries, which were identified using methylation-sensitive restriction enzymes. A close genomic relationship among four diploid species, viz., *E. indica, E. floccifolia, E. tristachya*, and *E. intermedia*, and also between these and a tetraploid species *E. coracana* was revealed by genomic *in-situ* hybridization (GISH) (Bisht and Mukai, 2001). Hiremath and Gowda (2018) used 25 primer pairs for varietal classification and genetic assessment in 32 finger millet accessions to examine the applicability of existing SSR markers for discovered polymorphisms. According to their study, polymorphism was demonstrated by all SSRs, of which 17 had a Polymorphic Information Content (PIC) value of more than 0.30. According to a recent study, when the finger millet genotypes PR 202 and IE 2606 were utilized for reciprocal crossing, 12 out of 28 RAPD markers, four out of 20 ISSR markers, and 20 out of 101 SSR markers revealed polymorphism (Krishna et al., 2019). In a diverse collection of finger millet, Sharma et al. (2018) reported for the first time 109 novel SNPs linked to significant agro-morphological traits using genome-wide association mapping utilizing the single locus single trait (SLST), multi-trait mixed model (MTMM), and multi-locus mixed model (MLMM) approaches. The study discovered nine common SNP markers for grain yield and its component traits using these three methods of marker-trait association. Of these, five were found to have orthologous candidate genes of rice and foxtail millet, which are crucial for plant growth, flowering, and grain yield.

3.6.2 GENOMIC RESOURCES AND OMICS APPROACHES

Genetic and genomic resources serve as a primary input for plant breeding practices and molecular breeding, respectively, by utilization in the identification of desirable genes, alleles, and QTLs. Although ample genetic resources are available for the other millet crops like pearl millet and foxtail millet, information on genomic sources, including physical/ chromosomal maps, genetic markers, and QTLs, is somewhat less for finger millet, which obstructs the further improvement of the finger millet crop. Despite its excellent source of nutrients and abiotic stress tolerance, very limited efforts have been reported due to its distant classification compared to other cereals and also its large genome size, which hampers the generation of genomic resources, including genome sequencing. Very few ESTs are available for finger millet when compared to those of rice and maize. With only 1934 ESTs, finger millet has a value nearly 100 times lower than that of rice and maize (Ceasar et al., 2019). Genomic sources would be inevitable for studies on large-scale genotyping studies such as genetic diversity analysis, construction of high-density linkage maps, and QTL mapping related to agronomic, stress-tolerant, and nutritional traits. Kumar et al. identified thousands of SSRs with microsatellite markers by sequencing the finger millet transcriptome of high and low-calcium-containing seeds (2015). Whole information pertaining to the nucleotide arrangements of a crop could be revealed by NGS technologies. It allows breeders to generate a wide range of genetic markers and QTLs required for genomics-assisted breeding of the respective crop. The subsequent development in finger millet research after the identification of SSR markers, identification of R-gene homologs from the NBS-LRR region, and comparative genomics followed by population structure analysis made a roadmap for enhancing omics technology for a clear understanding of nutritional variation, gene mining, drought tolerance, and transgenic development.

The more genetically diverse populations among finger millet core collections, along with the vast information on relevant genetic markers generated through NGS platforms, can directly contribute to enhancing the mapping resolution of traits, such as calcium content, through genome-wide association studies (GWAS) (Puranik et al., 2017). Grain calcium content phenotyping data from multi-location trials, together with a wide range of genetic sources from finger millet core collections, could serve as an efficient model for Ca homeostasis mechanisms as well as the identification of putative novel candidates. A sophisticated approach in the field of genomics-assisted breeding, known as an extension of marker-assisted selection with genomic selection (GS) (Meuwissen et al., 2001), provides genomic estimated breeding values (GEBV) from genome-wide marker data in conjunction with available phenotypic data for a tested (reference/training) population. Generation cycle time can be significantly reduced with this approach, allowing breeders to make appropriate and desired selections even earlier in the program (Heffner et al., 2010). Zhang et al. (2019) used the redundancy-reducing Evidential Gene pipeline in combination with Illumina technology to generate transcriptomes for six *Eleusine* species in order to examine their genetic relationships and evolutionary history. The sequencing-read data for the transcriptomes of all six species studied were deposited and made publicly available in the National Center for Biotechnology Information (NCBI) Sequence Read Archive (SRA) database. The extensive genetic resources of *Eleusine* generated through this research will be useful for continued studies of *Eleusine* evolution.

Recently, genome editing technologies, including protein-guided programmable nucleases such as mega-nucleases, zinc finger nucleases (ZFNs), transcription activator-like effector nucleases (TALENs), and RNA-guided engineered nucleases such as the CRISPR/Cas system, have emerged as effective tools to modify the genome with greater accuracy and predictability. Among all, the CRISPR/Cas9 system is a simple and adaptable mechanism for gene editing or genome editing guided by RNA or DNA (Sood et al., 2019). There are currently no reports on the modification of genes or genomes in finger millet. It is emphasized that there is a possibility to address many major production constraints in finger millet, such as yield and quality, as well as stress tolerance, by regulating gene expressions through gene editing technologies.

The omics approaches are also attempted for the biofortification of finger millet:

3.6.2.1 BIOFORTIFICATION OF FINGER MILLET THROUGH OMICS APPROACHES

Transcriptomics or gene expression profiling techniques such as serial analysis of gene expression studies (SAGE) have been used to reveal the gene expression patterns, especially genes involved in nutrient accumulation and distribution within the seed during the grain-filling stage. Nutrient composition and nutraceutical properties of two finger millet cultivars (Ravana and Osadha) were compared with two rice varieties (Bg300 and Basmati) to improve future uses and advantages of finger millet in the functional food and processing industry by evaluation (Priyanwada et al., 2020). The study also involved the identification of *in vitro* fermentation ability, α-amylase inhibitory activity, and antioxidant activity of the samples tested. Goel et al. (2012) reported a well-accepted hypothetical model for elucidating the transport and distribution of calcium within embryos during grain filling through meta-analysis of a publicly available finger millet database. These transcriptomic studies identified gene expression patterns of *calmodulin* and *Cax1* transporter genes that showed overexpression during the grain-filling stages in finger millet genotypes with high calcium content (Kumar et al., 2014; Mirza et al., 2014; Nath et al., 2010). Large-scale studies that combine the expression, function, and structure of proteins can be considered as proteomics (Kussmann et al., 2006). The transcriptome, along with the complete metabolome, represents a proteome. Although there are widely used techniques of proteomics, such as 2D gel electrophoresis and mass spectrometry tools for the separation and identification of proteins from complex protein mixtures (Kussmann et al., 2006; Mariman, 2006; Wang et al., 2006), very limited work has been done pertaining to the proteomics of finger millet. However, a new calcium-binding protein, calreticulin, has been identified in spikes of finger millet through peptide mass fingerprinting (Singh et al., 2016). Kumar et al. reported an association between elevated immune detection of *calmodulin* protein in the embryo along with the aleurone layer of high grain calcium finger millet genotype and stimulation of higher calcium accumulation at the grain filling stage (2014). A complete profiling of the metabolome of a particular species could make it possible for effective phenotyping to develop genetic markers for nutrient traits. It shows a wide range of variations due to differential imbibition, transport, and sequestration of minerals and ions (Kumar et al., 2016). In recent years, many finger millet researchers have focused on developing biofortified cultivars for human consumption through metabolic profile evaluation, including secondary substitutes such as polyphenols and ions. The metabolomics studies can be carried out by several techniques such as high-performance liquid chromatography (HPLC) and nuclear magnetic resonance (NMR). The combination of bioinformatics tools, functional analysis, and high-throughput approaches such as genomics, transcriptomics, proteomics, and metabolomics can enlighten the way to unravel the complex traits of physio-morphological and nutritionally important finger millet.

3.6.3 GENETIC IMPROVEMENT EFFORTS FOR ABIOTIC AND BIOTIC STRESS

Abiotic stresses are listed as the top major constraints to global food security. Besides affecting crop yield, they also affect the quality of harvested produce (Wang and Frei, 2011). Major abiotic stresses such as salinity, drought, flooding, and deficiencies of nutrients seem to affect the growth and development, which leads to yield reduction in finger millet (Maharajan et al., 2018; Ramakrishnan et al., 2017; Yamunarani et al., 2016). Among the majorly cultivated crops across the globe, millets are relatively more climate-resilient due to their high adaptability to a wide range of ecological and climatic conditions and less vulnerability to environmental stresses. The shorter lifecycle of millet crops allows the crop to escape from stress prominently, whereas other cereal crops require a long duration to mature. In particular, finger millet possesses several morpho-physiological, molecular, and biochemical behaviors that confer more tolerance to abiotic stresses than many other cereals. Several characteristic features such as short stature, less leaf area, thick cell walls, dense root systems, and more seed layers of coat assist in withstanding adverse conditions. The C_4 mechanism of the crop confers high water use efficiency and nutrient use efficiency, which enhances growth and development and ecological enactment under stress environments by flexible allocation patterns of biomass and controlled hydraulic conductivity per unit leaf area (Sage and Zhu, 2011). These attributes of the finger millet crop make it a next-generation crop by holding the potential for research to explore more climate-resilient traits. A wide range of genomic resources in millets, including finger millet, paved the way for the integration of omics technologies in the development of climate-resilient millet cultivars. However, the information regarding the application of genomics in finger millet is very limited (Gupta et al., 2017). There are just a few transcriptome sequences for finger millet under specific stress conditions and for the Ca content of the grain available. Association mapping of phosphorus response traits through mixed linear models was done by Ramakrishnan et al. (2017) using a germplasm assembly constituting 128 finger millet cultivars collected from 18 countries, which was evaluated for phosphorus (P) responses at the seedling stage by growing them in P-sufficient and P-deficient treatments. The study revealed four quantitative trait loci (QTLs) in which two QTLs, viz., *qLRDW.1* and *qLRDW.2*, were expressed for low P. Nevertheless, there hasn't been much work done on sequencing the transcriptome of *Eleusine* cultivars that have been exposed to several stresses, such as blast, salinity, and drought. With the increasing availability of molecular markers and transcriptome sequences, elevated progression in finger millet breeding is moving toward most of the stresses in general. Ramakrishna et al. (2017) identified a novel gene *EcbZIP17*, isolated from an endoplasmic reticulum membrane-tethered *bZIP* transcription factor from finger millet, which showed over-expression in transgenic tobacco plants with better vegetative growth and grain yield compared with wild type under optimal growth conditions, and the gene confirmed upregulation of brassinosteroid signaling genes (2018). These transgenic plants also showed better germination rates, biomass, primary and secondary roots, and recovery rates than wild types under various managed abiotic stresses, such as 250 mM NaCl, 400 mM mannitol, 10% PEG6000, water deficit, and heat stresses.

3.6.3.1 DROUGHT STRESS

Among all the abiotic stresses, drought is one of the main constraints of finger millet production. Drought stress resulted in wilting and leaf rolling, which caused the reduction of leaf-compatible solute potential and chlorophyll content due to enhanced production of many stress-responsive genes when compared to the control environment (Parvathi et al., 2013). To simulate field stress conditions, Parvathi et al. (2013) extracted transcriptome data from finger millet leaf tissue and subjected it to gravimetrically controlled drought stress. The finger millet cultivar GPU-28 was used in the study to assess how drought stress affected the expression of potential genes. By performing functional annotation and pathway analysis, the study was able to determine the activation of various genes involved in the drought-stress signaling cascade, such as PP2A (serine-threonine protein phosphatase 2A), FPS (farnesyl pyrophosphate synthase), CIPK31 (calcineurin B-like interacting protein kinase 31), and SRPR β (signal recognition particle receptor α), and so forth. Additionally, some regulatory genes, including TBP (TATA-binding protein-associated factors), were found to be drought-responsive, which can be evident that genes involved in basal regulatory processes are activated under water deficit conditions in finger millet. The possibility of allele mining of genes that are responsible for developing transgenic varieties by biotechnological tools was revealed by Bartwal et al. (2016) through the evaluation of two drought-tolerant finger millet genotypes, PRM6107 and PR202. The overexpression of finger millet TF genes *EcNAC1* (Ramegowda et al., 2012) and *EcNAC67* (Rehman et al., 2016) conferred abiotic stress tolerance, especially when imparted from tobacco and rice. An effort was made by Singh et al. to prospect the key role of the *EcDehydrin7* protein from the cDNA library of drought-tolerant finger millet in transgenic tobacco (2015). The relatively high expression of the *EcDehydrin7* protein, when compared to wild types under drought conditions, was identified through Northern and qRT-PCR analyses (Singh et al., 2015). According to a recently reported whole-genome sequencing (WGS) study (Hittalmani et al., 2017), 2,866 drought-responsive genes were associated with major transcription factor (TF) families across 19 Pfam domains (Ceasar et al., 2019). The transgenic finger millet plant carrying the bacterial mannitol-1-phosphate dehydrogenase (*mtlD*) gene showed greater osmotic-stress tolerance through chlorophyll retention (stay-green trait) under drought stress than wild-type plants (Hema et al., 2014). After evaluating a set of thirty-eight finger millet genotypes in field and mini-lysimeter settings under water stress and stress-free conditions, a recent physiological study found that variation in shoot biomass and grain yield under water stress varied by two and nine times, respectively (Talwar et al., 2020). The study also confirmed that reproductive growth is more sensitive to water stress than vegetative growth.

3.6.3.2 NUTRIENT DEFICIENCY

A few reports were available on the effects of nutrient deficiency, such as decreased tiller number due to nitrogen deficiency (Goron et al., 2015), reduced growth and biomass due to phosphorus deficiency (Ramakrishna et al., 2017), and stunted growth and delayed seed maturity due to zinc deficiency (Yamunarani et al., 2016). Genetic variation in grain Zn concentration among 319 finger millet germplasm, the effect of external application of Zn on grain Zn concentration, and Zn uptake and translocation patterns from plant parts were assessed by Yamunarani et al. (2016). The study reported that the genotypes, viz., GEC331 and GEC164, showed better responses to uptake and translocation under external Zn application, as indicated by molecular and biochemical studies.

3.6.3.3 SALINITY STRESS

Parvati and Natarajan reported a significant reduction in plant growth and biomass by evaluating finger millet seedlings of genotype GPU-28 under PEG and oxidative stress (2017). Water content, plant height, leaf expansion, length and width of a finger, grain weight, and flowering were affected by salinity stress (Anjaneyulu et al., 2014). In comparison to wild-type plants, the transgenic finger millet exhibited a greater degree of saline resistance. Mahalakshmi et al. (2006) reported that overexpression of the *Porteresia coarctata* serine-rich protein (*PcSrp*) gene in finger millet was induced under salinity conditions, and the plants showed normal growth and flowering behavior under 250 mM NaCl stress conditions. Agrobacterium-mediated gene transformation was used to create transgenic finger millet containing the mannitol-1-phosphate dehydrogenase (*mtlD*) gene, which grew more effectively than the wild type under salt- and drought-stress conditions (Hema et al., 2014). Rehman et al. (2014) reported two contrasting genes, viz., Co-12, which is susceptible, and *Trichy1*, which is tolerant under salinity conditions, through the RNAseq method. Transcriptome analysis of leaves from the salt-tolerant genotype Trichy1 showed the upregulation of various functional groups of genes belonging to transporter, transcription factor, osmotic-homeostasis, cell-signaling, and compatible-solute-biosynthesis families, whereas the downregulation of flavonoid biosynthesis was also observed in the salt-tolerant genotype (Rahman et al., 2014). A few efforts related to transgenic development for stress tolerance in finger millet are summarized in Table 3.3, along with the genes or proteins transferred and the methods of gene transformation (see Table 3.3 for details).

TABLE 3.3 Details of Transgenics Developed for Stress Tolerance in Finger Millet

Sl. No.	Gene/Protein Transferred or Identified	Trait	Gene Transformation Method	Reference
1.	Bar gene (pBar 35S) Pin gene (pPin 35S)	Leaf blast	Particle gun	Latha et al. (2005)
2.	*Porteresia coarctata* serine-rich-protein (PcSrp) encoding gene	Salt tolerance	Particle gun	Mahalaksmi et al. (2006)
3.	Rice chitinase gene	Leaf blast disease	Agrobacterium-mediated	Ceasar and Ignacimuthu (2011); Ignacimuthu and Ceasar (2012)
4.	*EcNAC1* expressed in Tobacco, a stress-responsive NAC gene derived from finger millet.	A combined stress, including drought and salt	Agrobacterium-mediated	Ramegowda et al. (2012)
5.	*Ex-apx1*	Drought stress		Bhatt et al. (2013)
6.	*Metallothionein*	Drought stress		Parvathi et al. (2013)
7.	*Protein phosphatase 2A*	Drought stress		Parvathi et al. (2013)
8.	*RISBZ4*	Drought stress		Parvathi et al. (2013)
9.	Vacuolar proton pyrophosphatase (*SbVPPase*) gene from *Sorghum bicolor*	Salt tolerance	Agrobacterium mediated	Anjaneyulu et al. (2014)
10.	C_2H_2 type of zinc finger transcription factors (TFs)	Salinity, dehydration and cold tolerance		Muthamilarasan et al. (2014)
11.	*mtlD*	Over expression stimulated multi-stress tolerance		Hema et al. (2014)
12.	*bHLH* transcription factor	Salinity, oxidative and drought stress	Agrobacterium mediated	Babitha et al. (2015)
13.	*Monodehydroascorbatereductase*	Drought, salt and UV radiation		Sudan et al. (2015)
14.	The protein *EcDehydrin7* from finger millet is expressed in tobacco	Drought stress	Agrobacterium mediated	Singh R. K. et al. (2015)
15.	NAC transcription factor	Salinity and drought stress		Rehman et al. (2016)
16.	*EcJAZ*	Biotic and biotic stress tolerance		Sen and Dutta (2016)
17.	Phosphate transporters (*Pt*)	Phosphate ion (Pi) stress		Pudake et al. (2017)

3.6.3.4 BIOTIC STRESSES

Ignacimuthu and Ceasar (2012) transferred the *Chi11* (*chitinase11*) gene from rice to finger millet using Agrobacterium-mediated transformation to provide leaf blast resistance. A transgenic finger millet that shows resistance to the leaf blast variety was developed by the introduction of an anti-fungal protein (PIN) gene from a prawn, which was chemically synthesized and cloned into plasmid *pPin35S* along with the *CaMV35S* promoter by the biolistic method of gene transformation (Latha et al., 2005). Apart from these, very limited work has been done in finger millet to develop transgenic varieties for blast resistance. Screening a large number of potential antifungal genes and gene pyramiding would be helpful in developing transgenic cultivars that are resistant to a broad spectrum of fungal diseases (Ceasar et al., 2019). Babu et al. (2014) carried out a comparative genomic study using SSR markers to identify genes/QTLs associated with the blast resistance genes. Fifty-eight genic SSR markers were developed and utilized to perform genetic analysis on a set of 190 finger millet accessions collected globally. Four QTLs linked to finger blast and one QTL for neck blast resistance were identified through association mapping by using 104 SSR markers. Blast-resistant genes were identified, mapped, and improved for future introgressed approaches to enhance blast resistance in finger millet by comparative genome analysis and *in silico* studies by Babu et al. (2014a). Three SSR markers, viz., RM23842, RM5963, and RM262, with tetra-, tri-, and di-repeat motifs respectively, were identified to be the most highly polymorphic loci for differentiating the 190 global collections of finger millet genotypes, which were classified into four major groups based on their response to blast resistance by using 58 genic SSR loci (Babu et al., 2014b). A high genetic diversity within a finger millet germplasm containing 67 cultivars, including resistant and susceptible to blast disease, was observed through Nei's gene diversity (h) index and analysis of molecular variance by using sequence-related amplified polymorphism (SRAP) and simple sequence repeat (SSR) markers (Saha et al., 2016).

3.7 GENOME SEQUENCING (WGS) STATUS

The rapid advancement of high-throughput techniques like Illumina sequencing has resulted in a notable acceleration in whole genome and transcriptome sequencing for various plant species in recent years (Bolger et al., 2014). Even with more complex genomes, WGS has already been accomplished for numerous cereals and a number of model plants. Undoubtedly, the WGS of finger millet has been delayed with respect to other major cereal crops (Figure 3.3). The polyploidy condition that contains homeologous genes is one of the major limitations for the genome assembly of finger millet, which delayed the finger millet genome sequencing. A purely African-driven, joint initiative between the African Orphan Crops Consortium (AOCC) and Bio-innovate, Africa, launched a WGS project for finger millet in 2014 with major objectives such as the creation of a reference genome, development of a robust set of molecular markers (microsatellite and SNPs), and capacity building in the application of genomics in finger millet breeding. The project submitted the first draft of the finger millet genome and made it available in public domains like NCBI and PubMed. Whole genome sequence information of finger millet has been made available recently, along with other millet crops like sorghum, pearl millet, and foxtail millets, which has been utilized efficiently to identify responsive genes for abiotic stress tolerance and for advancing breeding strategies such as GS in finger millet (Ceasar et al., 2018). With a genomic size of 1.5 Gb, finger millet has the largest genome of all the millets (Vetriventhan et al., 2020).

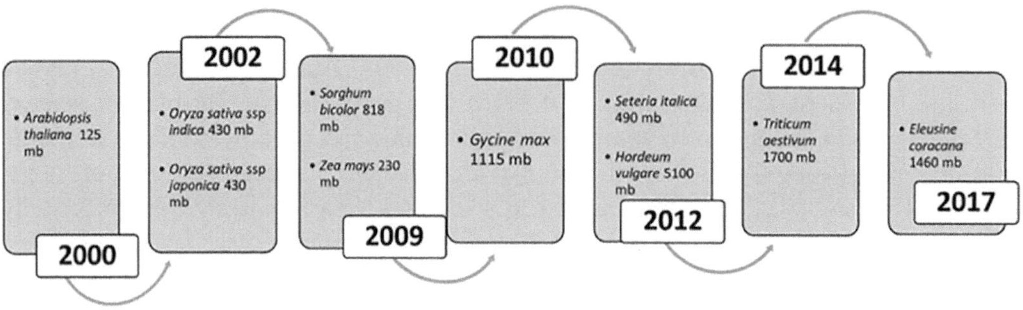

FIGURE 3.3 Timeline of genome sequence project.

3.7.1 WGS OF FINGER MILLET CULTIVAR ML-365

By Hittalmani et al., a whole genome sequence of genomic DNA extracted from a leaf sample of ML-365 (*Eleusine coracana* subsp. *coracana*) through a shotgun sequencing project was submitted to NCBI (2017). To overcome the hurdle of polyploidy, they sequenced the genome via a novel multiple hybrid assembly using diverse technologies with maximum accuracy, combining next-generation single-molecule sequencing followed by whole-genome sequencing through optical mapping. Very interestingly, it has been found to have maximum co-linearity with rice and foxtail millet, as associated with other species belonging to the Poaceae family. According to Hittalmani et al. (2017), this investigation also demonstrated that the genome sizes of the two species of finger millet, the African (*E. coracana* subspecies *africana*) and Asiatic (*E. coracana* subspecies *coracana*), were comparatively similar. This finding may be proof of the domestication described by Dida et al. (2008). This first version of the project has the accession number LXGH01000000 and consists of sequences LXGH01000001 to LXGH01525627. The project reported that the genome is tetraploid, with some duplicate scaffolds having a size of 1195.99 Mb and 44.80% GC content. Genome sequence technologies utilized in the project were Illumina NextSeq, Illumina HiSeq, and SOLiD. The gene ontology and pathway predictions, distribution of orthologous gene families among major cereal species, and the phylogenetic relationship of finger millet with six other Poaceae species, including rice, maize, sorghum, foxtail millet, and *Brachypodium* (revealed based on single-copy ortholog genes), syntenic genomic blocks of finger millet, gene ontology annotation of putative transcripts, differential gene expression in low moisture stress and control samples, and distribution of transcription factors within the finger millet genome were reported by Hittalmani et al. (2017). Twenty-eight genomic data projects pertaining to finger millet have been submitted to the NCBI platform, which includes different data types such as raw sequence reads, exome sequences, transcriptomes, and maps. The details are presented in Table 3.4.

TABLE 3.4 Genomic Resources Along with Project Details Available in NCBI

Sl. No.	Project Title	Goal/Object of the Project	Accession Number and Project ID	Data Type	Institute / University	Registration Date
1.	Integrated genomics-assisted breeding for efficient development of superior finger millet varieties for Karnataka	To identify superior breeding lines and develop new breeding lines by integrated molecular breeding through sequencing of multiple finger millet germplasm and subsequent identification of molecular markers for key traits by generating genomic resources.	Accession: PRJNA610152 ID: 610152	Raw sequence reads	Centre for Cellular and Molecular Platforms (C-CAMP) and The University of Trans-Disciplinary Health Sciences and Technology	4-Mar-2020
2.	*Eleusine coracana* subsp. *africana* synthetic B transcriptome	Chloroplast and mitochondrial genes were removed from E. indica from the *E. africana* reads. Then, the filtered *E. africana* reads they were mapped to *E. indica* transcriptome using bowtie2 and then the mapped and unmapped reads were extracted using same tools separately. Reads lengths below 90 were discarded after mapped and unmapped reads were filtered.	Accession: PRJNA473252 ID: 473252	Transcriptome or Gene expression	Auburn University	27-May-2018
3.	*Eleusine coracana* subsp. *africana* synthetic A transcriptome		Accession: PRJNA473251 ID: 473251			
4.	*Eleusine coracana* Transcriptome or Gene expression		Accession: PRJNA449118 ID: 449118			6-Apr-2018
5.	*E. coracana* Synthetic A genome sequence reads		Accession: PRJNA449117 ID: 449117			
6.	Finger millet F2 mapping population (MD20 X Okhale1) GBS sequencing data	Finger millet F2 mapping population (MD20 X Okhale1) GBS sequencing data, Genotyping by sequencing for library preparation, and PE150 for sequencing.	Accession: PRJNA445435 ID: 445435	Raw sequence reads	University of Georgia	23-Mar-2018
7.	*Eleusine coracana* subsp. *coracana* strain:PR202	de novo genome assembly and annotation for *Eleusine coracana* subsp. coracana, variety name:PR202 (Hatakeyama et al., 2017)	Accession: PRJDB5606 ID: 402014	Genome sequencing	Department of Evolutionary Biology and Environmental Studies, University of Zurich	6-Sep-2017
8.	Metagenome sequencing of blast disease susceptible and resistant finger millet varieties-Raw reads	Whole genome metagenome sequencing for blast disease susceptible and resistant finger millet varieties. Genomic DNA was pooled and constructed paired end Illumina library	Accession: PRJNA383952 ID: 383952	Raw sequence reads	University of agricultural sciences, Karnataka	24-Apr-2017
9.	*Eleusine coracana* Raw sequence reads	To enhance crop improvement efforts by generating whole genome and transcriptome sequences of finger millet genotype KNE796.	Accession: PRJNA377606 ID: 377606	Raw sequence reads, Genome sequencing	BGI-Shenzhen	2-Mar-2017

TABLE 3.4 *(Continued)*

Sl. No.	Project Title	Goal/Object of the Project	Accession Number and Project ID	Data Type	Institute / University	Registration Date
10.	*Eleusine coracana* Raw sequence reads	Application of crop genetics to improve Calcium content in millets and other crops for promoting health benefits in the prevention and treatment of osteoporosis	Accession: PRJNA374981 ID: 374981	Raw sequence reads	Aberystwyth University	16-Feb-2017
11.	*Eleusine coracana* subsp. *coracana* cultivar:ML-365 Transcriptome or Gene expression	RNA sequencing of moisture stressed and control plants of ML-365.	Accession: PRJNA339512 ID: 339512	Transcriptome or Gene expression	University of Agricultural Sciences	19-Aug-2016
12.	*Eleusine coracana* subsp. *coracana* cultivar: ML-365 Raw sequence reads	The reads of whole genome sequencing of ML-365, drought tolerant and high yielding finger millet variety developed at UAS, Bangalore (Hittalmani et al., 2017).	Accession: PRJNA338521 ID: 338521	Raw sequence reads	University of Agricultural Sciences	1-Aug-2016
13.	*Eleusine coracana* subsp. coracana cultivar:ML-365 Genome sequencing and assembly	Finger millet whole genome sequencing, assembly and annotation (Hittalmani et al., 2017)	Accession: PRJNA318349 ID: 318349	Genome sequencing and assembly	University of Agricultural Sciences	8-Jun-2017
14.	*Eleusine coracana* subsp. *coracana* cultivar: KNE796 Raw sequence reads	Finger millet sequences generated for high throughput marker development	Accession: PRJNA317618 ID: 317618	Raw sequence reads	Bio-Innovate (Damaris Odeny, ICRISAT, Nairobi, Kenya)	7-Apr-2016
15.	Comparative miRNA profiling of diverse finger millet (*E. coracana*) genotypes under heat stress	Molecular basis of heat tolerance and therefore enable the effective use of genetic and genomic approaches to improve heat tolerance in major cultivated crops	Accession: PRJNA312052 ID: 312052	miRNA-Seq	National Research Centre on Plant Biotechnology	15-Feb-2016
16.	Comparative miRNA profiling of contrasting genotypes of finger millet (*Eleusine coracana*) under heat stress	Molecular basis of heat tolerance and therefore enable the effective use of genetic and genomic approaches to improve heat tolerance in major cultivated crops.	Accession: PRJNA310802 ID: 310802	Exome	National Research Centre on Plant Biotechnology	4-Feb-2016
17.	*Eleusine coracana* Raw sequence reads.	Genotyping-by-Sequencing Analysis for determination of Population Structure amongst Finger Millet (*Eleusine coracana*) germplasm of diverse origin for its use in crop improvement	Accession: PRJNA306606 ID: 306606	Raw sequence reads	G. B. Pant University of Agriculture and Technology, Pantnagar.	21-Dec-2015
18.	*Eleusine coracana* cultivar: GPU 28 Transcriptome or Gene expression	To obtain a global view of the transcriptome of Finger millet (*Eleusine coracana* L.) under drought condition	Accession: PRJNA282860 ID: 282860	Transcriptome or Gene expression	University of Agricultural Sciences, Bangalore	2-May-2015

TABLE 3.4 *(Continued)*

Sl. No.	Project Title	Goal/Object of the Project	Accession Number and Project ID	Data Type	Institute / University	Registration Date
19.	*Eleusine coracana* Transcriptome or Gene expression	To obtain a global view of the transcriptome of Finger millet (*Eleusine coracana* L.) under drought and control conditions this study has performed.	Accession: PRJNA282578 ID: 282578	Transcriptome or Gene expression	University of Agricultural Sciences, Bangalore	29-Apr-2015
20.	*Eleusine coracana* subsp. coracana Transcriptome or Gene expression	Analysis of effect of water deficit on gene expression in two accessions of finger millet	Accession: PRJNA229808 ID: 229808	Transcriptome or Gene expression	University of Alberta	24-Nov-2013
21.	*Eleusine coracana* cultivar:MR1 Raw sequence reads.	To analyze the effect of drought stress on miRNA expression	Accession: PRJNA277250 ID: 277250	Raw sequence reads	Maharani Science College for Women Bangalore	5-Mar-2015
22.	*Eleusine coracana* Transcriptome or Gene expression	De novo transcriptome sequencing of finger millet cultivar for generating genomic resources for elucidating differential genes and markers involved in blast resistance in finger millet	Accession: PRJNA268401 ID: 268401	Transcriptome or Gene expression	National Bureau of Plant Genetic Resources	25-Nov-2014
23.	*Eleusine coracana* subsp. coracana Transcriptome or Gene expression	Transcriptome analysis of 2 contrasting genotypes of Finger Millet	Accession: PRJNA236796 ID: 236796	Transcriptome or Gene expression	G B Pant University and NxGenBio Life Sciences	30-Jan-2014
24.	*Eleusine coracana* strain:CO12, *Trichy 1* Transcriptome or Gene expression	To compare transcriptional response to salinity stress between susceptible CO 12 and tolerant genotype *trichy 1* of finger millet	Accession: PRJNA236733 ID: 236733	Transcriptome or Gene expression	Tamil Nadu Agricultural University	29-Jan-2014
25.	*Eleusine coracana* subsp. coracana Transcriptome or Gene expression	Analysis of effect of water deficit on gene expression in two accessions of finger millet	Accession: PRJNA229808 ID: 229808	Transcriptome or Gene expression	University of Alberta	24-Nov-2013
26.	*Eleusine coracana* Map	A mapping population of 15 F_2 progeny was developed from the cross *'MD-20'* × *'Okahle-1.'* The plants were genotyped with RFLP markers, SSR markers, SSCP markers, and AFLP markers.	Accession: PRJNA72291 ID: 72291	Map	Bisht and Mukai (2000); Dida et al. (2007);	-

3.7.2 *WGS OF FINGER MILLET CULTIVAR PR-202*

Whole genome sequence and assembly of finger millet cultivar PR-202 (IC: 479099) submitted by Hatakeyama et al. (2018) using a unique polyploidy genome assembly workflow has a genome size of 1.5 Gb, with an assembled genome size of 1189 Mb, which covered almost 80% of the genome. Based on these findings, finger millet included 2387 scaffolds with a maximum sequence length of 5 Mb and an N50 value of 905.318 Kb. More than 60 thousand genes were made available publicly on the NCBI platform, of which 96.5% are single-copy genes, whereas 91% are functionally annotated (Hatakeyama et al., 2018). The raw sequence data from each library and coverage depth of the genome were included in the report. The number of sequences of raw data from each library, including paired-end, mate-pair, and PacBio CLR, were obtained through Illumina NextSeq, Illumina MiSeq, and PacBio RS II sequencers, respectively. Since most of the homeologs were constructed separately, the suggested method can also be used for the assembly of additional allotetraploid genomes (Hatakeyama et al., 2018). This extensive work disclosed many interesting findings pertaining to the finger millet genome through gene annotation and assembly validation, polyploid assembly and phylogenetic analyses of donor genomes, and sequence analysis of drought-tolerance genes such as *EcNAC1* and Trichy1. The whole *de novo* assembly results generated using Bionano IrysSolve™ (CMAP format), along with all sequenced raw data (FASTQ format), scanned molecular data (BNX format), annotations (FASTA and GFF formats), and the final scaffolds, are available in the Bioinformation and DDBJ Sequence Read Archive (BI-DRA) database under genome assembly project accession number PRJDB5606 and BioSample number SAMD00076255. The availability of WGSs of these two finger millet cultivars could assist breeders effectively in further studies, such as linkage maps, association mapping, SNP identification, GWAS, and allele discovery by NGS techniques.

3.8 CONCLUSION

Millions of people who live in comparatively arid regions of Asia and Africa rely on finger millet, one of the world's oldest cereal staple crops. Even though finger millet has significantly improved genetically through conventional phenotype-based breeding, more work needs to be done to increase its economic value. This can be achieved by creating stable, widely adapted high-yielding varieties, which will give finger millet a competitive advantage over other similar cereals and encourage widespread cultivation. To accelerate and improve the effectiveness of finger millet breeding, various forms of marker-assisted selection such as F_2 enrichment, marker-assisted recurrent and GSs, and genome editing must be implemented. These methods require an understanding of the available genomic resources of the crop, including DNA markers, candidate genes, and transcription factors. The complex floral biology of finger millets impedes the artificial hybridization process, which leads to the discovery of a way to generate 100% crossed seeds in order to fully utilize the genetic potential. This chapter's main objective is to cover the cytogenetics, the genetics of qualitative and quantitative traits, breeding for resistance to biotic production constraints along with the origin, economic significance, and status of available finger millet germplasm, as well as its characterization and evaluation. Additionally, the chapter covers the utilization of genomic resources to improve the speed and accuracy of finger millet breeding for increased productivity. Recently revealed WGS may facilitate finger millet researchers to develop new cultivars that are more resilient to multiple nutrient stressors and also towards ever-changing climatic conditions through genetic and genomic investigations.

KEYWORDS

- **artificial hybridization process**
- **crop improvement**
- **DNA markers**
- **drought stress**
- **finger millet**
- **genome sequencing**
- **hybridization**
- **molecular breeding**
- **whole genome sequencing**

REFERENCES

Anjaneyulu, E., Reddy, P. S., Sunita, M. S., Kishor, P. B. K., & Meriga, B. (2014). Salt tolerance and activity of antioxidative enzymes of transgenic finger millet overexpressing a vacuolar H^+-pyrophosphatase gene (SbVPPase) from *Sorghum bicolor*. *Journal of Plant Physiology, 171*(10), 789–798. https://doi.org/10.1016/j.jplph.2014.02.001 (accessed on 7th November 2025).

Antony Ceasar, S., & Ignacimuthu, S. (2011). *Agrobacterium*-mediated transformation of finger millet (*Eleusine coracana* (L.) Gaertn.) using shoot apex explants. *Plant Cell Reports, 30*(9), 1759–1770. https://doi.org/10.1007/s00299-011-1084-0 (accessed on 7th November 2025).

Antony Ceasar, S., Maharajan, T., Ajeesh Krishna, T. P., Ramakrishnan, M., Victor Roch, G., Satish, L., & Ignacimuthu, S. (2018). Finger millet (*Eleusine coracana* (L.) Gaertn.) improvement: Current status and future interventions of whole genome sequence. *Frontiers in Plant Science, 9*, 1054. https://doi.org/10.3389/fpls.2018.01054 (accessed on 7th November 2025).

Babitha, K. C., Ramu, S. V., Nataraja, K. N., & Udayakumar, M. (2015). EcbZIP60, a basic leucine zipper transcription factor from *Eleusine coracana* L. improves abiotic stress tolerance in tobacco by activating unfolded protein response pathway. *Molecular Breeding, 35*(9). https://doi.org/10.1007/s11032-015-0374-6 (accessed on 7th November 2025).

Babitha, K. C., Vemanna, R. S., Nataraja, K. N., & Udayakumar, M. (2015). Overexpression of EcbHLH57 transcription factor from *Eleusine coracana* L. in tobacco confers tolerance to salt, oxidative and drought stress. *PLoS One, 10*(9), e0137098. https://doi.org/10.1371/journal.pone.0137098 (accessed on 7th November 2025).

Babu, A., Geetha, K., Manjunatha, V., & Shankar, A. (2012). An efficient high-throughput plant regeneration and transformation protocol for production of transgenics tolerant to salt in finger millet. *International Journal of Forest Crop Improvement, 3*, 16–20.

Babu, B. K., Agrawal, P. K., Pandey, D., & Kumar, A. (2014). Comparative genomics and association mapping approaches for opaque2 modifier genes in finger millet accessions using genic, genomic and candidate gene-based simple sequence repeat markers. *Molecular Breeding, 34*(3), 1261–1279. https://doi.org/10.1007/s11032-014-0115-2 (accessed on 7th November 2025).

Babu, B. K., Agrawal, P. K., Pandey, D., Sood, S., & Kumar, A. (2014). Molecular analysis of world collection of finger millet accessions for blast disease resistance using functional SSR markers. *SABRAO Journal of Breeding and Genetics, 46*(2), 202–216.

Babu, B. K., Dinesh, P., Agrawal, P. K., Sood, S., Chandrashekara, C., Bhatt, J. C., & Kumar, A. (2014). Comparative genomics and association mapping approaches for blast resistant genes in finger millet using SSRs. *PLoS One, 9*(6), e99182. https://doi.org/10.1371/journal.pone.0099182 (accessed on 7th November 2025).

Babu, B. K., Senthil, N., Gomez, S. M., Biji, K. R., Rajendraprasad, N. S., Kumar, S. S., & Babu, R. C. (2007). Assessment of genetic diversity among finger millet (*Eleusine coracana* (L.) Gaertn.) accessions using molecular markers. *Genetic Resources and Crop Evolution, 54*(2), 399–404. https://doi.org/10.1007/s10722-006-0002-8 (accessed on 7th November 2025).

Bhatt, D., Saxena, S. C., Jain, S., Dobriyal, A. K., Majee, M., & Arora, S. (2013). Cloning, expression and functional validation of drought inducible ascorbate peroxidase (Ec-Apx1) from *Eleusine coracana. Molecular Biology Reports, 40*(2), 1155–1165. https://doi.org/10.1007/s11033-012-2157-z (accessed on 7th November 2025).

Bisht, M. S., & Mukai, Y. (2001). Genomic *in situ* hybridization identifies genome donor of finger millet (*Eleusine coracana*). *Theoretical and Applied Genetics, 102*(6–7), 825–832. https://doi.org/10.1007/s001220000497 (accessed on 7th November 2025).

Bisht, M., & Mukai, Y. (2002). Genome organization and polyploid evolution in the genus *Eleusine* (Poaceae). *Plant Systematics and Evolution, 233*(3), 243–258.

Bolger, M. E., Weisshaar, B., Scholz, U., Stein, N., Usadel, B., & Mayer, K. F. X. (2014). Plant genome sequencing—applications for crop improvement. *Current Opinion in Biotechnology, 26*, 31–37. https://doi.org/10.1016/j.copbio.2013.08.019 (accessed on 7th November 2025).

Ceasar, S. A. (2019). Genome-wide identification and *in silico* analysis of PHT1 family genes and proteins in *Setaria viridis*: the best model to study nutrient transport in millets. *The Plant Genome, 12*(1), 180019. https://doi.org/10.3835/plantgenome2018.04.0019 (accessed on 7th November 2025).

Chandra, D., Chandra, S., Pallavi, & Sharma, A. K. (2016). Review of finger millet (*Eleusine coracana* (L.) Gaertn): A powerhouse of health benefiting nutrients. *Food Science and Human Wellness, 5*(3), 149–155. https://doi.org/10.1016/j.fshw.2016.05.004 (accessed on 7th November 2025).

Dewet, J. M. J. (1995). Finger millet, *Eleusine coracana*. In Smartt, J., & Simmonds, N. W. (Eds.), *Evolution of crop plants* (pp. 137–140). Longman, Singapore.

Dida, M. M. (1998). *Genome Studies in Finger Millet (Eleusine coracana L.)* (PhD dissertation). John Innes Centre.

Dida, M. M., Ramakrishnan, S., Bennetzen, J. L., Gale, M., & Devos, K. M. (2006). The genetic map of finger millet, *Eleusine coracana, Theoretical and Applied Genetics, 114*, 321–332

Dida, M. M., Wanyera, N., Harrison Dunn, M. L., Bennetzen, J. L., & Devos, K. M. (2008). Population structure and diversity in finger millet (*Eleusine coracana*) germplasm. *Tropical Plant Biology, 1*(2), 131–141. https://doi.org/10.1007/s12042-008-9012-3 (accessed on 7th November 2025).

Dodake, S., & Dhonukshe, B. L. (1998). Variability in floral structure and floral biology of finger millet (*Eleusine coracana* (L.) Gaertn.). *Indian Journal of Genetics, 58*, 107–112.

Dutta, S. K. (2016). Cloning, characterization, and subcellular localization of a novel JAZ repressor from *Eleusine coracana. Biologic Plantarum, 60*, 715–723.

Dwivedi, S. L., Upadhyaya, H. D., Senthilvel, S., Hash, C. T., Fukunaga, K., & Diao, X. (2012). Millets: Genetic and genomic resources. In Janick, J. (Ed.), *Plant Breeding Reviews* (pp. [not provided]). Wiley, Hoboken, NJ.

Fakrudin, B., Shashidhar, H., Kulkarni, R., & Hittalmani, S. (2004). Genetic diversity assessment of finger millet, *Eleusine coracana* (Gaertn.), germplasm through RAPD analysis. *PGR Newsletter, 138*, 50–54.

Frei, W. (2011). Stressed food: The impact of abiotic environmental stresses on crop quality. *Agriculture, Ecosystems & Environment, 141*(3, 4), 271–286.

Ganapathy, K. N. (2017). Improvement in finger millet: Status and future prospects. In *Millets and Sorghum* (pp. 87–111). John Wiley & Sons Ltd, Chichester, UK.

Gimode, D., Odeny, D. A., de Villiers, E. P., Wanyonyi, S., Dida, M. M., Mneney, E. E., Muchugi, A., Machuka, J., & de Villiers, S. M. (2016). Identification of SNP and SSR markers in finger millet using next generation sequencing technologies. *PLoS One, 11*(7), e0159437. https://doi.org/10.1371/journal.pone.0159437 (accessed on 7th November 2025).

Goel, A., Gaur, V. S., Arora, S., Gupta, S., & Kumar, A. (2012). *In silico* analysis of expression data for identification of genes involved in spatial accumulation of calcium in developing seeds of rice. *OMICS: A Journal of Integrative Biology, 16*(7–8), 402–413. https://doi.org/10.1089/omi.2012.0004 (accessed on 7th November 2025).

Goron, T. L., & Raizada, M. N. (2015). Genetic diversity and genomic resources available for the small millet crops to accelerate a new green revolution. *Frontiers in Plant Science, 6*, 157. https://doi.org/10.3389/fpls.2015.00157 (accessed on 7th November 2025).

Goron, T. L., Bhosekar, V. K., Shearer, C. R., Watts, S., & Raizada, M. N. (2015). Whole plant acclimation responses by finger millet to low nitrogen stress. *Frontiers in Plant Science, 6*, 652. https://doi.org/10.3389/fpls.2015.00652 (accessed on 7th November 2025).

Gowda, B. T. S. (1997). Genetic enhancement and breeding strategies in finger millet (*Eleusine coracana* Gaertn). *National Seminar on Small Millets*, 16–18.

Gowda, M., Pushpalatha, N., Jhadav, S. S., Satish, R. G., Boronayaka, M. B., Sujay, V., Pramila, C. K., Manasa, K. G., Gowda, J., Ravishankar, P., Krishnappa, M., & Narasimhamurthy, D. N. (2014). PS-1 (IC0598201; INGR14015), a finger millet (*Eleusine coracana*) germplasm with partial sterility, useful in hybridization and easy maintenance. *Indian Journal of Plant Genetic Resources, 27*.

Gupta, A., Sood, S., Agrawal, P. K., & Bhatt, J. C. (2012). Floral biology and pollination system in small millets. *European Journal of Plant Science and Biotechnology, 6*, 80–86.

Gupta, R., Verma, K., Joshi, D. C., Yadav, D., & Singh, M. (2010). Assessment of genetic relatedness among three varieties of finger millet with variable seed coat color using RAPD and ISSR markers. *Genetic Engineering and Biotechnology Journal, 2010*, 1–9.

Gupta, S. C., Muza, F. R., & Andrews, D. J. (1997). Registration of INFM 95001 finger millet genetic male-sterile line. *Crop Science, 37*(4), 1409. https://doi.org/10.2135/cropsci1997.0011183x003700040097x (accessed on 7th November 2025).

Gupta, S. M., Arora, S., Mirza, N., Pande, A., Lata, C., Puranik, S., Kumar, J., & Kumar, A. (2017). Finger millet: A "certain" crop for an "uncertain" future and a solution to food insecurity and hidden hunger under stressful environments. *Frontiers in Plant Science, 8*, 643. https://doi.org/10.3389/fpls.2017.00643 (accessed on 7th November 2025).

Harinarayana, G. (1986). Breeding and varietal improvement of small millets in India. In Seetharam, A., Riley, K. W., & Harinarayana, G. (Eds.), *Proceedings of the First International Small Millets Workshop*. Oxford & IBH Publishing Co. Pvt. Ltd, Bangalore, India.

Hatakeyama, M., Sirisha, A., Balachadran, M. T., Sivarajan, S. R., Patrignani, A., Gruter, S., Poveda, L., Shimizu-Inatsugi, R., Baeten, J., Francois, K., Nataraja, K. N., Reddy, Y. A. N., Phadnis, S., Ravikumar, R. L., Schlapbach, R., Sreeman, S. M., & Shimizu, K. K. (2017). Multiple hybrid de novo genome assembly of finger millet, an orphan allotetraploid crop. *DNA Research, 25*(1), 39–47.

Heffner, E. L., Lorenz, A. J., Jannink, J. L., & Sorrells, M. E. (2010). Plant breeding with genomic selection: Gain per unit time and cost. *Crop Science, 50*(5), 1681–1690. https://doi.org/10.2135/cropsci2009.11.0662 (accessed on 7th November 2025).

Hema, R., Vemanna, R. S., Sreeramulu, S., Reddy, C. P., Senthil-Kumar, M., & Udayakumar, M. (2014). Stable expression of mtlD gene imparts multiple stress tolerance in finger millet. *PLoS One, 9*(6), e99110. https://doi.org/10.1371/journal.pone.0099110 (accessed on 7th November 2025).

Hilu, K. W. (1988). Identification of the 'A' genome of finger millet using chloroplast DNA. *Genetics, 118*, 163–167.

Hilu, K. W., & Dewet, J. M. J. (1976). Domestication of *Eleusine coracana. Economic Botany, 30*, 199–208.

Hiremath, S. C., & Salimath, S. S. (1992). The 'A' genome donor of *Eleusine coracana* (L.) Gaertn. (Gramineae). *Theoretical and Applied Genetics, 84*, 747–754.

Hittalmani, S., Mahesh, H. B., Shirke, M. D., Biradar, H., Uday, G., Aruna, Y. R., Lohithaswa, H. C., & Mohanrao, A. (2017). Genome and transcriptome sequence of finger millet (*Eleusine coracana* (L.) Gaertn.) provides insights into drought tolerance and nutraceutical properties. *BMC Genomics, 18*(1), 465. https://doi.org/10.1186/s12864-017-3850-z (accessed on 7th November 2025).

Ignacimuthu, S., & Ceasar, S. A. (2012). Development of transgenic finger millet (*Eleusine coracana* (L.) Gaertn.) resistant to leaf blast disease. *Journal of Biosciences, 37*(1), 135–147. https://doi.org/10.1007/s12038-011-9178-y (accessed on 7th November 2025).

Jansen, P. C. M., & Ong, H. C. (1996). *Eleusine coracana* (L.) Gaertn. In Grubben, G., & Partohardjono, S. (Eds.), *Plant Resources of South-East Asia*. Backhuys Publishers, Leiden, The Netherlands.

Kalyana Babu, B., Agrawal, P. K., Pandey, D., Jaiswal, J. P., & Kumar, A. (2014). Association mapping of agro-morphological characters among the global collection of finger millet genotypes using genomic SSR markers. *Molecular Biology Reports, 41*(8), 5287–5297. https://doi.org/10.1007/s11033-014-3400-6 (accessed on 7th November 2025).

Klichowska, M. (1984). Plants of the Neolithic Kadera (Central Sudan): A palaeobotanical study of the plant impressions on pottery. In *Origins and early development of food producing cultures in North-Eastern Africa*. Warsaw.

Krishna, T. P. A., Maharajan, T., Roch, G. V., Ramakrishnan, M., Ceasar, S. A., & Ignacimuthu, S. (2020). Hybridization and hybrid detection through molecular markers in finger millet (*Eleusine coracana* (L.) Gaertn.). *Journal of Crop Improvement, 34*(3), 335–355. https://doi.org/10.1080/15427528.2019.1709596 (accessed on 7th November 2025).

Kumar, A., Gaur, V. S., Goel, A., & Gupta, A. K. (2015). De novo assembly and characterization of developing spikes transcriptome of finger millet (*Eleusine coracana*): A minor crop having nutraceutical properties. *Plant Molecular Biology Reporter, 33*(4), 905–922. https://doi.org/10.1007/s11105-014-0802-5 (accessed on 7th November 2025).

Kumar, A., Metwal, M., Kaur, S., Gupta, A. K., Puranik, S., & Singh, S. (2016). Nutraceutical value of finger millet (*Eleusine coracana* (L.) Gaertn.) and their improvement using omics approaches. *Frontiers in Plant Science, 7*, 934. https://doi.org/10.3389/fpls.2016.00934 (accessed on 7th November 2025).

Kumar, A., Yadav, S., Panwar, P., Gaur, V. S., & Sood, S. (2015). Identification of anchored simple sequence repeat markers associated with calcium content in finger millet (*Eleusine coracana*). *Proceedings of the National Academy of Sciences, India Section B: Biological Sciences, 85*(1), 311–317. https://doi.org/10.1007/s40011-013-0296-1 (accessed on 7th November 2025).

Kussmann, M., Raymond, F., & Affolter, M. (2006). OMICS-driven biomarker discovery in nutrition and health. *Journal of Biotechnology, 124*(4), 758–787. https://doi.org/10.1016/j.jbiotec.2006.02.014 (accessed on 7th November 2025).

Latha, A. M., Rao, K. V., & Reddy, V. D. (2005). Production of transgenic plants resistant to leaf blast disease in finger millet (*Eleusine coracana* (L.) Gaertn.). *Plant Science, 169*(4), 657–667. https://doi.org/10.1016/j.plantsci.2005.05.009 (accessed on 7th November 2025).

Mahalakshmi, S., Christopher, G., Reddy, T., Rao, K., & Reddy, V. (2006). Isolation of a cDNA clone (PcSrp) encoding serine-rich protein from *Porteresia coarctata* T. and its expression in yeast and finger millet (*Eleusine coracana* L.) affording salt tolerance. *Planta, 224*, 347–359. https://doi.org/10.1007/s00425-005-0218-4 (accessed on 7th November 2025).

Maharajan, T., Ceasar, S. A., Ajeesh Krishna, T. P., Ramakrishnan, M., Duraipandiyan, V., Abdulla, A. D. N., & Ignacimuthu, S. (2018). Utilization of molecular markers for improving the phosphorus efficiency in crop plants. *Plant Breeding, 137*(1), 10–26. https://doi.org/10.1111/pbr.12537 (accessed on 7th November 2025).

Mariman, E. C. M. (2006). Nutrigenomics and nutrigenetics: The 'omics' revolution in nutritional science. *Biotechnology and Applied Biochemistry, 44*(3), 119–128. https://doi.org/10.1042/ba20050112 (accessed on 7th November 2025).

Meuwissen, T. H., Hayes, B. J., & Goddard, M. E. (2001). Prediction of total genetic value using genome-wide dense marker maps. *Genetics, 157*(4), 1819–1829. https://doi.org/10.1093/genetics/157.4.1819 (accessed on 7th November 2025).

Mirza, N., Taj, G., Arora, S., & Kumar, A. (2014). Transcriptional expression analysis of genes involved in regulation of calcium translocation and storage in finger millet (*Eleusine coracana* L.). *Gene, 550*, 171–179.

Muthamilarasan, M., Bonthala, V. S., Mishra, A. K., Khandelwal, R., Khan, Y., Roy, R., & Prasad, M. (2014). C_2H_2 type of zinc finger transcription factors in foxtail millet define response to abiotic stresses. *Functional & Integrative Genomics, 14*(3), 531–543. https://doi.org/10.1007/s10142-014-0383-2 (accessed on 7th November 2025).

Muza, F. R., Lee, D. J., Andrews, D., & Gupta, S. C. (1995). Mitochondrial DNA variation in finger millet (*Eleusine coracana*). *Euphytica, 81*, 199–205.

Nagaraja, T. E., Nandini, C., Bhat, S., & Parveen, S. G. (2023). Artificial hybridization techniques in small millets–A review. *Frontiers in Plant Science, 14*, 1112117. https://doi.org/10.3389/fpls.2023.1112117 (accessed on 7th November 2025).

Nath, M., Goel, A., Taj, G., & Kumar, A. (2010). Molecular cloning and comparative *in silico* analysis of calmodulin genes from cereals and millets for understanding the mechanism of differential calcium accumulation. *Journal of Proteomics & Bioinformatics, 3*, 294–301.

Oduori, C., & Kanyenji, B. (2007). Finger millet in Kenya: Importance, advances in R&D, challenges and opportunities for improved production and profitability. In M. A. Mgonja, J. M. Lenné, E. Manyasa, & S. Sreenivasaprasad (Eds.), *Finger Millet Blast Management in East Africa* (pp. 10–22). ICRISAT.

Padulosi, S., Mal, B., Ravi, S. B., Gowda, J., Gowda, K. T. K., Shanthakumar, G., & Dutta, M. (2009). Food security and climate change: Role of plant genetic resources of minor millets. *Indian Journal of Plant Genetic Resources, 22*(1), 1–16.

Parvathi, M. S., & Nataraja, K. N. (2017). Discovery of stress responsive TATA-box binding protein associated factor6 (TAF6) from finger millet (*Eleusine coracana* (L.) Gaertn). *Journal of Plant Biology, 60*, 335–342. https://doi.org/10.1007/s12374-016-0574-6 (accessed on 7th November 2025).

Parvathi, M. S., Nataraja, K. N., Yashoda, B. K., Ramegowda, H. V., Mamrutha, H. M., & Rama, N. (2013). Expression analysis of stress responsive pathway genes linked to drought hardiness in an adapted crop, finger millet (*Eleusine coracana*). *Journal of Plant Biochemistry and Biotechnology, 22*(2), 193–201. https://doi.org/10.1007/s13562-012-0135-0 (accessed on 7th November 2025).

Priyanwada, L., Ruvini, L., Barana, J., & Janak, V. (2020). A comparative study on nutritional and nutraceutical properties of finger millet (*Eleusine coracana*) and rice (*Oryza sativa*). *World Journal of Bio-Pharma and Health Sciences, 1*(1), 017–024. https://doi.org/10.30574/wjbphs.2020.1.1.0008 (accessed on 7th November 2025).

Pudake, R. N., Mehta, C. M., Mohanta, T. K., Sharma, S., Varma, A., & Sharma, A. K. (2017). Expression of four phosphate transporter genes from finger millet (*Eleusine coracana* L.) in response to mycorrhizal colonization and Pi stress. *3 Biotech, 7*(1), 20. https://doi.org/10.1007/s13205-017-0609-9 (accessed on 7th November 2025).

Puranik, S., Kam, J., Sahu, P. P., Yadav, R., Srivastava, R. K., & Ojulong, H. (2017). Harnessing finger millet to combat calcium deficiency in humans: Challenges and prospects. *Frontiers in Plant Science, 8*, 131. https://doi.org/10.3389/fpls.2017.0131 (accessed on 7th November 2025).

Purseglove, J. W. (1972). *Tropical Crops: Monocotyledons*. Volumes I and II. Longman Scientific and Technical; John Wiley, Harlow, Essex, UK; New York.

Rachie, K. O., & Peters, L. V. (1977). *The Eleusines: A Review of the World Literature*. International Crops Research Institute for the Semi-Arid Tropics (ICRISAT).

Rahman, H. N., Jagadeeshselvam, R., & Valarmath, B. (2014). Transcriptome analysis of salinity responsiveness in contrasting genotypes of finger millet (*Eleusine coracana* L.) through RNA sequencing. *Plant Molecular Biology, 85*(4–5), 485–503.

Rahman, H., Ramanathan, V., Nallathambi, J., Duraialagaraja, S., & Muthurajan, R. (2016). Over-expression of a NAC 67 transcription factor from finger millet (*Eleusine coracana* L.) confers tolerance against salinity and drought stress in rice. *BMC Biotechnology, 16*(S1), 35. https://doi.org/10.1186/s12896-016-0261-1 (accessed on 7th November 2025).

Rajendran, H. A. D., Muthusamy, R., Stanislaus, A. C., Krishnaraj, T., Kuppusamy, S., Ignacimuthu, S., & Al-Dhabi, N. A. (2016). Analysis of molecular variance and population structure in southern Indian finger millet genotypes using three different molecular markers. *Journal of Crop Science and Biotechnology, 19*(4), 275–283. https://doi.org/10.1007/s12892-016-0015-6 (accessed on 7th November 2025).

Ramakrishnan, M., Ceasar, S. A., Vinod, K. K., Duraipandiyan, V., Ajeesh Krishna, T. P., Upadhyaya, H. D., Al-Dhabi, N. A., & Ignacimuthu, S. (2017). Identification of putative QTLs for seedling stage phosphorus starvation response in finger millet (*Eleusine coracana* L. Gaertn.) by association mapping and cross species synteny analysis. *PLoS One, 12*(8), e0183261. https://doi.org/10.1371/journal.pone.0183261 (accessed on 7th November 2025).

Ramegowda, V., Senthil-Kumar, M., Nataraja, K. N., Reddy, M. K., Mysore, K. S., & Udayakumar, M. (2012). Expression of a finger millet transcription factor, EcNAC1, in tobacco confers abiotic stress-tolerance. *PLoS One, 7*(7), e40397. https://doi.org/10.1371/journal.pone.0040397 (accessed on 7th November 2025).

Sage, R. F., & Zhu, X. G. (2011). Exploiting the engine of C4 photosynthesis. *Journal of Experimental Botany, 62*(9), 2989–3000. https://doi.org/10.1093/jxb/err179 (accessed on 7th November 2025).

Saha, D., Gowda, M. V. C., Arya, L., Verma, M., & Bansal, K. C. (2016). Genetic and genomic resources of small millets. *Critical Reviews in Plant Sciences, 35*(1), 56–79. https://doi.org/10.1080/07352689.2016.1147907 (accessed on 7th November 2025).

Sakamma, S., Umesh, K. B., Girish, M. R., Ravi, S. C., Satishkumar, M., & Bellundagi, V. (2017). Finger millet (*Eleusine coracana* L. Gaertn.) production system: Status, potential, constraints and implications for improving small farmer's welfare. *Journal of Agricultural Science, 10*(1), 162. https://doi.org/10.5539/jas.v10n1p162 (accessed on 7th November 2025).

Seetharam, A., & Halaswamy, J. H. (2003). *Nucleus and Breeder Seed Production Manual.* In Chowdhury, S. K. & Lal, S. K. (Eds.), New Delhi, India.

Sharma, D., Tiwari, A., Sood, S., Jamra, G., Singh, N. K., & Meher, P. K. (2018). Genome-wide association mapping of agro-morphological traits among a diverse collection of finger millet (*Eleusine coracana* L.) genotypes using SNP markers. *PLoS One, 13*(8), e0199444. https://doi.org/10.1371/journal.pone.0199444 (accessed on 7th November 2025).

Singh, M., Metwal, M., & Kumar, V. (2016). Identification and molecular characterization of 48 kDa calcium binding protein as calreticulin from finger millet (*Eleusine coracana*) using peptide mass fingerprinting and transcript profiling. *Journal of the Science of Food and Agriculture, 96*, 672–679.

Singh, R. K., Singh, V. K., Raghavendrarao, S., Phanindra, M. L. V., Venkat Raman, K., Solanke, A. U., Kumar, P. A., & Sharma, T. R. (2015). Expression of finger millet EcDehydrin7 in transgenic tobacco confers tolerance to drought stress. *Applied Biochemistry and Biotechnology, 177*(1), 207–216. https://doi.org/10.1007/s12010-015-1738-4 (accessed on 7th November 2025).

Singh, V., Lee, G., Son, H., Amani, S., Baunthiyal, M., & Shin, J. H. (2022). Anti-diabetic prospects of dietary bio-actives of millets and the significance of the gut microbiota: A case of finger millet. *Frontiers in Nutrition, 9*, 1056445. https://doi.org/10.3389/fnut.2022.1056445 (accessed on 7th November 2025).

Sood, S., Joshi, D. C., Chandra, A. K., & Kumar, A. (2019). Phenomics and genomics of finger millet: Current status and future prospects. *Planta, 250*(3), 731–751. https://doi.org/10.1007/s00425-019-03159-6 (accessed on 7th November 2025).

Talwar, H. S., Kumar, S., Madhusudhana, R., Nanaiah, G. K., Ronanki, S., & Tonapi, V. A. (2020). Variations in drought tolerance components and their association with yield components in finger millet (*Eleusine coracana*). *Functional Plant Biology, 47*(7), 659–674. https://doi.org/10.1071/FP19274 (accessed on 7th November 2025).

Vetriventhan, M., Azevedo, V. C. R., Upadhyaya, H. D., Nirmalakumari, A., Kane-Potaka, J., Anitha, S., Ceasar, S. A., Muthamilarasan, M., Bhat, B. V., Hariprasanna, K., Bellundagi, A., Cheruku, D., Backiyalakshmi, C., Santra, D., Vanniarajan, C., & Tonapi, V. A. (2020). Genetic and genomic resources, and breeding for accelerating improvement of small millets: Current status and future interventions. *Nucleus (Calcutta), 63*(3), 217–239. https://doi.org/10.1007/s13237-020-00322-3 (accessed on 7th November 2025).

Wang, J., Li, D., Dangott, L. J., & Wu, G. (2006). Proteomics and its role in nutrition research. *Journal of Nutrition, 136*(7), 1759–1762. https://doi.org/10.1093/jn/136.7.1759 (accessed on 7th November 2025).

Yamunarani, R., Govind, G., Ramegowda, V., Vokkaliga Thammegowda, H., & Ambarahalli Guligowda, S. (2016). Genetic diversity for grain Zn concentration in finger millet genotypes: Potential for improving human Zn nutrition. *Crop Journal, 4*(3), 229–234. https://doi.org/10.1016/j.cj.2015.12.001 (accessed on 7th November 2025).

Zhang, H., Hall, N., Goertzen, L. R., Chen, C. Y., Peatman, E., Patel, J., & McElroy, J. S. (2019). Transcriptome analysis reveals unique relationships among *Eleusine* species and heritage of *Eleusine coracana*. *G3: Genes, Genomes, Genetics, 9*(6), 2029–2036. https://doi.org/10.1534/g3.119.400214 (accessed on 7th November 2025).

CHAPTER 4

Genetic Improvement in Small Millets

GOVINDA RAI SARMA,[1,*] ADITHYA P. BALAKRISHNAN,[1] SUNAINA YADAV,[1] M. S. NIJI,[2] and IKKURTI GOPINATH[1]

[1]*Division of Genetics, ICAR–Indian Agricultural Research Institute, Pusa, New Delhi, Delhi, India*

[2]*College of Agriculture, Kerala Agricultural University, Vellayani, Thiruvananthapuram, Kerala, India*

*Corresponding author

4.1 INTRODUCTION

Small, also known as minor millets, are potentially highly nutrient-rich crops with great value for human and animal consumption. They are the small-seeded crops of the family Poaceae, comprising finger millet, foxtail millet, proso millet, kodo millet, little millet, barnyard millet, and other minor members such as teff, fonio, job's tears, guinea millet, and browntop millet. They are grown over wide agroecologies and included in conventional cropping systems. Millets are ethnic foods of many socio-geographical cultures, adding to their food diversity. Being a significant source of sustenance in dryland areas as well as in various tribal communities in plains and hills, they ensure the region's food security owing to the high resilience in the adaptation of small millets. The nutrient-rich small millets are whole grain foods with a higher content of quality protein, fiber, and several micronutrients, designated as Nutri-cereals along with other millets. Due to their immense health benefits and medicinal properties (Nithiyanantham et al., 2019), millets, including small millets, are undoubtedly regarded as nutraceuticals, and their new name in Indian languages, "Shree Anna," meaning "the best among all food grains," is self-explanatory. The concern about a healthy diet and the challenge of developing climate resilience among crop plants highlights the necessity of breeding efforts in small millets.

All the millets are grasses belonging to the Poaceae family. In the Poaceae family, pearl millet and all the small millets belong to the Paniceae tribe, the largest tribe of the Panicoideae subfamily. In contrast, sorghum and job's tears are members of the Andropogoneae tribe of the same subfamily. The prime member of small millet, finger millet, belongs to the Eragrostideae tribe of the Chloridoideae subfamily. Teff also belongs to the same (Figure 4.1). All small millets are C4 plants since the PACMAD (Aristidoideae, Panicoideae, Chloridoideae, Danthonioideae, Arundinoideae, Micrairoideae) clade of subfamilies of Poaceae

Millets in Sustainable Agriculture: Genetic Improvement, Nutritional Quality, Cultivation, and Integrated Pest–Disease Management. Roshni Vijayan, A. John Joel, R. Ambika Rajendran, and Adithya P. Balakrishnan (Eds.)

DOI: 10.1201/9781779642332-5

is the only group in Poaceae within which the C4 photosynthetic pathway has evolved (Grass Phylogeny Working Group II, 2012).

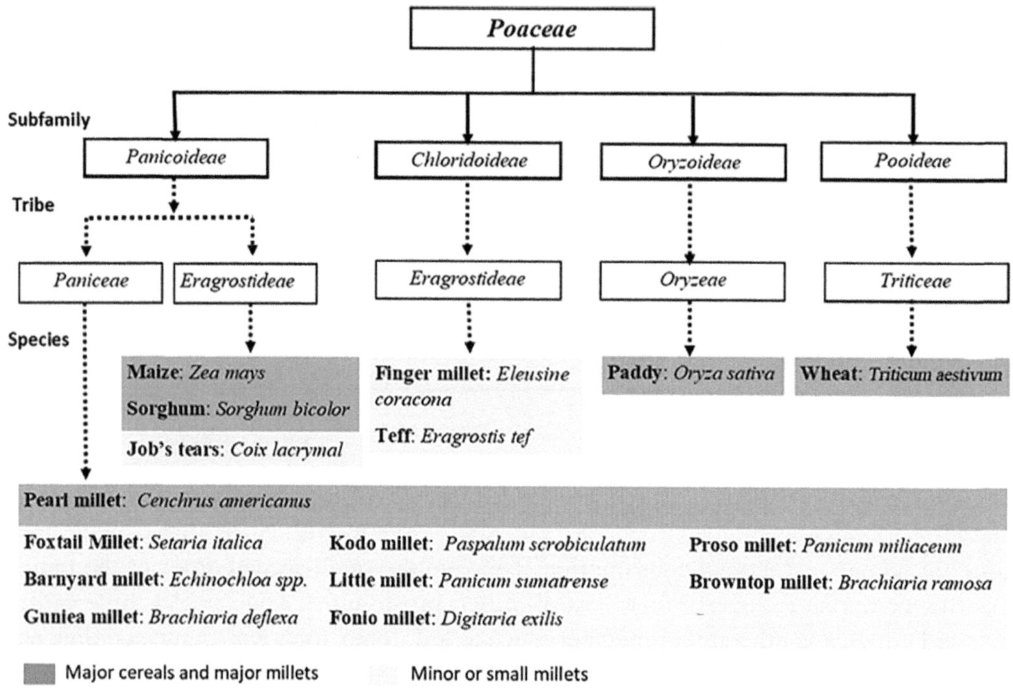

FIGURE 4.1 The taxonomic details of small millets and other associated cereals.
Source: Adapted from Vetriventhan et al. (2020).

Foxtail millet (*Setaria italica* (L.) P. Beauv.) is known worldwide as Italian/German millet; however, due to its small grain size, it is categorized as little millet. Foxtail millet is diploid (2n=2x=18), self-compatible, and has a substantially preserved genomic structure relative to its ancestral grass lineage (Devos et al., 1998). The green foxtail, *Setaria viridis*, is regarded as the wild ancestor of cultivated foxtail millet. It has higher crude fat (5–8%), seed protein (14–16%), and minerals than *Eleusine coracona* (Dwivedi et al., 2012; Ravindran, 1991). Also, foxtail protein contains seven essential amino acids, making its digestible protein's biological value even higher than that of major cereals, rice, and wheat (Zhang et al., 2007). The bran holds 9.4% crude oil supplemented with oleic (13.0%) and linoleic (66.5%) acids (Dwivedi et al., 2012; Liang et al., 2010), and grains hold 2.5 times more edible fiber (42.5%) than rice (Amadou et al., 2011). Foxtail millet is used as a forage crop throughout Asia's semi-arid tropical regions, as well as in North Africa, North America, Europe, and Australia. Because of its drought resistance, foxtail millet was previously a popular rainfed crop in India. However, during the 1990s, its acreage shrank by more than half, owing to the advent of more profitable and remunerative crops. Foxtail millet is also well known for its strong drought tolerance and resistance to disease and pests. Consequently, it has relatively low fertilizer and chemical input requirements, reducing production costs and making the grain more attractive as a healthy food.

Kodo millet *(Paspalum scrobiculatum* L.) or Indian crown grass is known as kodra (Marathi, Hindi, Punjabi, Gujarati), koden (Hindi), varagu (Tamil), arika (Telugu), harka (Kannada), koovaragu (Malayalam), and kodua (Odiya). It is an allotetraploid (2n = 2x = 40) tropical millet that originated in Africa but is mainly cultivated in the Indian subcontinent. Major growers in India include Madhya Pradesh, Maharashtra, Tamil Nadu, Karnataka, and Arunachal Pradesh (Jhum cultivation) (IIMR-RPOP). Kodo millet is a nutritionally rich and climate-resilient crop that can substitute staples like rice and wheat. The grain is an excellent source of carbohydrates (65.9%), crude fiber (9%), protein (8.3%), minerals (2.6%), vitamins like thiamin, niacin, riboflavin, and folic acid, and micronutrients like iron, zinc, and phosphorus (Deshpande et al., 2015). Supplementation with kodo millet reduces the effects of a high-fat diet and can thus be used as a functional element to manage obesity (Sharma et al., 2017).

Proso millet (*Panicum miliaceum* L.) is an allotetraploid (2n=4x=36), annual, self-pollinated, C4 crop brought under cultivation about 10,000 years ago in the semi-arid regions of China (Hunt et al., 2011; Lu et al., 2009). The crop thrives well in marginal lands with nominal water and nutrient requirements; cultivation is distributed across India, China, the USA, Russia, Japan, Mongolia, Iran, Iraq, and other East Asian countries. Proso is the pan-Slavic name for millet; different names are used for the crop depending on the geographic location, such as common and hog millet in the USA, broomcorn millet in China, common millet in Japan, Korea, and other Asia-Pacific countries, French white in France, and hersey millet in Germany (Rajput et al., 2014). In India, it is mainly grown in Tamil Nadu, Andhra Pradesh, Karnataka, Maharashtra, Madhya Pradesh, Uttar Pradesh, and Bihar and is known by different vernacular names such as variga, cheena, panivagaru, and bagaru. The crop has a very low glycemic index with high fiber, protein (>10%), vitamins, and minerals (Gomashe et al., 2017). Proso millet is mainly utilized as livestock and bird feed, with immense potential for developing superior varieties due to its highly nutritive and health-promoting benefits (Wang et al., 2016).

Barnyard millet (*Echinochloa* sp.) is one of the oldest domesticated millets. It is grown in warm and temperate regions of the world, including India, Japan, China, Nepal, Pakistan, and Africa, and is known by several names, viz., ooda, oadalu, sanwank, sawan, and sanwa. This allohexaploid (2n = 6x = 54) crop is the fourth most-produced small millet, providing food and nutritional security to many people across the globe (Renganathan et al., 2020). It has the ability to thrive in harsh conditions with meager input and can tolerate various biotic and abiotic stresses. The short life cycle adds to its climate resilience, which is particularly important in marginal rainfed regions. In the Indian Himalayan region, grain was used as a rice substitute in traditional systems. It is an excellent livestock feed apart from the major purpose of human consumption. The richness of protein (11%), carbohydrate (65%), crude fiber (13.6%), and, essentially, the iron and zinc micronutrients make it the best food and feed source (Manimekalai et al., 2018). All these characteristics of barnyard millet make it suitable for subsistence farming (Gupta et al., 2009).

Little millet (*Panicum sumatrense* Roth. ex. Roem. and Schult.) is a drought and waterlogging-resistant, fast-growing Poaceae member. It is also known as Indian millet, with various other common names such as Kutki (Hindi), Saame (Kannada), Saave (Kannada), Chama (Malayalam), Saamai (Tamil), Samalu (Telugu), Sava (Marathi), Halvi (Marathi), and Suan (Oriya). The allotetraploid crop (2n = 4x = 36) was domesticated in India roughly 5000 years ago (de Wet et al., 1983) and is primarily cultivated in the states of Jharkhand, Chhattisgarh, Madhya Pradesh, Andhra Pradesh, Karnataka, Maharashtra, Odisha, and Tamil Nadu. The crop is also distributed over Nepal, Pakistan, Sri Lanka, Indonesia, and Myanmar. The carbohydrate content of the grain is comparable to that of rice and sorghum. However, high levels of protein (8.7 g/100 g), fiber (8.6 g/100 g), fat (5.3 g/100 g), and mineral matter (1.7 g/100 g) make it nutritionally significant (Ganapathy, 2017). Moreover, the vitamins thiamine (0.2–0.48 mg/100 g), riboflavin (0.12 mg/100 g), and minerals potassium and magnesium are all abundant (Saleh et al., 2013).

All these small millets can potentially supplement the nutritional diet of human beings and animals. Their stress tolerance at various levels of environmental changes is incomparable or superior to that of other major crops. However, the lack of intense popularity of these crops and their derived products in consumer markets is a challenge. Despite these constraints, the declaration of 2023 as the International Year of Millets (UNO, 2022) and subsequent promotional programs hold promise for improving the overall status of small millets. In this review, we attempt to summarize the various aspects of crop improvement for important small millets undertaken thus far, categorized under distinct subheadings. Finger millet is well described in another chapter; hence, we are not discussing finger millet here. These crops will undoubtedly arise as climate-smart crops in the near future.

4.2 PRE-BREEDING, GERMPLASM CONSERVATION, AND VARIETAL DEVELOPMENT

Success in any crop improvement plan is largely determined by the degree of variability in the germplasm for the target traits. Therefore, it is important to focus on species diversity, germplasm collection, evaluation, utilization of unadapted accessions through pre-breeding, and conservation in small millets. These activities are not explored much in these crops compared to major staple crops like rice, wheat, and maize.

The genus *Setaria,* widely distributed throughout the world in both tropical and temperate climates, comprises 125 species. The green (*S. viridis*), yellow (*S. glauca*), giant (*S. faberi*), knotroot (*S. geniculata*), and bristly (*S. verticillata*) foxtails are the many wild relatives that contribute to the domestication of the *Setaria* genus along with *S. italica.* Wild foxtails are characterized by seed dormancy, strong phenotypic plasticity, and tolerance to both mechanical and chemical (herbicide) damages. Interestingly, green foxtail (*S. viridis*) is the ancestor of the domesticated form *S. italica*, and both species are still interfertile (Panaud, 2006). *Ex-situ* seed storage is the most widely used method to conserve the germplasm of foxtail millet. About 46,070 cultivated accessions and 906 wild accessions of foxtail millets made it one of the largest collections (Table 4.1), also considered a mine of phenotypic variations in plant height, growth habits, flowering time, inflorescence traits, and seed phenology (Upadhyaya et al., 2006). The wild relatives of *S. italica* have been found to be treasures of valuable genes for improving many traits of the cultivated gene pool of foxtail millet, for example, resistance to certain herbicides such as triazine, sethoxydim, dinitroaniline, and trifluralin. Herbicide resistance is a trait that has been effectively transferred into *S. italica* from *S. viridis* as it is controlled by one or two major genes with some modifier consequences (Darmency and Pernes, 1985; Darmency et al., 2011; Wang and Darmency, 1997; Wang et al., 1996; Wang et al., 2010). Similarly, Upadhyaya et al. (2009, 2011) and Lata et al. (2011a, b) reported core collections of 155 and 152 accessions, respectively. The 155 accessions (trait-specific germplasm) were examined, and accessions with characteristics such as early maturity, high grain yield, different panicle form, size, or seed color, or higher grain Fe, Zn, and (or) Ca concentrations were detected (Upadhyaya et al., 2011). Nirmalakumari and Vetriventhan (2010) assessed 741 foxtail millet germplasm accessions and found that genotypes with more productive tillers, intermediate panicle length, and medium flowering period had higher yield potential. *Pyricularia grisea* (Cooke) Sacc. (teleomorph: *Magnaporthe grisea*) causes blast disease in foxtail millet (*Setaria italica* [L.] P. Beauv.) in India and Africa, resulting in significant yield loss. Foxtail millet accessions resistant to neck blast (ISe 375, ISe 748, ISe 480, ISe 751, ISe 1037, ISe 769, ISe 1067, ISe 1204, ISe 1320, ISe 1593, ISe 1685, ISe 1335, ISe 1419, ISe 1387, ISe 1547, ISe 376, and ISe 1541) were identified from a foxtail millet core collection (ICRISAT Archival Report, 2009). In foxtail millet, ISe 1575 and ISe 1647 (23 days) were the earliest flowering accessions. ISe 1647 and ISe 1575 were also identified as high-yielding accessions (1.7 t ha^{-1}), whereas ISe 1851 and ISe 1789 were recognized for high inflorescence length (>250 mm) and width (>45 mm) in the foxtail millet core collection (ICRISAT Archival Report, 2008).

Kodo millet genetic research is still in its infancy. It is critical to focus on the available germplasm and its diversity to facilitate crop improvement in small millets such as kodo millet. With 8587 accessions stored in various gene banks, kodo millet germplasm accounts for around 7% of global small-millet germplasm (Upadhyaya et al., 2014). The majority of these accessions originate from India and the United States. The following numbers of accessions are stored in *ex situ* gene banks by different organizations: 2273 at the National Bureau of Plant Genetic Resources (NBPGR) in New Delhi, 1537 at the Project Coordinating Unit, All India Coordinated Research Project on Small Millets (AICRP-Small Millets) in Bangalore, 656 at the International Crops Research Institute for Semi-Arid Tropics (ICRISAT) in Patancheru, 427 at the Tamil Nadu Agricultural University (TNAU) in Coimbatore, and 336 at the USDA-ARS Plant Genetic Resources Conservation Unit in Griffin (Table 4.1). To improve and enhance the use of diverse available germplasm, a core collection was formed for Kodo millet. Upadhyaya et al. evaluated the entire germplasm collection from ICRISAT for morpho-agronomic traits. According to IBPGR's Kodo millet descriptors, 75 accessions representing 11% of 656 germplasm lines from all three races were created. These reduced sets of accessions can be used to find new sources of variation to improve the genetic potential (2014).

Subramanian et al. estimated the extent of variation in yield-contributing traits in 188 kodo millet accessions, including core collections, and observed the highest genotypic and phenotypic variance for dry fodder yield, plant height, and grain yield per plant (2010). Additive gene action is reported, and selection is suggested to improve grain and fodder yield per plant based on high heritability and observed genetic advance. Selection is also recommended for traits such as the number of productive tillers, the number of basal tillers, the length of the inflorescence, the length of the thumb, and so on (Nirubana et al., 2017). Variability in nutritional features is also documented since small millets such as kodo are nutritionally superior and regarded as naturally biofortified crops. About 30 varieties were reported to be rich in grain nutrient content from the ICRISAT gene bank, including IPs776 and IPs618 for iron and protein, Ips 344 for iron and zinc, and IPs 244 and IPs 5 for zinc and iron (Vetriventhan and Upadhyaya, 2019). Days to maturity, days to 50% flowering, and fodder yield per plot are relevant for parent selection and breeding program design since they contribute more towards the genetic divergence seen in kodo millet germplasm (Deepika et al., 2021; Kumari and Singh, 2015; Manimozhi, 2014; Thakur et al., 2020; Ulagnathan et al., 2013). Line RK96/1 was discovered as early as the flowering period (60 days) and yielded more based on Euclidean distance and cluster formation (Deepika et al., 2021). Based on association studies, the major yield components to enforce selection criteria are tillers per plant, panicle length, panicles per plant, days to flowering, days to maturity, fodder yield, and test weight (Sao et al., 2018; Thakur et al., 2020). Most of the varieties developed in kodo millets were by following conventional selection from germplasm lines and landraces. Being a self-pollinated crop, pureline selection is often utilized. Mutation breeding is also exploited with gamma (ɤ) radiation in this minor crop to generate variation. For proso millet, Northern China is the widely acknowledged center of origin; however, some evidence suggests China and Eastern Europe are two independent centers of origin (Lu et al., 2009; Zhang et al., 2010). *P. miliaceum* subsp. miliaceum is a cultivated species, and it is subdivided into five races based on panicle morphology, that is, *miliaceum, contractum, compactum, patentissimu*m, and *ovatum* (Lu et al., 2009). *P. miliaceum* subsp. ruderale includes weedy genotypes. Wild weedy relatives of *P. miliaceum* include *Panicum capillare* and *Panicum repens*. The crop germplasms maintained mainly in the gene banks of India, the USA, China, Ukraine, and Russia hold the wide genetic diversity of the crop (Upadhyaya et al., 2016). Among the 29,300 conserved accessions available all over the world (Table 4.1), the highest number of accessions is maintained in Europe (15,812), followed by Asia (12,110). Evaluation of the 850 accessions in the gene bank in ICRISAT derived a core collection comprising 106 accessions representing all five races (Upadhyaya et al., 2011). Efficient breeding programs based on a wide genetic variability of parental germplasm could be the most important strategy to improve crop yield and other agronomic traits. Evaluation of available germplasm collections could provide diversity scenarios and line-based trait profiles for further utilization in breeding programs for regional adaptation since assessment of agro-morphological traits diversity and identification of germplasm for breeding programs are limited.

Evaluation of 80 *P. miliaceum* accessions for important agronomic traits under Mediterranean pedoclimatic conditions exhibited a wide variation in grain yield, total dry biomass, harvest index, growing degree days, days to maturity, and plant height. This may provide significant information to breeders in developing superior varieties (Calamai et al., 2020). Vetriventhan and Upadhyaya studied 200 accessions of proso millet originating in 30 countries to assess phenotypic diversity to identify nutrient-rich, high-yielding accessions (2018). Trait-specific sources, including 18 high-yielding, 10 large-seeded, and 26 multiple nutrient-rich accessions, were identified for potential use in proso millet improvement. They included IPm 9 and IPm 2661 for high grain yield, and IPm 2069, IPm

2076, and IPm 2537 for high Fe, Zn, Ca, and protein contents. 364 proso millet germplasm accessions were evaluated by Salini et al. (2010), who observed high variability for most of the traits, including grain yield, signifying the scope for improvement of these traits. The principal component analysis identified characters that could be used to distinguish the accessions. The positive direct effect of plant height, number of productive tillers, and hundred seed weight suggests that direct selection for these characters would improve grain yield. Being a crop adapted to drought areas where water availability can be as low as 20–50 cm of rainfall per year, germplasm evaluation should be carried out through various breeding programs to identify lines suitable for various regional needs. Habiyaremye et al. identified Minsum and GR665 varieties of proso millet as good performers in non-irrigated conditions (2017). They reported high variability, genetic advance, and heritability for yield-contributing traits studied in the evaluations of 360 proso millet accessions. The information thus generated from these studies is of supreme importance for crop improvement and conservation purposes.

Proso millet is commonly used as livestock and bird feed despite its high nutritional value and impeccable environmental benefits. The inherent drought-tolerant nature of the crop makes it an ideal choice for the present climate change scenario. Recently, the potential health benefits of proso millet, attributed to its high content of different vitamins and minerals along with its low glycemic index and gluten-free property, have attracted food industries as well as scientific communities (Wang et al., 2016). Plant breeders across the globe are using both conventional and modern crop breeding strategies to deliver superior varieties. One of the major limitations of the crop is its lower yield in comparison with other major cereal crops. Hence, the development of proso millet cultivars with high yield, lodging and seed-shattering tolerance, and nutritional quality is crucial to promote large-scale commercial cultivation and adoption in the food industry. Currently, 24 varieties of proso millet have been released in India (Vetriventhan et al., 2020). Landraces have been the primary source of varietal development in proso millet, and a few have resulted from hybridization efforts. Proso millet is not easily amenable to hybridization due to the high rate of self-pollination and the smaller flower size. Out of all released varieties in India, only 25% are developed through hybridization. The majority of them are selections from landraces. The notable varieties include Co (PV), TNAU 151, TNAU 164, and TNAU 202.

Echinochloa frumentacea and *Echinochloa esculenta* are the most important species among the cultivated and wild species of the barnyard millet. *E. frumentacea* (syn *E. colona* var. *frumentacea*; Indian barnyard millet) originated from wild *E. colona* ((L.) Link) (Jungle rice) and exhibits an analogous line of evolution both in India and Africa (De Wet et al., 1983). *E. esculenta* (Japanese barnyard millet; syn. *E. utilis*) originated from wild *E. crusgalli* (L.) *P. Beauv.* (Barnyard grass) was domesticated around 4000 years ago in the Japanese temperate regions (Doggett, 1989). The allohexaploid species (2n = 6x = 54) are annual in growth habit with overlapping features. However, awned and large spikelets with papery glumes make Japanese barnyard millet different from Indian barnyard millet. *E. frumentacea* widely cultivated in tropical and subtropical regions, including India, Nepal, Pakistan, Malawi, Central Africa, and Tanzania, comprised four races: *stolonifera*, *robusta*, *intermedia*, and *laxa*. The *crusgalli*, *macrocarpa*, *utilis*, and *intermedia* are the races of *E.*

esculenta cultivated primarly in temperate countries, viz., Japan, Korea, China, Russia, and Germany (Renganathan et al., 2020; Upadhyaya et al., 2014).

Echinochloa comprises around 250 annual and perennial species (Bajwa et al., 2015). Both cultivated species of *Echinochloa,* Japanese and Indian barnyard millet, as mentioned earlier, have a close relationship with their wild ancestors, *E. crusgalli* and *E. colona,* respectively. Both cultivated species and their progenitors are hexaploid with $2n = 6x = 54$, where $x = 9$ (Yabuno, 1966). Cultivated and wild species of *Echinochloa* differ from each other in morphology, growth habitat, and other characteristics. *E. frumentacea* and *E. esculenta* showed a marked difference from their respective wild counterparts with respect to more compact growth habits, reduced vegetative branching, thick culms, larger inflorescences, reduced seed shattering, and larger seed size. Both species also differ in their genomic composition (Yabuno, 1966). The agronomic difference observed in the cultivated species includes higher seed-shattering behavior of *E. frumentacea* compared to *E. esculenta.* Differences are also observed in the color of caryopsis, with straw brown coloration getting darker in *E. crusgalli* and yellow-brown in *E. esculenta.* Plant height varies among the species, the tallest being *E. crusgalli* and *E. frumentacea,* with a height of around 250 cm, and *E. colona* usually reaches around 60 cm while *E. esculenta* shows intermediate stature (Liu et al., 2022).

In India, barnyard millet diversity is being quickly eroded due to changing socioeconomic and cultural dimensions of the farming community and a reduction in crop acreage (Maikhuri et al., 2001). Sizeable efforts have been taken to conserve the crop diversity *ex situ,* but very limited information is available regarding *in situ,* on-farm conservation (Padulosi et al., 2009). The worldwide *Echinochloa* germplasm collections total 11,268 accessions (http://apps3.fao.org). Core collections have been established after assessing diversity in more than 700 accessions, and core sets of 89 and 50 representative worldwide collections have been identified (Gowda et al., 2009; Upadhyaya et al., 2014). Major collections of barnyard millet are maintained in Asia (India and Japan), of which the Indian ICAR-NARS national collections and ICRISAT collections constitute 4734 and 749, respectively (Table 4.1). The prebreeding reports state that F1s derived from the crosses between *E. colona* x *E. frumentacea* and *E. crusgalli* x *E. esculenta* were found to be fertile. Meanwhile, progenies of *E. esculenta* x *E. frumentacea* as well as *E. crusgalli* x *E. colona* were sterile due to abnormal meiotic divisions (Yabuno, 1966) ascribable to the poor genomic affinity among *Echinochloa* species revealed through cytological studies. However, the current diversity within the genus is due to the rampant historical hybridization reported (Hilu, 1994; Yabuno, 1962; Yamaguchi et al., 2005).

Barnyard millet production has decreased worldwide due to intensive rice, wheat, and maize cultivation, which in turn reduced the crop diversity and led to inattention to its landraces (Sood et al., 2015). Comparable to the other minor crops, research and resources devoted to the study of barnyard millet were limited. In recent times, consumers have become more health-conscious, and barnyard millet and other small millets are getting much-needed attention owing to their nutritional benefits. In-depth studies in barnyard millet germplasm can provide various information for utilizing them in crop improvement programs. The Indian Himalayan regions harbor favorable genotypes for traits like tiller number, plant height, inflorescence length, and grain yield (Gupta et al., 2009). Barnyard millet germplasm evaluation observed considerable variation for yield-related traits (IIMR, 2018). Germplasm characterization has identified different sources of quality traits such as NDS6 and NDS2 (high protein content), NDS6 and NDS5 (high lysine content), NDS6 and NDS3 (low phytic acid and tannin), and B29 (easy de-hulling trait) (Gupta et al., 2015). Disease resistance sources include RBM77, RBM57, and RBM58 (grain smut resistance) and DHBM996 and TNEF-204 (shoot fly and stem borer resistance) (Rawat et al., 2019).

Barnyard millet cultivation in India is mainly confined to Tamil Nadu, Andhra Pradesh, Karnataka, and Uttarakhand. The inception of the All India Coordinated Millets Improvement Project in 1969 gained national-level interest in barnyard millet breeding (Seetharam, 1998). The breeding objectives include higher yield, nutritional value, pest and disease resistance, easy de-hulling, climate resilience, genotypes suitable for mechanical harvesting and post-harvest processing, dual-purpose genotypes, as well as early ground cover, chlorophyll retention, and stay-green traits contributing greatly to drought stress.

TABLE 4.1 Global Status of Small Millet Germplasm Collection

Crop Name	Asia			Europe	USA	Others	Total
	ICAR-NARS	ICRISAT	Others				
Foxtail millet (*Setaria italica*)	4573	1535		4548	1145	496	7724
Kodo millet (*Paspalum scrobiculatum*)	4237	656	_	354	14	3326	8587
Proso millet (*Panicum miliaceum*)	931	849	10330	15812	1147	239	29308
Barnyard millet (*Echinochloa colona & E. crus-galli*)	4734	749	4503	80	842	360	11268
Little millet (*Panicum sumatrense*)	1253	473	1104	_	226	8	3064

The rate of cross-pollination in the highly self-pollinated genus *Echinochloa* is sufficient to ensure gene exchange. Emasculation and artificial pollination are difficult due to small flower size, early flowering, reduced pollen viability, and non-availability of pollen grains (Nirmalakumari and Vetriventhan, 2010). Although some varieties developed through the pedigree method of selection in hybridization progenies have been released, mass selection and pure line selection are the most commonly used breeding methods due to their mode of pollination and floral characteristics. The barnyard millet growing regions of the country have more than 20 improved cultivars. A majority of them were developed through pure-line selection from local landraces/cultivars, accompanied by pedigree selection. The first released variety was K1. Other notable varieties released in India are PRJ1, MDU1, Co2, and VL207. Until 2012, selection from landraces has facilitated the release of nine varieties, of which three were developed through hybridization (https://www.millets.res.in/aicrp_small.php). Even though mutation breeding, which is pertinent to other self-pollinating crops, is equally applicable in barnyard millet, the development of mutant phenotypes is difficult due to the polyploid nature of the crop. Mutation breeding was initiated to induce genetic polymorphism for yield traits. Gamma irradiation treatment has been reported to improve genetic variability for seed yield and yield-attributing traits such as the number of tillers and ear heads (Mehra et al., 1985). Full waxy stable mutant lines were developed through γ-irradiation of the low amylase landrace "Nogehie," as most of the released cultivars were non-waxy, which gives a poor texture to food products (Hoshino et al., 2010). Increasing the grain size of barnyard millet will enhance its appeal among consumers. However, it would be challenging due to the narrow range of variability available for the trait.

Combining the wild grasses *Panicum psilopodium* and *Panicum sumatrense* based on morphological comparison, two subspecies are established in little millet, that is, *Panicum sumatrense* subsp. sumatrense and *Panicum sumatrense* subsp. psilopodium (de Wet, 1983). The *P. psilopodium*, with a basic chromosome number of x=10, has been proven to be the progenitor of cultivated species with x=9 through genomic homology studies of inter-specific crosses between *P. sumatrense* and *P. psilopodium* (Hiremath et al., 1990). For a better understanding of germplasm collection, conservation, and utilization, the chloroplast (cp) genome is utilized in little millet. Phylogenetic studies using the complete cp genome sequence indicated the similarity of *P. sumatrense* with *P. miliaceum* (common millet) and *P. capillare* (witchgrass) among *Poaceae* species (Sebastin et al., 2018). Nana and robusta are the two main races of *P. sumatrense*, with differences in height and inflorescence characteristics (Joshi et al., 2021). It is reported that the robusta accessions are high-yielding, while the nana accessions are high in grain nutrients (Vetriventhan et al., 2020). About 3,064 accessions are available in the global germplasm collection of little millet, of which 2,830 are from Asia, 226 from the USA, seven from Africa, and one from Oceania (Vetriventhan et al., 2020). More than 473 germplasm lines from various countries are conserved at the ICRISAT gene bank and 1,253 at NBPGR (Table 4.1). In 2014, ICRISAT created a core collection of 56 accessions to find trait-specific diverse lines to employ in crop improvement programs by screening for 21 morpho-agronomic traits. Variability and diversity studies on the little millet germplasm have been conducted as an initiation of crop improvement efforts. Due to additive gene effects, simple phenotypic selection on parameters such as the total number of tillers per plant, the number of productive tillers per plant, flag leaf width, and flag leaf sheath length is recommended to enhance grain yield (Nirmalakumari et al., 2010).

There is a controversy about the diversity present among landraces of little millet. The diversity present in the landraces of Kolli Hills is analyzed using participatory evaluation along with experimental methods (Arunachalam et al., 2005). The analysis using Randomly Amplified Polymorphic DNA (RAPD) markers showed not much genetic divergence in contrast to remarkable inter-group distances in biometrical studies. The studies conclude that the cause of genetic divergence cannot be attributed directly to the environment due to the considerable distance in maturity, flowering time, yield, and harvest index as found in the studies (Arunachalam et al., 2005).

4.3 MARKER DEVELOPMENT AND MAPPING OF AGRO-MORPHOLOGICAL TRAITS

The use of molecular markers to map desirable traits is common in modern crop improvement programs. This activity provides a starting point for marker-assisted breeding to accelerate the efficiency of crop breeding. In small millets, the lack of reference genome sequence hindered the development of high-density PCR and sequencing-based markers for mapping various traits to the respective linkage group. The inbreeding characteristic of foxtail millet and its small genome (515 Mb) with a limited number of chromosomes (2n = 2x = 18) make it a valuable model for investigating plant architecture, drought tolerance, and C4 photosynthesis in grain and bioenergy crops (Lata et al., 2013; Muthamilarasan and Prasad, 2015). Therefore, foxtail millet scientists are working hard to create high-yielding, high-quality, and stress-resistant cultivars. A foxtail millet (*S. italica*) draft genome (423 Mb) was created, which was anchored onto nine chromosomes and comprised 38,801 annotated genes (Zhang et al., 2012). In a study, a high-quality reference genome sequence encompassed 80% of the genome and >95% of the gene space in a 400-Mb assembly was assembled (Bennetzen et al., 2012). Post the assembly of the whole genome sequence of foxtail, several valuable mapping studies followed. Bennetzen et al. developed a genetic map with 992 SNP (Single Nucleotide Polymorphism) markers totaling 1416 cM in length (2012). About 21,294 microsatellite primer pairs based on the reference genome (https://phytozome.jgi.doe.gov/pz/portal.html) were produced, and a total of 15,573 markers were physically mapped on nine chromosomes of foxtail millet (Pandey et al., 2013). Jia et al. employed 2.58 million SNPs and 0.8 million common SNPs to create a haplotype map of the foxtail millet genome (2013). A hybrid population between *S. italica* acc. B100 and *S. viridis* acc. A10 led to the discovery of 25 quantitative trait loci (QTL) for vegetative branching and inflorescence architecture, as well as potential genes for branching control (Doust et al., 2004). Using an F2 population resulting from a cross between Zhang Gu 1 and A2, a high-density linkage map containing 118 SNP markers and 641 SVs (Structure Variations) markers was created (Zhang et al., 2012). Since the invention of next-generation sequencing (NGS), restriction site-associated DNA sequencing (RAD-seq) has been widely used in plants to generate thousands of Bins (derived from SNPs) for the production of high-density genetic maps (Peng et al., 2016; Xu et al., 2013). At present, many QTLs for agronomic and yield traits have been identified in foxtail millet, such as those related to drought stress (Qie et al., 2014), tiller number, plant height, panicle length, panicle weight, and panicle diameter (Wang et al., 2017; Zhang et al., 2017) and panicle branching (Masumoto et al., 2016). Using a Shen3 × Jinggu20 F2 population, Wang et al. discovered two QTLs associated with plant height, one related to panicle length, one related to panicle, and one connected to grain weight (2013). Based on the reference genome sequence, Fang et al. produced 10,598 SSR (Simple Sequence Repeats) markers (2016). These newly developed markers were not the same as those developed by Zhang et al. (2014) and Pandey et al. (2013) and, hence, beneficial for fine-mapping, map-based cloning, and molecular marker-aided breeding.

In foxtail, the best mapped and studied trait is the color of various plant organs. Resequencing-based InDel and SVs markers were used to map the white leaf sheath gene *SiWLS1* (Zhang et al., 2021). Yellow pigmentation of the kernel and its genetic variation are studied in Chinese accessions (Yang et al., 2017). A GWAS (Genome-wide Association Studies) and metabolite-based GWAS of 398 Chinese accessions reported the effects of domestication on the genetic architectures of various metabolites (Li et al., 2022). The study figured out the candidate gene for the yellow pigmentation of the kernel, found to be one of the domestication traits in foxtail millet, determined by *SiPSY1* (PHYTOENE SYNTHASE 1), and a KASP marker for the gene has been developed (Zuo et al., 2023). Similarly, hull genetics of hull color were also determined through QTL mapping as well as BSR-Seq and genome resequencing approaches (Guo et al., 2023; Tian et al., 2022). It proposed multiple genes involved in hull color regulation. In a combined GWAS, MWAS (microbiome-wide associated studies), and mGWAS (microbiome GWAS) analysis using 827 foxtail accessions for 12 different agronomic and yield traits and their associated rhizoplane microbiome species strains or, more specifically, biomarkers, association with six key agronomic traits were reported (Wang et al., 2023). This study reveals that most trait variations of top second leaf width, panicle diameter of the main stem, and main stem panicle weight are determined mainly by the rhizoplane microbiota alone. The growth and gene induction or suppression responses of foxtail cultivars are both host-genotype and strain-specific. The strain-specific responses of selected genotypes for precision cultivation were also validated. This research has highlighted the relationship between microbiomes and millet species. Whether the small millets' climate resilience and tolerance to various abiotic stresses are related to the specific microbiome they are associated with is a question of which significant answers are needed. All these mapping, sequencing, and marker studies added resources and methodologies for marker development and molecular breeding in *S. italica* for various objectives.

Genome sequencing has not yet been done in Kodo millet, which limits its application in developing sequence-based markers and applications through molecular breeding. However, PCR-based markers such as RAPD have been used to differentiate Indian and African cultivars to construct geographic groupings and study gene flow between them (M'Ribu and Hilu, 1996). The Indian Institute of Millets Research (IIMR) developed microsatellite (SSR) markers to utilize in the genetic analysis of Kodo millet. The very first genome-wide population structure analysis in Kodo millet concluded that at the SNP level, the existing grouping is not correlating (Johnson et al., 2019). Future research following this will create a scope for QTL identification and molecular breeding of Kodo millet. Validation of the SSR markers is currently a work in progress. The absence of a well-annotated reference genome sequence forces researchers to conduct tedious *de novo* analysis or comparative genomic analysis to develop valid molecular markers to begin molecular breeding. Molecular markers have been instrumental in the analysis of genetic diversity and population structure across diverse species (Rajput and Santra, 2016). Yet, molecular characterization of the proso millet gene pool remains difficult due to its complex tetraploid nature (Hunt et al., 2011). Nevertheless, molecular markers such as RAPD, AFLP (Amplified Fragment Length Polymorphism), ISSR (Inter and Simple Sequence Repeats), and SSR have been employed in proso millet genetic diversity analysis (Habiyaremye et al., 2017; Karam et al., 2004; Trivedi et al., 2015). M'Ribu and Hilu (1994) used RAPD markers to study the diversity among 13 accessions of *Panicum miliaceum*. The polymorphism observed in RAPD markers in the study demonstrated its effectiveness in determining intraspecific variation. The study separated the proso millet accessions into groups based on geographical origins. SSR markers for proso millet were extracted from the genomic sequences of related species such as switchgrass, barley, oat, and paddy in the initial period for the genetic diversity analysis (Rajput et al., 2014). SSR-based molecular studies identified numerous alleles across accessions and produced relative geographic structuring along with line-based DNA profiling (Cho et al., 2010; Rajput and Santra, 2016). A clustering of landraces based on geographical locations was restructured when markers derived from intron splice junctions were used for diversity analysis (Hu et al., 2008; Weining and Henry, 1995). A similar study of phylogenetic relationships clustered

Chinese and Russian accessions together and Korean accessions separately based on 5S ribosomal sequences (Pak et al., 2012).

Advancements in crop genomic resources contribute immensely to the development of molecular markers capable of providing genomic profiles of species as well as better screening methods for varietal identification and trait dissection. The proso millet genome, along with the complete chloroplast genome sequence, has been elucidated and made available. The landrace accession 00000390 was used to produce a 923 Mb long sequence (Kubesova et al., 2010) and 55,930 predicted genes (Zou et al., 2019). A sequenced genome always provides opportunities for resequencing with new annotations to avoid the present genome gaps, gene mining, and identifying various markers through comparative genomics. The genome sequence in proso millet enhanced the development of molecular and functional markers such as SSRs from related genomes or its own sequence, expressed sequence tag (EST) derived markers, and SNP markers, besides understanding the genome-level diversity in proso millet (Hu et al., 2009; Saha et al., 2016; Yue et al., 2016). Beyond SSR markers, the availability of specific miRNA sequences, waxy gene sequences, and genetic linkage maps added to the molecular tools (Wu et al., 2012). These molecular markers are significant in studying genetic relationships among germplasm, their characterization, classification, and species identification, and can be exploited for further expanding linkage maps to complex levels.

Cho et al. first developed the genome-derived SSR markers in proso millet (2010). They constructed an SSR-enriched library to delineate 25 polymorphic SSR markers. The SSR library was further utilized by Hunt et al. in the investigation of genetic diversity among 98 landraces from Eurasia, and they were able to make predictions regarding the domestication and dissemination of *P. miliaceum* (2011). Core sets were developed in USDA-ARS and ICRISAT using the species-specific and related genomic SSRs (Rajput and Santra, 2016; Upadhyaya et al., 2011). 500 proso millet specific-SSR primer pairs were developed by Liu et al. through high-throughput sequencing, and they screened eight accessions selected randomly from 73 germplasm collections using them (2016). 162 primers produced reproducible polymorphic fragments in the screening. As combining ability studies have not been reported so far in proso millet, the development of hybrids in the future could be derived from the clusters classified using molecular markers.

Gene-based markers used in proso millet provided an understanding of the mutations and gene-based diversity. Hunt et al. sequenced the Granule-bound starch synthase I (GBSSI) gene (waxy gene), the prime enzyme for amylose synthesis in the endosperm of the grain, and reported two homologues, L and S forms (2010). They identified a 15-bp deletion in the S homologue and two sequence polymorphisms in the L homologue of the gene coding sequence. The deletion was reported to cause the loss of function of the proteins, resulting in 100% correlation with the waxy phenotype. The mutant alleles thus identified were further scrutinized in Japanese and Eurasian landraces for *Waxy* gene diversity, and accessions with high and low waxiness were identified (Araki et al., 2012; Hunt et al., 2013). 20 common proso millet cultivars were compared with medieval millet landraces using ISSR, simple SSR, and cleaved amplified polymorphic DNA (CAPS) markers by Lágler et al. (2005).

The first robust genetic linkage map constructed using 2412 SNPs in proso millet identified 18 QTLs for agro-morphological traits on 14 linkage groups, with a phenotypic variance explained ranging from 13 to 34% (Rajput et al., 2016). 10 marker-trait associations (MTAs) of several key seed morphology traits like perimeter, length, width, and circularity of the seeds were established using 88 highly diverse accessions with significant levels of admixture of proso millet. The development of linked markers from the MTAs and their validation will be helpful in their utilization for marker-assisted selection (MAS).

In proso millet, the development of contrasting mutants, their characterization, and the generation of functional genomic information for yield was conducted by Francis et al. (2023). A gamma-irradiated mutant population was evaluated up to the M4 generation to identify high-yielding mutants. Sequencing of three mutants including ATL_hy (dense panicle and high yield), ATL_ly (lax panicle and low yield), and the wild type using the genotyping-by-sequencing approach detected SNPs. A total of 84 and 171 functional SNPs were identified in high-yielding and low-yielding mutants, respectively, on annotation with respect to the reference genome. The SNPs, two in ATL_hy and one in ATL_ly, were functional corresponding to the "E3 ubiquitin-protein ligase UPL7" coding gene (Francis et al., 2023). 1882 SNPs were identified in GWAS of 190 proso millet accessions that classified them into eight putative subpopulations (Johnson et al., 2019). Zou et al. reported a high-quality, chromosome-scale genome assembly of proso millet using a combination of short-read sequencing, single-molecule real-time sequencing, Hi-C, and a high-density genetic map (2019). The millet contains 55,930 protein-coding genes and 339 microRNA genes, and phylogenetic analyses revealed two sets of homologous chromosomes that may have merged around 5.6 million years ago.

Molecular marker application in barnyard millet closely followed other small millet crops such as finger and foxtail millet. Genetic diversity analysis in barnyard millet using molecular markers was initiated using RAPD markers to decipher the phylogenetic relationship among *Echinochloa* species (Hilu et al., 1994; Rutledge et al., 2000). Indian barnyard millet was shown to be high in genetic diversity compared to *E. esculenta*, consistent with a high degree of morphological variability observed in *E. frumentacea*, proving that RAPD markers are operative in differentiating various *Echinochloa* species at the genomic level. However, further validation revealed that RAPD markers were inefficient in representing the actual diversity of barnyard millet germplasm (Prabha et al., 2010). Later, the low polymorphic information generated by RAPD markers was overcome by employing AFLP markers in the investigation of genetic divergence, which developed a higher ability to reveal the genetic diversity in *Echinochloa* species (Tabacchi et al., 2009). Polymerase chain reaction-restriction fragment length polymorphism (PCR-RFLP) technique was successfully explored for species differentiation in *E. crusgalli* and *E. oryzicola* (Yasuda et al., 2006). InDel markers identified in *E. colona* for rbcL, matK, and ITS genes have been widely used in the DNA barcoding of *E. crus-galli*, *E. colona*, and *E. oryzicola* as a cost-effective approach (Lee et al., 2014). In the case of insufficient sequence information, the aforementioned markers are the lone choice in diversity studies and the predecessors of molecular breeding. Later, with significant development of sequencing technologies, the limitations present in the RAPD, RFLP, and AFLP techniques were eliminated. Sequence-based markers such as SSRs, EST-SSRs, and SNPs are more desirable in genetic diversity analyses due to their co-dominance, reproducibility, and high level of polymorphism (Lin et al., 2011; Nozawa et al., 2006). Although the importance of a sequenced genome is known, there is no annotated genome sequence available in barnyard millet, but it is available in the ancestral wild species, *Echinochloa crusgalli* (Gou et al., 2017; Vetriventhan et al., 2020), which could provide insights into Japanese barnyard millet compared to Indian species. Chloroplast genome sequencing of seven *Echinochloa* species provided information about cp-genome structure, organization, and conserved gene order (Perumal et al., 2016; Ye et al., 2014). Among the seven species, chloroplast genome size ranged between 139,592-139,851 bp with an identity of 99.5%. The chloroplast genome has the presence of 61% AT-rich regions and 38.6% GC-rich regions with a maximum of 131 genes (Sebastin et al., 2019). Molecular breeding techniques such as genetic diversity analysis, linkage mapping, and identification of QTLs demand molecular markers in bulk. In the absence of a sequenced genome, comparative genomics plays an important role, as there is limited genomic information available in barnyard millet. Genomic resources from closely related species, such as foxtail millet (Kumari et al., 2013; Muthamilarasan et al., 2014; Pandey et al., 2013; Yadav et al., 2014), can be exploited in this regard. Regardless, the genetic diversity of Indian barnyard millet germplasm was analyzed using independent sequencing attempts such as transcriptome sequencing and genotyping-by-sequencing (GBS)-based EST-SSR markers developed from the transcriptome sequence of cultivated *E. frumentacea* (Babu et al., 2017; Murukarthick et al., 2019). A total of 21,000 SNPs were generated through the whole-genome GBS of a core collection comprising 95 barnyard millet accessions (Wallace et al., 2015).

Recent advances have identified numerous SSR and SNP markers recognized in barnyard millet (Manimekalai et al., 2018; Wallace et al., 2015). This offers breeders the opportunities to delineate novel QTLs and genes responsible for important traits. However, no genetic map or QTL mapping has been reported so far. Ishikawa et al. performed molecular characterization of three waxy genes, namely, *EeWx1, EeWx2*, and *EeWx3*, determining the amylose level in the endosperm (2013). Plants having functional alleles at all three loci produced normal amylose content (wild type), while mutant forms in any loci (natural mutant) exhibited low amylose content. Homozygous mutated alleles in all three waxy loci showed waxy (amylose-free) phenotypes. The study developed useful markers for selecting low amylose and waxy lines in barnyard millet. Renganathan et al. employed bulk segregant analysis (BSA) to study marker-trait associations (MTAs) in barnyard millet and reported that the SSR marker, BMESSR39, is closely linked to anthocyanin pigmentation in barnyard millet (2019). Like kodo millet, little millet lacks a sequenced genome. The scarcity of molecular markers greatly restricts the use of advanced genomic techniques in little millet breeding. RAPD markers are being used in this crop to begin the various marker analyses for genetic characterization. 32 genotypes and thirty-six RAPD primers were used to amplify 155 polymorphic sites (Tiwari et al., 2018). Recently, about 4443 novel potential expressed gene-derived SSR markers were developed using transcriptome data. Most unigenes harboring these microsatellites were discovered to be related to thiamine, purine metabolism, and signaling pathways, highlighting their importance (Desai et al., 2021). The absence of a reference genome sequence is a prime hindrance to beginning deep and engaging research activities for future breeding in any crop. However, sequencing and annotating a well-characterized genotype will be a research objective before the scientific community to begin the exploration towards omics of little millet.

4.4 MOLECULAR AND TRANSGENIC BREEDING

Nowadays, crop breeding programs rely more on molecular and genetic engineering techniques than conventional breeding. In fact, the molecular breeding methods are much more rewarding than the conventional methods. They add pace to the genetic gain in neglected crops like small millets to address food and nutritional security. Efforts are still sparse in small millets, particularly barnyard and little millets.

Foxtail millet has a greater genetic and physiological likeness to the potential biofuel crops, which makes it the main target for genetic transformation. Ban et al. published the first report on tissue culture of foxtail millet, using anthers as explants (1971). The development of foxtail millet transgenic plants by pollen and inflorescence particle bombardment has also been described, although their transformation efficiency is low (Dong and Duan, 2000; Wang et al., 2011). Agrobacterium-mediated transformation is the most common method to overexpress any desired gene to generate transgenics, which is also more useful than particle bombardment (Ceasar and Ignacimuthu, 2009). Only a few reports of *Agrobacterium*-mediated transformation in foxtail millet have been published so far. The first *Agrobacterium*-mediated transformation system was created by Liu et al. in foxtail millet, with a transformation efficiency of 6.6% (2005). To convert foxtail millet using *Zm13:Si401*, Qin et al. used the same technique, where *Zm13* was a pollen-specific maize promoter, and the plasmid vector used was pBIZm13Si401 (2008). Chen et al. have developed methods for callus initiation, somatic embryo development, and plantlet regeneration from explants of Yugu1, a foxtail millet genotype (2011). Foxtail millet has also been recognized as an effective system for virus-induced gene silencing for functional assessment of genes in bioenergy crops that are more technically complicated (Doust et al., 2009).

Bhatt et al. proposed the first successful genetic transformation process in Kodo millet in order to improve the orphan crop in terms of quality genes. The protocol was standardized by optimizing the MS medium components and external supplementation of acetosyringone in co-culture media (2021). *Agrobacterium*-mediated gene transfer is a well-established method for incorporating any desired gene into a crop.

Despite limited genomic resources, ESTs were used in proso millet to mine miRNAs, where 43 miRNAs were identified, and 12 were subsequently validated (Wu et al., 2012). These identified miRNAs were found to target genes associated with development, stress response, and metabolism. The subsequent study of drought-tolerant and sensitive proso millet cultivars using transcriptome sequencing identified 833 and 2166 differentially expressed genes, respectively. The study summarized that induced water stress triggered differential responses due to the ROS scavenging system and the jasmonic acid signal transduction pathway. Notably, no reports of transgenics are available in this crop.

Marker-assisted breeding and other transgenic breeding techniques have yet to be initiated in barnyard and little millet due to diverse reasons. The advent of whole genome sequencing in these crops can bring drastic changes. The bulk generation of molecular markers in the newly sequenced genomes can bring new breeding programs to light. However, independent transcriptome sequencing and gene expression studies are still lacking due to the crop's lack of wide economic importance.

4.5 ABIOTIC AND BIOTIC STRESS RESISTANCE

One of the most important areas of research right now is the production of cultivars with improved nutritional value and stress tolerance. The demand for climate-resilient cultivars is increasing due to changes in the peculiar characteristics of each season in different parts of the world. Abiotic stresses like drought, heat, cold, nutritional deficiency, excess salt, and toxic metals, as well as biotic stresses like diseases and pest attacks, can hinder plant growth and development. Small millets are adapted to arid, semi-arid, marginal, and dry regions, and they are excellent sources of stress tolerance.

Blast (*Pyricularia setariae*), rust (Uromyces setariae-italiae), downy mildew (*Sclerospora graminicola*), and smut (*Ustilago crameri*) are the most significant diseases of foxtail millet (Dwivedi et al., 2012). ICRISAT investigated foxtail millet core collection accessions for blast resistance in the field and greenhouse under artificial inoculation circumstances, finding 21 accessions resistant to neck and head blast in the field and 11 accessions resistant to seedling leaf blast in the greenhouse. Following further testing against four isolates of the blast pathogen, 16 accessions were found to be resistant to at least one isolate for leaf, sheath, neck, and head blast, and two accessions (ISe 1181 and ISe 1547) were found to be free of head blast infection and resistant to neck, leaf, and sheath blast against four isolates (Sharma et al., 2014).

In comparison to other grains, foxtail millet is a drought-tolerant crop. Agronomic traits like grain weight per panicle, panicle weight, plant height, length of rachis, and 1000-grain weight, as well as physiological parameters like relative chlorophyll, soluble protein, malondialdehyde (MDA), and superoxide dismutase (SOD), were found to have significant correlations with the drought-resistant index (DRI) under drought conditions (Zhang et al., 2012). Seedling survival following repeated drought stress was used to classify a total of 17,799 foxtail millet accessions, including 486 elite cultivars and 17,313 landraces, into five grades of drought tolerance, with grade 1 accessions being the most drought tolerant and containing the most elite cultivars. Lata et al. investigated transcriptome alterations in the tolerant cultivar "Prasad" after 0.5 and 6 hours of dehydration stress, discovering that there are two different waves of transcript modifications in foxtail millet at the early and late phases of stress treatment (2010). After dehydration stress, expression analysis of nine up-regulated transcripts (around 2.5-fold changes) between drought-tolerant and susceptible cultivars reveals that they play a role in dehydration adaptation and tolerance in foxtail millet. Using mannitol or polyethylene glycol (PEG-6000) assays, a quick and easy screening technique for foxtail millet drought tolerance was developed, with relative water content and germination rate under osmotic stress suggested as indicators of drought resistance at the seedling stage (Zhang et al., 2005; Zhu et al., 2008). Lata et al. screened 107 cultivars using lipid peroxidation as a biochemical marker to determine membrane integrity under stress and classified the genotypes as very tolerant, tolerant, sensitive, and highly sensitive (2011a).

Krishnamurthy et al. used pot culture to screen the foxtail millet core collection for salinity tolerance, which indicated a wide range of tolerance and discovered salinity-tolerant accessions (2014). Sreenivasulu et al. used a 711 cDNA insert macro array filter representing 620 unigenes of a barley EST collection to isolate differentially expressed ESTs in the salt-tolerant cultivar Prasad and salt-sensitive cultivars cv. Lepakshi, discovering that at high salinity, the transcripts of H_2O_2 scavenging enzymes, as well as those of some cellular metabolism genes, were particularly induced in the tolerant cultivar (2004). Proline accumulation (increased GS and P5C reductase activities) correlated with salinity tolerance in foxtail millet, as suggested by Veeranagamallaiah et al. (2007). Tian et al. proposed the lodging coefficient as a viable indicator for field selection for lodging resistance (2010). China has reported landraces and improved cultivars of foxtail millet that are resistant to lodging. In finger millet, prolonged waterlogging reduced the total root number but increased it in foxtail millet (Kono et al., 1988). Under waterlogging, Zegada-Lizarazu and Iijima (2005) found a significant reduction in water use efficiency (WUE) in these crops, while under drought, significant reductions in shoot dry weight and leaf area were seen. Jinan 8337 (Zhang and Guan, 1995) and Lugu 7 (Chen and Qi, 1993) are a few varieties reported as tolerant to waterlogging in foxtail millet.

Culm strength is one of the major focuses of researchers in kodo millet because of the crop's high susceptibility to lodging during maturity. The association between culm lignin content and lodging resistance is well established in different crops (Ma, 2010; Ookawa and Ishihara, 1993). Sreeja et al. (2016) confirmed the association of the *Flexible Culm 1 (FC1)* gene, identified from the *fc1* mutant in rice, with low lignin content in the Adari and Aamo 10 genotypes of kodo millet through differential expression studies. Studies based on RNA-seq have been conducted to identify stress-response genes in kodo millet. Characterization of downregulated ABA-responsive genes such as TY_DN18151_c0_g1 and TRINITY_DN12488_c0_g1, as well as transcription factors such as bZIP12 and bZIP46, was observed in the dehydration-susceptible kodo millet cultivar "CO3." This represents a potential avenue for developing climate-resilient cultivars in kodo millet and other crops through gene transfer (Suresh et al., 2022). It can also contribute to the understanding of stress-tolerance mechanisms.

Foxtail millet can tolerate drought and intense heat stress or escape these stresses through early maturity (Baltensperger, 2002). However, water stress influences the crop characteristics at different growth stages. Drought and salinity tolerance are important for the success of the crop in different agroecosystems. Seghatoleslami et al. studied the responses of proso millet to drought stress in different growth stages using five proso millet genotypes (K-C-M.6, K-C-M.4, K-4C-M.2, K-C-M.9, and a local landrace) under five irrigation treatments. The treatments included drought at vegetative, ear emergence, seed filling, and control (irrigated) stages (2008). A great reduction in grain yield was recorded at the ear emergence stage due to increased floret death and reduced seed size in drought stress (Matsuura et al., 2012; Seghatoleslami et al., 2008). A comparison of genotypes indicated that two genotypes, K-C-M.9 and K-C-M.4, had the highest WUE and relatively higher harvest indices; they can be recommended in drought-prone regions (Seghatoleslami et al., 2008). Thus, the direct selection of the best-yielding drought-tolerant accessions for each agroecological region will be the best criterion when considering the present breeding activities in the crop. Framing selection indices using water stress-related traits, including biochemical activities of enzymes such as superoxide dismutase (SOD) and peroxidase (POD), which are antioxidative, can yield the identification of accessions showing a high degree of drought tolerance due to high osmoregulation during stress in proso millet (Karyudi and Fletcher, 2002; Zhang et al., 2012). The studies on the genetics of drought stress effects in the crop suggest the S-adenosylmethionine synthase gene as one of the crucial genes whose expression declines in drought conditions and increases in rehydration. Besides this, 32 ESTs associated with abiotic stress response were also identified through the suppressive subtraction hybridization technique (Lin et al., 2006).

Like other crops, soil salinity induces osmotic and ionic stresses, affecting the growth and development of proso millet. As a mitigation effort for salinity stress-induced effects, exogenous application of selenium was found to improve the fresh weight, shoot and root length, and biomass accumulation of proso millet plants under salinity stress (Rasool et al., 2023). Selenium enhanced the activity of antioxidant enzymes like superoxide dismutase and catalase and improved the levels of osmolytes such as proline, glycine betaine, and total soluble sugar. Hence, selenium in low concentration was found to improve the defense mechanisms of the crop by boosting the antioxidant machinery and regulating Na+ transporters, thereby mitigating the adverse effects of NaCl stress (Mushtaq et al., 2023; Rasool et al., 2023).

The primary diseases of proso millet are of fungal origin, mainly head smut (*Sporisorium destruens*) and kernel smut (*Ustilago crameri*). The bacterial stripe disease (*Pseudomonas avenae*) also adds to the damage. Konstantinov et al. conducted breeding of proso millet for resistance to head smut (*Sporisorium destruens*; Syn. *Sphacelotheca destruens*) and melanosis (*Pseudomonas syringae* and *Xanthomonas campestris* pv. *holcicola*) and developed the smut-resistant variety Khar'kovskoe 86 (Kh86) and Kh22, with combined resistance to smut and melanosis (1991). Mutation breeding for smut resistance identified 79-6554 as the most resistant (1.8% infection) line. Considerable diversity is reported in *Sporisorium destruens* isolates affecting various cultivars, with three pathotypes and 10 differential hosts for pathotype identification in proso millet (Zhou et al., 2016). The genetic and physiological response of crop plants to the disease pathogen is now well understood (Jin et al., 2020; Jin et al., 2023). Blast disease is another fungal disease affecting all growing stages. Subedi et al. assessed 14 proso millet genotypes against blast disease caused by *Pyricularia grisea* under natural epiphytotic conditions (2022). Although the experiment reported that most of the genotypes showed moderately susceptible and susceptible reactions to the leaf blast disease, genotype C04654 was found to be resistant with the highest yield of 2.6 t/ha in blast incidence. Humla-239 and C04651 genotypes were found to be moderately susceptible, with 2.5 and 2.4 t/ha of yield, respectively. Considering pests, shoot fly (*Atherigona pulla*) is also prominent in most of the crop-growing belt (Lyon et al., 2008). The monsoon crop is mostly affected by the pest. Resistant cultivars for insect pest infestation are seldom identified in proso millet; however, the cultivar Huntsman was identified to be resistant to the Russian wheat aphid (Baltensperger et al., 1995)

Being a hardy crop, barnyard millet can withstand a range of biotic and abiotic stresses with minimum yield losses. However, the identification of the most resistant lines and the variability in various stresses welcomes new benefits in cultivation. The crop species usually have higher WUE and dry matter production compared to other millets (Zegada-Lizarazu and Iijima, 2005). Similarly, salinity-tolerant accessions, CO(Kv)2, TNEF204, and MDU 1 were identified (Arthi et al., 2019). Nevertheless, the characterization of genes related to abiotic stress resistance is still to be reported to exploit for introgression in popular high-yielding lines.

E. frumentacea is usually vulnerable to many biotic stresses at different stages during crop growth (Jain et al., 1997; Jagadish et al., 2008). Major biotic stresses in barnyard millet are grain smut, loose smut, sheath blight, shoot fly, stem borer, and aphid. Grain smut caused by *Ustilago panici-frumentacei* Brefeld. is the most important disease and the main productivity constraint in barnyard millet, resulting in up to 60.8% yield reduction (Jain et al., 1997). The resistant cultivar for grain smut has not been reported so far in *E. frumentacea*, whereas *E. esculenta* confers strong resistance against the fungus (Joshi et al., 2021). However, hybridization barriers between the two species make it difficult to transfer the resistance from *E. esculenta* to *E. frumentacea* (Sood et al., 2014). Hence, *E. colona*, the wild progenitor of Indian barnyard millet, the cross-compatible wild species, could be used as an alternative source of grain smut resistance. Joshi et al. investigated grain smut disease in 49 Indian landraces of barnyard millet and reported the presence of resistance in some lines among them, viz., RBM57, RBM77, RBM78, RBM81, and RBM82 (2015). Two accessions of barnyard millet from species *E. esculenta*, PRB 9402 and PRB 9602, were observed to be immune to grain smut after phenotypic screening (AICSMIP, 2001). Loose smut caused by the fungus *Ustilago tritici* is another main disease affecting grain development in both cultivated species (Gupta et al., 2010; Jain et al., 1997). Subsequently, barnyard accessions resistant to both grain smut and loose smut were identified (Gupta et al., 2010; Nagaraj and Mantur, 2008). Shoot fly (*Atherigona falcata*) and pink stem borer (*Sesamia inferens*) are the major pests damaging the crop. Rawat et al. identified resistant genotypes DHBM996 and TNEF204 against shoot fly and stem borer (2019).

Drought, salt, and high temperatures are among the abiotic stress conditions in which little millet thrives. Genotypes like RLM 37 and OLM 20 are identified as sources of drought stress tolerance by analyzing the adaptive mechanisms, such as the accumulation of leaf proline content (Patil et al., 2018). The genes for amino-transferase, thionin, aquaporin, synaptotagmin, CDPK, cytho protein, TaNAC-2, EcNAC-67, NAM, U2-SnRNP, HvNAC, OsNAC-29-2, MPK17-1, DQP1, DQP2, DQP3, DQP4, and DQP6 are identified as key genes for moisture stress in little millet due to the up-regulation in tolerant genotypes like RLM-37, MM-23, MM-10, BL-4, BL-8, and BL-15-1 (Sushmitha et al., 2018). The role of polyamine coding genes in drought stress tolerance was established in transcriptome sequencing of the drought-subjected cultivar OLM 203 or "Tarini" (Narayanrao et al., 2023).

4.6 CONCLUSION

After decades of research focusing solely on the major staples, plant breeders are emphasizing the genetic improvement of neglected and underutilized crops. Minor or small millets are one such category. The national and international stakeholders have been encouraged to promote the research, use, and consumption of millets since the declaration of International Millet Year. The nutraceutical properties of the crops necessitate studies on the critical genes and metabolites providing these properties. However, the basic need of this hour is to promote the cultivation of small millets that are necessary to exploit their uses and meet the demands. Crop improvement research is essential to develop ideal varieties to tailor to various emerging consumer needs and provide substantial income to farmers. The pre-breeding germplasm conservation-evaluation studies in each small millet are very elaborate and depict the variability for various traits for the breeding programs to begin with. However, many undiscovered variations exist for several traits and breeding objectives. Accelerated genetic gains are of immediate need to cater to the upcoming demand for climate resilience. However, to provide molecular breeding assistance, the omics of various crops are still in their infancy, except for foxtail millet, pertaining to its small genome size and ploidy, which are similar to rice. Although whole genome sequences are available for foxtail millet and proso millet, and complete chloroplast genome sequences are available for foxtail millet, proso millet, barnyard millet, and little millet, there are several milestones to complete for the effective utilization of these valuable data. Hence, most of the marker development and mapping works so far have focused on finger millet and foxtail millet, where the omics data are much more developed. More efforts should be directed to reveal the genomic structure of kodo, barnyard, and little millet using NGS techniques, and the implementation of molecular breeding and genome editing techniques is essential to promote small millet as a dependable source for achieving nutritional security and climate resilience. The allopolyploid nature of small millets, except for foxtail millet, hints at their tolerance to various abiotic stresses; however, the extent of variation persisting in each trait of stress tolerance has not been studied and exploited until now. Although most of the breeding efforts are diverted towards foxtail millet, it is also lacking in several attributes. This comprehensive report summarizes that interdisciplinary and multidisciplinary efforts are essential for these small orphan crops to exploit their potential for climate change and to start a "New Green Revolution."

KEYWORDS

- **allopolyploidy**
- **artificial pollination**
- **DNA profiling**
- **Echinochloa crusgalli**
- **gene-based markers**
- **genome sequencing**
- **minor millets**
- **molecular breeding**

REFERENCES

Amadou, I., Amza, T., Shi, Y. H., & Le, G. W. (2011). Chemical analysis and antioxidant properties of foxtail millet bran extracts. *Songklanakarin Journal of Science and Technology*, *33*, 509–515. https://www.thaiscience.info/journals/Article/SONG/10773514.pdf (accessed on 7th November 2025).

Araki, M., Numaoka, A., Kawase, M., & Fukunaga, K. (2012). Origin of waxy common millet, *Panicum miliaceum* L. in Japan. *Genetic Resources and Crop Evolution*, *59*(7), 1303–1308. https://doi.org/10.1007/s10722-011-9755-9 (accessed on 7th November 2025).

Arthi, N., Rajagopal, B., Geethanjali, S., Nirmalakumari, A., & Senthil, N. (2019). Screening of barnyard millet (*Echinochloa frumentacea*) germplasm for salinity tolerance. *Electronic Journal of Plant Breeding*, *10*(2), 659. https://doi.org/10.5958/0975-928x.2019.00083.8 (accessed on 7th November 2025).

Arunachalam, V., Rengalakshmi, R., & Kubera Raj, M. S. (2005). Ecological stability of genetic diversity among landraces of little millet (*Panicum sumatrense*) in South India. *Genetic Resources and Crop Evolution*, *52*(1), 15–19. https://doi.org/10.1007/s10722-005-6693-4 (accessed on 7th November 2025).

Babu, B. K., & Chauhan, R. (2017). *In-silico* identification of EST based microsatellite markers and SNPs, and comparative genomic analysis of ESTs in barnyard millet for their omics applications. *Current Agriculture Research Journal*, *5*(3), 279–287. https://doi.org/10.12944/carj.5.3.03 (accessed on 7th November 2025).

Baltensperger, D. D., Lyon, D. J., Anderson, R., Holman, T., Stymieste, C., Shanahan, J., Nelson, L. A., DeBoer, K. L., Hein, G. L., & Krall, J. (1995). *Producing and Marketing Proso Millet in the High Plains, EC195-137*. University of Nebraska, Lincoln, NE.

Ban, Y., Kokuba, T., & Miyaji, Y. (1971). Production of haploid plant by anther culture of *Setaria italica*. *Bulletin of the Faculty of Agriculture, Kagoshima University*, *21*, 7.

Bennetzen, J. L., Schmutz, J., Wang, H., Percifield, R., Hawkins, J., Pontaroli, A. C., et al. (2012). Reference genome sequence of the model plant *Setaria*. *Nature Biotechnology*, *30*(6), 555–561. https://doi.org/10.1038/nbt.2196 (accessed on 7th November 2025).

Bhatt, R., Asopa, P. P., Jain, R., Kothari-Chajer, A., Kothari, S. L., & Kachhwaha, S. (2021). Optimization of *Agrobacterium* mediated genetic transformation in *Paspalum scrobiculatum* L. (Kodo millet). *Agronomy*, *11*(6), 1104. https://doi.org/10.3390/agronomy11061104 (accessed on 7th November 2025).

Calamai, A., Masoni, A., Marini, L., Dell'acqua, M., Ganugi, P., Boukail, S., Benedettelli, S., & Palchetti, E. (2020). Evaluation of the agronomic traits of 80 accessions of proso millet (*Panicum miliaceum* L.) under Mediterranean pedoclimatic conditions. *Agriculture*, *10*(12), 578. https://doi.org/10.3390/agriculture10120578 (accessed on 7th November 2025).

Ceasar, S. A., & Ignacimuthu, S. (2009). Genetic engineering of millets: Current status and future prospects. *Biotechnology Letters*, *31*(6), 779–788. https://doi.org/10.1007/s10529-009-9933-4 (accessed on 7th November 2025).

Chen, J., & Qi, Y. (1993). Recent developments in foxtail millet cultivation and research in China. In K. W. Riley, S. C. Gupta, A. Seetharam, & J. N. Mushonga (Eds.), *Advances in Small Millets* (pp. 101–107). Oxford and IBH Publishing Co: New Delhi.

Chen, X., Zale, J., & Chen, F. (2011). The regeneration and transformation of foxtail millet (*Setaria italica*), a model biofuel crop. In *Joint USDA-DOE Plant Feedstock Genomics for Bioenergy Awardee Meeting*, Crystal City, Virginia, USA, April 10–13, 2011 (p. 278). USDA & USDOE: Washington, D.C.

Cho, Y. I., Chung, J. W., Lee, G. A., Ma, K. H., Dixit, A., Gwag, J. G., & Park, Y. J. (2010). Development and characterization of twenty-five new polymorphic microsatellite markers in proso millet (*Panicum miliaceum* L.). *Genes & Genomics*, *32*(3), 267–273. https://doi.org/10.1007/s13258-010-0007-8 (accessed on 7th November 2025).

Darmency, H., & Pernes, J. (1985). Use of wild *Setaria viridis* (L.) Beauv. to improve triazine resistance in cultivated *S. italica* (L.) by hybridization. *Weed Research*, *25*(3), 175–179. https://doi.org/10.1111/j.1365-3180.1985.tb00633.x (accessed on 7th November 2025).

Darmency, H., Picard, J. C., & Wang, T. (2011). Fitness costs linked to dinitroaniline resistance mutation in *Setaria*. *Heredity*, *107*(1), 80–86. https://doi.org/10.1038/hdy.2010.169 (accessed on 7th November 2025).

de Wet, J. M. J., Brink, D. E., Rao, K. E. P., & Mengesha, M. H. (1983). Diversity in kodo millet, *Paspalum scrobiculatum*. *Economic Botany*, *37*(2), 159–163. https://doi.org/10.1007/bf02858779 (accessed on 7th November 2025).

Deepika, C. (2021). Genetic enhancement of grain yield and contributing traits in kodo millet (*Paspalum scrobiculatum* L.). *Annual Report*. Indian Institute of Millet Research: Hyderabad, India.

Desai, H., Hamid, R., Ghorbanzadeh, Z., Bhut, N., Padhiyar, S. M., Kheni, J., & Tomar, R. S. (2021). Genic microsatellite marker characterization and development in little millet (*Panicum sumatrense*) using transcriptome sequencing. *Scientific Reports*, *11*(1). https://doi.org/10.1038/s41598-021-00100-4 (accessed on 7th November 2025).

Deshpande, S. S., Mohapatra, D., Tripathi, M. K., & Sadvatha, R. H. (2015). Kodo millet–nutritional value and utilization in Indian foods. *Journal of Grain Processing and Storage*, *2*(2), 16–23.

Devos, K. M., Wang, Z. M., Beales, J., Sasaki, T., & Gale, M. D. (1998). Comparative genetic maps of foxtail millet (*Setaria italica*) and rice (*Oryza sativa*). *Theoretical and Applied Genetics*, *96*(1), 63–68. https://doi.org/10.1007/s001220050709 (accessed on 7th November 2025).

Doggett, H. (1989). Small millets–a selective overview. In A. Seetharam, K. W. Riley, & G. Harinarayana (Eds.), *Small Millets in Global Agriculture* (pp. 3–18). Oxford and IBH Publications: New Delhi.

Dong, Y., & Duan, S. (2000). Production of transgenic millet plants via particle bombardment. *Acta Botanica Boreali-Occidentalia Sinica*, *20*, 175–178.

Doust, A. N., Devos, K. M., Gadberry, M. D., Gale, M. D., & Kellogg, E. A. (2004). Genetic control of branching in foxtail millet. *Proceedings of the National Academy of Sciences of the United States of America*, *101*(24), 9045–9050. https://doi.org/10.1073/pnas.0402892101 (accessed on 7th November 2025).

Doust, A. N., Kellogg, E. A., Devos, K. M., & Bennetzen, J. L. (2009). Foxtail millet: A sequence-driven grass model system. *Plant Physiology*, *149*(1), 137–141. https://doi.org/10.1104/pp.108.129627 (accessed on 7th November 2025).

Dwivedi, S. L., Upadhyaya, H. D., Senthilvel, S., Hash, C. T., Fukunaga, K., Diao, X., Santra, D., Baltensperger, D., & Prasad, M. (2012). Millets: Genetic and genomic resources. *Plant Breeding Reviews*, *35*, 247–375.

Fang, F. Q., Qian, Z., Ao, G. M., & Yu, J. J. (2008). Co-suppression of Si401, a maize pollen specific *Zm401* homologous gene, results in aberrant anther development in foxtail millet. *Euphotic*, *163*, 103–111.

Fang, X., Dong, K., Wang, X., Liu, T., He, J., Ren, R., Zhang, L., Liu, R., Liu, X., Li, M., Huang, M., Zhang, Z., & Yang, T. (2016). A high-density genetic map and QTL for agronomic and yield traits in foxtail millet [*Setaria italica* (L.) P. Beauv.]. *BMC Genomics*, *17*(1). https://doi.org/10.1186/s12864-016-2628-z (accessed on 7th November 2025).

Francis, N., Rajasekaran, R., Rajagopalan, V. R., Bakya, S. V., Muthurajan, R., Kumar, A. G., Alagarswamy, S., Krishnamoorthy, I., & Thiyagarajan, C. (2023). Molecular characterization and SNP identification using genotyping-by-sequencing in high-yielding mutants of proso millet. *Frontiers in Plant Science*, *14*. https://doi.org/10.3389/fpls.2023.1108203 (accessed on 7th November 2025).

Ganapathy, K. N. (2017). Genetic improvement in little millet. In J. V. Patil (Ed.), *Millets and Sorghum: Biology and Genetic Improvement* (pp. 170–183). John Wiley & Sons: New York. https://doi.org/10.1002/9781119130765.ch6 (accessed on 7th November 2025).

Gomashe, S. S. (2017). Proso millet, *Panicum miliaceum* (L.): Genetic improvement and research needs. In J. V. Patil (Ed.), *Millets and Sorghum: Biology and Genetic Improvement* (pp. 150–169). John Wiley & Sons: New York. https://doi.org/10.1002/9781119130765.ch5 (accessed on 7th November 2025).

Gowda, J., Bharathi, S., Somu, G. G., Krishnappa, M. M., & Rekha, D. (2009). Formation of core set in barnyard millet [*Echinochloa frumentacea* (Roxb.) Link] germplasm using data on twenty-four morpho-agronomic traits. *International Journal of Plant Sciences*, *4*(1), 1–5.

Grass Phylogeny Working Group II. (2012). New grass phylogeny resolves deep evolutionary relationships and discovers C4 origins. *New Phytologist*, *193*(2), 304–312. https://doi.org/10.1111/j.1469-8137.2011.03972.x (accessed on 7th November 2025).

Guo, L., Qiu, J., Ye, C., Jin, G., Mao, L., Zhang, H., et al. (2017). *Echinochloa crus-galli* genome analysis provides insight into its adaptation and invasiveness as a weed. *Nature Communications*, *8*(1). https://doi.org/10.1038/s41467-017-01067-5 (accessed on 7th November 2025).

Guo, S., Chai, S., Guo, Y., Shi, X., Han, F., Qu, T., Xing, L., Yang, Q., Gao, J., Gao, X., Feng, B., Song, H., & Yang, P. (2023). Mapping of major QTL and candidate gene analysis for hull colour in foxtail millet (*Setaria italica* (L.) P. Beauv.). *BMC Genomics*, *24*(1). https://doi.org/10.1186/s12864-023-09517-9 (accessed on 7th November 2025).

Gupta, A., Joshi, D., Mahajan, V., & Gupta, H. S. (2010). Screening barnyard millet germplasm against grain smut (*Ustilago panici-frumentacei* Brefeld). *Plant Genetic Resources*, *8*(1), 52–54. https://doi.org/10.1017/s1479262109990141 (accessed on 7th November 2025).

Gupta, A., Mahajan, V., Kumar, M., & Gupta, H. S. (2009). Biodiversity in the barnyard millet (*Echinochloa frumentacea* Link, Poaceae) germplasm in India. *Genetic Resources and Crop Evolution*, *56*(6), 883–889. https://doi.org/10.1007/s10722-009-9462-y (accessed on 7th November 2025).

Gupta, A., Sood, S., Agrawal, P. K., & Bhatt, J. C. (2015). B 29: An easy dehulling barnyard millet (*Echinochloa frumentacea* Link) genotype. *National Academy Science Letters*, *38*(1), 21–24. https://doi.org/10.1007/s40009-014-0295-7 (accessed on 7th November 2025).

Habiyaremye, C., Matanguihan, J. B., D'Alpoim Guedes, J., Ganjyal, G. M., Whiteman, M. R., Kidwell, K. K., & Murphy, K. M. (2017). Proso millet (*Panicum miliaceum* L.) and its potential for cultivation in the Pacific Northwest, U.S.A review. *Frontiers in Plant Science*, *7*. https://doi.org/10.3389/fpls.2016.01961 (accessed on 7th November 2025).

Hilu, K. W. (1994). Evidence from RAPD markers in the evolution of *Echinochloa* millets (Poaceae). *Österreichische Botanische Zeitschrift*, *189*(3–4), 247–257. https://doi.org/10.1007/bf00939730 (accessed on 7th November 2025).

Hiremath, S. C., Patil, G. N. V., & Salimath, S. S. (1990). Genome homology and origin of *Panicum sumatrense* (Gramineae). *Cytologic (Tokyo)*, *55*(2), 315–319. https://doi.org/10.1508/cytologia.55.315 (accessed on 7th November 2025).

Hoshino, T., Nakamura, T., Seimiya, Y., Kamada, T., Ishikawa, G., Ogasawara, A., Sagawa, S., Satto, M., Shimizu, H., Nishi, M., Watanabe, M., Takeda, J., & Takahata, Y. (2010). Production of a full waxy line and analysis of waxy genes in the allohexaploid crop, Japanese barnyard millet. *Plant Breeding*, *129*, 349–355. https://doi.org/10.1111/j.1439-0523.2009.01668.x (accessed on 7th November 2025).

Hu, X., Wang, J., Lu, P., & Zhang, H. (2009). Assessment of genetic diversity in broomcorn millet (*Panicum miliaceum* L.) using SSR markers. *Journal of Genetics and Genomics*, *36*(8), 491–500. https://doi.org/10.1016/s1673-8527(08)60139-3 (accessed on 7th November 2025).

Hu, Y. G., Zhu, J., Liu, F., Zhang, Z., Chai, Y., & Weining, S. (2008). Genetic diversity among Chinese landraces and cultivars of broomcorn millet (*Panicum miliaceum*) revealed by the polymerase chain reaction. *Annals of Applied Biology*, *153*(3), 357–364. https://doi.org/10.1111/j.1744-7348.2008.00263.x (accessed on 7th November 2025).

Hunt, H. V., Campana, M. G., Lawes, M. C., Park, Y. J., Bower, M. A., Howe, C. J., & Jones, M. K. (2011). Genetic diversity and phylogeography of broomcorn millet (*Panicum miliaceum* L.) across Eurasia. *Molecular Ecology*, *20*(22), 4756–4771. https://doi.org/10.1111/j.1365-294x.2011.05318.x (accessed on 7th November 2025).

Hunt, H. V., Denyer, K., Packman, L. C., Jones, M. K., & Howe, C. J. (2010). Molecular basis of the waxy endosperm starch phenotype in broomcorn millet (*Panicum miliaceum* L.). *Molecular Biology and Evolution* *27*(7), 1478–1494. https://doi.org/10.1093/molbev/msq040 (accessed on 7th November 2025).

Hunt, H. V., Moots, H. M., Graybosch, R. A., Jones, H., Parker, M., Romanova, O., Jones, M. K., Howe, C. J., & Trafford, K. (2013). Waxy phenotype evolution in the allotetraploid cereal broomcorn millet: Mutations at the *GBSSI* locus in their functional and phylogenetic context. *Molecular Biology and Evolution*, *30*(1), 109–122. https://doi.org/10.1093/molbev/mss209 (accessed on 7th November 2025).

Ishikawa, G., Seimiya, Y., Saito, M., Nakamura, T., & Hoshino, T. (2013). Molecular characterization of spontaneous and induced mutations in the three homoeologous waxy genes of Japanese barnyard millet [*Echinochloa esculenta* (A. Braun) H. Scholz]. *Molecular Breeding*, *31*(1), 69–78. https://doi.org/10.1007/s11032-012-9769-9 (accessed on 7th November 2025).

Jagadish, P. S., Mohapatra, H. K., Chakravarthy, M. K., Srivastava, N., & Nangia, N. (2008). *A Compendium of Insect Pests of Finger Millet and Other Small Millets*. All India Coordinated Small Millets Improvement Project.

Jain, A. K., Jain, S. K., & Yadava, H. S. (1997). Assessment of yield losses due to grain smut in barnyard millet. *Indian Phytopathology, 50*, 49–52.

Jia, G., Huang, X., Zhi, H., Zhao, Y., Zhao, Q., Li, W., et al. (2013). A haplotype map of genomic variations and genome-wide association studies of agronomic traits in foxtail millet (*Setaria italica*). *Nature Genetics, 45*(8), 957–961. https://doi.org/10.1038/ng.2673 (accessed on 7th November 2025).

Jin, F., Liu, J., Wu, E., Yang, P., Gao, J., Gao, X., & Feng, B. (2020). Leaf transcriptome analysis of broomcorn millet uncovers key genes and pathways in response to *Sporisorium destruens*. *International Journal of Molecular Sciences, 22*(17), 9542. https://doi.org/10.3390/ijms22179542 (accessed on 7th November 2025).

Jin, F., Xu, L., Lv, Z., Zhang, Y., Yang, Q., Han, Q., & Feng, B. (2023). Combined physiological and pathway analysis revealed the effect of *Sporisorium destruens* on photosynthesis in broomcorn millet (*Panicum miliaceum* L.). *Journal of Integrative Agriculture*. https://doi.org/10.1016/j.jia.2023.12.027 (accessed on 7th November 2025).

Johnson, M., Deshpande, S., Vetriventhan, M., Upadhyaya, H. D., & Wallace, J. G. (2019). Genome-wide population structure analyses of three minor millets: Kodo millet, little millet, and proso millet. *The Plant Genome, 12*(3). https://doi.org/10.3835/plantgenome2019.03.0021 (accessed on 7th November 2025).

Joshi, D. C., Meena, R. P., & Chandora, R. (2021). Genetic resources: Collection, characterization, conservation, and documentation. In M. Singh & S. Sood (Eds.), *Millets and Pseudo Cereals* (pp. 19–31). Elsevier.

Joshi, R. P., Jain, A. K., Chauhan, S. S., & Singh, G. (2015). Characterization of barnyard millet (*Echinochloa frumentacea* (Roxb.) Link.) landraces for agro-morphological traits and disease resistance. *Electronic Journal of Plant Breeding, 6*(4), 888–898.

Karam, D., Westra, P., Nissen, S. J., Ward, S. M., & Figueiredo, J. E. F. (2004). Genetic diversity among proso millet (*Panicum miliaceum*) biotypes assessed by AFLP technique. *Planta Daninha, 22*(2), 167–174. https://doi.org/10.1590/s0100-83582004000200001 (accessed on 7th November 2025).

Karyudi, & Fletcher, R. J. (2002). Osmoregulation in birdseed millet under conditions of water stress. I. Variation in *Setaria italica* and *Panicum miliaceum*. *Euphytica, 125*(3), 337–348. https://doi.org/10.1023/a:1016073910886 (accessed on 7th November 2025).

Kono, Y., Yamauchi, A., Nonoyama, T., & Tatsumi, J. (1988). Comparison of growth responses to waterlogging of summer cereals with special reference to rooting ability. *Japanese Journal of Crop Science, 57*(2), 321–331.

Konstantinov, S. I., Linnik, V. M., Ya Shapina, L., & Grigorashchenko, L. V. (1991). *Breeding Proso Millet for Resistance to Diseases* (pp. 112–117). Ukrainian Academy of Sciences: Kiev, Ukraine.

Krishnamurthy, L., Upadhyaya, H. D., Gowda, C. L. L., Kashiwagi, J., Purushothaman, R., Singh, S., & Vadez, V. (2014). Large variation for salinity tolerance in the core collection of foxtail millet (*Setaria italica* (L.) P. Beauv.) germplasm. *Crop and Pasture Science, 65*(4), 353–353. https://doi.org/10.1071/cp13282 (accessed on 7th November 2025).

Kubešová, M., Moravcová, L., Suda, J., Jarošík, V., & Pyšek, P. (2010). Naturalized plants have smaller genomes than their non-invading relatives: A flow cytometric analysis of the Czech alien flora. *Preslia, 82*, 81–96.

Kumari, K., Muthamilarasan, M., Misra, G., Gupta, S., & Alagesan, S. (2013). Development of eSSR markers in *Setaria italica* and their applicability in studying genetic diversity, cross-transferability and comparative mapping in millet and non-millet species. *PLoS ONE, 8*. https://doi.org/10.1371/journal.pone.0067742 (accessed on 7th November 2025).

Kumari, S., & Singh, S. K. (2015). Assessment of genetic diversity in promising finger millet [*Eleusine coracana* (L.) Gaertn] genotypes. *The Bioscan, 10*(2), 825–830.

Lágler, R., Gyulai, G., Humphreys, M., Szabó, Z., Horváth, L., Bittsánszky, A., Kiss, J., Holly, L., & Heszky, L. (2005). Morphological and molecular analysis of common millet (*Panicum miliaceum*) cultivars compared to an aDNA sample from the 15th century (Hungary). *Euphotic, 146*(1–2), 77–85. https://doi.org/10.1007/s10681-005-5814-7 (accessed on 7th November 2025).

Lata, C., Bhutty, S., Bahadur, R. P., Majee, M., & Prasad, M. (2011b). Association of an SNP in a novel DREB2-like gene *SiDREB2* with stress tolerance in foxtail millet [*Setaria italica* (L.)]. *Journal of Experimental Botany, 62*(10), 3387–3401. https://doi.org/10.1093/jxb/err016 (accessed on 7th November 2025).

Lata, C., Gupta, S., & Prasad, M. (2013). Foxtail millet: A model crop for genetic and genomic studies in bioenergy grasses. *Critical Reviews in Biotechnology*, *33*(3), 328–343. https://doi.org/10.3109/07388551.2012.716809 (accessed on 7th November 2025).

Lata, C., Jha, S., Dixit, V., Sreenivasulu, N., & Prasad, M. (2011a). Differential antioxidative responses to dehydration-induced oxidative stress in core set of foxtail millet cultivars [*Setaria italica* (L.)]. *Protoplasm*, *248*(4), 817–828. https://doi.org/10.1007/s00709-010-0257-y (accessed on 7th November 2025).

Lata, C., Sahu, P. P., & Prasad, M. (2010). Comparative transcriptome analysis of differentially expressed genes in foxtail millet (*Setaria italica* L.) during dehydration stress. *Biochemical and Biophysical Research Communications*, *393*(4), 720–727. https://doi.org/10.1016/j.bbrc.2010.02.068 (accessed on 7th November 2025).

Lee, J., Park, K., Lee, I., Kim, C., Kown, O., & Park, T. (2014). Simple sequence repeats analysis of genetic diversity among acetyl-CoA carboxylase inhibitor-resistant and -susceptible *Echinochloa crus-galli* and *E. oryzicola* populations in Korea. *Weed Research*, *55*, 90–100.

Li, X., Gao, J., Song, J., Guo, K., Hou, S., Wang, X., et al. (2022). Multi-omics analyses of 398 foxtail millet accessions reveal genomic regions associated with domestication, metabolite traits, and anti-inflammatory effects. *Molecular Plant*, *15*(8), 1367–1383. https://doi.org/10.1016/j.molp.2022.07.003 (accessed on 7th November 2025).

Liang, S., Yang, G., & Ma, Y. (2010). Chemical characteristics and fatty acid profile of foxtail millet bran oil. *Journal of the American Oil Chemists' Society*, *87*(1), 63–67. https://doi.org/10.1007/s11746-009-1475-3 (accessed on 7th November 2025).

Lin, F., Hu, Y., Song, G., Zhang, H., Liu, T., & He, B. (2006). Isolation and analysis of genes induced by rehydration after serious drought in broomcorn millet (*Panicum miliaceum* L.) by SSH. *Chinese Journal of Agricultural Biotechnology*, *3*(3), 237–242. https://doi.org/10.1079/cjb2006119 (accessed on 7th November 2025).

Lin, H. S., Chiang, C. Y., Chang, S. B., & Kuoh, C. S. (2011). Development of simple sequence repeats (SSR) markers in *Setaria italica* (Poaceae) and cross amplification in related species. *International Journal of Molecular Sciences*, *12*, 7835–7845. https://doi.org/10.3390/ijms12117835 (accessed on 7th November 2025).

Liu, M., Xu, Y., He, J., Zhang, S., Wang, Y., & Lu, P. (2016). Genetic diversity and population structure of broomcorn millet (*Panicum miliaceum* L.) cultivars and landraces in China based on microsatellite markers. *International Journal of Molecular Sciences*, *17*(3), 370. https://doi.org/10.3390/ijms17030370 (accessed on 7th November 2025).

Liu, R., Singh, V., Abugho, S., Lin, H. S., Zhou, X. G., & Bagavathiannan, M. (2022). Morphophysiological diversity and its association with herbicide resistance in *Echinochloa* ecotypes. *Weed Science*, *70*(1), 26–35. https://doi.org/10.1017/wsc.2021.64 (accessed on 7th November 2025).

Liu, Y. H., Yu, J. J., Zhao, Q., & Ao, G. M. (2005). Genetic transformation of millet (*Setaria italica*) by *Agrobacterium*-mediated. *Agricultural Biotechnology Journal*, *13*, 32–37.

Lu, H., Zhang, J., Liu, K. B., Wu, N., Li, Y., Zhou, K., Ye, M., Zhang, T., Zhang, H., Yang, X., Shen, L., Xu, D., & Li, Q. (2009). Earliest domestication of common millet (*Panicum miliaceum*) in East Asia extended to 10,000 years ago. *Proceedings of the National Academy of Sciences of the United States of America*, *106*(18), 7367–7372. https://doi.org/10.1073/pnas.0900158106 (accessed on 7th November 2025).

M'Ribu, H. K., & Hilu, K. W. (1996). Application of random amplified polymorphic DNA to study genetic diversity in *Paspalum scrobiculatum* L. (kodo millet, Poaceae). *Genetic Resources and Crop Evolution*, *43*(3), 203–210. https://doi.org/10.1007/bf00123272 (accessed on 7th November 2025).

Maikhuri, R. K., Rao, K. S., & Semwal, R. S. (2001). Changing scenario of Himalayan agro-ecosystem: Loss of agro-biodiversity an indicator of environment changes in Central Himalaya, India. *Environmentalist*, *21*, 23–29. https://doi.org/10.1023/A:1010638104135 (accessed on 7th November 2025).

Manimekalai, M., Dhasarathan, M., Karthikeyan, A., Murukarthick, J., Renganathan, V. G., Thangaraj, K., Vellaikumar, S., Vanniarajan, C., & Senthil, N. (2018). Genetic diversity in the barnyard millet (*Echinochloa frumentacea*) germplasms revealed by morphological traits and simple sequence repeat markers. *Current Plant Biology*, *14*, 71–78. https://doi.org/10.1016/j.cpb.2018.09.006 (accessed on 7th November 2025).

Manimozhi, S. V. (2014). *Characterization and Evaluation of Little Millet (Panicum sumatrense Roth ex Roem. and Schultz) Genotypes for Nutritional Traits* (PhD dissertation). Tamil Nadu Agricultural University, Coimbatore, India.

Masumoto, H., Takagi, H., Mukainari, Y., Terauchi, R., & Fukunaga, K. (2016). Genetic analysis of *NEKODE1* gene involved in panicle branching of foxtail millet, *Setaria italica* (L.) P. Beauv., and mapping by using QTL-Seq. *Molecular Breeding*, *36*(5). https://doi.org/10.1007/s11032-016-0481-z (accessed on 7th November 2025).

Matsuura, A., Tsuji, W., An, P., Inanaga, S., & Murata, K. (2012). Effect of pre- and post-heading water deficit on growth and grain yield of four millets. *Plant Production Science*, *15*(4), 323–331. https://doi.org/10.1626/pps.15.323 (accessed on 7th November 2025).

Mehra, H. S., Joshi, H. C., & Chikara, J. (1985). Induced mutations in Japanese millet. *Indian Journal of Agricultural Sciences*, *55*, 294–295.

Murukarthick, J., Manimekalai, M., Karthikeyan, A., Perumal, S., Dhasarathan, M., Kandasamy, T., Yang, T. J., & Natesan, S. (2019). Transcriptomes of Indian barnyard millet and barnyard grass reveal putative genes involved in drought adaptation and micronutrient accumulation. *Acta Physiologiae Plantarum*, *41*. https://doi.org/10.1007/s11738-019-2855-4 (accessed on 7th November 2025).

Mushtaq, N. U., Alghamdi, K. M., Saleem, S., Tahir, I., Bahieldin, A., Henrissat, B., Alghamdi, M. K., Rehman, R. U., & Hakeem, K. R. (2023). Exogenous zinc mitigates salinity stress by stimulating proline metabolism in proso millet (*Panicum miliaceum* L.). *Frontiers in Plant Science*, *14*. https://doi.org/10.3389/fpls.2023.1053869 (accessed on 7th November 2025).

Muthamilarasan, M., & Prasad, M. (2015). Advances in *Setaria* genomics for genetic improvement of cereals and bioenergy grasses. *Theoretical and Applied Genetics*, *128*(1), 1–14. https://doi.org/10.1007/s00122-014-2399-3 (accessed on 7th November 2025).

N/A. (2009). *Archival Report 2008*. International Crop Research Semi-Arid Reports: Patancheru, Telangana, India. https://oar.icrisat.org/1995/1/ICRISA-Archival-Report_2008.pdf (accessed on 7th November 2025).

N/A. (2010). *Archival Report 2009*. International Crops Research Institute for the Semi-Arid Tropics: Patancheru, Telangana, India. https://oar.icrisat.org/2037/1/ICRISAT-Arch.Rep.pdf (accessed on 7th November 2025).

N/A. (2018). *MILLETs, Annual Report 2017–2018*. Hyderabad, India. https://www.millets.res.in/ar_page.php (accessed on 7th November 2025).

N/A. (2023). *International Year of Millets 2023–Building Momentum for the Year*. United Nations Web TV. https://webtv.un.org/en/asset/k1h/k1h0tqi7qq (accessed on 7th November 2025).

N/A. (2023). *Recommended Package of Practices in Kodo Millet (Paspalum scrobiculatum L.)*. https://www.millets.res.in/technologies/3-Recommended_Package_of_Practices-Kodo_Millet.pdf (accessed on 7th November 2025).

Nagaraja, A., & Mantur, S. G. (2008). Evaluation of barnyard millet entries for grain smut resistance and yield. *Mysore Journal of Agricultural Sciences*, *42*, 375–377.

Narayanrao, D. R., Tomar, R. S., Sm, P., Jasminkumar, K., Ashish, G., Chauhan, N. M., Singh, S. C., Upadhye, V., Kuddus, M., Kamble, L., & Hajare, S. T. (2023). De novo transcriptome sequencing of drought tolerance–associated genes in little millet (*Panicum sumatrense* L.). *Functional & Integrative Genomics*, *23*(4). https://doi.org/10.1007/s10142-023-01221-x (accessed on 7th November 2025).

Nirmalakumari, A., & Vetriventhan, M. (2010). Characterization of foxtail millet germplasm collections for yield contributing traits. *Electronic Journal of Plant Breeding*, *1*, 140–147.

Nirmalakumari, A., Salini, K., & Veerabadhiran, P. (2010). Morphological characterization and evaluation of little millet (*Panicum sumatrense* Roth. ex Roem. and Schultz.) germplasm. *Electronic Journal of Plant Breeding*, *1*(2), 148–155.

Nirubana, V., Ganesamurthy, K., Ravikesavan, R., & Chitdeshwari, T. (2017). Genetic diversity studies in kodo millet (*Paspalum scrobiculatum* L.) germplasm accessions based on biometrical and nutritional quality traits. *International Journal of Current Microbiology and Applied Sciences*, *6*(10), 832–839. https://doi.org/10.20546/ijcmas.2017.610.099 (accessed on 7th November 2025).

Nithiyanantham, S., Kalaiselvi, P., Mahomoodally, M. F., Zengin, G., Abirami, A., & Srinivasan, G. (2019). Nutritional and functional roles of millets—A review. *Journal of Food Biochemistry*, *43*(7). https://doi.org/10.1111/jfbc.12859 (accessed on 7th November 2025).

Nozawa, S., Takahashi, M., Nakai, H., & Sato, Y. I. (2006). Difference in SSR variations between Japanese barnyard millet (*Echinochloa esculenta*) and its wild relative *E. crus-galli*. *Breeding Science*, *56*(4), 335–340. https://doi.org/10.1270/jsbbs.56.335 (accessed on 7th November 2025).

Ookawa, T., & Ishihara, K. (1993). Varietal difference of the cell wall components affecting the bending stress of the culm in relation to the lodging resistance in paddy rice. *Japanese Journal of Crop Science*, *62*(3), 378–384. https://doi.org/10.1626/jcs.62.378 (accessed on 7th November 2025).

Padulosi, S., Mal, B., Ravi, S., Gowda, J., Gowda, K. T. K., Shanthakumar, G., Yenagi, N., & Dutta, M. (2009). Food security and climate change: Role of plant genetic resources of minor millets. *Indian Journal of Plant Genetic Resources*, *22*, 1–16.

Pak, J. H., Kim, M. J., Kim, H. J., Shin, S. H., Seo, M. C., Oh, I. S., Kim, K. Y., Jung, T. W., & Chung, Y. S. (2012). Phylogenetic analysis of common millet (*Panicum miliaceum* L.) using NTS of 5S ribosomal DNA. *Korean Journal of Breeding Science*, *44*(4), 470–475. https://doi.org/10.9787/kjbs.2012.44.4.470 (accessed on 7th November 2025).

Panaud, O. (2006). Foxtail millet. In C. Kole (Ed.), *Cereals and Millets. Genome Mapping and Molecular Breeding in Plants* (Vol. 1, pp. 325–334). Springer: Berlin, Germany. https://doi.org/10.1007/978-3-540-34389-99 (accessed on 7th November 2025).

Pandey, G., Misra, G., Kumari, K., Gupta, S., Parida, S. K., Chattopadhyay, D., & Prasad, M. (2013). Genome-wide development and use of microsatellite markers for large-scale genotyping applications in foxtail millet [*Setaria italica* (L.)]. *DNA Research*, *20*(2), 197–207. https://doi.org/10.1093/dnares/dst002 (accessed on 7th November 2025).

Patil, A. H., Dubey, M., Sao, A., & Chandel, G. (2018). Characterization of minor millets (*Panicum sumatrense* and *Eleusine coracana*) for trait related to moisture stress tolerance. *International Journal of Bio-Resource and Stress Management*, *9*(2), 224–230. https://doi.org/10.23910/ijbsm/2018.9.2.1867 (accessed on 7th November 2025).

Peng, Y., Hu, Y., Mao, B., Xiang, H., Shao, Y., Pan, Y., Sheng, X., Li, Y., Ni, X., Xia, Y., Zhang, G., Yuan, L., Quan, Z., & Zhao, B. (2016). Genetic analysis for rice grain quality traits in the YVB stable variant line using RAD-Seq. *Molecular Genetics and Genomics*, *291*(1), 297–307. https://doi.org/10.1007/s00438-015-1104-9 (accessed on 7th November 2025).

Pernes, J. (1985). *The Evolution of Cultivated Plants: The Example of Cereals*. Comptes Rendus de l'Academie des Sciences La Vie des Sciences: Sdriegendrale, France.

Perumal, S., Jayakodi, M., Kim, D. S., Yang, T. J., & Natesan, S. (2016). The complete chloroplast genome sequence of Indian barnyard millet, *Echinochloa frumentacea* (Poaceae). *Mitochondrial DNA Part B: Resources*, *1*(1), 79–80. https://doi.org/10.1080/23802359.2015.1137832 (accessed on 7th November 2025).

Prabha, D., Negi, Y. K., & Khanna, V. K. (2010). Morphological and isozyme diversity in the accessions of two cultivated species of barnyard millet. *Natural Science*, *8*, 71–76.

Qie, L., Jia, G., Zhang, W., Schnable, J., Shang, Z., Li, W., Liu, B., Li, M., Chai, Y., Zhi, H., & Diao, X. (2014). Mapping of quantitative trait locus (QTLs) that contribute to germination and early seedling drought tolerance in the interspecific cross *Setaria italica* × *Setaria viridis*. *PLoS ONE*, *9*(7), e101868. https://doi.org/10.1371/journal.pone.0101868 (accessed on 7th November 2025).

Qin, F. F., Zhao, Q., Ao, G. M., & Yu, J. J. (2008). Co-suppression of *Si401*, a maize pollen specific *Zm401* homologous gene, results in aberrant anther development in foxtail millet. *Euphytica*, *163*(1), 103–111. https://doi.org/10.1007/s10681-007-9610-4 (accessed on 7th November 2025).

Rajput, S. G., & Santra, D. K. (2016). Evaluation of genetic diversity of proso millet germplasm available in the United States using simple-sequence repeat markers. *Crop Science*, *56*(5), 2401–2409. https://doi.org/10.2135/cropsci2015.10.0644 (accessed on 7th November 2025).

Rajput, S. G., Plyler-Harveson, T., & Santra, D. K. (2014). Development and characterization of SSR markers in proso millet based on switchgrass genomics. *American Journal of Plant Sciences*, *5*(1), 175–186. https://doi.org/10.4236/ajps.2014.51023 (accessed on 7th November 2025).

Rajput, S. G., Santra, D. K., & Schnable, J. (2016). Mapping QTLs for morpho-agronomic traits in proso millet (*Panicum miliaceum* L.). *Molecular Breeding*, *36*(4). https://doi.org/10.1007/s11032-016-0460-4 (accessed on 7th November 2025).

Rasool, A., Hafiz Shah, W., Padder, S. A., Tahir, I., Alharby, H. F., Hakeem, K. R., & ul Rehman, R. (2023). Exogenous selenium treatment alleviates salinity stress in proso millet (*Panicum miliaceum* L.) by enhancing the antioxidant defense system and regulation of ionic channels. *Plant Growth Regulation*, *100*(2), 479–494. https://doi.org/10.1007/s10725-022-00826-9 (accessed on 7th November 2025).

Ravindran, G. (1991). Studies on millets: Proximate composition, mineral composition, and phytate and oxalate contents. *Food Chemistry*, *39*(1), 99–107. https://doi.org/10.1016/0308-8146(91)90088-6 (accessed on 7th November 2025).

Rawat, L., Nautiyal, A., Bisht, T. S., Prasad, S., Naithani, D., Makhloga, K., & Tewari, A. (2019). Screening of barnyard millet germplasm against shoot fly and stem borer damage under field conditions. *International Journal of Current Microbiology and Applied Sciences*, *8*(02), 1221–1226. https://doi.org/10.20546/ijcmas.2019.802.142 (accessed on 7th November 2025).

Renganathan, V. G., Vanniarajan, C., & Ramalingam, J. (2019). Genetic analysis and identification of molecular markers linked to the anthocyanin pigmentation in barnyard millet [*Echinochloa frumentacea* Roxb. (Link)]. In *Proceedings of the Neglected and Underutilized Crop Species for Food, Nutrition, Energy and Environment* (p. 43). NIPGR: New Delhi, India.

Renganathan, V. G., Vanniarajan, C., & Ramalingam, J. (2020). Barnyard millet for food and nutritional security: Current status and future research direction. *Frontiers in Genetics*, *11*. https://doi.org/10.3389/fgene.2020.00500 (accessed on 7th November 2025).

Rutledge, J., Talbert, R. E., & Sneller, C. H. (2000). RAPD analysis of genetic variation among propanil-resistant and -susceptible *Echinochloa crus-galli* populations in Arkansas. *Weed Science*, *48*(6), 669–674. https://doi.org/10.1614/0043-1745(2000)048[0669:raogva]2.0.co;2 (accessed on 7th November 2025).

Saha, D., Gowda, M. V. C., Arya, L., Verma, M., & Bansal, K. C. (2016). Genetic and genomic resources of small millets. *Critical Reviews in Plant Sciences*, *35*(1), 56–79. https://doi.org/10.1080/07352689.2016.1147907 (accessed on 7th November 2025).

Saleh, A. S. M., Zhang, Q., Chen, J., & Shen, Q. (2013). Millet grains: Nutritional quality, processing, and potential health benefits. *Comprehensive Reviews in Food Science and Food Safety*, *12*(3), 281–295. https://doi.org/10.1111/1541-4337.12012 (accessed on 7th November 2025).

Salini, K., Nirmalakumari, A., Muthiah, A., & Senthil, N. (2010). Evaluation of proso millet (*Panicum miliaceum* L.) germplasm collections. *Electronic Journal of Plant Breeding*.

Sao, A., Singh, P., Kumar, P., & Pradhan, A. (2017). Estimates of genetic parameters for yield and contributing traits in kodo millet (*Paspalum scrobiculatum* L.). *Research Journal of Agricultural Sciences*, *8*(1), 120–122.

Sebastin, R., Lee, G. A., Lee, K. J., Shin, M. J., Cho, G. T., Lee, J. R., Ma, K. H., & Chung, J. W. (2018). The complete chloroplast genome sequences of little millet (*Panicum sumatrense* Roth ex Roem. and Schult.) (Poaceae). *Mitochondrial DNA Part B: Resources*, *3*(2), 719–720. https://doi.org/10.1080/23802359.2018.1483771 (accessed on 7th November 2025).

Sebastin, R., Lee, K. J., Cho, G. T., Lee, J. R., Shin, M. J., Kim, S. H., Lee, G. A., Chung, J. W., & Hyun, D. Y. (2019). The complete chloroplast genome sequence of Japanese millet *Echinochloa esculenta* (A. Braun) H. Scholz (Poaceae). *Mitochondrial DNA Part B: Resources*, *4*(1), 1392–1393. https://doi.org/10.1080/23802359.2019.1598787 (accessed on 7th November 2025).

Seghatoleslami, M. J., Kafi, M., & Majidi, E. (2008). Effect of drought stress at different growth stages on yield and water use efficiency of five proso millet (*Panicum miliaceum*) genotypes. *Pakistan Journal of Botany*, *40*(4), 1427–1432.

Sharma, R., Girish, A. G., Upadhyaya, H. D., Humayun, P., Babu, T. K., Rao, V. P., & Thakur, R. P. (2014). Identification of blast resistance in a core collection of foxtail millet germplasm. *Plant Disease*, *98*(4), 519–524. https://doi.org/10.1094/pdis-06-13-0593-re (accessed on 7th November 2025).

Sharma, S., Sharma, N., Handa, S., & Pathania, S. (2017). Evaluation of health potential of nutritionally enriched kodo millet (*Eleusine coracana*) grown in Himachal Pradesh, India. *Food Chemistry*, *214*, 162–168. https://doi.org/10.1016/j.foodchem.2016.07.086 (accessed on 7th November 2025).

Sood, S., Khulbe, R., Saini, N., Gupta, A., & Agrawal, P. (2014). Interspecific hybrid between *Echinochloa esculenta* (Japanese barnyard millet) and *E. frumentacea* (Indian barnyard millet)–A new avenue for genetic enhancement of barnyard millet. *Electronic Journal of Plant Breeding*, *5*(2), 248–253.

Sreeja, R., Balaji, S., Arul, L., Nirmala Kumari, A., Kannan Bapu, J. R., & Subramanian, A. (2016). Association of lignin and FLEXIBLE CULM 1 (FC1) ortholog in imparting culm strength and lodging resistance in kodo millet (*Paspalum scrobiculatum* L.). *Molecular Breeding, 36*(11). https://doi.org/10.1007/s11032-016-0577-5 (accessed on 7th November 2025).

Sreenivasulu, N., Miranda, M., Prakash, H. S., Wobus, U., & Weschke, W. (2004). Transcriptome changes in foxtail millet genotypes at high salinity: Identification and characterization of a PHGPX gene specifically up-regulated by NaCl in a salt-tolerant line. *Journal of Plant Physiology, 161*(4), 467–477. https://doi.org/10.1078/0176-1617-01112 (accessed on 7th November 2025).

Subramanian, A., Nirmalakumari, A., & Veerabadhiran, P. (2010). Trait based selection of superior kodo millet (*Paspalum scrobiculatum* L.) genotypes. *Electronic Journal of Plant Breeding, 1*(4), 852–855.

Suresh, B. V., Choudhary, P., Aggarwal, P. R., Rana, S., Singh, R. K., Ravikesavan, R., Prasad, M., & Muthamilarasan, M. (2022). De novo transcriptome analysis identifies key genes involved in dehydration stress response in kodo millet (*Paspalum scrobiculatum* L.). *Genomics, 114*(3), 110347. https://doi.org/10.1016/j.ygeno.2022.110347 (accessed on 7th November 2025).

Sushmitha, B. (2018). Transcript analysis of the known moisture stress responsive gene orthologs among different genotypes of little millet, *Panicum sumatrense*. *Biosciences Biotechnology Research Communications, 11*(2), 335–346. https://doi.org/10.21786/bbrc/11.2/21 (accessed on 7th November 2025).

Tabacchi, M., Mantegazza, R., Spada, A., & Ferrero, A. (2006). Morphological traits and molecular markers for classification of *Echinochloa* species from Italian rice fields. *Weed Science, 54*(6), 1086–1093. https://doi.org/10.1614/ws-06-018r1.1 (accessed on 7th November 2025).

Thakur, J., Kanwar, R. R., & Shyam, K. K. (2020). Studies of correlation between yield and yield attributing characters of kodo millet (*Paspalum scrobiculatum* L.). *Journal of Pharmacognosy and Phytochemistry, 9*(6), 1919–1922.

Tian, B., Wang, J., Zhang, L., Li, Y., Wang, S., & Li, H. (2010). Assessment of resistance to lodging of landrace and improved cultivars in foxtail millet. *Euphotic, 172*(3), 295–302. https://doi.org/10.1007/s10681-009-9999-z (accessed on 7th November 2025).

Tian, B., Zhang, L., Hu, J., Liu, Y., Zhou, L., Ping, W., Zou, J., & Li, H. (2022). Genetic characterization of hull color using BSR-Seq and genome re-sequencing approaches in foxtail millet. *Frontiers in Plant Science, 13*. https://doi.org/10.3389/fpls.2022.1019496 (accessed on 7th November 2025).

Tiwari, N., Tiwari, S., & Tripathi, N. (2018). Genetic characterization of Indian little millet (*Panicum sumatrense*) genotypes using random amplified polymorphic DNA markers. *Agriculture and Natural Resources, 52*(4), 347–353. https://doi.org/10.1016/j.anres.2018.10.008 (accessed on 7th November 2025).

Trivedi, A. K., Arya, L., Verma, M., Verma, S. K., Tyagi, R. K., & Hemantaranjan, A. (2015). Genetic variability in proso millet [*Panicum miliaceum*] germplasm of Central Himalayan region based on morpho-physiological traits and molecular markers. *Acta Physiologies Plantarum, 37*(2). https://doi.org/10.1007/s11738-014-1770-y (accessed on 7th November 2025).

Ulaganathan, V. (2013). *Evaluation and Characterization of Genetic Resources and Stability Analysis of Selected Genotypes in Finger Millet (Eleusine coracana (L.) Gaertn.)* (PhD dissertation). Tamil Nadu Agricultural University, Coimbatore, India.

Upadhyaya, H. D., Dwivedi, S. L., Singh, S. K., Singh, S., Vetriventhan, M., & Sharma, S. (2014). Forming core collections in barnyard, kodo, and little millets using morpho-agronomic descriptors. *Crop Science, 54*(6), 2673–2682. https://doi.org/10.2135/cropsci2014.03.0221 (accessed on 7th November 2025).

Upadhyaya, H. D., Pundir, R. P. S., Gowda, C. L. L., Gopal Reddy, V., & Singh, S. (2009). Establishing a core collection of foxtail millet to enhance the utilization of germplasm of an underutilized crop. *Plant Genetic Resources, 7*(2), 177–184. https://doi.org/10.1017/s1479262108178042 (accessed on 7th November 2025).

Upadhyaya, H. D., Ravishankar, C. R., Narasimhudu, Y., Sarma, N. D. R. K., Singh, S. K., Varshney, S. K., Reddy, V. G., Singh, S., Parzies, H. K., Dwivedi, S. L., Nadaf, H. L., Sahrawat, K. L., & Gowda, C. L. L. (2011). Identification of trait-specific germplasm and developing a mini core collection for efficient use of foxtail millet genetic resources in crop improvement. *Field Crops Research, 124*(3), 459–467. https://doi.org/10.1016/j.fcr.2011.08.004 (accessed on 7th November 2025).

Upadhyaya, H. D., Reddy, V. G., & Gowda, C. L. L. (2006). Characterization of world's foxtail millet germplasm collections for morphological traits. *SAT eJournal, 2*(1), 1–3.

Upadhyaya, H. D., Vetriventhan, M., Dwivedi, S. L., Pattanashetti, S. K., & Singh, S. K. (2016). Proso, barnyard, little, and kodo millets. In M. Singh & H. D. Upadhyaya (Eds.), *Genetic and Genomic Resources for Grain Cereals Improvement* (pp. 321–343). Elsevier. https://doi.org/10.1016/B978-0-12-802000-5.00008-3 (accessed on 7th November 2025).

Veeranagamallaiah, G., Jyothsnakumari, G., Thippeswamy, M., Chandra Obul Reddy, P., Surabhi, G. K., Sriranganayakulu, G., Mahesh, Y., Rajasekhar, B., Madhurarekha, C., & Sudhakar, C. (2008). Proteomic analysis of salt stress responses in foxtail millet (*Setaria italica* L. cv. Prasad) seedlings. *Plant Science, 175*(5), 631–641. https://doi.org/10.1016/j.plantsci.2008.06.017 (accessed on 7th November 2025).

Vetriventhan, M., & Upadhyaya, H. D. (2018). Diversity and trait-specific sources for productivity and nutritional traits in the global proso millet (*Panicum miliaceum* L.) germplasm collection. *Crop Journal, 6*(5), 451–463. https://doi.org/10.1016/j.cj.2018.04.002 (accessed on 7th November 2025).

Vetriventhan, M., & Upadhyaya, H. D. (2019). Variability for productivity and nutritional traits in germplasm of kodo millet, an underutilized nutrient-rich climate smart crop. *Crop Science, 59*(3), 1095–1106. https://doi.org/10.2135/cropsci2018.07.0450 (accessed on 7th November 2025).

Vetriventhan, M., Azevedo, V. C. R., Upadhyaya, H. D., Nirmalakumari, A., Kane-Potaka, J., Anitha, S., et al. (2020). Genetic and genomic resources, and breeding for accelerating improvement of small millets: Current status and future interventions. *Nucleus (Calcutta), 63*(3), 217–239. https://doi.org/10.1007/s13237-020-00322-3 (accessed on 7th November 2025).

Wallace, J. G., Upadhyaya, H. D., Vetriventhan, M., Buckler, E. S., Hash, C. T., & Ramu, P. (2015). The genetic makeup of a global barnyard millet germplasm collection. *Plant Genome, 8*(1). https://doi.org/10.3835/plantgenome2014.10.0067 (accessed on 7th November 2025).

Wang, C., Chen, J., Zhi, H., Yang, L., Li, W., Wang, Y., Li, H., Zhao, B., Chen, M., & Diao, X. (2010). Population genetics of foxtail millet and its wild ancestor. *BMC Genetics, 11*(1). https://doi.org/10.1186/1471-2156-11-90 (accessed on 7th November 2025).

Wang, J., Wang, Z., Du, X., Yang, H., Han, F., Han, Y., Yuan, F., Zhang, L., Peng, S., & Guo, E. (2017). A high-density genetic map and QTL analysis of agronomic traits in foxtail millet (*Setaria italica* (L.) P. Beauv.) using RAD-Seq. *PLoS One, 12*(6), e0179717. https://doi.org/10.1371/journal.pone.0179717 (accessed on 7th November 2025).

Wang, M. Z. (2011). Culturing of immature inflorescences and *Agrobacterium*-mediated transformation of foxtail millet (*Setaria italica*). *African Journal of Biotechnology, 10*(73). https://doi.org/10.5897/ajb10.2330 (accessed on 7th November 2025).

Wang, R., Hunt, H. V., Qiao, Z., Wang, L., & Han, Y. (2016). Diversity and cultivation of broomcorn millet (*Panicum miliaceum* L.) in China: A review. *Economic Botany, 70*(3), 332–342. https://doi.org/10.1007/s12231-016-9357-8 (accessed on 7th November 2025).

Wang, T., & Darmency, H. (1997). Inheritance of sethoxydim resistance in foxtail millet, *Setaria italica* (L.) Beauv. *Euphytica, 94*(1), 69–73. https://doi.org/10.1023/a:1002989725995 (accessed on 7th November 2025).

Wang, T., Fleury, A., Ma, J., & Darmency, H. (1996). Genetic control of dinitroaniline resistance in foxtail millet (*Setaria italica*). *Journal of Heredity, 87*(6), 423–426. https://doi.org/10.1093/oxfordjournals.jhered.a023031 (accessed on 7th November 2025).

Wang, X. Y., Diao, X. M., Wang, J. Z., Wang, C. F., Wang, G. Q., Hao, X. F., Liang, Z. H., Wang, X. M., & Zhao, F. F. (2013). Construction of genetic map and QTL analysis of some main agronomic traits in millet (*Setaria italica*). *Journal of Plant Genetic Resources, 14*(5), 871–878.

Wang, Y., Wang, X., Sun, S., Jin, C., Su, J., Wei, J., Luo, X., Wen, J., Wei, T., Sahu, S. K., Zou, H., Chen, H., Mu, Z., Zhang, G., Liu, X., Xu, X., Gram, L., Yang, H., Wang, E., & Liu, H. (2023). GWAS, mWAS and mGWAS provide insights into precision agriculture based on genotype-dependent microbial effects in foxtail millet. *Nature Communications, 13*(1). https://doi.org/10.1038/s41467-022-33238-4 (accessed on 7th November 2025).

Weining, S., & Henry, R. J. (1995). Molecular analysis of the DNA polymorphism of wild barley (*Hordeum spontaneum*) germplasm using the polymerase chain reaction. *Genetic Resources and Crop Evolution, 42*(3), 273–280. https://doi.org/10.1007/bf02431262 (accessed on 7th November 2025).

Wu, Y., Du, J., Wang, X., Fang, X., Shan, W., & Liang, Z. (2012). Computational prediction and experimental verification of miRNAs in *Panicum miliaceum* L. *Science China Life Sciences, 55*(9), 807–817. https://doi.org/10.1007/s11427-012-4367-y (accessed on 7th November 2025).

Xu, X., Zeng, L., Tao, Y., Vuong, T., Wan, J., Boerma, R., Noe, J., Li, Z., Finnerty, S., Pathan, S. M., Shannon, J. G., & Nguyen, H. T. (2013). Pinpointing genes underlying the quantitative trait loci for root-knot nematode resistance in paleopolyploid soybean by whole genome resequencing. *Proceedings of the National Academy of Sciences of the United States of America, 110*(33), 13469–13474. https://doi.org/10.1073/pnas.1222368110 (accessed on 7th November 2025).

Yabuno, T. (1962). Cytotaxonomic studies on the two cultivated species and the wild relatives in the genus *Echinochloa*. *Cytologic (Tokyo), 27*(3), 296–305. https://doi.org/10.1508/cytologia.27.296 (accessed on 7th November 2025).

Yabuno, T. (1966). Biosystematics study of the genus *Echinochloa*. *Japanese Journal of Botany, 19*, 277–323.

Yadav, S., Gaur, V. S., Jaiswal, J. P., & Kumar, A. (2014). Simple sequence repeat (SSR) analysis in relation to calcium transport and signaling genes reveals transferability among grasses and a conserved behavior within finger millet genotypes. *Österreichische Botanische Zeitschrift, 300*(6), 1561–1568. https://doi.org/10.1007/s00606-014-0982-3 (accessed on 7th November 2025).

Yamaguchi, H., Utano, A., Yasuda, K., Yano, A., & Soejima, A. (2005). A molecular phylogeny of wild and cultivated *Echinochloa* in East Asia inferred from non-coding region sequences of trnT-L-F. *Weed Biology and Management, 5*(4), 210–218. https://doi.org/10.1111/j.1445-6664.2005.00185.x (accessed on 7th November 2025).

Yang, Y. B., Jia, G. Q., Deng, L. G., Qin, L., Chen, E. Y., Cong, X. J., Zou, R. F., Wang, H. L., Zhang, H. W., Liu, B., Guan, Y. A., Diao, X. M., & Yin, Y. P. (2017). Genetic variation of yellow pigment and its components in foxtail millet (*Setaria italica* (L.) P. Beauv.) from different eco-regions in China. *Journal of Integrative Agriculture, 16*(11), 2459–2469. https://doi.org/10.1016/s2095-3119(16)61598-8 (accessed on 7th November 2025).

Yasuda, K., Yano, A., Nakayama, Y., & Yamaguchi, H. (2006). Molecular identification of *Echinochloa oryzicola* Vasing. and *E. crusgalli* (L.) Beauv. using a polymerase chain reaction restriction fragment length polymorphism technique. *Biological Management, 2*, 11–17.

Ye, C. Y., Lin, Z., Li, G., Wang, Y. Y., Qiu, J., Fu, F., Zhang, H., Chen, L., Ye, S., Song, W., Jin, G., Zhu, J., Lu, Y., Guo, L., & Fan, L. (2014). *Echinochloa* chloroplast genomes: Insights into the evolution and taxonomic identification of two weedy species. *PLoS One, 9*(11), e113657. https://doi.org/10.1371/journal.pone.0113657 (accessed on 7th November 2025).

Yue, H., Wang, L., Liu, H., Yue, W., Du, X., Song, W., & Nie, X. (2016). De novo assembly and characterization of the transcriptome of broomcorn millet (*Panicum miliaceum* L.) for gene discovery and marker development. *Frontiers in Plant Science, 7*. https://doi.org/10.3389/fpls.2016.01083 (accessed on 7th November 2025).

Zegada-Lizarazu, W., & Iijima, M. (2005). Deep root water uptake ability and water use efficiency of pearl millet in comparison to other millet species. *Plant Production Science, 8*(4), 454–460. https://doi.org/10.1626/pps.8.454 (accessed on 7th November 2025).

Zhang, D., Chen, F., Bettinger, R. L., Barton, L., Ji, D., Morgan, C., Wang, H., Cheng, X., Dong, G., Guilderson, T. P., & Zhao, H. (2010). Archaeological records of Dadiwan in the past 60 ka and the origin of millet agriculture. *Chinese Science Bulletin, 55*(16), 1636–1642. https://doi.org/10.1007/s11434-010-3097-4 (accessed on 7th November 2025).

Zhang, G., Liu, X., Quan, Z., Cheng, S., Xu, X., Pan, S., et al. (2012). Genome sequence of foxtail millet (*Setaria italica*) provides insights into grass evolution and biofuel potential. *Nature Biotechnology, 30*(6), 549–554. https://doi.org/10.1038/nbt.2195 (accessed on 7th November 2025).

Zhang, H., Tang, S., Schnable, J. C., He, Q., Gao, Y., Luo, M., Jia, G., Feng, B., Zhi, H., & Diao, X. (2021). Genome-wide DNA polymorphism analysis and molecular marker development for the *Setaria italica* variety "SSR41" and positional cloning of the *Setaria* white leaf sheath gene *SiWLS1*. *Frontiers in Plant Science, 12*. https://doi.org/10.3389/fpls.2021.743782 (accessed on 7th November 2025).

Zhang, J. P., Wang, M. Y., Bai, Y. F., Jia, J. P., & Wang, G. Y. (2005). Rapid evaluation on drought tolerance of foxtail millet at seedling stage. *Journal of Plant Genetic Resources (Chinese, English abstract), 6*(1), 59–62.

Zhang, J., Liu, T., Fu, J., Zhu, Y., Jia, J., Zheng, J., Zhao, Y., Zhang, Y., & Wang, G. (2007). Construction and application of EST library from *Setaria italica* in response to dehydration stress. *Genomics, 90*(1), 121–131. https://doi.org/10.1016/j.ygeno.2007.03.016 (accessed on 7th November 2025).

Zhang, J., Lu, H., Wu, N., Yang, X., & Diao, X. (2011). Phytolith analysis for differentiating between foxtail millet (*Setaria italica*) and green foxtail (*Setaria viridis*). *PLoS One, 6*(5), e19726. https://doi.org/10.1371/journal.pone.0019726 (accessed on 7th November 2025).

Zhang, K., Fan, G., Zhang, X., Zhao, F., Wei, W., Du, G., Feng, X., Wang, X., Wang, F., Song, G., Zou, H., Zhang, X., Li, S., Ni, X., Zhang, G., & Zhao, Z. (2017). Identification of QTLs for 14 agronomically important traits in *Setaria italica* based on SNPs generated from high-throughput sequencing. *G3 (Bethesda), 7*(5), 1587–1594. https://doi.org/10.1534/g3.117.041517 (accessed on 7th November 2025).

Zhang, P., Feng, B., Wang, P., Gao, X., J., G., H., S., Zhang, X., & Chai, Y. (2012). Study on identification of drought-resistance indexes at seedling state in broomcorn millet under PEG stress. *Journal of China Agricultural University*, 1–10.

Zhang, S., Tang, C., Zhao, Q., Li, J., Yang, L., Qie, L., et al. (2014). Development of highly polymorphic simple sequence repeat markers using genome-wide microsatellite variant analysis in foxtail millet [*Setaria italica* (L.) P. Beauv.]. *BMC Genomics, 15*(1). https://doi.org/10.1186/1471-2164-15-78 (accessed on 7th November 2025).

Zhang, W. L., & Guan, Y. A. (1995). Summer millet Jinan 8337 with high yield and quality. *Crop Genetic Resources, 28*(3).

Zhang, W. Y., Liu, B. H., Xie, J. X., Li, J. M., Gu, G. Q., Wang, Y. F., Li, H. Q., Li, Y. Q., & Diao, X. M. (2012). Screening of indexes for drought tolerance test at booting stage in foxtail millet. *Journal of Plant Genetic Resources, 13*(5), 765–772.

Zhou, Y., Qu, Y., Zhu, M., Liu, J., Wang, Y., Song, H., & Feng, B. (2016). Genetic diversity and virulence variation of *Sporisorium destruens* isolates and evaluation of broomcorn millet for resistance to head smut. *Euphotic, 211*, 59–70. https://doi.org/10.1007/s10681-016-1723-1 (accessed on 7th November 2025).

Zhu, X. H., Song, Y. C., Zhao, Z. H., Shi, Y. S., Liu, Y. H., Li, Y., & Wang, T. Y. (2008). Methods for identification of drought tolerance at germination period of foxtail millet by osmotic stress. *Journal of Plant Genetic Resources, 9*(1), 62–67.

Zou, C., Li, L., Miki, D., Li, D., Tang, Q., Xiao, L., et al. (2019). The genome of broomcorn millet. *Nature Communications, 10*(1). https://doi.org/10.1038/s41467-019-08409-5 (accessed on 7th November 2025).

Zuo, R., Zhang, Y., Yang, Y., Wang, C., Zhi, H., Zhang, L., Tang, S., Guan, Y., Li, S., Cheng, R., Shang, Z., Jia, G., & Diao, X. (2023). Haplotype variation and KASP markers for *SiPSY1*–A key gene controlling yellow kernel pigmentation in foxtail millet. *Crop Journal*. https://doi.org/10.1016/j.cj.2023.09.008 (accessed on 7th November 2025).

PART II
Nutritional Quality Enhancement

CHAPTER 5

Nutritional Quality Enhancement in Sorghum

NEETHU FRANCIS,[1,*] S. M. INDHU,[2] B. MOHANAPRIYA,[3] P. VINOTH,[2] and R. JAISHREEPRIYANKA[3]

[1]*School of Agricultural Sciences, Karunya Institute of Technology and Sciences, Karunya Nagar, Coimbatore, Tamil Nadu, India*

[2]*Department of Plant Biotechnology, Center of Plant Molecular Biology and Biotechnology, Tamil Nadu Agricultural University, Coimbatore, Tamil Nadu, India*

[3]*Department of Millets, Center for Plant Breeding and Genetics, Tamil Nadu Agricultural University, Coimbatore, Tamil Nadu, India*

*Corresponding author

5.1 INTRODUCTION

Sorghum (*Sorghum bicolor* (L.) Moench) is a hardy cereal crop with low water requirements and a short growing period. It is the fifth most important cereal crop and is an important food crop for future agriculture. Sorghum has been used in food products such as thin and thick porridges, fermented and unfermented breads, lactic acid fermented and alcoholic beers, non-alcoholic beverages, and malted flours (Are et al., 2019). It is one of the cheapest sources of energy, protein, iron, and zinc, and meets more than 50% of iron and zinc requirements in low-income populations in India. It is mainly eaten in the form of unleavened flatbread or roti (Kulamarva et al., 2009). The bran, endosperm, and germ are the major parts of sorghum grain. The bran layer is made of pericarp and testa. This layer has a high quantity of health-protective phenolic compounds such as flavonoids, 3-deoxyx-anthocyanins, and condensed tannins, which can be used in functional foods (Awika and Rooney, 2004; Burdette et al., 2010; Xiong et al., 2019). The endosperm and germ fractions contain starch, proteins, lipids, vitamins, and minerals. The durra and bicolor-durra races of the sorghum germplasm panel had the highest protein (12.6%) and fat (3%) and the lowest starch (66.1%), while the kafir race had the lowest protein content (10.9%) and the highest starch content (67.6%). Red sorghum has high levels of antioxidants with high anti-inflammatory and anti-cancerous activities (Rhodes et al., 2017). However, a high concentration of antinutritional substances such as tannins, phytic acids, and oxalates (Ojha et al., 2018) interferes with the digestion of proteins and the bioavailability of certain minerals when used directly as food (Keyata et al., 2021; Popova and Mihaylova, 2019;). The poor digestibility of starch and protein and the bioavailability of minerals reduce the

Millets in Sustainable Agriculture: Genetic Improvement, Nutritional Quality, Cultivation, and Integrated Pest–Disease Management. Roshni Vijayan, A. John Joel, R. Ambika Rajendran, and Adithya P. Balakrishnan (Eds.)

DOI: 10.1201/9781779642332-7

nutritional value and quality of sorghum-based foods. See Table 5.1. A study about the production constraints of sorghum and farmer preferences in Uganda highlighted the need for improved nutritious varieties with low tannin levels (Andiku et al., 2021). Protein content and quality, Fe and Zn content and bioavailability, Vitamin A content enhancement, sugar and starch content, seed shelf life, and processing traits improvement are a few of the important areas that need to be addressed for overall quality enhancement (Figure 5.1).

TABLE 5.1 Nutritional Profile of Sorghum in Comparison With Major Food Crops

Food Crops	Energy (Kcal)	CHO (g)	Protein (g)	Fat (g)	Ash (g)	Fiber (g)	Fe (mg)	Ca (mg)
Rice	362	71	11.6	2	1.3	2	1.8	33
Wheat	348	74.8	8.5	2.6	1.6	0.9	3.5	30
Maize	358	73	9.2	4.6	1.2	2.8	2.7	26
Sorghum	329	76	9.5	2.5	1.6	2.7	5.4	25

Source: Juliano 1979; Kulamarva et al. (2009; Leder 2004; Subramanian and Metta (2000).

Various quality improvement aspects and genetic, breeding and biotechnological approaches to achieve this are discussed in this chapter.

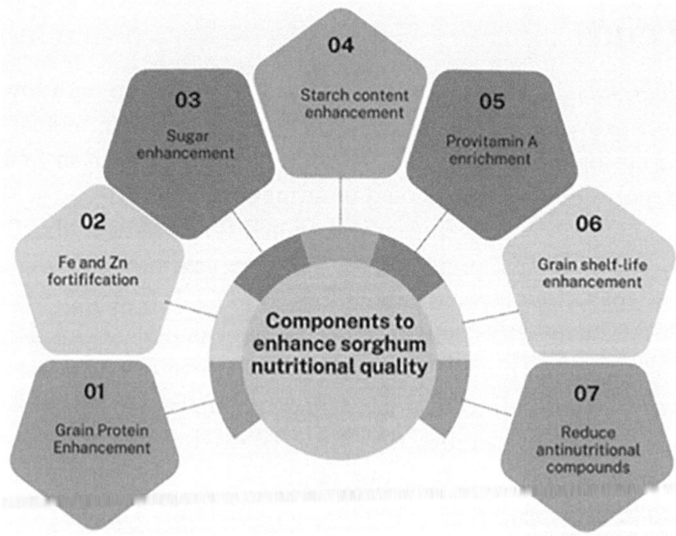

FIGURE 5.1 Components for nutritional quality enhancement in sorghum.

5.2 GRAIN PROTEIN ENHANCEMENT

Sorghum is being recognized for drought hardiness over common staple crops and is an important crop for food security in arid and semi-arid areas. However, the protein content, quality, digestibility, and functionality need improvement to be efficiently utilized as a protein source in diets. Drought stress is known to affect the protein content in sorghum; however, the Sorghum Payam genotype, which is a drought-tolerant genotype, produced the highest protein content even under high stress (Safian et al., 2022). In a study to identify the nutritionally superior genotype for the preparation of a sorghum snack in India, that is, *Sorghum Chakali*, the variety M 35-1 emerged, contributing higher protein levels (Chavan et al., 2016). The grain protein lacks the essential amino acids, lysine and tryptophan, and has poor digestibility (Bressani and Rios, 1962; Mertz et al., 1984). Higher protein sorghum grains most often have a lower lysine content. Moreover, poor protein digestibility in sorghum can also impact starch digestion, as starch granules are entrapped in the protein matrix (Zarei et al., 2022). In the sorghum grain, the endosperm harbors the maximum protein content, that is, up to 80% of the total protein content. Protein content in sorghum grains is in the range of 8–17% (Rooney and Serna-Saldivar, 2000) with an average of 11% (Lasztity, 2017). The storage protein in sorghum is called Kafirin, as it was the first and major protein isolated from the kafir sorghum. Kafirins are alcohol-soluble proteins that belong to the class of prolamins. The composition of this protein resembles the maize protein zein (Johns and Brewster, 1916; Labuschagne, 2018). Kafirin proteins make up about 70–80% of the total protein in the sorghum endosperm (Hamaker et al., 1995; Wong et al., 2010). Kafirins are further subdivided into α-, β-, γ-, and δ-kafirins based on characteristics such as molecular weight, solubility, and structure (Shewry and Halford, 2002). The α-kafirin fraction constitutes 80% of the total kafirin protein. The β- and γ-kafirins constitute about 5% and 15%, respectively, of the total kafirins in sorghum (Hamaker and Bugusu, 2003). These proteins are present as spherical protein bodies in the endoplasmic reticulum (Shull et al., 1992). Although α-kafirins are digestible, their internal localization within the protein bodies makes them the last kafirins to be digested. Although β- and γ-kafirins are arranged peripherally, they exhibit greater resistance to proteases because they are cross-linked by intra- and intermolecular disulfide bonds. These prolamins, or kafirins, are deficient in lysine, are less digestible, and reduce the nutritional quality of sorghum protein (Li et al., 2018). The non-prolamin proteins or non-kafirins are albumins, globulins, and glutelins, which are rich in lysine but constitute only 30% of the protein. Kafirins are encapsulated into the protein bodies around which the non-kafirin proteins are arranged. These proteins are arranged around the starch granules to form a matrix, reducing the digestibility of the sorghum endosperm (Hamaker and Bugusu, 2003). Genetic enhancement of the quality of sorghum protein can be achieved by increasing the lysine-rich non-kafirins and the digestibility of the kafirins. Understanding the natural variation existing among the germplasm sources of sorghum can be useful in utilizing them as donors in crop improvement programs. There have been various attempts to understand the genes and alleles controlling kafirin proteins. Six alleles were reported for β- and γ-kafirins among sorghum genotypes (Laidlaw et al., 2010). Genes encoding α-kafirins and their subunits in different sorghum genotypes have also been identified (Song et al., 2004; Xu and Messing, 2008). Xu and Messing (2009) found that chromosome 5 harbored 20 genes encoding α-kafirins in the BTx623 genome of sorghum. In "P721Q," an EMS mutant of sorghum, the protein structure was modified from the natural spherical form to a reticulate form, which made it more accessible to protease enzymes, thereby increasing protein digestibility. The invaginations, or folded protein structure, made the centrally arranged α-kafirins more accessible to gastric enzymes and also increased the enzyme-exposed surface area of the protein bodies. The mutation, which created a single amino acid substitution in the gene encoding a 22-kDa α-kafirin, also increased the proportion of non-kafirin proteins (Oria et al., 2000; Wu et al., 2013).

In another approach to improve the digestibility of sorghum protein, RNA interference (RNAi) technology was used to silence the 29-kDa α-kafirin. This reduced the 29-kDa α-kafirin content, modified protein morphology, and enhanced protein digestibility (Kumar et al., 2012). Li and co-workers used the CRISPR/Cas9 approach to edit the K1C family genes that encode kafirin proteins. The seeds of the transgenic lines exhibited reduced levels of α-kafirin and increased digestibility and lysine content (Li et al., 2018). Proteomics approaches have also shed light on the expressed proteins of the genes encoding kafirins and non-kafirins in sorghum. Liquid chromatography and mass spectrometry techniques identified various proteins involved in the protein-starch matrix formation and protein-protein interactions in sorghum grain. As the kafirins of sorghum and zein in maize are similar, findings about maize protein are cross-transferable to sorghum (Labuschagne, 2018).

5.3 FE AND ZN FORTIFICATION

Micronutrients are necessary for human growth, development, and immune system maintenance. A deficiency of these minerals deteriorates physical as well as mental health and results in diseases that can even lead to death. Iron (Fe) and Zinc (Zn) are the major contributors to micronutrient malnourishment and related health problems. Nearly 3 billion people worldwide suffer from Fe deficiency. Similarly, Zn deficiency poses a threat to one-third of the population in developing nations (Çakmak et al., 2004; Hotz and Brown, 2004). Lack of diversified diets, particularly in poor countries, is a major cause of this rampant malnourishment. Staple cereals that are widely consumed are low in Fe and Zn content. Nutrient supplementation, food fortification, biofortification, and diet diversification are different ways to improve this micronutrient intake. Biofortification involves genetically increasing the micronutrient content in food crops through plant breeding or genetic engineering (Saltzman et al., 2013). It is a practical and affordable way to address micronutrient deficiencies, particularly in developing countries. Hardy crops like sorghum are important for future agriculture, and improving their nutritional profile is a sustainable approach to addressing food and nutritional insecurity.

Various methods are being used for estimating Fe and Zn content and phenotyping the accessions. The most common methods are colorimetric analysis (semi-quantitative), atomic absorption spectrometer (AAS), inductively coupled plasma-optical emission spectrometer (ICP-OES), inductively coupled plasma-mass spectrometer (ICP-MS), and X-ray fluorescence (XRF). The method of choice can be decided based on the purpose, budget, and accuracy of results desired. Colorimetric analysis can be used to screen a large number of genotypes at minimum cost and time. Staining of seeds with Prussian blue for Fe content and diphenyl thiocarbazone-based dithizone (DTZ) for Zn content has been used as a rapid screening protocol in various cereals (Shobhana et al., 2013). Staining methods can be used to identify high and low genotypes, which can be further confirmed using more precise methods. XRF is a non-destructive method that can be used for mass screening of genotypes. AAS, ICP-OES, and ICP-MS give more precise quantitative estimations of micronutrients. The accuracy of nutritional content estimations and repeatability of the results also depend on the nutrients applied during growth and post-harvest handling procedures. Extreme caution should be exercised to prevent contamination of the seed sample with dust, soil minerals, etc., that can tamper with the results. Sorghum grain Fe content ranged from 26 mg/Kg to 61 mg/Kg, with a mean of 46 mg/Kg, and Zn content ranged from 21 mg/Kg to 57 mg/Kg among the germplasm accessions of the sorghum core collection maintained at ICRISAT. An accession from India of race Dura had the highest Fe and Zn content. Among the commercial sorghum cultivars (hybrids, varieties, and landraces) in India, Zn and Fe content was in the range of 22–44 mg/Kg for Fe and 22–33 mg/Kg for Zn content (Ashok Kumar et al., 2012). The aleurone layer in the seed coat, which will be removed during the milling process, has the highest Fe and Zn content. Decortication reduces these micronutrient contents by half (Chavan and Patil, 2010; Hama et al., 2011). Additionally, the bioavailability of these nutrients is reduced by the presence of antinutritional factors like phytates and tannins. The average phytate content in sorghum is 7 mg/g. Fe availability is severely hampered if the phytate/Fe ratio goes above 14 (Kayodé et al., 2006). Tannins are the primary anti-nutritional factor and impede its utilization in animal feed and the livestock industry. Brown sorghum is known to have high tannin levels compared to white sorghum, which is tannin-free. These tannin-rich brown sorghum types are known to reduce feed efficiency by up to 30% (Zarei et al., 2022).

Micronutrient genetic enhancement can be accomplished through biotechnological and traditional plant breeding techniques. Sorghum's research programs aim to increase the Fe concentration by 30 mg/kg and zinc concentration by 12 mg/kg over the base levels (30 mg/kg and 20 mg/kg for Fe and Zn, respectively). These targets are set, taking into consideration the daily intake, nutrient absorption, daily nutrient requirement, storage losses, etc. (Kumar et al., 2015). Characterization and identification of potential donors from the available germplasm is essential to exploit the available variability for the traits. Variability studies among 84 sorghum parental lines revealed considerable genetic variation for Fe, Zn, and phytate content. Fe content was in the range of 20 to 37 mg/Kg, and Zn content ranged between 13.4 and 31 mg/Kg. Phytate content was in the range of 3.8 to 13.5 mg/g. New variability can be generated through induced mutagenesis. Reducing anti-nutritional components in the grain can indirectly increase the bioavailable Fe and Zn content. Genotypes with high Fe and Zn content and low phytate content can be utilized as donors in crop improvement programs to develop nutritionally superior cultivars. Sorghum with white pericarp (white sorghum) is known to have low phenols and tannins (Awika and Rooney, 2004). Fe and Zn showed a positive association and high heritability (Reddy et al., 2005). Hence, efforts to improve one can indirectly select for the other positively. Establishing associations between nutritional content and easily observable and measurable agronomic parameters can be beneficial in identifying nutritionally superior genotypes.

5.4 SUGAR ENHANCEMENT AND BIOFUEL APPLICATION

Sweet sorghum can accumulate high levels of fermentable sugars in its stalks and grains and adapt quickly to adverse climatic conditions. This makes it a potential feedstock crop for new-age biofuels. Specific sweet sorghum genotypes accumulate about 78% of plant biomass as juice (Vinutha et al., 2014). Numerous methods have been suggested to maximize sugar content in sweet sorghum, viz., by modulating the plant growing environments (Li et al., 2018), use of plant growth regulators (Macedo et al., 2017), and genetic enhancement

(Liu et al., 2021). Physical methods like the removal of other sinks for carbohydrates, that is, inflorescence before anthesis, can also maximize sugar yield in stalks (Jebril et al., 2021; Kering et al., 2017). However, genetic improvement is the most durable and effective way to maximize the sugar yield and reduce the overall cost of biofuel production. The ideal cultivar suitable for biofuel production should possess high biomass and high sugar content. However, the genotypes exhibit trade-offs between these two traits.

Downregulation of sucrose transporters, SUT1 and SUT4, was found to be associated with high sugar accumulation (Qazi et al., 2012), but understanding their accurate function and role in sugar accumulation is limited (Bihmidine et al., 2015). In addition, the reduced expression of transcripts linked with cell wall activities in sweet sorghum over grain sorghum proves that carbon partitioning in the stalk might be the mechanism that imparts variation in monomeric carbohydrates among the genotypes (Calviño and Messing, 2012). The plant cell wall is made up of cellulose, hemicellulose, and lignin. Higher lignin content not only affects the fermentable sugars from the cell wall but also correlates with inferior forage quality. Novel mutations in the phenylpropanoid pathway result in the brown mid-rib (*bmr*) mutants characterized by low lignin content and increased yield of fermentable sugars.

The miRNAs also play a vital role in sugar accumulation. Higher expression of miR169 shows a negative association with brix value (Calviño et al., 2011; Yu et al., 2015). Another miRNA, miR127, was identified as a potential candidate for regulation of stem-sugar accumulation and serves as the target for cellulase-synthase-like genes (Yu et al., 2015). Apart from this, more than 850 genomic loci associated with sugar metabolism and biomass-related traits, viz., plant height, leaf architecture, delayed flowering, stalk thickness, stalk diameter, sugar, cellulose, lignin content, etc., have been reported in different mapping populations (Rosas-Quijano et al., 2021). Fine mapping of genes present within these QTLs can aid in further improvement of sorghum for biofuel production. Cooper et al. compared the genomic sequence of BTx623 (grain sorghum) with Rio (sweet sorghum) and found that the gene content and genomic structure are almost similar (2019). He also reported a loss of function mutation, especially among the sugar metabolism genes. Many transgenic approaches are also reported to be successful in increasing the sugar content in sorghum. Isomaltulase acts as an acariogenic sweetener in plants. Isomaltulase can be produced by the overexpression of sucrose isomerase (SI) genes using strong stalk-specific promoters. Liu et al. (2021) targeted the stalk vacuole in sorghum and achieved a several-fold increase in sugar content over the control in transgenic lines. Sugar content in the transgenic lines and their F_1 and F_2 progenies was higher than that in field-grown sugarcane. The regulatory network underlying the accumulation of high sugar is complex and remains ambiguous. Addressing this issue will require more knowledge of available genetic and genomic resources in sorghum and other closely related species and collaborative breeding efforts, that is, extensive screening of germplasm along with the use of new biotechnological approaches. This will help utilize the unexplored regions of the genome and identify unique and novel allelic variants for sugar accumulation and biomass-related traits.

5.5 STARCH CONTENT ENHANCEMENT

Sorghum is a significant food in the human diet as well as for ruminants, pigs, and poultry. Carbohydrates in food supply the necessary dietary energy for both humans and animals.

Plant carbohydrates, such as sugars, cellulose, gums, and starches, are adequate for all essential nutrients of a regular diet. Among them, the primary source of nutritional energy for humans is starch (storage carbohydrate). Starch granules are accumulated in semi-spherical or oval shapes at the roots, corms, tubers, stems, seeds, and fruits of plants. The key sources of starch in human diets are cereal grains (barley, corn, and sorghum) and tuber crops (Theurer, 1986). Among these sources, sorghum is the economically cheaper alternative to maize starch (Dayakar et al., 2003). About 80% of starch in sorghum is present in seed dry matter (Rooney and Waniska, 2004) and 9% in stem dry matter (McKinley et al., 2018). Profiling of sorghum landraces in Tigray, Ethiopia, for starch, protein, and other nutritional and anti-nutritional components revealed the rich genetic variability for these traits (Semere et al., 2023). The starch is mainly composed of 25.5% amylose content and 75% amylopectin, with an amylose/amylopectin ratio of 0.35 (Silva et al., 2019). Amylose is a linear polymer of α-1→4 linked D-glucose units (Galicia-García et al., 2007) and is vital to the properties and uses of starch. Wide genetic variation for amylose content in sorghum has been reported. It ranged from 24% to 33% in Zimbabwean landraces (Beta et al., 2001) and 16.1% to 55.8% in another set of 55 genotypes (Hill et al., 2012). Amylopectin is a branched polymer with linear chains of α-1→4 linked D-glucose and α-1→6 linked D-glucose, with branching points every 20 to 25 glucose residues (French, 1984; Blennow et al., 2002). Starch granule size varied from 5.5 to 30 μm, with an average size of 15.5 μm (Silva et al., 2019). The physical properties of starch during processing depend on the amylose-amylopectin ratio, structure of amylose and amylopectin, size and distribution of granules, water content, etc. (Zhu, 2014).

Starches are classified based on rates of digestion as resistant, slowly digestible, and easily digestible starch. Among them, the resistant starch (RS) is not digested in the upper gastrointestinal tract. It is not hydrolyzed by digestive enzymes during incubation (Englyst et al., 1992). Though starch is the primary source of nutritional energy for humans in routine life, the presence of resistant type starch is not desirable. Sorghum starch contains a higher level of resistant starch than maize starch and other cereal starches (Niba and Hoffman, 2003). It ranges from 12% to 21.5% (Sang et al., 2008). In general, cereal starches are easily digested when compared to root and tuber starches. However, in sorghum, the hard peripheral endosperm layer makes the starch resistant to digestive action. The starch digestibility is inversely proportional to the amylose content and starch/protein proportion. In general, to improve starch digestion, processing methods such as pressure cooking, puffing, steaming, and flaking are effective methods (Alsaffar, 2011). In crop improvement programs, resistant starch must be taken into account when assessing the nutritional value of sorghum.

Starch Synthesis-Related Genes (SSRGs) are involved in the starch synthesis process, and they are divided into four classes: ABP-glucose pyrophosphorylase (AGPase), starch synthase (SS), starch branching enzyme (SBE), and starch de-branching enzyme (DBE). The α-amylase (Amy), β-amylase (Bmy), isoamylase (ISA), limit dextrinase (LD), α-glucosidase (AGL), and glucan-water dikinase are involved in the starch degradation process. ADP-glucose and cytosolic heteroglucan metabolism may play a role in stem sucrose/starch accumulation. There is a positive correlation between grain yield and grain starch content in sorghum (Murray et al., 2008). Understanding the inheritance and genetic control of starch is beneficial in enhancing starch yield. Molecular markers and quantitative genetics will help dissect the genetic control of these complex traits through marker-assisted breeding. Table 5.2 gives an overview of several previous studies that have identified quantitative trait loci (QTLs) for sorghum starch content. These identified genes and QTLs can be used for marker-assisted breeding, genome-wide association studies, and RNA-seq techniques to improve the digestible starch content along with economic yield.

TABLE 5.2 List of QTLs and Genes Related to Starch Content in Sorghum

QTL/Gene Name	Chromo. No.	Position/ Gene location	LOD Score	Reference
qSTR1.1	1	134.8	7.45	Ayalew et al. (2022)
qSTR2.1	2	11.00	3.2	
qSTR2.2	2	125.80	3.08	
qSTR3.1	3	22.00	4.55	
qSTR3.2	3	72.20	3.58	
qSTR5.1	5	30.40	3.16	
qSTR7.1	7	105.50	4.17	
qSTR9.1	9	70.50	4.56	
qSTR10.1	10	27.00	4.02	
qAMY1.1	1	134.80	9.21	
qAMY2.1	2	119.20	4.36	
qAMY3.1	3	86.90	3.8	
qAMY3.2	3	48.30	3.22	
qAMY4.1	4	30.80	5.2	
qAMY4.2	4	68.30	3.02	
qAMY7.1	7	80.00	3.59	
qAMY9.1	9	23.50	5.7	
qAMY10.1	10	48.50	4.01	
qAMY10.2	10	92.30	3.72	
Sobic.001G475600	1	74788578–74792376		Chen et al (2019)
Sobic.005G084100	5	11366416–11370715		
Sobic.006G036500.1	6	12446395–12449252		
Sobic.006G234100	6	57574088–57577745		
Sobic.006G234700	6	57643053–57648782		
Sobic.006G235300	6	57694800–57696355		
Sobic.006G235350	6	57696523–57699060		
Sobic.006G236700	6	57815633–57822442		
Sobic.006G237200	6	57856599–57860447		
Sobic.009G071800	9	8621987–8624768		
Sobic.009G072400	9	8881404–8887144		
Sobic.009G072700	9	8916935–8923148		
Sobic.001G054500	1	4066711–4067588		Sapkota et al. (2020)
Sobic.008G111500	8	51714673–51715254		
Sobic.008G111600	8	51719209–51726960		
Sobic.008G158332	8	59121190–59129810		
SB00115.3	3	58,000,108		Sukumaran et al. (2012)
SB00086.1	4	45,665,466		

5.6 PROVITAMIN A ENRICHMENT IN SORGHUM

Globally, vitamin A deficiency is more prevalent among nutrient deficiencies (WHO, 2009). Vitamin A is a group of fat-soluble nutritionally unsaturated hydrocarbons, which include retinal, retinoic acid, retinol, and several provitamin A carotenoids. Preformed vitamin A is found in animal products, whereas provitamin A is found in plant-based food products like fruits and vegetables. Carotenoid is the most common form of provitamin A. Provitamin A carotenoids include beta-carotene, beta-cryptoxanthin, and alpha-carotene (Cruet-Burgos et al., 2020). Cereals are the primary source of nutrients in many regions where vitamin A deficiency is more prevalent. They especially lack provitamin A carotenoids (Zhai et al., 2016). Hence, an increase in provitamin A carotenoids could alleviate vitamin A deficiency problems in areas where cereals serve as their primary dietary source. Sorghum is a staple food for millions of people in semi-arid regions of sub-Saharan Africa and Asia (FAO, 1995), but it is deficient in provitamin A.

Genetic variation is reported for provitamin A carotenoids in sorghum. The variability for β-carotene concentration was low, including for yellow endosperm lines of sorghum, whose maximum was 1.13 mg/kg in IS 26886 (Reddy et al., 2005). Recently, Cruet-Burgos et al. assessed variability for provitamin A carotenoids in a diverse germplasm accession using GWAS (2020). There was moderate variation for β-carotene (0.8 μg/g), lutein (0.3–9.4 μg/g), and zeaxanthin (0.2–9.1 μg/g), but β-cryptoxanthin and α-carotene were nearly undetectable. GWAS revealed significant associations between carotenoids and SNPs, which were co-localized with known carotenoid genes. A notable SNP was found inside the putative ortholog of maize (ZEP). Based on association with zeaxanthin and with the lutein-zeaxanthin ratio, zeaxanthin epoxidase was found to be the major gene that controls carotenoid variation. This trait is known to have an oligogenic inheritance. To date, no biofortified variety for provitamin A has been released in sorghum. A transgenic sorghum with high levels of provitamin A carotenoids was developed by the Africa Biofortified Sorghum project in 2012 but was prevented from public dissemination due to regulatory interventions (Zhao, 2007). Genomics-assisted molecular tools can be used as selection tools in traditional sorghum biofortification programs to fasten the breeding cycle.

5.7 SORGHUM GRAIN SHELF-LIFE ENHANCEMENT

Shelf life is an important parameter that promotes sorghum and other millets in the context of nutri-cereals on a commercial scale. Sorghum grains are mostly stored as whole grains or flour. Millets like sorghum and pearl millet, when stored as flour, become rancid after a period. Higher lipid content is present in the germ and endosperm of sorghum grain (Sruthi and Rao, 2021). Lipolytic enzymes like lipases released during milling degrade fat and release free fatty acids, which result in flour rancidity and poor quality (Dayakar et al., 2015). This, in turn, affects the palatability and quality of the products made from them.

The important indices that can be used to determine the quality of sorghum flour under storage are alcoholic acidity and water activity. Alcoholic acidity is defined as mg of H_2SO_4 required for 100 g of the sample to have the same alcohol-soluble acids. It should not exceed 0.18 to avoid deterioration of flour during storage. Water activity is considered an essential and useful expression for microbial growth and enzyme activity. The water activity level in food decides and controls the onset and extent of mold attack. Food that spoils easily is likely to deteriorate due to biochemical changes resulting from high moisture content. In a recent study by Nanaiah et al., raw sorghum and parboiled sorghum were observed over 90 days of storage (2019). Alcoholic acidity increases in raw sorghum during storage, which affects the shelf life of sorghum. Meanwhile, in parboiled sorghum, it remains the same, allowing it to be stored for a more extended period of 90 days. Hence, parboiling prolongs the storage period. Parboiled sorghum was found to have more water activity at the initial stages of storage due to biological and chemical changes during parboiling. Meanwhile, in raw sorghum, there was no such change. Breeding efforts to identify and exploit the variation for these traits among the adapted cultivars and germplasm sources can improve the shelf life of sorghum grain and flour. Though it is one of the most critical post-harvest issues to be addressed, research on this aspect is scarce.

5.8 POST-HARVEST SEED PROCESSING TRAIT ENHANCEMENT IN SORGHUM

The sorghum grain can be cooked similarly to rice, cracked similarly to oats for porridge, malted similarly to barley for beer, baked similarly to wheat to make flatbreads, and can also be popped similarly to popcorn for snacks. Dehulling and milling the grain into flour and semolina are the first steps in processing. After the primary process, the raw material is used to create ready-to-cook and ready-to-eat goods through secondary processing, including flaking, extrusion, popping, baking, and combining with other cereals. The development of food-grade sorghum varieties takes into account several important characteristics, such as white pericarp color, thin mesocarp, standard endosperm type, low tannin content, tan-plant necrotic lesion color (Tuinstra, 2008; Waniska and Rooney, 2002), grain size, shape, and endosperm texture. Endosperm texture is divided into vitreous (hard), intermediate, and floury (soft) kinds depending on the ratio of floury to vitreous endosperm in the grains. Based on endosperm quality, sorghum grains are categorized as waxy, heterodoxy, and non-waxy. The amylose/amylopectin ratio and endosperm type can be modified by altering genes like waxy (wx), which is involved in starch production. Endosperm hardness and density are positively correlated with better milling yield. Hard endosperm cultivars are more resistant to grain molds than other cultivars (Jambunathan et al., 1992). According to Rooney and Waniska, the grains with good milling qualities contain spherical kernels with a high proportion of firm endosperm, thick white pericarps, and non-pigmented tests that yield the most decorticated grain with the least amount of breakage during milling (2000). Compared to products made with whole wheat, high-tannin sorghum brans can be added to bread and cookies in small amounts without significantly changing the texture or flavor. Cookies taste better when made with malted and steam-treated sorghum flour, which includes more damaged starch (Hugo et al., 2000). Additionally, it has been noted that soft endosperm sorghum flour produces biscuits of better quality (Aruna and Visarada, 2019). Grains with heterodoxy endosperm texture are suitable for unfermented bread making and tortillas, while grains with waxy endosperm are suitable for making fermented bread. Amylose content and the interactions of protein and starch inside the endosperm cells are two elements that determine the quality of the porridge (Bello et al., 1990).

Grains with hard and non-waxy endosperm textures are preferred for thick porridge and couscous making. For brewing grains, a soft endosperm is required that can quickly absorb water for enzyme mobilization during grain germination to yield malt (House et al., 2000). Additionally, *in vitro* research revealed that waxy sorghum differs from standard sorghum in that it contains a significantly higher proportion of quickly digestible starch and a smaller amount of resistant starch (Sang et al., 2008), which allows for rapid digestion with amylases and is suitable for brewing. The waxy and heteroxy waxy lines had the maximum efficiency in the synthesis of bioethanol (Taylor et al., 2014; Wu et al., 2010). Compared to other unpigmented cereals, sorghum is dark in color and less palatable. Breeding sorghum for decreased polyphenol oxidase (PPO) activity can reduce this dark coloration of grain (Hikeezi et al., 2013). Food-grade sorghum cultivars with white pericarp, tan secondary-plant color, and light-colored glumes are used to make the finest quality sorghum flours and food products. Other traits that enhance grain and flour color characteristics, like glume color, can also have a significant impact on the grain's quality. The glumes can be black, dark brown, or purple, or they might be light tan or straw-colored in sorghum. Light-colored cultivars typically yield grain and flour with excellent color qualities. Research groups in ICRISAT and India have released many food-grade sorghum hybrids. They are tall, late duration, and high-yielding in nature. To make beer, colored sorghum is preferable over white grain sorghum because it imparts the desired dark color. Sorghums with red pericarps, notably those with colored testas, are frequently utilized to make malt, as well as local alcoholic beverages and porridge (Waniska and Rooney, 2002). The size, hardness, and decortication yield of the kernels are additional crucial quality factors. Small or soft grain varieties are not favored, whereas varieties with large or bold grains often produce more flour (Tuinstra, 2008). Amylose content and thick pericarp genotypes were found to pop well (Mishra et al., 2015). It has been claimed that popping sorghum grain will reduce its phytic acid level and improve the digestion of its starch and protein (Saravanabavan et al., 2013). Popping efficiency and expansion ratio were both influenced by both genotype and environment, but the genotype effect was significantly larger than the environment effect, and both traits have high heritability, suggesting that selection for improved popping is possible (Pugh et al., 2017; Rooney and Rooney, 2013). *Sorghum roxburghii* var. *hians*, which is the group of sorghum varieties known as the *Talavirchina*, is known for its tiny grain and corneous endosperm and is suitable for popping (Ayyangar and Ayyar, 1936). Sorghum is a starchy cereal with outstanding extrusion cooking capabilities, making it comparable to corn and rice in terms of extrusion properties (Llopart et al., 2014). Hard endosperm sorghum generates the greatest noodles in terms of texture after cooking, but soft texture sorghum is needed for excellent pasta products. Intermediate to harder endosperm texture with high recovery is suitable for semolina making (Ganapathy and Dayakar, 2016). Most of the breeding programs in sorghum hitherto were aimed at improving yield characteristics and biotic and abiotic stress tolerance. Though some works were focused on the improvement of processing traits, more attention has to be directed to meeting the end-use-based breeding. Significant diversity in kernel size, structure, shape, texture, endosperm hardness, pigmentation, starch recovery, peripheral endosperm fraction, protein, and oil is present in the germplasm, which can be extensively screened and utilized in crop improvement programs for improved milling and processing in sorghum.

5.9 CONCLUSION

Present-day agriculture is challenged by the varying climatic conditions, which reduce the performance of major food crops like rice and wheat. Climate-resilient food crops like sorghum are the hope for future food and nutritional security. Though sorghum is a hardy crop, many nutritional and quality parameters need further improvement to exploit its full potential. Protein content and quality, Fe and Zn content and bioavailability, Vitamin A content enhancement, sugar and starch content, seed shelf life, and processing traits improvement are a few of the important areas that need to be addressed for overall quality enhancement. Understanding the genetics and regulatory network behind these traits and their interrelationships can better enhance them through crop improvement and molecular tools. Characterization of the already existing genetic variation and ways to incorporate it into the cultivars need to be focused upon. However, research into these areas is limited. Intensive efforts towards sorghum quality improvement, integrating both nutritional and quality aspects of sorghum-based food and bio-products, are essential.

KEYWORDS

- **Africa biofortified sorghum**
- **food fortification**
- **fortification**
- **grain protein**
- **provitamin A carotenoids**
- **sorghum nutritional quality**
- **starch content**
- **sugar content**

REFERENCES

Alsaffar, A. A. (2011). Effect of food processing on the resistant starch content of cereals and cereal products—a review. *International Journal of Food Science & Technology, 46*(3), 455–462.

Andiku, C., Shimelis, H., Laing, M., Shayanowako, A. I. T., Adrogu Ugen, M., Manyasa, E., & Ojiewo, C. (2021). Assessment of sorghum production constraints and farmer preferences for sorghum variety in Uganda: Implications for nutritional quality breeding. *Acta Agriculture Scandinavica, Section B—Soil & Plant Science, 71*(7), 620–632. https://doi.org/10.1080/09064710.2021.1944297 (accessed on 7th November 2025).

Are, A. K., Srivastava, R. K., Mahalingam, G., Gorthy, S., Gaddameedi, A., Kunapareddy, A., Kotla, A., & Jaganathan, J. (2019). Application of plant breeding and genomics for improved sorghum and pearl millet grain nutritional quality. In *Sorghum and Millets* (pp. 51–68). Elsevier.

Aruna, C., & Visarada, K. (2019). Sorghum grain in the food and brewing industry. In *Breeding Sorghum for Diverse End Uses*. Elsevier.

Awika, J. M., & Rooney, L. W. (2004). Sorghum phytochemicals and their potential impact on human health. *Phytochemistry, 65*(9), 1199–1221. https://doi.org/10.1016/j.phytochem.2004.04.001 (accessed on 7th November 2025).

Ayalew, H., Peiris, S., Chiluwal, A., Kumar, R., Tiwari, M., Ostmeyer, T., Bean, S., & Jagadish, S. V. K. (2022). Stable sorghum grain quality QTL was identified using SC35× RTx430 mapping population. *The Plant Genome*.

Ayyangar, G. N. R., & Ayyar, M. A. S. (1936). Sorghum for popping. *Madras Agricultural Journal, 25*, 323–328.

Bello, A. B., Rooney, L. W., & Waniska, R. D. (1990). Factors affecting quality of sorghum tô, a thick porridge. *Cereal Chemistry, 67*(1), 20–25.

Beta, T., Obilana, A. B., & Corke, H. (2001). Genetic diversity in properties of starch from Zimbabwean sorghum landraces. *Cereal Chemistry, 78*(5), 583–589.

Bihmidine, S., Baker, R. F., Hoffner, C., & Braun, D. M. (2015). Sucrose accumulation in sweet sorghum stems occurs by apo plasmic phloem unloading and does not involve differential sucrose transporter expression. *BMC Plant Biology, 15*(1), 186. https://doi.org/10.1186/s12870-015-0572-8 (accessed on 7th November 2025).

Blennow, A., Nielsen, T. H., Baunsgaard, L., Mikkelsen, R., & Engelsen, S. B. (2002). Starch phosphorylation: A new front line in starch research. *Trends in Plant Science, 7*(10), 445–450. https://doi.org/10.1016/s1360-1385(02)02332-4 (accessed on 7th November 2025).

Bressani, R., & Rios, B. J. (1962). The chemical and essential amino acid composition of twenty-five selections of grain sorghum. *Cereal Chemistry, 39*, 50–58.

Brown, K. H., Rivera, J. A., Bhutta, Z., Gibson, R. S., King, J. C., Lönnerdal, B., Ruel, M. T., Sandtröm, B., Wasantwisut, E., & Hotz, C. (2004). International Zinc Nutrition Consultative Group (IZiNCG) technical document #1. Assessment of the risk of zinc deficiency in populations and options for its control. *Food and Nutrition Bulletin, 25*(1 Suppl 2), S99–S203.

Burdette, A., Garner, P. L., Mayer, E. P., Hargrove, J. L., Hartle, D. K., & Greenspan, P. (2010). Anti-inflammatory activity of select sorghum (*Sorghum bicolor*) brans. *Journal of Medicinal Food, 13*(4), 879–887. https://doi.org/10.1089/jmf.2009.0147 (accessed on 7th November 2025).

Cakmak, I., Torun, A., Millet, E., Feldman, M., Fahima, T., Korol, A., Nevo, E., Braun, H. J., & Özkan, H. (2004). *Triticum dicoccoides*: An important genetic resource for increasing zinc and iron concentration in modern cultivated wheat. *Soil Science and Plant Nutrition, 50*(7), 1047–1054. https://doi.org/10.1080/00380768.2004.10408573 (accessed on 7th November 2025).

Calviño, M., & Messing, J. (2012). Sweet sorghum as a model system for bioenergy crops. *Current Opinion in Biotechnology, 23*(3), 323–329. https://doi.org/10.1016/j.copbio.2011.12.002 (accessed on 7th November 2025).

Calviño, M., Bruggmann, R., & Messing, J. (2011). Characterization of the small RNA component of the transcriptome from grain and sweet sorghum stems. *BMC Genomics, 12*(1), 356. https://doi.org/10.1186/1471-2164-12-356 (accessed on 7th November 2025).

Chavan, U. D., & Patil, J. V. (2010). *Grain Sorghum Processing*. IBDC Publishers.

Chavan, U. D., Jagtap, Y. K., Shinde, M. S., & Patil, J. V. (2016). Preparation and nutritional quality of sorghum chakali. *International Journal of Recent Scientific Research, 7*(1), 8404–8411.

Chen, B. R., Wang, C., Ping, W., Zhu, Z., Ning, X., Shi, G., Miao, Y., Nai, W., Li, J., & Hou, J. (2019). Genome-wide association study for starch content and constitution in sorghum (*Sorghum bicolor* (L.) Moench). *Journal of Integrative Agriculture, 18*(11), 2446–2456.

Cooper, E. A., Brenton, Z. W., Flinn, B. S., Jenkins, J., Shu, S., Flowers, D., Luo, F., Wang, Y., Xia, P., Barry, K., Daum, C., Lipzen, A., Yoshinaga, Y., Schmutz, J., Saski, C., Vermerris, W., & Kresovich, S. (2019). A new reference genome for *Sorghum bicolor* reveals high levels of sequence similarity between sweet and grain genotypes: Implications for the genetics of sugar metabolism. *BMC Genomics, 20*(1), 420. https://doi.org/10.1186/s12864-019-5734-x (accessed on 7th November 2025).

Cruet-Burgos, C., Cox, S., Ioerger, B. P., Perumal, R., Hu, Z., Herald, T. J., Bean, S. R., & Rhodes, D. H. (2020). Advancing provitamin A biofortification in sorghum: Genome-wide association studies of grain carotenoids in global germplasm: Cruet-Burgos et al. *Plant Genome, 13*(1), e20013. https://doi.org/10.1002/tpg2.20013 (accessed on 7th November 2025).

Englyst, H. N., Kingman, S. M., & Cummings, J. H. (1992). Classification and measurement of nutritionally important starch fractions. *European Journal of Clinical Nutrition, 46*(Suppl 2), S33–S50.

French, D. (1984). Organization of starch granules. In *Starch: Chemistry and Technology* (pp. 183–247). Elsevier.

Galicia-García, T., Martínez-Bustos, F., Jiménez-Arévalo, O., & Aguilar-Palazuelos, E. (2007). Mechanical properties of extruded biodegradable films of native starch and sugar cane fibre. *Food Engineering Research Developments*, 263–278.

Ganapathy, K. N., & Dayakar, B. R. (2016). End product specific breeding in sorghum—Quality and nutritional considerations. *Indian Farm, 65*, 18–20.

Hama, F., Icard-Vernière, C., Guyot, J. P., Picq, C., Diawara, B., & Mouquet-Rivier, C. (2011). Changes in micro- and macronutrient composition of pearl millet and white sorghum during in-field versus laboratory decortication. *Journal of Cereal Science, 54*(3), 425–433.

Hamaker, B. R., & Bugusu, B. A. (2003). Overview: Sorghum proteins and food quality. In *Workshop on the Proteins of Sorghum and Millets: Enhancing Nutritional and Functional Properties for Africa*. Pretoria, South Africa.

Hamaker, B. R., Mohamed, A. A., Habben, J. E., Huang, C. P., & Larkins, B. A. (1995). Efficient procedure for extracting maize and sorghum kernel proteins reveals higher prolamin contents than the conventional method. *Cereal Chemistry*.

Hikeezi, D. M., Duodu, K. G., Chisi, M., Rooney, L. W., & Taylor, J. R. N. (2013). Polyphenol oxidase activity in white tan-plant-type sorghums: An important determinant of the relatively dark color of their porridges. *International Journal of Food Science and Technology, 48*(5), 941–946. https://doi.org/10.1111/ijfs.12045 (accessed on 7th November 2025).

Hill, H., Slade Lee, L., & Henry, R. J. (2012). Variation in sorghum starch synthesis genes associated with differences in starch phenotype. *Food Chemistry, 131*(1), 175–183. https://doi.org/10.1016/j.foodchem.2011.08.057 (accessed on 7th November 2025).

House, L. R., Gomez, M., Murty, D. S., Sun, Y., & Verma, B. N. (2000). Development of some agricultural industries in several African and Asian countries. In *Sorghum: Origin, History, Technology and Production* (pp. 131–190).

Hugo, L. F., Rooney, L. W., & Taylor, J. R. N. (2000). Malted sorghum as a functional ingredient in composite bread. *Cereal Chemistry, 77*(4), 428–432. https://doi.org/10.1094/cchem.2000.77.4.428 (accessed on 7th November 2025).

Jambunathan, R., Kherdekar, M. S., & Stenhouse, J. W. (1992). Sorghum grain hardness and its relationship to mold susceptibility and mold resistance. *Journal of Agricultural and Food Chemistry, 40*(8), 1403–1408. https://doi.org/10.1021/jf00020a023 (accessed on 7th November 2025).

Jebril, J., Wang, D., Rozeboom, K., & Tesso, T. (2021). Grain sink removal increases stalk juice yield, sugar accumulation, and biomass in sweet sorghum (*Sorghum bicolor* (L.) Moench). *Industrial Crops and Products, 173*(114089), 114089. https://doi.org/10.1016/j.indcrop.2021.114089 (accessed on 7th November 2025).

Johns, C. O., & Brewster, J. F. (1916). Kafirin, an alcohol-soluble protein from kafir, *Andropogon sorghum*. *Journal of Biological Chemistry, 28*(1), 59–65. https://doi.org/10.1016/s0021-9258(18)86843-0 (accessed on 7th November 2025).

Juliano, B. O. (1979). The chemical basis of rice grain quality. In *Chemical Aspects of Rice Grain Quality* (p. 69).

Kayodé, A. P. P., Linnemann, A. R., Hounhouigan, J. D., Nout, M. J. R., & van Boekel, M. A. J. S. (2006). Genetic and environmental impact on iron, zinc, and phytate in food sorghum grown in Benin. *Journal of Agricultural and Food Chemistry, 54*(1), 256–262. https://doi.org/10.1021/jf0521404 (accessed on 7th November 2025).

Kering, M. K., Temu, V. W., & Rutto, L. K. (2017). Nitrogen fertilizer and panicle removal in sweet sorghum production: Effect on biomass, juice yield and soluble sugar content. *Journal of Sustainable Bioenergy Systems, 7*(1), 14–26. https://doi.org/10.4236/jsbs.2017.71002 (accessed on 7th November 2025).

Keyata, E. O., Tola, Y. B., Bultosa, G., & Forsido, S. F. (2021). Premilling treatments effects on nutritional composition, antinutritional factors, and *in vitro* mineral bioavailability of the improved Assosa I sorghum variety (*Sorghum bicolor* L). *Food Science & Nutrition, 9*(4), 1929–1938.

Kulamarva, A. G., Sosle, V. R., & Raghavan, G. S. V. (2009). Nutritional and rheological properties of sorghum. *International Journal of Food Properties, 12*(1), 55–69. https://doi.org/10.1080/10942910802252148 (accessed on 7th November 2025).

Kumar, A. A., Anuradha, K., Ramaiah, B., Grando, S., Frederick, H., Rattunde, W., Virk, P., & Pfeiffer, W. H. (2015). Recent advances in sorghum biofortification research. In *Plant Breeding Reviews: Volume 39* (pp. 89–124). John Wiley & Sons, Inc.

Kumar, A., Reddy, A., Ramaiah, B. V. S., Sahrawat, B., & Pfeiffer, K. L. (2012). Genetic variability and character association for grain iron and zinc contents in sorghum germplasm accessions and commercial cultivars. *European Journal of Plant Science and Biotechnology, 6*(1), 1–5.

Kumar, T., Dweikat, I., Sato, S., Ge, Z., Nersesian, N., Chen, H., Elthon, T., Bean, S., Ioerger, B. P., Tilley, M., & Clemente, T. (2012). Modulation of kernel storage proteins in grain sorghum (*Sorghum bicolor* (L.) Moench): Modulation of sorghum kernel proteins. *Plant Biotechnology Journal, 10*(5), 533–544. https://doi.org/10.1111/j.1467-7652.2012.00685.x (accessed on 7th November 2025).

Labuschagne, M. T. (2018). A review of cereal grain proteomics and its potential for sorghum improvement. *Journal of Cereal Science, 84*, 151–158. https://doi.org/10.1016/j.jcs.2018.10.010 (accessed on 7th November 2025).

Laidlaw, H. K. C., Mace, E. S., Williams, S. B., Sakrewski, K., Mudge, A. M., Prentis, P. J., Jordan, D. R., & Godwin, I. D. (2010). Allelic variation of the β-, γ- and δ-kafirin genes in diverse sorghum genotypes. *Theoretical and Applied Genetics, 121*(7), 1227–1237.

Lasztity, R. (2017). *The Chemistry of Cereal Proteins* (2nd ed.). Routledge.

Li, A., Jia, S., Yobi, A., Ge, Z., Sato, S. J., Zhang, C., Angelovici, R., Clemente, T. E., & Holding, D. R. (2018). Editing of an alpha-kafirin gene family increases digestibility and protein quality in sorghum. *Plant Physiology, 177*(4), 1425–1438.

Li, Y., Yuan, F., & Wang, B. (2018). Changes in the sugar content of sweet sorghum stems under natural conditions during winter in saline soil of the Yellow River Delta. In *IOP Conference Series: Earth and Environmental Science*. IOP Publishing.

Liu, G., Zhang, Y., Gong, H., Li, S., Pan, Y., Davis, C., Jing, H. C., Wu, L., & Godwin, I. D. (2021). Stem vacuole-targeted sucrose isomerase enhances sugar content in sorghum. *Biotechnology for Biofuels, 14*(1), 1–14.

Llopart, E. E., Drago, S. R., De Greef, D. M., Torres, R. L., & González, R. J. (2014). Effects of extrusion conditions on physical and nutritional properties of extruded whole grain red sorghum (*Sorghum spp*). *International Journal of Food Sciences and Nutrition, 65*(1), 34–41. https://doi.org/10.3109/09637486.2013.836737 (accessed on 7th November 2025).

Macedo, W. R., Araújo, D. K., Santos, V. M., Castro, P. R. de C. e., & Fernandes, G. M. (2017). Plant growth regulators on sweet sorghum: Physiological and nutritional value analysis. *Comunicações em Ciências, 8*(1), 170. https://doi.org/10.14295/cs.v8i1.1315 (accessed on 7th November 2025).

McKinley, B. A., Casto, A. L., Rooney, W. L., & Mullet, J. E. (2018). Developmental dynamics of stem starch accumulation in *Sorghum bicolor*. *Plant Direct, 2*(8), e00074. https://doi.org/10.1002/pld3.74 (accessed on 7th November 2025).

Mertz, E. T., Hassen, M. M., Cairns-Whittern, C., Kirleis, A. W., Tu, L., & Axtell, J. D. (1984). Pepsin digestibility of proteins in sorghum and other major cereals. *Proceedings of the National Academy of Sciences of the United States of America, 81*(1), 1–2. https://doi.org/10.1073/pnas.81.1.1 (accessed on 7th November 2025).

Mishra, G., Joshi, D. C., Mohapatra, D., & Babu, V. B. (2015). Varietal influence on the microwave popping characteristics of sorghum. *Journal of Cereal Science, 65*, 19–24. https://doi.org/10.1016/j.jcs.2015.06.001 (accessed on 7th November 2025).

Murray, S. C., Sharma, A., Rooney, W. L., Klein, P. E., Mullet, J. E., Mitchell, S. E., & Kresovich, S. (2008). Genetic improvement of sorghum as a biofuel feedstock: I. QTL for stem sugar and grain nonstructural carbohydrates. *Crop Science, 48*(6), 2165–2179. https://doi.org/10.2135/cropsci2008.01.0016 (accessed on 7th November 2025).

Nanaiah, G. K., Benhur, D. R., Rakshit, S., Devender, V., Chinnegowda, V. S., & Tonapi, V. A. (2019). Variation in nutritional and shelf-life parameters among Rabi sorghum cultivars and effect of processing on these parameters. *Agricultural Research, 8*(4), 513–522. https://doi.org/10.1007/s40003-018-0393-8 (accessed on 7th November 2025).

Niba, L. L., & Hoffman, J. (2003). Resistant starch and β-glucan levels in grain sorghum (*Sorghum bicolor* M.) are influenced by soaking and autoclaving. *Food Chemistry, 81*(1), 113–118. https://doi.org/10.1016/s0308-8146(02)00386-2 (accessed on 7th November 2025).

Ojha, P., Adhikari, R., Karki, R., Mishra, A., Subedi, U., & Karki, T. B. (2018). Malting and fermentation effects on antinutritional components and functional characteristics of sorghum flour. *Food Science & Nutrition, 6*(1), 47–53. https://doi.org/10.1002/fsn3.525 (accessed on 7th November 2025).

Oria, M. P., Hamaker, B. R., Axtell, J. D., & Huang, C. P. (2000). A highly digestible sorghum mutant cultivar exhibits a unique folded structure of endosperm protein bodies. *Proceedings of the National Academy of Sciences, 97*(10), 5065–5070.

Popova, A., & Mihaylova, D. (2019). Antinutrients in plant-based foods: A review. *The Open Biotechnology Journal.*

Pugh, N. A., Awika, J. M., & Rooney, W. L. (2017). Heritability of popping characteristics in sorghum grain. *Crop Science, 57*(1), 71–77.

Qazi, H. A., Paranjpe, S., & Bhargava, S. (2012). Stem sugar accumulation in sweet sorghum—Activity and expression of sucrose metabolizing enzymes and sucrose transporters. *Journal of Plant Physiology, 169*(6), 605–613.

Rao, D., Kalpana, B., Srinivas, K., & Patil, K. (2015). Development and standardization of sorghum-rich multigrain flour and assessment of its storage stability with addition of TBHQ. *Journal of Food Processing and Preservation, 39*(5), 451–457.

Rao, D., Mathew, B., Bharath Kumar, B., Karthikeyan, K. A., Hymajyothi, K., Parwez, S., Ratnavathi, S., & Seetharama, C. (2003). Industrial utilization of sorghum in India—Status and prospects. *NATP (NRCS) Series.*

Reddy, B. V. S., Ramesh, S., & Longvah, T. (2005). Prospects of breeding for micronutrients and β-carotene-dense sorghums. *International Sorghum and Millets Newsletter, 46*, 10–14.

Rhodes, D. H., Hoffmann, L., Rooney, W. L., Herald, T. J., Bean, S., Boyles, R., Brenton, Z. W., & Kresovich, S. (2017). Genetic architecture of kernel composition in global sorghum germplasm. *BMC Genomics, 18*(1). https://doi.org/10.1186/s12864-016-3403-x (accessed on 7th November 2025).

Rooney, L. W., & Serna-Saldivar, S. O. (2000). Sorghum. In *Handbook of Cereal Science and Technology* (2nd ed.). In Kulp, K., & Ponte, J. G. (Eds.).

Rooney, L. W., & Waniska, R. D. (2000). Sorghum food and industrial utilization. In *Sorghum: Origin, History, Technology, And Production* (pp. 689–729).

Rooney, L. W., & Waniska, R. D. (2004). Crop utilization and marketing: Food and nutritional quality of sorghum and millet, 72474–77843.

Rooney, T. E., & Rooney, W. L. (2013). Genotype and environment effects on the popping characteristics of grain sorghum. *Journal of Crop Improvement, 27*(4), 460–468. https://doi.org/10.1080/15427528.2013.794181 (accessed on 7th November 2025).

Rosas-Quijano, R., Ontiveros-Cisneros, A., Montes-García, N., Díaz-Franco, A., Vázquez-Ovando, A., & Gálvez-López, D. (2021). A general overview of sweet sorghum genomics. In *Application of Biotechnology for Biomass*. IntechOpen.

Safian, N., Naderi, M. R., Torabi, M., Soleymani, A., & Salemi, H. R. (2022). Corn (*Zea mays* L.) and sorghum (*Sorghum bicolor* (L.) Moench) yield and nutritional quality affected by drought stress. *Biocatalysis and Agricultural Biotechnology, 45*(102486), 102486. https://doi.org/10.1016/j.bcab.2022.102486 (accessed on 7th November 2025).

Saltzman, A., Birol, E., Bouis, H. E., Boy, E., De Moura, F. F., Islam, Y., & Pfeiffer, W. H. (2013). Biofortification: Progress toward a more nourishing future. *Global Food Security, 2*(1), 9–17. https://doi.org/10.1016/j.gfs.2012.12.003 (accessed on 7th November 2025).

Sang, Y., Bean, S., Seib, P. A., Pedersen, J., & Shi, Y. C. (2008). Structure and functional properties of sorghum starches differing in amylose content. *Journal of Agricultural and Food Chemistry, 56*(15), 6680–6685. https://doi.org/10.1021/jf800577x (accessed on 7th November 2025).

Sapkota, S., Boatwright, J. L., Jordan, K., Boyles, R., & Kresovich, S. (2020). Identification of novel genomic associations and gene candidates for grain starch content in sorghum. *Genes (Basel), 11*(12), 1448. https://doi.org/10.3390/genes11121448 (accessed on 7th November 2025).

Semere, T., Tsehaye, Y., Tareke, L., Westengen, O. T., & Fjellheim, S. (2023). Nutritional and antinutritional potentials of sorghum: A comparative study among different sorghum landraces of Tigray, Northern Ethiopia. *Agriculture, 13*(9).

Shewry, P. R., & Halford, N. G. (2002). Cereal seed storage proteins: Structures, properties and role in grain utilization. *Journal of Experimental Botany, 53*(370), 947–958. https://doi.org/10.1093/jexbot/53.370.947 (accessed on 7th November 2025).

Shobhana, V. G., Senthil, N., Kalpana, K., Abirami, B., Sangeetha, J., Saranya, B., Jeevan, R. J., Tamilkumar, P., Nagarajan, P., Velu, V., Arumugachamy, S., Joel, A. J., Vellaikumar, S., & Raveendran, M. (2013). Comparative studies on the iron and zinc contents estimation using atomic absorption spectrophotometer and grain staining techniques (Prussian blue and DTZ) in maize germplasms. *Journal of Plant Nutrition, 36*(2), 329–342. https://doi.org/10.1080/01904167.2012.744419 (accessed on 7th November 2025).

Shull, J. M., Watterson, J. J., & Kirleis, A. W. (1992). Purification and immunocytochemical localization of kafirins in *Sorghum bicolor* (L. Moench) endosperm. *Protoplasm, 171*(1–2), 64–74. https://doi.org/10.1007/bf01379281 (accessed on 7th November 2025).

Silva, E. M. S., Peres, A. E. C., Silva, A. C., Leal, M. C. D. M., Liao, L. M., & De Almeida, V. O. (2019). Sorghum starch as depressant in mineral flotation: Part 1—Extraction and characterization. *Journal of Materials Research and Technology, 8*(1), 396–402.

Song, R., Segal, G., & Messing, J. (2004). Expression of the sorghum 10-member kafirin gene cluster in maize endosperm. *Nucleic Acids Research, 32*(22), e189. https://doi.org/10.1093/nar/gnh183 (accessed on 7th November 2025).

Sorghum and millet in cultivated plants, primarily as food sources. In *Encyclopedia of Life Support Systems (EOLSS)*, developed under the auspices of the UNESCO.

Sruthi, N. U., & Rao, P. S. (2021). Effect of processing on storage stability of millet flour: A review. *Trends in Food Science & Technology, 112*, 58–74. https://doi.org/10.1016/j.tifs.2021.03.043 (accessed on 7th November 2025).

Subramanian, V., & Metta, V. C. (2000). Sorghum grain for poultry feed. In A. Chandrasher, R. Bandyopadhayi, & A. J. Hall (Eds.), *Technical and Institutional Options for Sorghum Grain Mold Management* (pp. 502–504). International Crops Research Institute for the Semi-Arid Tropics (ICRISAT), Patancheru.

Sukumaran, S., Xiang, W., Bean, S. R., Pedersen, J. F., Kresovich, S., Tuinstra, M. R., Tesso, T. T., Hamblin, M. T., & Yu, J. (2012). Association mapping for grain quality in a diverse sorghum collection. *Plant Genome, 5*(3), 126–135. https://doi.org/10.3835/plantgenome2012.07.0016 (accessed on 7th November 2025).

Taylor, J. R. N., Belton, P. S., Beta, T., & Duodu, K. G. (2014). Increasing the utilization of sorghum, millets and pseudocereals: Developments in the science of their phenolic phytochemicals, biofortification and protein functionality. *Journal of Cereal Science, 59*(3), 257–275. https://doi.org/10.1016/j.jcs.2013.10.009 (accessed on 7th November 2025).

Theurer, C. B. (1986). Grain processing effects on starch utilization by ruminants. *Journal of Animal Science, 63*(5), 1649–1662. https://doi.org/10.2527/jas1986.6351649x (accessed on 7th November 2025).

Tuinstra, M. R. (2008). Food-grade sorghum varieties and production considerations: A review. *Journal of Plant Interactions, 3*(1), 69–72. https://doi.org/10.1080/17429140701722770 (accessed on 7th November 2025).

Vinutha, K. S., Rayaprolu, L., Yadagiri, K., Umakanth, A. V., Patil, J. V., & Srinivasa Rao, P. (2014). Sweet sorghum research and development in India: Status and prospects. *Sugar Tech, 16*(2), 133–143. https://doi.org/10.1007/s12355-014-0302-9 (accessed on 7th November 2025).

Waniska, R. D., & Rooney, L. K. (2008). Sorghum grain quality for increased utilization. In *Sorghum and Millets Diseases* (pp. 327–335). Iowa State Press, Ames, Iowa, USA.

Wong, J. H., Marx, D. B., Wilson, J. D., Buchanan, B. B., Lemaux, P. G., & Pedersen, J. F. (2010). Principal component analysis and biochemical characterization of protein and starch reveal primary targets for improving sorghum grain. *Plant Science, 179*(6), 598–611. https://doi.org/10.1016/j.plantsci.2010.08.020 (accessed on 7th November 2025).

World Health Organization. (2009). *Global Prevalence of Vitamin A Deficiency in Populations at Risk 1995–2005: WHO Global Database on Vitamin A Deficiency.*

Wu, X., Jampala, B., Robbins, A., Hays, D., Yan, S., Xu, F., Rooney, W., Peterson, G., Shi, Y. C., & Wang, D. (2010). Ethanol fermentation performance of grain sorghums (*Sorghum bicolor*) with modified endosperm matrices. *Journal of Agricultural and Food Chemistry, 58*(17), 9556–9562. https://doi.org/10.1021/jf101555d (accessed on 7th November 2025).

Wu, Y., Yuan, L., Guo, X., Holding, D. R., & Messing, J. (2013). Mutation in the seed storage protein kafirin creates a high-value food trait in sorghum. *Nature Communications, 4*(1), 2217. https://doi.org/10.1038/ncomms3217 (accessed on 7th November 2025).

Xiong, Y., Zhang, P., Warner, R. D., & Fang, Z. (2019). Sorghum grain: From genotype, nutrition, and phenolic profile to its health benefits and food applications. *Comprehensive Reviews in Food Science and Food Safety, 18*(6), 2025–2046. https://doi.org/10.1111/1541-4337.12506 (accessed on 7th November 2025).

Xu, J. H., & Messing, J. (2008). Organization of the prolamin gene family provides insight into the evolution of the maize genome and gene duplications in grass species. *Proceedings of the National Academy of Sciences of the United States of America, 105*(38), 14330–14335. https://doi.org/10.1073/pnas.0807026105 (accessed on 7th November 2025).

Xu, J. H., & Messing, J. (2009). Amplification of prolamin storage protein genes in different subfamilies of the Poaceae. *Theoretical and Applied Genetics, 119*(8), 1397–1412. https://doi.org/10.1007/s00122-009-1143-x (accessed on 7th November 2025).

Yu, H., Cong, L., Zhu, Z., Wang, C., Zou, J., Tao, C., Shi, Z., & Lu, X. (2015). Identification of differentially expressed microRNA in the stems and leaves during sugar accumulation in sweet sorghum. *Gene, 571*(2), 221–230. https://doi.org/10.1016/j.gene.2015.06.056 (accessed on 7th November 2025).

Zarei, M., Amirkolaei, A. K., Trushenski, J. T., Sealey, W. M., Schwarz, M. H., & Ovissipour, R. (2022). Sorghum as a potential valuable aquafeed ingredient: Nutritional quality and digestibility. *Agriculture, 12*(5), 669. https://doi.org/10.3390/agriculture12050669 (accessed on 7th November 2025).

Zhai, S., Xia, X., & He, Z. (2016). Carotenoids in staple cereals: Metabolism, regulation, and genetic manipulation. *Frontiers in Plant Science, 7*, 1197. https://doi.org/10.3389/fpls.2016.01197 (accessed on 7th November 2025).

Zhao, Z. (2007). The Africa Biofortified Sorghum Project—Applying biotechnology to develop nutritionally improved sorghum for Africa. In *Biotechnology and Sustainable Agriculture 2006 and Beyond* (pp. 273–277). Springer.

Zhu, F. (2014). Structure, physicochemical properties, modifications, and uses of sorghum starch: Sorghum starch review. *Comprehensive Reviews in Food Science and Food Safety, 13*(4), 597–610. https://doi.org/10.1111/1541-4337.12070 (accessed on 7th November 2025).

CHAPTER 6

Nutritional Quality Enhancement in Pearl Millet

V. KRISHNASREE,[1,*] ROSHNI VIJAYAN,[2] A. V. S. DURGA PRASAD,[3] and GOVINDA RAI SARMA[4]

[1]*Department of Community Science, College of Agriculture, Kerala Agricultural University, Padannakkad, Kerala, India*

[2]*Division of Pulses, Regional Agricultural Research Station–Pattambi, Kerala Agricultural University, India.*

[3]*Department of Plant Breeding and Genetics, Acharya N. G. Ranga Agricultural University, Vizianagaram, Andhra Pradesh, India*

[4]*Division of Genetics, ICAR–Indian Agricultural Research Institute, Pusa, New Delhi, Delhi, India*

*Corresponding author

6.1 INTRODUCTION

The FAO and UN have designated 2023 as the "International Year of Millets" to accelerate the accomplishment of the Sustainable Development Goals (SDGs) as delineated in the 2030 Agenda. These goals encompass the eradication of hunger and food insecurity (SDG 2.1) as well as the elimination of all types of malnutrition (SDG 2.2). Cereals like wheat, rice, and barley have consistently topped the list of domesticated grains since the inception of agriculture due to their efficacy in providing good yields under intensive cultivation. However, findings from recent climate modeling projects envisaged sweeping changes in seasons concerning delayed monsoons and less summer along with an increase in temperature. Considering these changing climatic conditions, there is an urge to revamp the existing choice of cereals from wheat and rice to millets, which could be cultivated in arid and marginal lands to ensure sustainable agriculture. On the other hand, the latest comprehensive national nutritional survey (2016–18) conducted in India revealed that the country's younger generation faces the double burden of malnutrition, viz., obesity and micronutrient deficiency. Micronutrient deficiencies, viz., zinc (32%), iron (22%), vitamin A (16%), B12 (31%), and folate (37%) are prominent among adolescents along with the onset of diabetes, where one in 10 children of this group have fasting plasma glucose >100 mg/dl & ≤ 126 mg/dl or glycosylated Hb levels between 5.7%–6.4% (MoHFW, 2019).

Worldwide, over 155 million children are chronically undernourished, and 45% of deaths in children under five years are due to nutritional deficiencies (UNICEF, WHO, and

Millets in Sustainable Agriculture: Genetic Improvement, Nutritional Quality, Cultivation, and Integrated Pest–Disease Management. Roshni Vijayan, A. John Joel, R. Ambika Rajendran, and Adithya P. Balakrishnan (Eds.)

DOI: 10.1201/9781779642332-8

World Bank Group, 2017). Considering these health statistics, one of the prime research foci in the current scenario is to augment the nutritional quality of staples. Nutritional insecurity in a predominantly cereal-consuming population is a significant threat, emphasizing the necessity to include energy-rich and nutritionally superior grains like millets into the diet. The nutritional profile of millets is exemplary of the existing major cereals, which contain amino acids, vitamins, and minerals apart from high protein content. Considering the average iron requirement of vulnerable populations like adolescent girls and young women, 20% of it could be covered with dietary inclusion of 100 grams of millet (Figure 6.1).

Millets are rich in minerals, dietary fiber, and vitamins and have a lower glycemic index than other cereals that are widely consumed today. Thus, millets have the capacity to address nutritional deficiencies among vulnerable populations. The genetic diversity of millets offers myriad opportunities for farmers and consumers, as well as the ability to contribute to nutrition, health, and environmental sustainability. *Pennisetum glaucum*, a highly nutritious cereal crop, is widely cultivated on more than 30 million hectares globally and is commonly known as pearl millet. The majority of this land is located in Africa, encompassing more than 18 million hectares, and Asia, encompassing more than 10 million hectares. The crop contains 36.11 μg/100 g folate, 28.23 μg/100 g β-carotene, 27.35 mg/100 g calcium, 6.42 mg/100 g iron, 2.76 mg/100 g zinc, and 30.04 μg/100 g selenium. According to Raheem et al. (2021), pearl millet is potentially one of the most nutritious food choices for addressing malnutrition. The average production of pearl millet during 2010–12 was 9.4 million metric tons (Directorate of Millets Development, 2020). Nutritional-quality enhancement of foods for alleviating hidden hunger and malnutrition has been implemented since the 1990s. Various approaches, viz., genetic improvement, agronomic intervention, food fortification, dietary diversification, and supplementation with pharmacological doses, are being employed (Figure 6.2).

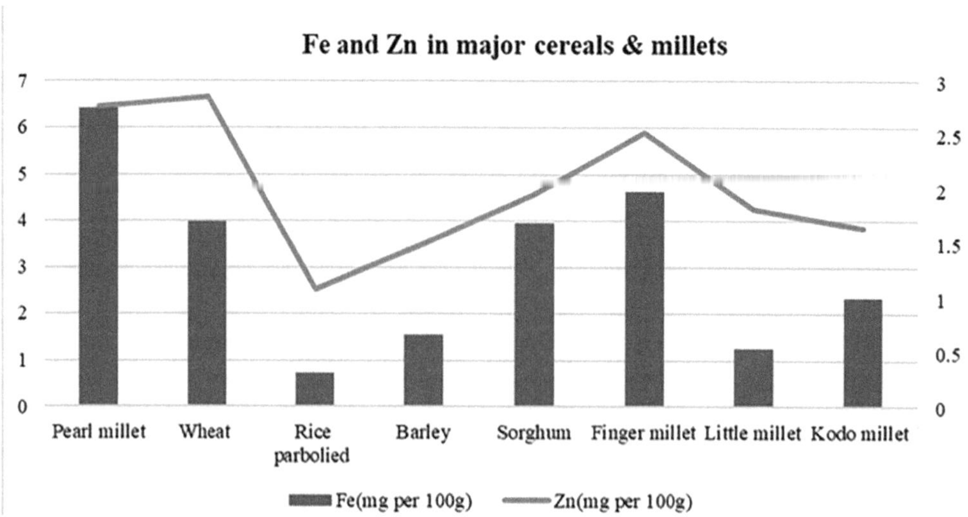

FIGURE 6.1 Iron and Zinc in major cereals and millets.

Source: Adapted Longvah et al. (2017).

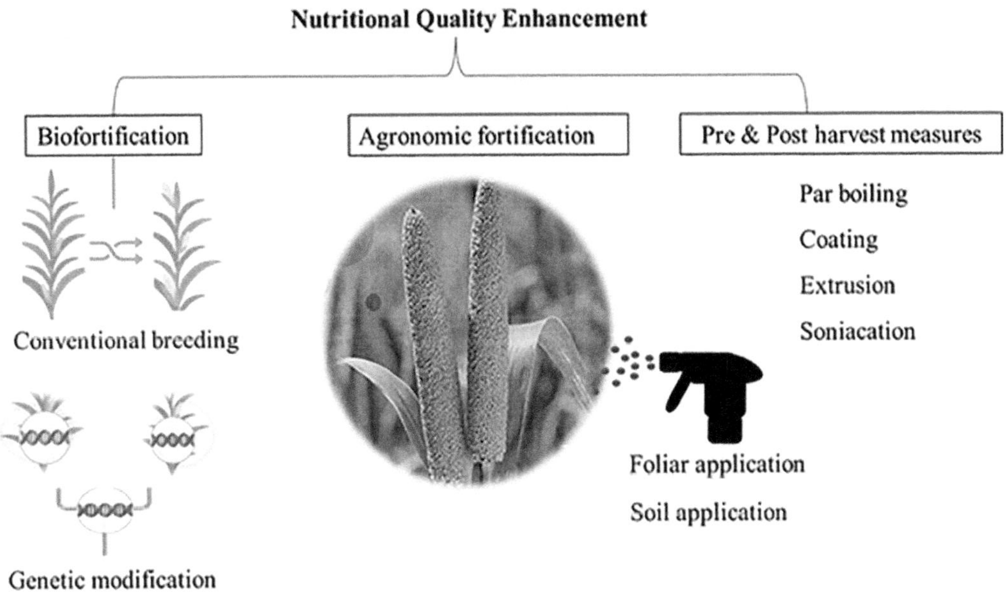

FIGURE 6.2 Nutritional quality enhancement of pearl millet.
Source: Srivastav et al. (2020); Satyavathi et al. (2021).

6.2 BIOFORTIFICATION OF PEARL MILLET

6.2.1 GERMPLASM SCREENING

Prior to introducing the most appropriate cultivars for biofortification, it is crucial to evaluate the genetic resources of high-quality cultivars that are rich in vitamins, cereal protein, and micronutrients and have a low glycemic index. The "Harvest Plus" project of the Consultative Group of International Agricultural Research (CGIAR) is actively searching for pearl millet genotypes that have elevated levels of vital nutrients. In 1985, the Indian Council of Agricultural Research (ICAR) founded the All India Coordinated Pearl Millet Improvement Project (AICPMIP) in Jodhpur. This effort is responsible for the cultivation of 50% of the country's pearl millet (Figure 6.3).

ICRISAT, located in Patancheru, India, developed a pearl millet breeding line that has a significant amount of carotenoids (137 µg/100 g) (Hash et al., 1996).

To develop superior cultivars with increased mineral content, numerous studies have been conducted to evaluate and assess pearl millet germplasms with elevated iron (Fe) and zinc (Zn) levels. This study references research conducted by Kanatti et al. (2014), Abdalla et al. (1998), Malik (1999), and Jambunathan and Subramanian (1988). The germplasms exhibit significant differences in their Fe (30–140 ppm) and Zn (20–90 ppm) levels, as reported by Govindaraj et al. (2019). According to Rai et al. (2015), most accessions with elevated iron (Fe) and zinc (Zn) contents—95 to 121 ppm and 59 to 87 ppm, respectively—were sourced from Ghana and Togo. This demonstrates that Indian genetic material can be successfully used to genetically increase the levels of iron (Fe) and zinc (Zn). The primary

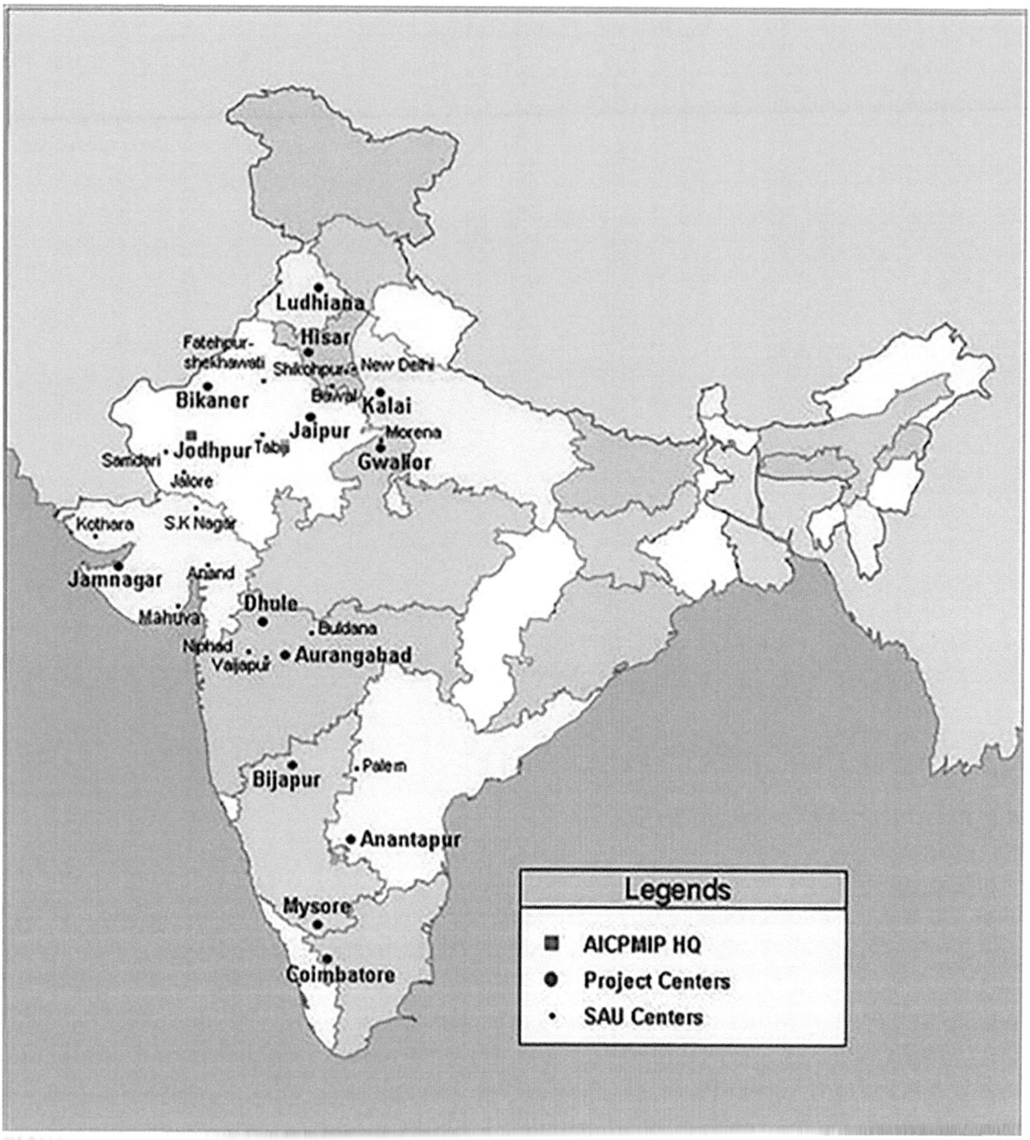

FIGURE 6.3 AICPMIP centers across India.

Source: Adopted from: http://www.aicpmip.res.in/aicpmipcentersmap.html.

pearl millet germplasm repositories are in Griffin, United States, at the International Crops Research Institute for the Semi-Arid Tropics (ICRISAT). The collections consist of 23,092 accessions from 52 countries, 238 mini core sets based on the entire collection, and a core set of 2094 accessions. The USDA-National Plant Germplasm System Pearl Millet Collection is another exceptional collection, consisting of 1297 distinct genotypes belonging to 31 countries.

6.2.2 GRAIN PROTEIN ENHANCEMENT

Pearl millet exhibits higher nutrient density compared to routinely consumed principal crops, viz., maize, wheat, rice, and jowar (Kumar et al., 2021). Pearl millet has a higher protein level compared to rice, although it is lower in protein content compared to wheat, with an approximate protein value of 11 g per 100 g. According to Hoseney et al., Rooney, and McDonough (1987), pearl millet grain has a more constant leucine/isoleucine ratio compared to sorghum or maize, despite being similar to small grains. Pearl millet has roughly 40% higher levels of lysine, methionine, and threonine compared to maize. According to Burton et al.'s findings in 1972, it is possible for it to have a crude protein content that is 60% higher. Research indicates that 100 grams of pearl millet grain typically contain a lysine content ranging from 1.9 to 3.9 grams. Pearl millet is a highly suitable option for persons with gluten intolerance because it has a low level of prolamin and maintains its alkaline properties even when heated.

Even though the protein content of pearl millet is 11.6%, it is devoid of important amino acids like tryptophan and lysine. Consequently, it is advisable to initiate research on quality protein pearl millet (QPPM) after conducting research on quality protein maize (QPM). It is crucial to highlight the significance of integrating desired features from end users or consumers into superior cultivars.

6.2.3 FE AND ZN FORTIFICATION

The initiative launched by ICRISAT and financed by HarvestPlus to alleviate the iron and zinc levels in pearl millet has demonstrated a viable approach that can be replicated to enhance the nutritional quality of other significant food crops. Pearl millet is identified as the primary product for increasing iron and zinc levels, as per the Biofortification Priority Index (BPI) established by the IFPRI. Since none of the widely recognized pearl millet varieties possess grain Fe levels beyond the Harvest Plus baseline of 77 ppm, they are not classified as High Fe pearl millet kinds. The concentrations of iron (Fe) and zinc (Zn) vary significantly, with values ranging from 31 to 125 parts per million (ppm) for Fe and 35 to 82 ppm for Zn, respectively. Genome-wide association studies (GWAS) and quantitative trait loci (QTL) analysis are effective strategies for identifying possible loci that influence the micronutrient content of pearl millet (Table 6.1). By utilizing these potential genetic locations to enhance the micronutrient composition of cereals, it is possible to improve the quality of commonly cultivated varieties.

A purposeful breeding program was undertaken with the aim of producing a cultivar that had a high iron content. This program exploits the genetic variation found in ICTP 8203, an open-pollinated cultivar known for its excellent productivity, early maturation, large seed size, and disease tolerance. The pearl millet variety "Dhanashakti" was introduced by ICRISAT in 2014 and is the first of its kind worldwide with a high iron content. Since its introduction, farmers have enthusiastically adopted it. The concentration of Zn remained unchanged, even though the ICTP 8203 formulation contains 71 mg/kg of Fe. On the other hand, the more advanced version of ICMV 221, known as ICMV 221Fe11-2, was specifically designed to have much higher levels of zinc (51 mg/kg) and iron (81 mg/kg).

TABLE 6.1 QTLs and Genes Identified for Fe and Zn Traits in Pearl Millet

Trait	QTL locus/markers	LG	LOD Value	PVE / Effect	References
Iron (Fe)	Xipes017-pgpb12900	1	6.22	9.0/4.0	Kumar et al. (2016, 2017, 2018), Anuradha et al. (2017)
	pgpb10531-pgpb9130	1	25.36	31.9/9.7	
	Xpsmp2261, Xipes0180, Xipes0096				
	Xipes188-pgpb6069	3	6.59	9.5/0.4	
	pgpb9502-pgpb6039	4B	6.87	10.4/−0.6	
	pgpb11956-pgpb9273	7	7.25	12.5/−1.9	
	pgpb8427-pgpb13221	7	8.58	12.2/5.3	
	pgpb11938-pgpb8987	7	8.83	12.5/4.9	
	pgpb6825-Xipes195	7	9.70	14.0/0.1	
	pgpb8445-pgpb11206	A	7.67	12.4/4.0	
	pgpb10660-pgpb8626	D	7.00	11.6/1.2	
	pgpb10727-Xipes179	E 3	9.364.68	14.3/3.1/8.5	
Zinc (Zn)	Xpsmp2214-Xipes142	3	9.66	32.3/8.5	Kumar et al. (2016, 2017, 2018)
	pgpb10531-pgpb9130	1	23.93	30.4/6.7	
	pgpb10397-pgpb10394	1	6.50	9.4/1.7	
	pgpb9502-pgpb6039	4B	7.33	11.1/−0.6	
	pgpb10483-pgpb11463	4B	6.68	11.6/−2.2	
	pgpb13229-pgpb12681	5	8.17	11.6/2.7	
	Xipes198-pgpb8427	7	7.16	10.2/2.7	
	pgpb12329-pgpb9721	7	7.58	10.9/2.8	
	pgpb8779-pgpb12691	H	6.68	11.6/2.1	

PVE = Phenotypic value explained (%).

The All India Coordinated Pearl Millet Improvement Project (AICPMIP) has made significant progress in the assessment and spread of biofortified pearl millet varieties in India. In addition, the ICAR has set specific benchmarks of 32 parts per million (ppm) for zinc and 42 ppm for iron as the minimum specifications for the upcoming pearl millet types that will be disseminated nationwide. The main goal of developing biofortified hybrids and variations with increased iron and zinc content was to boost the nutritional quality of pearl millet. Consequently, the productivity of pearl millet experienced growth. Combining new breeding procedures with good regulatory measures can greatly boost the success rates of high-Fe hybrids in pearl millet. The various hybrids and variations of pearl millet that have been systematically created and distributed are listed in Table 6.2.

Through diligent and prolonged dedication, it is feasible to alleviate the nutritional content of any crop through biofortification. Studies by Tripathi et al. in 2010 revealed that the zinc content of pearl millet flour was enhanced by adding zinc stearate and the metal chelator EDTA at a molar ratio of Zn:EDTA (1:1). ICRISAT has proven that traditional breeding methods may effectively merge cereal yield and micronutrient density thus far. The implementation of the fast-track strategy described in Figure 6.4 involves utilizing breeding lines that have moderate to elevated concentrations

TABLE 6.2 Catalogue of Biofortified Pearl Millet Varieties and Hybrids

Variety	Institute	Nutrient content	Salient Features	Reference
ABV 04	ANGRAU-ARS, Ananthapuram, under ICAR-AICRP on Pearl Millet, year of release: 2018	Fe: 70.0 ppm, Zn: 63.0 ppm	GY: 28.6 q/ha, DFY: 58.0 q/ha, Maturity: 86 days, Adaptation: Kharif season in Maharashtra, Karnataka, Andhra Pradesh, Telangana and Tamil Nadu	Devendra Kumar, 2020,
HHB 299	CCSHAU, Hisar & ICRISAT, year of release – 2017.	Fe: 73.0 ppm Zn: 41.0 ppm	GY: 32.7 q/ha, DFY: 73.0 q/ha, Maturity: 81 days, Adaptation: Kharif season in Haryana, Rajasthan, Gujarat, Punjab, Delhi, Maharashtra and Tamil Nadu	Yadava et al., 2017
HHB 311	CCSHAU, Hisar & ICRISAT	Fe: 83.0 ppm	GY: 31.7 q/ha, DFY: 72.0 q/ha, Maturity: 81 days, Adaptation: Kharif season in Rajasthan, Gujarat, Haryana, Punjab, Delhi, Maharashtra and Tamil Nadu	ICRISAT, 2014
AHB 1200Fe	VNMKV, Parbhani & ICRISAT, Year of release: 2018	Fe: 73.0 ppm	GY: 32.0 q/ha, DFY: 70.0 q/ha, Maturity: 78 days, Adaptation: Kharif season in Haryana, Rajasthan, Gujarat, Punjab, Delhi, Maharashtra and Tamil Nadu	Yadava et al., 2017
AHB 1269Fe		Fe: 91.0 ppm, Zn: 43.0 ppm	GY: 31.7 q/ha, DFY: 74.0 q/ha, Maturity: 82 days, Adaptation: Kharif season in Gujarat, Haryana, Punjab, Delhi, Maharashtra, Telangana and Tamil Nadu	Yadava et al., 2020
RHB 233	SKNAU, Rajasthan & ICRISAT, year of release-2019	Fe: 83.0 ppm, Zn: 46.0 ppm	GY: 31.6 q/ha, DFY: 74.0 q/ha, Maturity: 80 days, Adaptation: Kharif season in Rajasthan, Gujarat, Haryana, Madhya Pradesh, Delhi, Maharashtra and Tamil Nadu	ICRISAT, 2019
RHB 234		Fe: 84.0 ppm, Zn: 46.0 ppm	GY: 31.7 q/ha, DFY: 70.0 q/ha, Maturity: 81 days, Adaptation: Kharif season in Rajasthan, Gujarat, Haryana, Madhya Pradesh, Delhi, Maharashtra and Tamil Nadu	Yadava et al., 2020
Phule Mahashakti	MPKV, Dhule under ICAR-AICRP on Pearl Millet, year of release- 2018	Fe: 87.0 ppm, Zn: 41.0 ppm	GY: 29.3 q/ha, DFY: 56.0 q/ha, Maturity: 88 days, Adaptation: Kharif season in Maharashtra	Yadava et al., 2020
Dhanashakti	ICRISAT &	Fe: 71.0ppm	GY: 22.0 q/ha, DFY: 53.0 q/ha,	ICRISAT, 2014
Shakti-1201 (ICMH-1201)	Harvest Plus	Fe: 75 ppm Zn: 40 ppm		

GY: grain yield; DFY: dry fodder yield.

of iron (Fe) and zinc (Zn), and a population that has not been specifically bred for micronutrient content. To reach the ultimate goal of creating hybrids in the presence of high levels of iron (Fe), one can breed progenitor lines (consisting of seed and restorer parents) that possess a high density of Fe as a desired trait. A total of 174 high-Fe early-generation progenies, resulting from the crossbreeding of B × B and R × R, were generated and assessed at Patancheru. The progenies exhibited Fe densities surpassing 90 mg kg^{-1}, while the Zn densities varied between 36 and 72 mg kg^{-1}. The co-localization and overlap of QTL for zinc (Zn) and iron (Fe) concentrations on chromosome 3 (LG3) have been confirmed. As a result, there is a considerable difference in the concentrations of Fe and Zn in the top breeding populations. This calls for more research to create diagnostic procedures that may be used in the future to screen and classify materials (Govindaraj et al., 2019).

Apart from enhancing the concentration of targeted nutrients in seed grains, their bioavailability, consumer acceptance, and compactness to edaphic, climatic, and agronomic factors also need to be analyzed. Likewise, metabolomics related to the uptake of these beneficial nutrients should be carefully studied to avoid the undesirable outcomes of biofortification. For instance, genetic enhancement of Zn content in grains is attained by modulating and enhancing the transport proteins, which aids in the uptake, translocation, and storage of Zn. However, several studies also reported that the enhancement of ZIP family proteins like IRT1 facilitates the uptake of Cd along with zinc, which could result

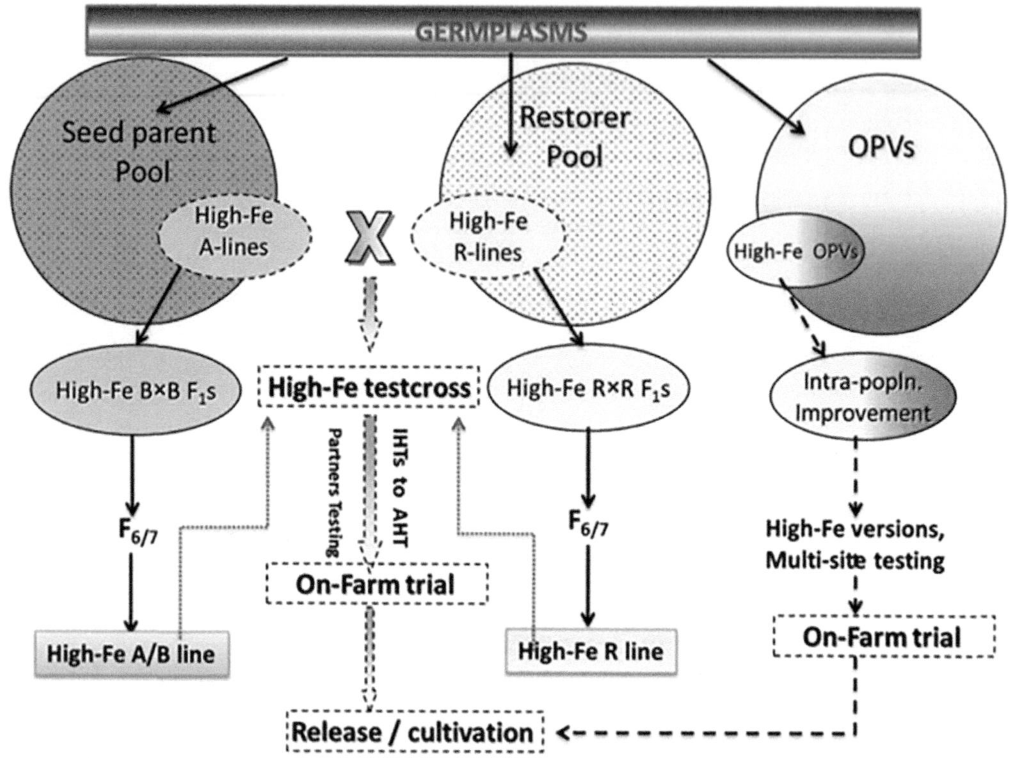

FIGURE 6.4 Instant breeding approach adopted at ICRISAT for the development of biofortified hybrids in India and OPVs in West Africa.

Source: Govindaraj et al. (2019).

in toxic outcomes (Guerinot, 2000; Zhao and McGrath, 2009). Further, the bioavailability of minerals always poses a problem for crops like pearl millet due to the existence of anti-nutritional components like phytates. Even though breeding selective lines of pearl millet with little phytate content is possible, the phytate content is also determined by environmental factors (Simwemba et al., 1984). Therefore, several post harvest measures are also adopted to reduce the anti-nutritional factors while maintaining the nutrient content (Kumar et al., 2018).

6.2.4 STARCH CONTENT ENHANCEMENT AND BIOFUEL APPLICATION

Wu et al. (2006) found that the fermentation efficiency of pearl millet is similar to that of corn and cereal sorghum. Pearl millet has a higher protein and fat content compared to maize or grain sorghum. Consequently, the levels of lipids and proteins in the dried grains with soluble (DDGS) produced by distillers, obtained from pearl millet, are increased. Pearl millet offers a feasible substitute for the manufacture of petroleum ethanol in areas where maize and cereal sorghum are not grown to the appropriate extent.

6.2.4.1 *SEED SHELF-LIFE ENHANCEMENT*

Although pearl millet has outstanding nutritious characteristics, the durability and ability to be absorbed by the body of its flour pose difficulties in promoting pearl millet products. The main goals should be to enhance the grain with zinc and iron by biofortification, gather accurate data on these elements' nutritional benefits and bioavailability, and prolong the shelf life of pearl millet flour. Developing flour variations that increase the product's shelf life while decreasing the amount of oil and phenolic compounds is crucial. Furthermore, further investigation is required to determine its nutraceutical value and potential health advantages to increase its attractiveness to consumers.

6.3 POST-HARVEST MEASURES TO MAINTAIN/ENHANCE THE NUTRITIONAL QUALITY

6.3.1 *GRAIN PROTEIN ENHANCEMENT*

Soaking, roasting, fermentation, etc., reduce the anti-nutritional factors in pearl millet. Soaking enhances the leaching of tannins from grains (Handa et al., 2017; Hussain et al., 2011). Pearl millet is a good source of protein; however, the digestibility of protein varies depending on several exogenous and endogenous factors. The preliminary processing methods, viz., roasting and germination, improved the protein digestibility by 65% and 88% in Kalukombu, a traditional pearl millet (Pushparaj and Urooj, 2011). Fermentation enhances the protein digestibility of pearl millet by reducing the antinutrient factors, such as tannin content, from 0.49 to 0.29 tannic acid equivalents per 100 grams (Rathore and Singh, 2018).

6.3.2 *ENHANCEMENT OF IRON BIOAVAILABILITY*

Malting is a traditional method adopted to enhance the nutritional value of millets, as it increases access to nutrients (iron by 30% and manganese by 17%) by hydrolyzing the anti-nutritional factors like phytate (Platel et al., 2010).

Malting of pearl millets → hydrolysis of phytate phosphorus → inositol monophosphate → reduces the phytic acid.

6.3.3 *STARCH PROFILE OF PEARL MILLETS*

Apart from cereals, millets are the next major source of starch. The quality and type of starch make millets a healthier alternative when compared to cereals. The more complex the starch is, the lower the rate of *in vitro* digestibility, thereby reducing the risk of diabetes. The amount of rapidly digestible starch in pearl millet is 46.30–50.10 g/100g starch, and slowly digestible starch is 37.20–38.70 g/100g starch (Sandhu and Siroha, 2017; Suma and Urooj, 2015). Likewise, the glycemic index for pearl millet is 54.15 (Sharma and Gujral, 2019). The modification of available starch in pearl millets to make it more resistant in nature is carried out through various methods like thermal, chemical, biochemical, or a combination of any of these treatments. Hydrothermal processing at 110°C for 8 hours and further drying at 40°C to limit the moisture content to 10% enhanced the resistant starch content of pearl millet from 2.4 g to 7.07 g/100 g of starch (Sharma et al., 2015). Esterification of pearl millet starch using 3% octenyl succinate anhydride resulted in comparatively higher resistant starch content, that is, 13.7 g/100 g of starch (Sharma et al., 2016).

6.4 BENEFITS OF NUTRITIONAL QUALITY ENHANCEMENT

We are in an era of nutritional transition, as witnessed in the latest NFHS survey 2019–21; around 32.1% of children under five are underweight, and 35.5% are stunted. Concurrently, the same survey indicated elevated levels of obesity, affecting 24% of women and 22.9% of men, along with the prevalence of non-communicable diseases. As per the NFHS survey, around 6.1% of women and 7.3% of men have high blood sugar levels, while 21.3% of women and 24% of men are detected to have elevated blood pressure (NFHS-5, 2019–21).

Repurposing agriculture by encompassing the production, collection, storage, and distribution of nutrient-rich crops can positively impact public health and the overall nutritional requirements of the population (FAO, 2022). Several policy initiatives have been proposed in this regard, from time to time, by the major international and national agencies, viz., WHO, FAO, UNICEF, and respective central and state governments.

6.4.1 INITIATIVES OF WHO IN SUSTAINABLE NUTRITIONAL SECURITY

The World Health Organization is dedicated to improving global nutritional security by addressing different aspects that lead to the vicious cycle of malnutrition. Encouraging individuals to adopt a balanced and nutritious diet is the central dogma of WHO for nutritional security. Diet improvisation by developing a regulatory framework and implementing school-based health programs are some of the key areas covered under the WHO's initiatives (Figure 6.5) to combat nutrition deficiency.

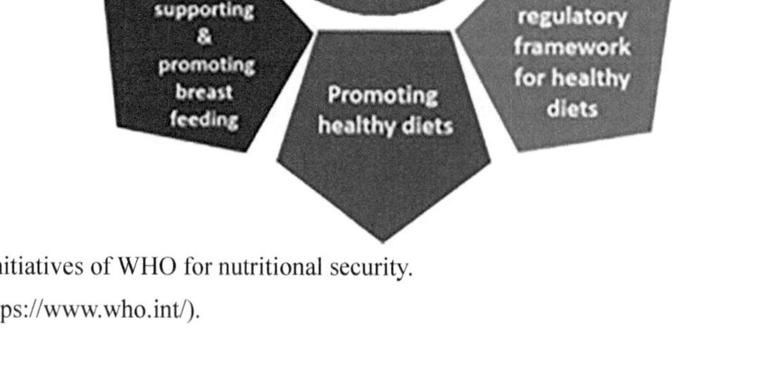

FIGURE 6.5 Initiatives of WHO for nutritional security.

(Data source: https://www.who.int/).

6.4.1.1 *PROMOTING HEALTHY DIETS*

The development of dietary recommendations and counseling on diverse and healthy food choices are the key initiatives adopted by the WHO to promote healthy diets. The observation from the Western Pacific region with respect to the causes of obesity identified a diet shift from fresh, locally grown, and traditional foods to calorie-dense foods with poor nutritional value. A diverse diet has been identified as the key point for reducing obesity and non-communicable diseases (Overweight and obesity in the Western Pacific Region, 2017).

6.4.1.2 *DEVELOPING REGULATORY FRAMEWORKS FOR HEALTHY DIETS*

In collaboration with stakeholders and governments, WHO established policies, viz., regulations for food labeling, advertising, and marketing, and advocating fiscal measures like imposing taxes on unhealthy foods and beverages (Overweight and obesity in the Western Pacific Region, 2017).

6.4.1.3 *IMPROVING SCHOOL-BASED HEALTH PROGRAMS*

Supporting school meal programs and integrating nutrition education into curricula by partnering with educational institutions paves a conducive environment for promoting positive nutritional and lifestyle changes among the population (Overweight and obesity in the Western Pacific Region, 2017).

6.4.1.4 *PROMOTING BREASTFEEDING*

Breastfeeding is the foundation of infant nutrition; hence, WHO takes active measures to support and promote breastfeeding by conducting awareness campaigns and providing guidance to healthcare systems (Overweight and obesity in the Western Pacific Region, 2017).

Apart from the above-mentioned key initiatives, WHO also organizes ongoing research and data collection to understand the regional variations for planning and executing area-specific nutritional interventions.

6.4.2 *INITIATIVES OF FAO*

FAO also emphasizes the promotion of healthy diets to ensure nutritional security. FAO focuses on aligning food safety and dietary diversification by including whole cereals, millets, legumes, fruits, and vegetables for sustainable nutritional security (https://www.fao.org/millets-2023/about/en). In this regard, FAO endorsed India's proposal for the commemoration of the International Year of Millets to create awareness across the globe concerning the nutritional and therapeutic potency of millets and their adaptability towards fluctuating weather conditions. Several outreach activities engaging youth were organized globally as part of IYM 2023. This also set the path for enhancing legislative actions to encourage millet cultivation, processing, and trade, envisaging climate resilience and dietary diversity (https://www.fao.org/millets-2023/about/en).

6.4.3 POSHAN ABHIYAN – A GOVERNMENT OF INDIA INITIATIVE

Poshan Abhiyan is a flagship program of the Government of India launched in 2018 with the aim of overarching holistic nutrition by suppressing malnutrition in a mission mode. One of the key highlights of this scheme in India was the inclusion of millets in the supplementary nutrition program (http://niti.gov.in/sites/default/files/2023-02/Poshan-Abhiyaan-Monitoring.pdf). The case study of Odisha tells us that the legislative decision to include millet in SNP had increased the area under cultivation of finger millet from 2949 hectares to 5182 hectares and the number of farmers cultivating millets (finger millet) from 7014 in 2016–17 to 8,596 in 2017–2018, thus resulting in a 120% increase in the yield. Since millets were abundantly available, the same was included in the ICDS scheme and distributed to the population at one rupee per kg through the public distribution system, thus ensuring a wide coverage of healthy diets (https://dmeo.gov.in/sites/default/files/2021-08/Package2_WCD_CaseStudy3.pdf).

The Government of India also included "Shree Anna," which translates to "mother of all grains," as part of the "Poshan Pakhwada" initiative to promote the benefits of millets among the general public by organizing diet camps, conducting home visits, and encouraging the inclusion of millets in SNP. The program also published a millet recipe book to enhance product diversification using millets (Millets Recipe Book 2023_Low Res_V5-min.pdf [wcd.nic.in]).

6.5 SIGNIFICANCE OF PEARL MILLET – IN A GLOBAL SCENARIO

From a global perspective, where we are witnessing a nutritional transition among the population, it is not only necessary to provide sufficient food for the burgeoning population but also to ensure its nutritional quality for preventing micronutrient deficiencies, hidden hunger, and non-communicable diseases (Kumar et al., 2020). Likewise, pearl millet has wide crop diversity, climate resiliency, and industrial applications. This represents the primary reason for including pearl millet in the biofortification program, and it is also being popularized through various government schemes.

Pearl millet is a nutritious grain rich in essential nutrients such as protein, fiber, vitamins, and minerals. This is particularly important in regions where malnutrition is a problem, as it can contribute to improving dietary diversity and overall health. It is known for its resistance to harsh environmental conditions, including drought and high temperatures. This makes it a vital crop in arid and semi-arid tracts where other crops may fail (Satyavathi et al., 2021).

In many parts of Africa and Asia, pearl millet is a staple food for smallholder farmers, contributing to their income and well-being. Its ability to grow in challenging conditions contributes to food security and stability in these regions. Thus, pearl millet cultivation provides livelihood opportunities to farmers, especially in marginalized and rural communities. Pearl millet has various industrial uses, including as animal feed and in the production of alcoholic beverages (Dwivedi et al., 2012). It also has commercial value as a grain for human consumption and for processing into products such as wheat flour and breakfast cereals (Satyavathi et al., 2021).

6.6 CONCLUSION

Since the year 2023 was announced as the International Year of Millets by UNGA to popularize millets as nutri-cereals upon request by the Government of India, it has paved the way for more research in reinvigorating millets, in general, to global dietary habits. Extensive research has been undertaken with respect to the biofortification of pearl millet for enhanced Fe and Zn content. Nevertheless, the bioavailability and consumer acceptance of these developed crops must be assessed to understand their wide acceptance among the public. In summary, pearl millet plays a key role in global agriculture, nutrition, and food security, making it an important crop that can have a significant influence on the overarching efforts to combat hunger and malnutrition globally.

KEYWORDS

- **bioavailability**
- **biofortification priority index**
- **biofortified varieties**
- **Fe and Zn content**
- **micronutrient deficiencies**
- **nutritional quality**
- **pearl millet**
- **sustainable development goals**

REFERENCES

Abdalla, A. A., El Tinay, A. H., Mohamed, B. E., & Abdalla, A. H. (1998). Proximate composition, starch, phytate and mineral contents of 10 pearl millet genotypes. *Food Chemistry, 63*(2), 243–246. https://doi.org/10.1016/s0308-8146(97)00228-8 (accessed on 7th November 2025).

Anuradha, N., Satyavathi, C. T., Bharadwaj, C., Nepolean, T., Sankar, S. M., Singh, S. P., Meena, M. C., Singhal, T., & Srivastava, R. K. (2017). Deciphering genomic regions for high grain iron and zinc content using association mapping in pearl millet. *Frontiers in Plant Science, 8*, 412. https://doi.org/10.3389/fpls.2017.00412 (accessed on 7th November 2025).

Burton, G. W., Wallace, A. T., & Rachie, K. O. (1972). Chemical composition and nutritive value of pearl millet (*Pennisetum typhoides* (Burm.) Stapf and *E. C. Hubbard*) grain. *Crop Science, 12*(2), 187–188. https://doi.org/10.2135/cropsci1972.0011183x001200020009x (accessed on 7th November 2025).

Collectives, P. F. (n.d.). Odisha Millet Mission (OMM). *Gov.in*. https://dmeo.gov.in/sites/default/files/2021-08/Package2_WCD_CaseStudy3.pdf (accessed on 7th November 2025).

Dwivedi, S., Upadhyaya, H., Senthilvel, S., & Hash, C. (2012). Millets: Genetic, and genomic resources. In J. Janick (Ed.), *Plant Breeding Reviews* (pp. 247–375). John Wiley and Sons, Inc.

Ejeta, G., Hansen, M. M., & Mertz, E. T. (1987). *In vitro* digestibility and amino acid composition of pearl millet (*Pennisetum typhoides*) and other cereals. *Proceedings of the National Academy of Sciences of the United States of America, 84*, 6016–6019.

Govindaraj, M., Rai, K. N., Cherian, B., Pfeiffer, W. H., Kanatti, A., & Shivade, H. (2019). Breeding biofortified pearl millet varieties and hybrids to enhance millet markets for human nutrition. *Agriculture, 9*(5), 106. https://doi.org/10.3390/agriculture9050106 (accessed on 7th November 2025).

Guerinot, M. L. (2000). The ZIP family of metal transporters. *Biochemical et Biophysical Acta (BBA)–Bio Membranes, 1465*(1–2), 190–198. https://doi.org/10.1016/s0005-2736(00)00138-3 (accessed on 7th November 2025).

Handa, V., Kumar, V., Panghal, A., Suri, S., & Kaur, J. (2017). Effect of soaking and germination on physicochemical and functional attributes of horsegram flour. *Journal of Food Science and Technology, 54*(13), 4229–4239. https://doi.org/10.1007/s13197-017-2892-1 (accessed on 7th November 2025).

Hash, C. T., Bhaskar Raj, A. G., Rao, S. A., & Singh, U. (1996). New sources of yellow endosperm and β-carotene in pearl millet. In *Proceedings of the International Conference on Genetic Improvement of Sorghum and Pearl Millet* (Vol. 650). TX, USA.

Herrington, C., Lividini, K., & Angel, M. D. (2019). Prioritizing countries for biofortification interventions: Biofortification priority index second edition (BPI 2.0). In *Harvest Plus Working Paper*.

Hoseney, R. C., Andrews, D. J., & Clark, H. (1987). *Nutritional Quality of Cereal Grains: Genetic and Agronomic Improvement*. In R. A. Olsen & J. K. Frey (Eds.), American Society of Agronomy.

Hussain, I., Uddin, M. B., & Aziz, M. G. (2011). Optimization of anti-nutritional factors from germinated wheat and mung bean by response surface methodology. *Food Research International Journal, 18*, 957–963.

Jambunathan, R., & Subramanian, V. (1988). Grain quality and utilization in sorghum and pearl millet. In *Proceedings of Workshop on Biotechnology for Tropical Crop Improvement* (pp. 133–139). Patancheru, India.

Kanatti, A., Rai, K. N., Radhika, K., Govindaraj, M., Sahrawat, K. L., & Rao, A. S. (2014). Grain iron and zinc density in pearl millet: Combining ability, heterosis and association with grain yield and grain size. *Springer Plus, 3*(1). https://doi.org/10.1186/2193-1801-3-763 (accessed on 7th November 2025).

Kumar, A., Tomer, V., Kaur, A., Kumar, V., & Gupta, K. (2018). Millets: A solution to agrarian and nutritional challenges. *Agriculture & Food Security, 7*(1). https://doi.org/10.1186/s40066-018-0183-3 (accessed on 7th November 2025).

Kumar, R. R., Bhargava, D. V., & Pandit, K. (2021a). Lipase—the fascinating dynamics of enzyme in seed storage and germination—a real challenge to pearl millet. *Food Chemistry, 361*.

Kumar, R. R., Rai, G. K., Goswami, S., Singh, S. P., Satyavathi, C. T., & Praveen, S. (2020). *Millets the Orphan Crop for Uncertain Future*. Scientific International Pvt. Ltd.

Kumar, S., Hash, C. T., Nepolean, T., Satyavathi, C. T., Singh, G., Mahendrakar, M. D., Yadav, R. S., & Srivastava, R. K. (2017). Mapping QTLs controlling flowering time and important agronomic traits in pearl millet. *Frontiers in Plant Science, 8*. (accessed on 7th November 2025).

Kumar, S., Hash, C. T., Thirunavukkarasu, N., Singh, G., Rajaram, V., Rathore, A., Senapathy, S., Mahendrakar, M. D., Yadav, R. S., & Srivastava, R. K. (2016). Mapping quantitative trait loci controlling high iron and zinc content in self and open pollinated grains of pearl millet (*Pennisetum glaucum* (L.) R. Br.). *Frontiers in Plant Science, 7*, 1636. https://doi.org/10.3389/fpls.2016.01636 (accessed on 7th November 2025).

Kumar, S., Hash, C., Nepolean, T., Mahendrakar, M., Satyavathi, C., Singh, G., Rathore, A., Yadav, R., Gupta, R., & Srivastava, R. (2018). Mapping grain iron and zinc content quantitative trait loci in an Iniadi-derived immortal population of pearl millet. *Genes (Basel), 9*(5), 248. https://doi.org/10.3390/genes9050248 (accessed on 7th November 2025).

Longvah, T., Ananthan, R., Bhaskarachary, K., & Venkaiah, K. (2017). *Indian food composition tables*. National Institute of Nutrition, Indian Council of Medical Research, Department of Health Research, Ministry of Health & Family Welfare, Government of India; Telangana, India.

Malik, M. (1999). *Functional and Nutritional Properties of Newly Developed Hybrids and Varieties of Pearl Millet*. Hisar, Haryana, India.

Manwaring, H. R., Bligh, H. F. J., & Yadav, R. (2016). The challenges and opportunities associated with biofortification of pearl millet (*Pennisetum glaucum*) with elevated levels of grain iron and zinc. *Frontiers in Plant Science, 7*. https://doi.org/10.3389/fpls.2016.01944 (accessed on 7th November 2025).

N/A. (2009). The World Health Organization (WHO). https://www.who.int/ (accessed on 7th November 2025).

N/A. (2017). *Levels and Trends in Child Malnutrition. UNICEF/WHO/World Bank Group Joint Malnutrition Estimates. Key Findings of the 2017 Edition*. New York/Geneva/Washington DC.

N/A. (2017). *Overweight and Obesity in the Western Pacific Region*. Manila, Philippines.

N/A. (2019). Ministry of Health and Family Welfare (MoHFW). *National Family Health Survey, Compendium of Fact Sheets, India and 14 States/UTs (Phase-II)*. Ministry of Health & Family Welfare, UNICEF and Population Council.

N/A. (2022). *Repurposing Agriculture's Public Budget to Accelerate Transformation in Ethiopia*.

N/A. (2023). *Millets Recipe Book 2023_Low Res_V5-Min.pdf*. https://wcd.nic.in/sites/default/files/Millets%20Recipe%20Book%202023_Low%20Res_V5-min.pdf (accessed on 7th November 2025).

N/A. (2023). *Preserving Progress on Nutrition in India: Gov.in*. http://niti.gov.in/sites/default/files/2023-02/Poshan-Abhiyaan-Monitoring.pdf (accessed on 7th November 2025).

N/A. (n.d.). About International Year of Millets. https://www.fao.org/millets-2023/about/en (accessed on 7th November 2025).

N/A. (n.d.). https://www.icrisat.org/africas-first-biofortified-pearl-millet-variety-aims-to-combat-anaemia/ (accessed on 7th November 2025).

N/A. (n.d.). Proceedings of the 52nd annual pearl millet workshop All India Coordinated Research Project on Pearl Millet. *Res.in*. http://www.aicpmip.res.in/pw2017.pdf (accessed on 7th November 2025).

Platel, K., Eipeson, S. W., & Srinivasan, K. (2010). Bioaccessible mineral content of malted finger millet (*Eleusine coracana*), wheat (*Triticum aestivum*), and barley (*Hordeum vulgare*). *Journal of Agricultural and Food Chemistry, 58*(13), 8100–8103. https://doi.org/10.1021/jf100846e (accessed on 7th November 2025).

Pushparaj, F. S., & Urooj, A. (2011). Influence of processing on dietary fiber, tannin and *in vitro* protein digestibility of pearl millet. *Food and Nutrition Sciences, 2*(8), 895–900. https://doi.org/10.4236/fns.2011.28122 (accessed on 7th November 2025).

Raheem, D., Dayoub, M., Birech, R., & Nakiyemba, A. (2021). The contribution of cereal grains to food security and sustainability in Africa: Potential application of UAV in Ghana, Nigeria, Uganda, and Namibia. *Urban Science, 5*(1), 8. https://doi.org/10.3390/urbansci5010008 (accessed on 7th November 2025).

Rai, K. N., Velu, G., Govindaraj, M., Upadhyaya, H. D., Rao, A. S., Shivade, H., & Reddy, K. N. (2015). Iniadi pearl millet germplasm as a valuable genetic resource for high grain iron and zinc densities. *Plant Genetic Resources, 13*(1), 75–82. https://doi.org/10.1017/S1479262114000665 (accessed on 7th November 2025).

Rai, K., Patil, H., Yadav, O., Khairwal, I., Rajpurohit, B., Rao, A., & Kulkarni, M. (2014). Dhanashakti: A high-iron pearl millet variety. *Indian Farming, 64*(7), 32–34.

Rathore, S., & Karunakar, S. (2018). Analysis of the effects of natural and pure culture fermentation for the qualitative enhancement of pearl millet flour. *Nutra foods, 17*, 145–153. https://doi.org/10.17470/NF-017-1013-3 (accessed on 7th November 2025).

Rooney, L. W., & Donough, M. (1987). Food quality and consumer acceptance in pearl millet. In *Proceedings in International Pearl Millet Workshop, ICRISAT* (pp. 43–61). Patancheru, India.

Sandhu, K. S., & Siroha, A. K. (2017). Relationships between physicochemical, thermal, rheological and *in vitro* digestibility properties of starches from pearl millet cultivars. *Lebensmittel-Wissenschaft & Technologie, 83*, 213–224. https://doi.org/10.1016/j.lwt.2017.05.015 (accessed on 7th November 2025).

Satyavathi, C. T., Ambawat, S., Khandelwal, V., & Srivastava, R. K. (2021). Pearl millet: A climate-resilient Nutri cereal for mitigating hidden hunger and provide nutritional security. *Frontiers in Plant Science, 12*. https://doi.org/10.3389/fpls.2021.659938 (accessed on 7th November 2025).

Sharma, B., & Gujral, H. S. (2019). Characterization of thermo-mechanical behavior of dough and starch digestibility profile of minor millet flat breads. *Journal of Cereal Science, 90*(102842), 102842. https://doi.org/10.1016/j.jcs.2019.102842 (accessed on 7th November 2025).

Sharma, M., Singh, A. K., Yadav, D. N., Arora, S., & Vishwakarma, R. K. (2016). Impact of octenyl succinylation on rheological, pasting, thermal and physicochemical properties of pearl millet (*Pennisetum typhoides*) starch. *Lebensmittel-Wissenschaft & Technologie, 73*, 52–59. https://doi.org/10.1016/j.lwt.2016.05.034 (accessed on 7th November 2025).

Sharma, M., Yadav, D. N., Singh, A. K., & Tomar, S. K. (2015). Rheological and functional properties of heat moisture treated pearl millet starch. *Journal of Food Science and Technology, 52*(10), 6502–6510. https://doi.org/10.1007/s13197-015-1735-1 (accessed on 7th November 2025).

Simwemba, C. G., Hoseney, R. C., Varriano-Marston, E., & Zeleznak, K. (1984). Certain B vitamin and phytic acid contents of pearl millet (*Pennisetum americanum* (L.) Leeke). *Journal of Agricultural and Food Chemistry, 32*(1), 31–34. https://doi.org/10.1021/jf00121a008 (accessed on 7th November 2025).

Srivastav, P., & Bharat Bhushan Rao, C. (2020). Effect of agronomic biofortification of pearl millet (*Pennisetum glaucum* L.) cultivars with iron on growth parameters and quality. *Journal of Pharmacognosy and Phytochemistry, 9*(5), 395–397.

Suma, P. F., & Urooj, A. (2015). Isolation and characterization of starch from pearl millet (*Pennisetum typhoidium*) flours. *International Journal of Food Properties, 18*(12), 2675–2687. https://doi.org/10.1080/10942912.2014.981640 (accessed on 7th November 2025).

Tripathi, B., & Patel, C. (2010). Fortification of sorghum (*Sorghum vulgare*) and pearl millet (*Pennisetum glaucum*) flour with zinc. *Journal of Trace Elements in Medicine and Biology, 24*, 257–262.

Velu, G., Rai, K. N., Muralidharan, V., Kulkarni, V. N., Longvah, T., & Raveendran, T. S. (2007). Prospects of breeding biofortified pearl millet with high grain iron and zinc content. *Plant Breeding, 126*, 182–185. https://doi.org/10.1111/j.1439-0523.2007.01322.x (accessed on 7th November 2025).

Wu, X., Wang, D., Bean, S., & Wilson, J. (2006). Ethanol production from pearl millet by using *Saccharomyces cerevisiae*. *Cereal Chemistry*, 127–131.

Yadava, D. K., Choudhury, P. R., Hossain, F., & Kumar, D. (2017). *Biofortified Varieties: Sustainable Way to Alleviate Malnutrition*. Indian Council of Agricultural Research, New Delhi.

Yadava, D. K., Choudhury, P. R., Hossain, F., Kumar, D., & Mohapatra, T. (2020). *Biofortified Varieties: Sustainable Way to Alleviate Malnutrition* (Third Edition). Indian Council of Agricultural Research, New Delhi.

Zhao, F. J., & McGrath, S. P. (2009). Biofortification and phytoremediation. *Current Opinion in Plant Biology, 12*(3), 373–380. https://doi.org/10.1016/j.pbi.2009.04.005 (accessed on 7th November 2025).

CHAPTER 7

Nutritional Quality Enhancement in Finger Millet

C. NANDINI,[1] SUJATA BHAT,[2] T. V. KRISHNA,[3] T. E. NAGARAJA,[2] and J. SOWJANYA[4,*]

[1]*Regional Agricultural Research Station-Central Zone, Kerala Agricultural University, Pattambi, Palakkad, Kerala, India*

[2]*Project Coordinating Unit, ICAR–AICRP on Small Millets, University of Agricultural Sciences, GKVK, Bangalore, Karnataka, India*

[3]*ICAR–AICRP on Kharif Pulses, Zonal Agricultural Research Station, University of Agricultural Sciences, GKVK, Bangalore, Karnataka, India*

[4]*ICAR–Central Research Institute for Dryland Agriculture, Santhoshnagar, Hyderabad, Telangana, India*

*Corresponding author

7.1 INTRODUCTION

The nutritional value of food is the most important aspect of maintaining good human health and physical well-being (Ansari et al., 2023). Millets provide numerous nutritional and physiological benefits (Amadou et al., 2013) and are regarded as a superfood. Most millets are far superior to other grains like rice and wheat. Millets are climate-resilient and sustainable crops that require minimal input. Millets are small-seeded cereal crops from the Poaceae family (Verma et al., 2012). They are pre-green revolution crops that have been farmed for millennia. Eating millets improves the immune system and lowers cholesterol and blood sugar. They are appropriate for heart conditions and can be consumed by those with type 2 diabetes (Jayanta et al., 2023).

One such nutrient-rich millet crop, finger millet, is utilized as animal feed (straw) in industrialized nations and as food (grains) in underdeveloped ones, demonstrating that it is regarded as food for the poor (Ceasar et al., 2018; Wambi et al., 2020). *Eleusine coracana* L., commonly referred to as finger millet, is also known by the following names: bulo (Uganda); kambale, lupoko, mawele, amale, bule (Zambia); poho, rapoko, zviyo, njera, mazhovole (Zimbabwe); finger millet, African millet, koracan (England); dagussa, tokuso, barankiya (Ethiopia); wimbi, mugimbi (Kenya). It is a major crop in the developing nations of Asia and Africa's dry and semi-arid regions (Ceasar et al., 2018; Krishna et al., 2020). According to Maharaj et al. (2019), finger millet is the fourth most produced millet worldwide, behind sorghum (*Sorghum bicolor),* pearl millet *(Cenchrus americanus*; formerly

Millets in Sustainable Agriculture: Genetic Improvement, Nutritional Quality, Cultivation, and Integrated Pest–Disease Management. Roshni Vijayan, A. John Joel, R. Ambika Rajendran, and Adithya P. Balakrishnan (Eds.)

DOI: 10.1201/9781779642332-9

known as *Pennisetum glaucum*), and foxtail millet (*Setaria italica*). Because the grains of finger millet contain significant levels of vitamins, iron, carbohydrates, calcium, potassium, zinc, phosphorus, magnesium, and vital amino acids, it is a more nutritious food than rice and wheat (Saleh et al., 2013). It is a staple food of major importance in many parts of eastern and central Africa, as well as India. With an approximate 40.62% (28.37 million) contribution, India is the leading producer of finger millet (FAOSTAT, 2019).

Compared to other cereals like barley, rice, maize, and wheat, it is a small-seeded minor grain with a light brown to brick red seed coat rich in phyto-chemicals, including dietary fiber and polyphenols. This indigenous small millet is used to make health meals and baby food in both natural and malted versions. This millet flour is used in many different recipes, such as porridge, puddings, pancakes, biscuits, bread, roti, and other snacks. It is considered a healthy food for diabetics and is used as a nourishing food for babies when malted.

7.2 NUTRITIONAL COMPOSITION OF FINGER MILLET

Similar to other cereals like rice, wheat, and other millets, finger millet has a high amount of diverse nutrients, including minerals like calcium (364 mg), zinc (2.53 mg), and iron (4.62 mg), as well as a significant amount of carbohydrates (66.7%), protein (7.1%), and dietary fiber (11.18%) (Font et al., 2020). Although the protein content of finger millet varies greatly, ranging from 5.6% to 12.70%, it is approximately 7% (Bhatt et al., 2003; Gautam, 2000). According to Mbithi et al. (2000), finger millet has a larger percentage of total amino acids (44.7%) than the FAO reference protein, which has 33.9% of the necessary amino acids (FAO, 1991). Furthermore, an adequate proportion of essential amino acids is provided by the finger millet amino acid profile. Because finger millet has higher levels of lysine, threonine, and valine than other millets, it has a relatively better balance of essential amino acids (Ravindran, 1992). According to reports, finger millet's crude fat content ranges from 1.3% to 1.9% (Bhatt et al., 2003). Oleic acid was the main fatty acid found in finger millet, with palmitic and linoleic acids coming in second and third. There was also a trace amount of linolenic acid in it. According to the fatty acid profile, unsaturated fatty acids make up 74.4% of total fatty acids, while saturated fatty acids make up 25.6% (Sridhar and Lakshminarayana, 1994).

Finger millet is the most plentiful source of calcium and iron. Inadequate iron intake causes anemia, which can be solved by including finger millet in our regular diet; calcium insufficiency causes problems with the bones, teeth, and blood vessels. Finger millet's high fiber content is responsible for its hypoglycemic effect. Complex carbohydrates, particularly those found in fiber-rich foods, are absorbed slowly and broken down gradually, lowering postprandial glucose levels and reducing the risk of constipation, high cholesterol, and intestinal cancer (Geetha and Parvathy, 1990).

7.3 NUTRITIONAL INHIBITORS AND TOXIC FACTORS

Finger millet contains phytates, tannins, polyphenols, trypsin inhibitory factors, and dietary fiber. Because of their ability to chelate metals and block enzymes, these were formerly regarded as "anti-nutrients," but today they are known as nutraceuticals. The antioxidant activity of millet meals is influenced by phytates, polyphenols, and tannins. This has a major effect on health, aging, and metabolic illnesses. According to Dharmaraj and Malleshi (2010), finger millet has a number of anti-nutritional substances, including tannins, phytic acid, and trypsin inhibitors. As reported by Anthony and Chandra (1998), finger millet has a high amount of tannins, in the range of 0.04% to 3.74% of equivalents of catechin. Brown finger millet contains 360 mg of tannins per 100 grams, according to Rao and Prabhavati (1982). Additionally, they discovered that 50% of the iron in food may be tannic. The tannin content of finger millet cultivars has been found to be adversely correlated with *in vitro* protein digestibility (Ramachandra et al., 1977). Tannin content has been reported to be decreased by soaking, roasting, boiling, germination, and fermentation (Rao and Prabhavathi, 1982). Brown finger millet's tannin concentration was reduced by 54% after malting (Rao, 1994).

Finger millet contains 41% of its total phosphorus in the form of phytic phosphorus (Deosthale, 2000). Phytic acid, present in the diet, binds not only to minerals from the seed but also to other minerals encountered in the digestive system (Raboy, 2000). Malting the grain significantly reduces its phytin phosphorus content, with a decrease of 58% to 65% reported for brown and white finger millet varieties, respectively (Rao, 1994). Processing methods like germination and fermentation have also been shown to significantly lower phytic acid levels, with reductions of 49.2% after germination and 66.5% after fermentation (Mammiro et al., 2001). Phytic acid decreased in finger millet by 84.7% when combinations of processing methods were used. The partial retention of phytates is beneficial for their contribution to health benefits such as antidiabetic, antioxidant, and anticancer effects, which have been recently recognized (Thompson, 1993).

Finger millet has drawn consumer interest recently because of its increased polyphenolic compound content and antioxidant activity (Kumari et al., 2019). Furthermore, finger millet is now favored by consumers over regularly consumed staples due to its higher resistant starch (RS) and slowly digested starch (SDS) content when compared to wheat and rice (Sharma and Gujral, 2019).

Compared to many other cereals like barley, rice, maize, and wheat, finger millet grain has a polyphenol-rich dark brown seed covering (Viswanath et al., 2009). These phenolics are not evenly distributed throughout the grain; instead, they are primarily concentrated in the outer layers, which make up the bran fraction's primary constituents: the testa, pericarp, and aleurone layer. Grain phenolic compounds come in two forms: free conjugates that are soluble and insoluble bound forms. Regular consumption of finger millet, an abundant source of dietary fiber and polyphenols, has demonstrated a protective effect against gastrointestinal disorders and type 2 diabetes. According to Tharanathan and Mahadevamma (2003), the health benefits of finger millet include:

Finger millet has been found to exhibit several health benefits, such as hypercholesterolemia, hypoglycemic effects, anti-ulcerative, antimicrobial, anti-inflammatory, antiviral, and anticancer activity. Several researchers have explored supplementing the major cereals with finger millet in a wide variety of food products, such as baked bread, biscuits, extruded pasta, and noodles (Shukla and Srivastava, 2014), besides traditional foods.

7.4 PROCESSING TECHNOLOGIES OF FINGER MILLET TO ENHANCE NUTRITIONAL QUALITY AND BIOAVAILABILITY

From a nutritional perspective, there are difficulties because of the low mineral content of cereals, which limits their bioavailability. Additionally, the presence of phytic acid and other anti-nutritional elements further reduces the bioavailability of cereals to 5–15%. Although grains include anti-nutritional components like tannin in greater levels with substantial negative correlations with iron and zinc, soaking, roasting, boiling, germination, fermentation, and malting are some examples of processing methods that lower tannin concentration and increase micronutrient availability.

7.4.1 SOAKING OR COOKING

The best approach for lowering the percentage of anti-nutrients in food is soaking. For a duration of one to two days, ragi was immersed in water at a ratio of 1:10 at room temperature. Trypsin inhibitors, polyphenols, phytates, saponins, and oxalate levels are all decreased throughout the soaking process (Hotz and Gibson, 2007). Soaking enhances the nutrient content, bioavailability, and bioaccessibility of minerals in millets. Soaking finger millets in distilled water or NaOH solution for eight hours greatly lowers their tannin content. Additionally, phytic acid concentrations are said to have decreased from 39.47% to 24.17%. The heat-labile anti-nutrients in finger millets can be rendered inactive through steaming or cooking (Kakade, 2015).

7.4.2 FERMENTATION

The oldest and best processing technique for food preservation is fermentation. The fermenting procedure concurrently lowers the anti-nutrient contents while improving the finger millet's biological value, net protein usage value, and percentage values of niacin, riboflavin, and thiamin. According to Anthony and Chandra (1998), there is a significant decrease in anti-nutrient contents such as tannins by 52%, trypsin inhibitor activity (TIA) by 32%, and phylate quantity by 20% when endogenous grain microflora is employed in the method of fermentation of finger millet flour. The fermenting process also increases the mineral contents and accessibility of finger millets, such as calcium (20%), ferrous (20%), zinc (20%), and phosphorus (26%) (Singh and Raghuvanshi, 2012). Millets undergo fermentation, which raises the lysine and niacin levels by 4.2 mg and 0.2 mg, respectively, per 100g (Chinenye et al., 2017). For the production of thick and thin porridge, baked and fried pancakes, and beverages made from finger millet—all of which have been enjoyed in the aforementioned form since ancient times—fermentation is a crucial and fundamental stage. The complex storage protein in fermentation can be broken down by microbial enzymes, which increases protein digestibility. Since acid generation and soaking boost *in vitro* digestibility after fermentation, an amylolytic attack is also possible.

7.4.3 *MILLING OR GRINDING*

Before consuming finger millets, it is necessary to remove the majority that is present as husk and coarse bran with a seed coat by the dehusking and debranning processes (Kakade and Hathan, 2015). The grain's germ and aleurone layers, which are nutrient-rich sections, are somewhat displaced, rendering them low in nutrients. By adding 10% water to the grain and moistening it, the fibrous husk can be removed (Hotz and Gibson, 2007). The process of moistening, grinding, and clarifying finger millet yields flour. The kernel is rich in calcium, polyphenols, and dietary fiber; seed coatings make up 15% of the kernel (Antony and Chandra, 1998). Generally, the entire meal or the ground millet is used in the manufacturing of food products. Finger millet's seed coat is edible, but when it's added to food, it takes on a chewy, black feel. It ought to be divided as a result during the processing phase. Finger millet is subjected to many processes to separate the seed coat: hydrothermal decortications, sifting and pulverizing, malting, and wet conditioning. The undamaged millet starch granules exhibit a small resistance to α and β amylase, while the granules that have been scratched are vulnerable to enzyme attack. Even though starch enhances the water binding capacity, damage but excess value causes the water binding ability of flour to decrease.

7.4.4 *ROASTING*

Grain taste is the main benefit of traditional roasting, but there are other advantages as well, such as decreased anti-nutritional components and longer shelf life. The grain is made edible, and its nutrients are preserved during the grinding and roasting processes. Although the processes of roasting and puffing are nearly identical, puffing involves a greater volume expansion. According to Kakade and Hathan (2015), roasting removes anti-nutritional compounds like hemagglutinin, goitrogenic agents, cyanogenic glycosides, alkaloids, saponins, and trypsin inhibitors. This procedure also extends the storage life of the product and is frequently used in rural and domestic settings. When finger millet is roasted, it increases the bioavailability of iron in weaning diets.

7.4.5 *PUFFING*

Puffing or popping is a simplified method of preparing grains to make them suitable for consumption. After popping, the finger millet grain becomes a porous, crisp, and easy-to-prepare meal that is precooked and has a decent flavor and aroma. In a nutrition intervention program, popped finger millet is considered a food component that is enhanced with nutrients (Ray et al., 2016). Malleshi and Desikachar (1986) indicated that processing temperatures of 250°C and moisture contents of approximately 19% are ideal for obtaining fully expanded millets. Because it is not advantageous to de-bran the grain after popping, the popped grains are quite fibrous. Popped millets are products that are porous, have a distinct flavor, an appealing texture, and a low bulk density. Puffing increases the nitrogen concentration and starch-digesting capacity *in vitro*. A high degree of starch gelatinization and the de-bonding of starch granules from protein are attributed to the improved starch digestion ability, which renders the grain more amenable to enzymatic digestion.

7.4.6 MALTING OR GERMINATION

According to Singh and Raghuvanshi (2012), germination is the most popular technique used in India to improve the nutritional profile of finger millet, and it has a higher nutritional value than both germinated sorghum and germinated maize. Finger millet is said to have a higher amylase activity value than sorghum, but other millets require four to five days to reach their maximal amylase activity (Malleshi and Desikachar, 1986). Germination enhances the nutritional content linked to biochemical changes and prevents fungal infection in grains. The concentration of β and α amylase increases steadily during the germination process, which will improve the aroma of malted meals throughout the roasting and kilning stages. The terms "malting," "sprouting," and "germination" are most frequently used in reference to the soaking process. Trypsin inhibitor, tannin, and phytic acid are examples of anti-nutritional contents that are simultaneously decreased and enhanced by germination. Additionally, it improves the bio-accessibility function of minerals found in finger millet, including calcium, iron, and zinc (Krishnan et al., 2012; Suma and Uroo, 2012). If finger millet is allowed to germinate for 48 hours, the ability to absorb iron, zinc, and calcium *in vitro* can be greatly enhanced (Mammiro, 2001). During the malting process, finger millets' nutritional and sensory qualities, as well as their digestibility and nutrient bioavailability, can all be altered to an extremely high degree. Finger millet, in its malted form, is regarded as a healthy food for diabetics and is used as a nourishing food for babies. Tryptophan, methionine, cysteine, and lysine content all rise when finger millet sprouts. Because the seed coat is removed during germination and malting, finger millet exhibits a drop in iron, calcium, and phosphorus. According to Chenenye (2017), malting brown and white finger millet dramatically raises the amounts of ionizable iron and soluble zinc while concurrently lowering the tannin and phytin phosphorus. During sprouting, microflora and lactic acid bacteria are said to proliferate, which is advantageous for the preparation of traditional foods (Krishnan et al., 2012).

7.5 VALUE-ADDED PRODUCTS TO ENHANCE FINGER MILLET CONSUMPTION

Finger millet is a versatile grain that works well in place of other grains, including rice and other starchy grains. These products have been shown to be effective in increasing finger millet consumption, or they are currently in use.

7.5.1 MULTI-GRAIN FLOUR/COMPOSITE FLOUR

The idea of composite or multigrain flour is not new to humankind. It has long been customary to combine two or three different kinds of grains or grains and pulses, depending on local commodity availability and dietary preferences, but in these situations, knowledge of nutritional security is not always connected. No Indian meal is complete without Indian-style bread or roti; therefore, multigrain flour, which is made by blending wheat and finger millet in a 7:3 (wheat: finger millet) ratio, is one of the basic semi-finished goods suitable for making chapatti (roti). Even though the suggested blend has much lower gluten content, the pleasing texture of the chapatti remains unaltered. However, the chapatti's color becomes a little darker. Incorporating finger millet into chapattis not only enhances their flavor but also effectively regulates blood sugar levels in those with diabetes (Kang et al., 2008). We feel fuller on fewer calories due to the fibers' bulkiness and the slower digestion rates, which may assist us in avoiding overindulging in calories. Those who struggle with constipation can benefit further from its high fiber content (Cade et al., 2007).

7.5.2 PAPAD

In millet-growing regions of South India, it has become customary to incorporate finger millet as one of the basic ingredients, in the range of 15–20% (w/w), in addition to other essential components including rice, spices, and either black or green gram. It has been reported that in certain regions of Karnataka, finger millet can be added to papad up to 60% (Begum et al., 2007).

Additionally, papad is made with finger millet flour, which serves as a basic material and is combined with salt and spices. Before making dough, flour is boiled in water until it becomes gelatinous. After rolling and cutting the dough into the appropriate shapes and sizes, a thin sheet is created. These papad pieces are then dried until the proper moisture content of 7–8% (db) is reached. Because the pericarp and starch are combined, the papad retains some of its black color when fried or roasted, but it also becomes lighter and is more palatable to consumers.

7.5.3 NOODLES – VERMICELLI

Noodles are in high demand both in India and outside due to the shifting eating patterns of adolescent and child demographics. As more people are becoming aware of the health advantages of millet, there is an increasing demand for finger millet noodles in particular. Noodles are pasta products—also referred to as convenience foods—that are made using a cold extrusion process and, once dried, are brittle and hard. Different combinations of noodles are made: finger millet-only noodles, finger millet and wheat in a 1:1 ratio, and finger millet combined with soy flour and wheat in a 5:4:1 ratio. When making noodles made only from millet, the millet flour is pre-treated to help with extrusion and maintain a smooth texture throughout drying. When making noodles, wheat flour is typically a key component of the mixture due to the additional benefit of wheat gluten, which not only facilitates efficient extrusion but also provides the finished product a smooth, fracture-free texture. While preparing noodles, there are numerous alternative mix combinations that may be tried, keeping in mind the nutritional content and accessibility of the ingredients.

7.5.4 EXTRUDED PRODUCTS

Another cutting-edge technique for turning ingredients into products with added value is extrusion technology. These days, extruded foodstuffs made from various grains are highly sought after by people of all ages. "Kurkure," for instance, is a popular product among kids. Food preferences are also changing dramatically as a result of the changed lifestyle, and as extruded foods are RTE products, they are now a great option for snacks.

Following the necessary conditioning and flour-making steps, all cereals with significant starch content can be extruded. Grits or flour made from finger millet have good extrusion properties. Extrusion cooking allows the product to be fully cooked and gelatinized, allowing it to be used as an RTE food. Similar to other preparations, finger millet flour can be combined in the right amounts with other flours made from legumes, along with additional fortification of vitamins and minerals, to create an extruded food that is nutritionally complete. As an alternative, the extrudates can be ground up and mixed with a measured quantity of additional cooked or pre-prepared food items to make additional food mixes for nursing moms, newborns, etc. To further enhance the value of extrudates made from finger millets and draw in kids, coat them in sweet or savory coatings.

7.5.5 BAKERY PRODUCTS

While there have been attempts to prepare bread, muffins, and biscuits with finger millet flour, standardizing the recipe and product quality is being worked on. In addition to offering excellent fiber and micronutrient content, the usage of millets in baked goods will also open up new opportunities for millets to be used in a variety of value-added products. According to Singh et al. (2012), finger millet has gained popularity recently, and initiatives are being made to make it easily accessible to consumers.

7.5.6 FERMENTED FOODS

Fermented foods like dosa and idli are popular in many parts of India. In the southern portion of the nation, they are frequently eaten for breakfast or even dinner. One common ingredient for these kinds of fermented dishes is finger millet. It tastes better and increases the nutritional value by providing more protein, calcium, and fiber. Depending on preference and flavor, finger millet or malted grains can also be sprouted for fermented dishes. After additional steps, ragi flour is combined with the other basic ingredients for fermented dishes.

7.5.7 RAGI SOUP

Blending 1 part ragi flour with 2.5 parts water will make one serving of ragi soup. Proper care is taken when stirring or mixing to prevent lumps from forming and to leave a thick and smooth mixture. After that, this mixture is cooked by heating it over medium heat for 15 to 17 minutes. Stirring must be done continuously during heating in order to prevent additional lump development. Once cooked, take the mixture off the heat and stir in half a teacup of curd and salt to taste. It's ready to be served warm after an additional five minutes. If serving cold, then chilling down much more is needed.

7.5.8 BREADS

In Indian kitchens, bread is a popular morning item. Although multigrain bread has gained popularity in recent years due to its nutritional benefits, some people find the taste to be too strong. For those individuals, adding ragi flour to wheat flour keeps the bread's flavor the same while increasing its nutritional content.

7.5.9 FINGER MILLET FLAKES

Finger millet flakes are added with corn flakes and can be consumed at breakfast by children and adults, as growing children need a higher amount of dietary fiber, fats, carbohydrates, and protein. Meanwhile, adults are conscious of their health problems, so the addition of ragi flakes will help adults to control their cholesterol levels and may reduce the chance of diabetes mellitus.

7.6 ENHANCEMENT OF NUTRIENTS USING BIOFORTIFICATION

Bio-fortification is the process of developing a crop with bioavailable micronutrients in its edible parts. The development of such crops can either be achieved by selective breeding or through modern biotechnological approaches (Garg et al., 2018).

7.7 ICAR'S INITIATIVE TO DEVELOP BIOFORTIFIED VARIETIES

The Consortia Research Platform (CRP) on bio-fortification was established by ICAR in 2014 with the goal of expediting the creation of bio-fortified varieties and improving the nutritional value of widely consumed staples. The CRP focused on rice, wheat, maize, sorghum, pearl millet, and small millets. Research is underway to create small-millet cultivars that are high in zinc and iron. The Prime Minister, Shri Narendra Modi, dedicated 17 bio-fortified varieties of eight crops to the country on the occasion of World Food Day 2020. Among these were the calcium, iron, and zinc-rich finger millet cultivars CFMV 1 and CFMV 2. The details are presented in Table 7.1.

TABLE 7.1 Bio-Fortified Varieties of Finger Millet

Sl. No.	Crop	Variety	Grain Yield (kg/ha)	Calcium (ppm)	Iron (ppm)	Zinc (ppm)	Developer	Area of Adoption
1	Finger millet	CFMV 1 (Indravathi)	3110	4280	58	44	ARS, Vizianagaram	AP, Karnataka, Tamil Nadu, Puducherry, Odisha
2	Finger millet	CFMV 1 (Gira)	2950	4740	39	25	HMRS, Waghai	AP, Chhattisgarh, Gujarat, Maharashtra, Odisha

7.8 HEALTH BENEFITS OF FINGER MILLET USAGE

Finger millet, which is high in polyphenols, has health benefits and can be used to make healthy dietary supplements. The brown kind of millet has higher levels of lysine, valine, and threonine, as well as 96% phenolic acid compared to white and other varieties.

7.8.1 FINGER MILLET AS AN ADDITIVE IN THE DIET FOR DIABETES

Phytochemicals in finger millet slow down digestion, which aids in blood sugar regulation and boosts antioxidant levels (Hegde et al., 2005). Finger millet has more fiber than wheat or rice, and its moderate glycemic response means it has less potential to elevate blood sugar and enhance the absorption of carbohydrates (Kumari and Sumathi, 2002). The phenolic seed coat of finger millet lowers postprandial hyperglycemia by inhibiting the activity of enzymes needed for the digestion of complex carbohydrates, such as amylase and alpha-glucosidase (Singh, 2016).

7.8.2 FINGER MILLET TO DECREASE CARDIOVASCULAR DISEASES

Methionine, lecithin, and threonine amino acids present in finger millet help to remove unnecessary fat from the liver, decrease cholesterol levels, and increase fat formation. Again, finger millet possesses a low level of serum triglycerides. The ingestion of finger millet lowers plasma triglycerides, which reduces the risk of cardiovascular diseases (Lee, 2010).

7.8.3 FINGER MILLET FOR CELIAC DISEASE

Due to its gluten-free nature, finger millet is a good choice for those with celiac disease, those who are sensitive to gluten, and those who generally detest the gluten found in wheat and other common cereal grains (Saleh et al., 2013).

7.8.4 FINGER MILLET AS A SOURCE FOR PROTEIN / AMINO ACIDS

Amino acids are essential for tissue repair and the functioning of the body. Tryptophan, valine, isoleucine, methionine, and threonine are abundant in finger millet. Amino acids enhance the body's metabolism, support the equilibrium of nitrogen levels in the body, and aid in the coordination of muscle and tissue repair. Isoleucine helps to create blood, which aids in the creation of bones and heals muscular diseases. Blood production aids in the development and restoration of healthy skin, whereas isoleucine aids in the healing of muscles. Methionine is an amino acid that is necessary for all bodily activities. It also helps the body eliminate extra fat and provides sulfur, which is needed for the body to produce glutathione, one of the body's natural antioxidants.

7.8.5 FINGER MILLET AS A MEDICATION FOR ANEMIA

Finger millet is high in iron and is used in diets to treat malnutrition and anemia.

7.8.6 FINGER MILLETS AS A WAY TO RELAXATION

Consumption of finger millet naturally induces relaxation of the body. Ragi has been utilized to treat anxiety, headaches, depression, and sleeplessness.

7.8.7 FINGER MILLETS: A WAY TO REDUCE AGING

Antioxidants and phenolics, which are important parameters of health, aging, and the metabolic syndrome, are abundant in finger millets.

7.9 CONCLUSION

A significant staple food crop in India and eastern and central Africa is finger millet. It is a simple food to digest and does not generate acids. Finger millet has high dietary fiber and offers several health benefits like anti-diabetic properties, chronic disease prevention, cardiac health, and hypocholesterolemic effects. Because the grain is coarse and difficult to process, this crop has a limited shelf life. It also has a low sensory quality. There are many technologies developed to make composite flour and value-added products from finger millet, but developed countries are not aware of this. Hence, information needs to be spread around the world about this crop, its nutritional properties, and its health benefits. In this regard, the year 2023 is declared as the International Year of Millet to increase the production and productivity of millet throughout the globe. Developing countries also face numerous challenges related to production, processing, value addition, marketing, and consumption. This has hindered overall production and consumption. In the future, there is a need to develop high-yielding varieties specific to different locations and value-added products from finger millet to make this crop suitable for sustainable world development. The food processing industries and biotechnology must play a vital role in the development of new value-added products to increase their consumption popularity across the globe.

KEYWORDS

- **bio-fortification**
- **Finger millet (Eleusine coracana**
- **Poaceae family**
- **processing technologies**
- **value-added products**

REFERENCES

Amadoubr, I., & Le, M. (2013). Millets: Nutritional composition, some health benefits and processing–a review. *Emirates Journal of Food and Agriculture, 25*(7), 501. https://doi.org/10.9755/ejfa.v25i7.12045 (accessed on 7th November 2025).

Ansari, M. A., Ravisankar, N., Ansari, M. H., Babu, S., Layek, J., & Panwar, A. S. (2023). Integrating conservation agriculture with intensive crop diversification in the maize-based organic system: Impact on sustaining food and nutritional security. *Frontiers in Nutrition, 10*. https://doi.org/10.3389/fnut.2023.1137247 (accessed on 7th November 2025).

Antony Ceasar, S., Maharajan, T., Ajeesh Krishna, T. P., Ramakrishnan, M., Victor Roch, G., Satish, L., & Ignacimuthu, S. (2018). Finger millet (*Eleusine coracana* (L.) Gaertn.) improvement: Current status and future interventions of whole genome sequence. *Frontiers in Plant Science, 9*. https://doi.org/10.3389/fpls.2018.01054 (accessed on 7th November 2025).

Antony, U., & Chandra, T. S. (1998). Antinutrient reduction and enhancement in protein, starch and mineral availability in fermented flour of finger millet (*Eleusine coracana* L.). *Journal of Agricultural and Food Chemistry, 46*(7), 2578–2582.

Begum, J. M. (2007). Refined processing and products for commercial use and health benefits from finger millet. In K. T. Krishne Gowda & A. Seetharam (Eds.), *All India Coordinated Small Millets Improvement Project.*

Bhatt, A., Singh, V., Shrotria, P. K., & Baskheti, D. C. (2003). Coarse grains of Uttaranchal: Ensuring sustainable food and nutritional security. *Indian Farmer's Digest*, 34–38.

Cade, J. E., Berley, V. J., & Greenwood, D. C. (2007). Dietary fibre and risk of breast cancer in the UK women's cohort study. *International Journal of Epidemiology, 36*, 431–438.

Chinenye, O. E., Ayodeji, O. A., & Baba, A. J. (2017). Effect of fermentation (natural and starter) on the physicochemical, anti-nutritional and proximate composition of pearl millet used for flour production. *American Journal of Bioscience and Bioengineering, 5*(1), 12–16.

Dharmaraj, U., & Malleshi, N. G. (2011). Changes in carbohydrates, proteins and lipids of finger millet after hydrothermal processing. *Lebensmittel-Wissenschaft & Technologie, 44*(7), 1636–1642. https://doi.org/10.1016/j.lwt.2010.08.014 (accessed on 7th November 2025).

Garg, M., Sharma, N., Sharma, S., Kapoor, P., Kumar, A., Chunduri, V., & Arora, P. (2018). Biofortified crops generated by breeding, agronomy, and transgenic approaches are improving lives of millions of people around the world. *Frontiers in Nutrition, 5*. https://doi.org/10.3389/fnut.2018.00012 (accessed on 7th November 2025).

Gautam, R. (2000). Bole-bisrae motae anaaj, *Kethi, 10*, 14–17.

Geetha, C., & Parvathi, P. (1990). Hypoglycemic effect of millet incorporated breakfast items on selected non-insulin dependent diabetic patients. *Indian Journal of Nutrition and Dietetics, 27*, 316–320.

Hegde, P. S., Rajasekaran, N. S., & Chandra, T. S. (2005). Effects of the antioxidant properties of millet species on oxidative stress and glycemic status in alloxan-induced rats. *Nutrition Research, 25*(12), 1109–1120. https://doi.org/10.1016/j.nutres.2005.09.020 (accessed on 7th November 2025).

Hotz, C., & Gibson, R. S. (2007). Traditional food-processing and preparation practices to enhance the bioavailability of micronutrients in plant-based diets. *Journal of Nutrition, 137*(4), 1097–1100. https://doi.org/10.1093/jn/137.4.1097 (accessed on 7th November 2025).

Kakade, S. B., & Hathan, B. S. (2015). Finger millet processing. *International Journal of Agricultural Innovations and Research, 3*(4), 1003–1008.

Kang, R. K., & Jain, R. (2008). Impact of indigenous fiber rich premix supplementation on blood glucose levels in diabetics. *American Journal of Food Technology, 3*(1), 50–55.

Krishna, T. P. A., Maharajan, T., Roch, G. V., Ramakrishnan, M., Ceasar, S. A., & Ignacimuthu, S. (2020). Hybridization and hybrid detection through molecular markers in finger millet (*Eleusine coracana* (L.) Gaertn.). *Journal of Crop Improvement, 34*(3), 335–355. https://doi.org/10.1080/15427528.2019.1709596 (accessed on 7th November 2025).

Krishnan, R., Dharmaraj, U., & Malleshi, N. G. (2012). Influence of decortication, popping and malting on bio accessibility of calcium, iron and zinc in finger millet. *LWT - Food Science and Technology, 48*(2), 169–174.

Kumari, D., Shahidi, F., & Chandrashekara, A. (2019). Bio accessibility and antioxidant activities of finger millet food phenolics. *Journal of Food Bioactive, 6*, 100–109.

Kumari, P. L., & Sumathi, S. (2002). Effect of consumption of finger millet on hyperglycemia in non-insulin dependent diabetes mellitus (NIDDM) subjects. *Plant Foods for Human Nutrition, 57*(3), 205–213.

Layek, J., Rangappa, K., Das, A., Ansari, M. A., Choudhary, S., Rajbonshi, N., Patra, S., Kumar, A., Mishra, V. K., Ravisankar, N., Kumar, S., Hazarika, S., Dutta, S. K., Babu, S., Tahasildar, M., & Shettigar, N. (2023). Evaluation of millets for physio-chemical and root morphological traits suitable for resilient farming and nutritional security in Eastern Himalayas. *Frontiers in Nutrition, 10*. https://doi.org/10.3389/fnut.2023.1198023 (accessed on 7th November 2025).

Lee, S. H., Chung, I. M., Cha, Y. S., & Park, Y. (2010). Millet consumption decreased serum concentration of triglyceride and C-reactive protein but not oxidative status in hyperlipidemic rats. *Nutrition Research, 30*(4), 290–296.

Maharajan, T., Ceasar, S. A., Krishna, T. P. A., & Ignacimuthu, S. (2019). Phosphate supply influenced the growth, yield and expression of PHT1 family phosphate transporters in seven millets. *Planta, 250*(5), 1433–1448. https://doi.org/10.1007/s00425-019-03237-9 (accessed on 7th November 2025).

Malleshi, N. G., & Desikachar, H. S. R. (1986). Influence of malting conditions on quality of finger millet malt. *Journal of the Institute of Brewing, 92*(1), 81–83. https://doi.org/10.1002/j.2050-0416.1986.tb04377.x (accessed on 7th November 2025).

Mamiro, P. R. S., Van, J., Mwikya, S. M., & Huyghebaert, A. (2001). *In vitro* extractability of calcium, iron, and zinc in finger millet and kidney beans during processing. *Journal of Food Science, 66*(9), 1271–1275. https://doi.org/10.1111/j.1365-2621.2001.tb15200.x (accessed on 7th November 2025).

N/A. (1991). *Protein Quality Evaluation*. FAO/WHO Food and Nutrition Paper. Italy, 12–24.

N/A. (n.d.). United Nations University. http://www.unu.edu (accessed on 7th November 2025).

Raboy, V. (2000). Low phytic acid grains. *Food & Nutrition Bulletin, 21*(4), 423–427.

Ramachandra, G., Virupaksha, T. K., & Shadaksharaswamy, M. (1977). Relationship between tannin levels and *in vitro* protein digestibility in finger millet (*Eleusine coracana* Gaertn). *Journal of Agricultural and Food Chemistry, 25*, 1101–1104.

Rao, B. S. N., & Prabhavati, T. (1982). Tannin content of foods consumed in India and its influence on ionizable iron. *Journal of the Science of Food and Agriculture, 33*, 89–96.

Rao, P. U. (1994). Evaluation of protein quality of brown and white ragi (*Eleusine coracana*) before and after malting. *Food Chemistry, 51*(4), 433–436. https://doi.org/10.1016/0308-8146(94)90198-8 (accessed on 7th November 2025).

Ravindran, G. (1992). Seed proteins of millets: Amino acid composition, proteinase inhibitors and *in vitro* digestibility. *Food Chemistry, 44*(1), 13–17.

Ray, M., Ghosh, K., Singh, S., & Chandra Mondal, K. (2016). Folk to functional: An explorative overview of rice-based fermented foods and beverages in India. *Journal of Ethnic Foods, 3*(1), 5–18. https://doi.org/10.1016/j.jef.2016.02.002 (accessed on 7th November 2025).

Saha, S., Gupta, A., Singh, S. R. K., Bharti, N., Singh, K. P., Mahajan, V., & Gupta, H. S. (2011). Compositional and varietal influence of finger millet flour on rheological properties of dough and quality of biscuit. *Lebensmittel-Wissenschaft & Technologie, 44*(3), 616–621. https://doi.org/10.1016/j.lwt.2010.08.009 (accessed on 7th November 2025).

Saleh, A. S. M., Zhang, Q., Chen, J., & Shen, Q. (2013). Millet grains: Nutritional quality, processing, and potential health benefits. *Comprehensive Reviews in Food Science and Food Safety, 12*(3), 281–295. https://doi.org/10.1111/1541-4337.12012 (accessed on 7th November 2025).

Sharma, B., & Gujral, H. S. (2019). Influence of nutritional and antinutritional components on dough rheology and *in vitro* protein & starch digestibility of minor millets. *Food Chemistry, 299*(125115), 125115. https://doi.org/10.1016/j.foodchem.2019.125115 (accessed on 7th November 2025).

Shukla, K., & Srivastava, S. (2014). Evaluation of finger millet incorporated noodles for nutritive value and glycemic index. *Journal of Food Science and Technology, 51*(3), 527–534. https://doi.org/10.1007/s13197-011-0530-x (accessed on 7th November 2025).

Singh, E. (n.d.). Potential functional implications of finger millet (*Eleusine coracana*) in nutritional benefits, processing, health and diseases: A review. *International Journal of Home Science, 2*(21), 151–155.

Singh, P. (2012). Finger millet for food and nutritional security. *African Journal of Food Science, 6*(4). https://doi.org/10.5897/ajfsx10.010 (accessed on 7th November 2025).

Sridhar, R., & Lakshminarayana, G. (1994). Content of total lipids and lipid classes and composition of fatty acids in small millets: Foxtail (*Setaria italica*), proso (*Panicum miliaceum*), and finger (*Eleusine coracana*). *Cereal Chemistry, 71*(4), 355–358.

Suma, P. F., & Urooj, A. (2012). Antioxidant activity of extracts from foxtail millet (*Setaria italica*). *Journal of Food Science and Technology, 49*(4), 500–504. https://doi.org/10.1007/s13197-011-0300-9 (accessed on 7th November 2025).

Tharanathan, R. N., & Mahadevamma, S. (2003). Grain legumes—a boon to human nutrition. *Trends in Food Science & Technology, 14*(12), 507–518. https://doi.org/10.1016/j.tifs.2003.07.002 (accessed on 7th November 2025).

Thompson, L. U. (1993). Potential health benefits and problems associated with anti-nutrients. *Food Research International, 26*, 131–149.

Urooj, A., & Malleshi, N. G. (2009). Evaluation of antioxidant and antimicrobial properties of finger millet polyphenols (*Eleusine coracana*). *Food Chemistry, 114*(1), 340–346.

Van Cam, J., & Nagundi, D. (2000). Amino acid profile after sprouting, autoclaving and lactic acid fermentation of finger millet (*Eleusine coracana*) and kidney beans (*Phaseolus vulgaris* L.). *Journal of Agricultural and Food Chemistry, 48*(8), 3081–3085.

Verma, V., & Patel, S. (2012). Nutritional security and value-added products from finger millets (ragi). *Journal of Applied Chemistry, 1*, 485–489.

Wambi, W., Otienno, G., Tumwesigye, W., & Mulumba, J. (2021). Genetic and genomic resources for finger millet improvement: Opportunities for advancing climate-smart agriculture. *Journal of Crop Improvement, 35*(2), 204–233. https://doi.org/10.1080/15427528.2020.1808133 (accessed on 7th November 2025).

Wen, Y., Liu, J., Meng, X., Zhang, D., & Zhao, G. (2014). Characterization of proso millet starches from different geographical origins of China. *Food Science and Biotechnology, 23*(5), 1371–1377. https://doi.org/10.1007/s10068-014-0188-z (accessed on 7th November 2025).

CHAPTER 8

Nutritional Quality Enhancement in Small Millets

U. KRISHNAJA,* P. SHRUTHI, and GOPIKA P. HASSAN

Department of Community Science, College of Agriculture, Kerala Agricultural University, Vellayani, Thiruvananthapuram, Kerala, India

*Corresponding author

8.1 INTRODUCTION

"Millets" are small-seeded grains named after the French word "mille," which means thousands of handfuls of grains (Kaur et al., 2023). They are considered among the oldest cultivated crops known to humans and are drought-tolerant, nutrient-dense, and rich sources of proteins, vitamins, and minerals; they are also referred to as "famine reserves." Millets belong to the family Poaceae and can be grouped into major millets (sorghum and pearl millet) and minor millets based on grain size. Both major and minor millets have been important components of the traditional food basket of poor populations around the world. Minor millets are better suited for cultivation in arid and semi-arid regions because they can grow in barren, sparsely irrigated areas with low soil fertility. A total of 74.00 million hectares of land were used to produce 89.17 million metric tons of millets globally. With a 12.45-million-acre cultivated area and a production of 15.53 million metric tons, India is the world's largest millet producer (Banshidhar et al., 2023). Minor millets comprise six species, namely finger millet (*Eleusine coracana* (L.)), little millet (*Panicum sumatranse*), kodo millet (*Paspalum scrobiculatum* (L.)), foxtail millet (*Setaria italica* (L.)), barnyard millet (*Echinochloa frumentacea* (L.)), and proso millet (*Panicum miliaceum* (L.)). These crops have traditionally been an indispensable component of dry farming systems in India and elsewhere. See Figure 8.1. They are commonly referred to as nutri-cereals or small-seeded grains that are nutritionally superior to rice and wheat and are considered gluten-free (Srivastava and Reddy, 2023).

8.2 NUTRITIONAL COMPOSITION OF MINOR MILLETS

Minor millets are small-seeded grains that resemble paddy or rough rice in the morphological features of the kernel. The kernel of the grain is generally composed of distinct husk, bran, and endosperm tissues. The unique tissue embryo makes up around 2% of the

Millets in Sustainable Agriculture: Genetic Improvement, Nutritional Quality, Cultivation, and Integrated Pest–Disease Management. Roshni Vijayan, A. John Joel, R. Ambika Rajendran, and Adithya P. Balakrishnan (Eds.)

DOI: 10.1201/9781779642332-10

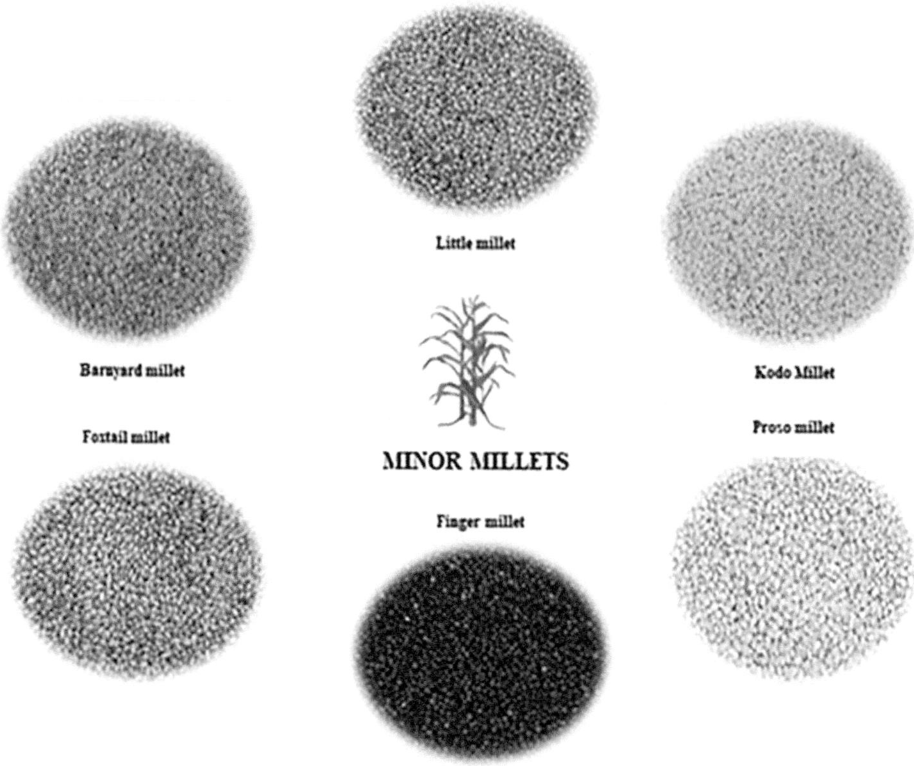

FIGURE 8.1 Minor millets.

grain. The inedible millet husk resembles the paddy husk; however, bran may be converted into an edible component and separated to develop milled millets for food uses. Typically, the endosperm makes up 75% of the kernel, with the husk making up 15% to 20%, and the rest 5% is bran. These grains are ovoid to spherical in shape, and studies have shown that their 1000-kernel weight and volume range from 1.9 to 5.5 g and 1.3 to 3.8 ml, respectively. The nutritional profile of minor millet is briefed in Tables 8.1–8.4.

TABLE 8.1 Nutrient Composition of Minor Millets (per 100 g)

Minor Millets	Energy (KJ)	Carbohydrates (g)	Protein(g)	Fat(g)	Crude Fiber(g)	Ca(mg)	P(mg)	Fe (mg)
Finger millet	328	72	7.3	1.3	3.6	344	283	3.9
Little millet	341	67.0	7.7	4.7	7.6	17	220	9.3
Kodo millet	309	65.9	8.3	1.4	9.0	27	188	0.5
Foxtail millet	331	60.9	12.3	4.3	8.0	31	290	2.8
Barnyard millet	307	65.5	6.2	2.2	9.8	20	280	5.0
Proso millet	341	70.4	12.5	1.1	2.2	14	206	0.8

Source: Adapted from a manual, *National Institute of Nutrition, Hyderabad* (1998).

TABLE 8.2 Essential Amino Acid Content of Minor Millets (mg/g of N)

Minor Millets	Arginine	Histidine	Lysine	Tryptophan	Phenyl Alanine	Tyrosine	Methionine	Cystine	Threonine	Leucine	Isoleucine	Valine
Finger millet	300	130	220	100	310	220	210	140	240	690	400	480
Little millet	250	120	110	60	330	-	180	90	190	760	370	350
Foxtail millet	220	130	140	60	420	-	180	100	190	1040	480	430
Barnyard millet	270	120	150	50	430	-	180	110	200	650	360	410
Proso millet	290	110	190	50	310	-	160	-	150	760	410	410

Source: Adapted from ICAR (2021).

TABLE 8.3 Vitamin Composition of Minor Millets (per 100 g)

Minor Millets	Thiamin (mg)	Niacin (mg)	Riboflavin	Vit A (carotene) (mg/100g)	Folic Acid (mg/100g)	Vit B5 (mg/100g)	Vit E (mg/100g)
Finger millet	0.42	1.1	0.19	42	18.3	-	22.0
Little millet	0.3	3.2	0.09	-	9.0	-	-
Kodo millet	0.15	2.0	0.09	-	23.1	-	-
Foxtail millet	0.59	3.2	0.11	32	15.0	0.82	31.0
Barnyard millet	0.33	4.2	0.1	-	-	-	-
Proso millet	0.41	4.5	0.28	-	-	1.2	-

Source: Adapted from a manual, National Institute of Nutrition, Hyderabad (1998); http://www.sahajasamrudha.org; Kulkarni et al. (2018).

TABLE 8.4 Mineral Composition of Minor Millets (Per 100 g)

Minor Millets	Magnesium (Mg) (mg)	Sodium (Na) (mg)	Potassium (K) (mg)	Copper (Cu) (mg)	Manganese (Mn) (mg)	Zinc (Zn) (mg)
Finger millet	137	11.0	408	0.47	5.49	2.3
Little millet	133	8.1	129	1.00	0.68	3.7
Kodo millet	147	4.6	144	1.60	1.10	0.7
Foxtail millet	81	4.6	250	1.40	0.60	2.4
Barnyard millet	82	-	-	0.60	0.96	3
Proso millet	153	8.2	113	1.60	0.60	1.4

Source: Adapted from http://www.sahajasamrudha.org; Kulkarni et al. (2018) and Longva et al. (2017).

Little, foxtail, and proso millet often have a single, glossy seed coat and husk. Kodo and barnyard millet, in contrast, have multi-layered seed coatings. The kodo millet typically has a brown seed coat, the foxtail millet has a yellowish seed coat, and the other millets have greyish seed coats. Similar to how rice is milled, the husk and bran of millet are separated to produce edible products. Hulling has no impact on the nutritional content because the germ is retained throughout (FAO, 1995).

Finger millet can be considered one of the richest sources of calcium (Ramashia et al., 2019). Phosphorus and iron are likewise abundant in the other minor millets. The protein content of millet ranges from 7 to 12%, and the fat content from 1 to 5.0%. The millet's protein content has a harmonious amino acid profile and is an excellent source of methionine, cystine, and lysine. For those who primarily rely on plant foods for their protein dietary needs, these essential amino acids offer unique benefits. Nearly 65% of the millet grain's calories come from carbohydrates, a large amount of which comes in the form of dietary fiber and non-starchy polysaccharides (NSP), which also decrease cholesterol and slow the release of glucose into the bloodstream during digestion. People who frequently eat millet have been shown to have lower rates of cardiovascular disease, duodenal ulcers, and diabetes. Additionally, millet grains are a good source of thiamine, riboflavin, folic acid, and niacin. These B-complex vitamins are necessary for the synthesis of nucleic acids, neurological chemicals, energy, and normal brain function. Although millets are rich in several minerals and fatty acids, they are comparable to rice and wheat in terms of most nutrients. Millets differ significantly in their carbohydrate makeup, with amylose and amylopectin contents ranging from 16 to 28 and from 74 to 84%, respectively. Resistant starch is thought to be present in minor amounts in millet grains, and these minor starches derived from millet can operate as prebiotics or as a food source for probiotics in the gut. The polyphenolic concentration of millet grains is the cause of their antioxidant activity. Cinnamic acid derivatives and benzoic acids combine to produce millet polyphenol, which has anticancer and enzyme-inhibitory properties (Kaur et al., 2023). Phytic acid, β-glucan, phenolic acids, fatty acids, and dietary fiber are among the functional components found in minor millet grains. They offer defense against lifestyle disorders, such as diabetes, cancer, and cardiovascular conditions (ICAR-IIMR, 2021).

Nutritional quality enhancement of millet can be done in different ways (*see* Figure 8.2).

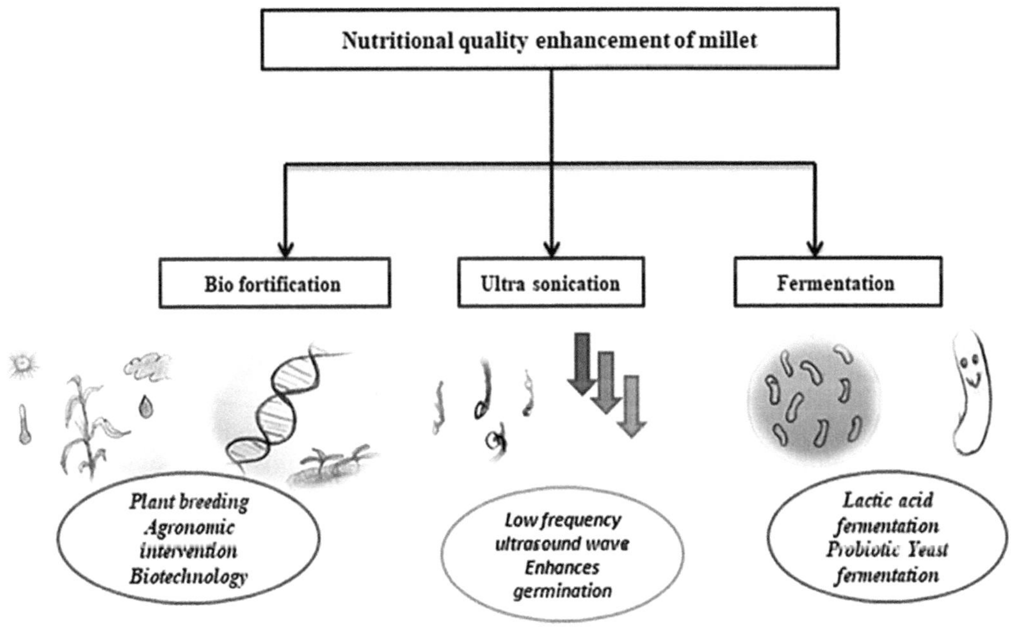

FIGURE 8.2 Ways for nutritional enhancement in millets.

8.3 BIOFORTIFICATION IN MILLETS

Biofortification is an act of enhancing the nutritional value of millet grains by possible practices like agronomic interventions, traditional plant breeding, and cutting-edge biotechnology techniques. It is a food-based strategy for overcoming nutrient deficiency by delivering nutrient-dense crops to poor people's doorsteps. Biofortified millet grains are a nutritious powerhouse and aid in the fight against malnutrition (Ramadas et al., 2020). Millets can be biofortified in two ways: (i) by increasing the accumulation of nutrients in milled grains and (ii) by lowering antinutrients to boost the bioavailability of minerals. The use of genetic engineering and genome editing techniques allows nutrients to accumulate in edible proportions while blocking antinutrient production. Plant genetic resources (PGRs) conservation ensures a steady supply of raw material for crop advancement. The success of a biofortification program depends on the long-term use of PGRs for nutritional enhancement. Due to a lack of genetic diversity, millet collections do not have enough beta carotene. Therefore, traditional millet breeding for high provitamin A levels is not feasible. In such cases, a transgenic strategy for vitamin production in plants supports metabolic engineering. Golden rice is a famous example of provitamin A genetic modification in grains. PGRs conservation ensures a steady stream of raw materials for crop innovation. Utilizing PGRs for nutritional augmentation in a sustainable manner is key to the success of biofortification programs. The International Crop Research Institute for the Semi-Arid Tropics (ICRISAT) has the largest collection of millet germplasm in the genebank, accounting for 27.4% of all crop accessions (Vinoth and Ravindhran, 2017).

8.4 LACTIC ACID FERMENTATION OF MILLETS

The fermentation process aids in the conversion of large organic molecules into simpler ones via microbial action (Sharma et al., 2020). The most well-known and time-tested method of enhancing the functionality, nutritional content, flavor, aesthetics, and safety of cereal comestibles is lactic acid fermentation. Millets are full of prebiotic dietary fiber, which serves as a food source for lactic acid bacteria, which in turn ferments the cereal grains to produce a salubrious food. Lactic acid bacteria aid in the production and absorption of vitamins (B, K) and micronutrients like iron, zinc, and calcium. Lactic acid fermentation plays a critical role in the elimination of aflatoxin and also enhances the grain substrates with protein, necessary amino acids, and fatty acids. Additionally, it enhances the organoleptic qualities, elevates the nutritional value of raw materials containing starch, functions as the main way of food preservation, and acts as a tactic for lowering harmful bacterial contamination (Petrova and Petrov, 2020). Little millet was successfully fermented by the addition of probiotic yeast (*S. boulardii*) and lactic acid bacteria (LAB; *L. acidophilus*) inoculums. This process improved the protein content, enhanced the bioavailability of micronutrients, and decreased the amount of phytic acid (Pampangouda et al., 2015). Millets undergoing lactic acid fermentation exhibited effective polyphenol content and the lowest antinutritional values, making them potentially useful for use in food items and preservation (Thamburaj et al., 2023).

8.5 LOW-FREQUENCY ULTRASOUND TECHNOLOGY

Ultrasound (US) is a mechanical wave that can be distinguished under three frequency ranges beyond the human audible range: conventional high-power or low-frequency ultrasound (LFUS) at frequencies between 20 and 100 kHz; sonochemistry at frequencies between 100 and 2 MHz; and medical or analytical ranges between 5 and 10 MHz. This technique has already been used to address issues like boosting yield and speeding up production. LFUS increases the extraction of polysaccharides, phenolic compounds, and other dietary ingredients; alters the structure of starches, changing their pasting properties; and partially denatures proteins, enhancing their interfacial properties and peptide availability. The technical, functional, and biological qualities of cereal and pseudocereal macromolecules, such as starch and protein, can often be improved through the use of low-frequency ultrasonic technology (Estivi et al., 2022). Finger millet can be processed more effectively with ultrasound-assisted hydration, resulting in fewer antinutrients and higher-quality starch. The procedure also aids in producing isolated starch with improved structure, water-binding ability, and water solubility (Yadav et al., 2021). Improved gel formation and stability can be achieved by subjecting foxtail millet starch to annealing and sonication treatments (Babu et al., 2019). Resistant starch concentration in pearl millet flours can be increased by using basic, ecologically friendly technologies like dry heat and ultrasonication to change the properties of starch (Vidhyalakshmi and Meera, 2023).

8.6 FINGER MILLET

Finger millet, scientifically known as *Eleusine coracana*, locally called panjapullu, kooravu, muthari, and ragi, is a staple cereal grain in low-income regions of the world and is the oldest cultivated crop in India. It is referred to as nrttakondaka, which also goes by the names "rajika" or "markataka," signifying dancing grain. This gluten-free grain is high in carbohydrates, dietary fiber, phytochemicals, and other essential nutrients and is the prime ingredient in weaning foods. The grain is utilized using various processing techniques to improve the sensorial and nutritional properties and to reduce the anti-nutritional factors (Ramashia et al., 2019). The process of producing nutrient-dense staple crops through the application of contemporary biotechnology and the best conventional breeding techniques, all while maintaining key consumer-preferred features and agronomic performance, is known as biofortification. To improve the dietary intake of calcium in humans and to combat hidden hunger, finger millet crops can be biofortified. Adding the appropriate mineral as an organic compound to fertilizer, enhancing plant nutrition through conventional breeding combined with mutagenesis, and using marker-assisted selection to introduce such traits into widely cultivated, adapted genotypes are some strategies for biofortifying crops with essential nutrients. It's important to maximize the absorption of grains that have been biofortified while reducing the absorption of anti-nutrients. Certain vitamins, inulin, cysteine amino acid, and provitamins are among the substances that aid in absorption and bioavailability (Sharma et al., 2017). An alternative option to generate high calcium finger millet varieties through conventional or contemporary breeding approaches and transformation-based methods is to mine markers strongly connected to other variables determining grain calcium content and uncover underlying genes (Puranik et al., 2017). Processing techniques include popping, decorting, malting, fermenting, parboiling, roasting, and milling, which can enhance the nutritional value of finger millet. The anti-nutritional content can be reduced by soaking and cooking (Rathore et al., 2019; Vagdevi et al., 2023). Millets are traditionally stored in underground pits, knotted bundles, earthen pots, mud rhombuses, etc., for long-term storage (Mobolade et al., 2019). Using dried red chilies and fresh neem leaves in storage bins and containers helps reduce insect infestation and is an affordable and sustainable method of millet storage. The storage stability of millet flours can be increased by using both cutting-edge technologies like microwave processing, infrared heating, gamma radiation, and high-pressure processing, as well as conventional processing techniques like mechanical processing, fermentation, germination, thermal treatment, and adding preservatives (Kaur et al., 2023). Starch is the main carbohydrate that is widely present in finger millet grains. It usually ranges in size from 1 to 120 mm and can take on a variety of morphologies. Because of starch's exceptional functional properties, the demand for it in a wide range of applications—such as adhesives, diluents, thickeners, emulsion stabilizers, fillers, bulking and suspending agents, and more in both the food and non-food industries—has dramatically increased (Gautam et al., 2021).

8.7 BARNYARD MILLET

Barnyard millet is widely grown throughout Asia, particularly in India, China, Japan, and Korea. It is the fourth most extensively cultivated minor millet in the world, providing millions of famished individuals with food security. Common names for barnyard millet include cockspur grass, Korean native millet, prickly millet, sawa millet, and watergrass. When grown naturally, this annual summer crop serves as both a food source and a source of animal feed. Due to its quick growth, resilience to drought, and ability to thrive in challenging circumstances, it is an important crop for famine areas. It is mostly grown in two agro-ecologies in India: one in the mid-hills of Uttarakhand's Himalayan region in the north, and another in Tamil Nadu's Deccan plateau region in the south. It has a harvesting period of six weeks and is among the fastest-growing millets. As well as providing rich amounts of fiber (13.6%) and iron (186 mg/kg dry matter), barnyard millet grains contain other beneficial compounds, such as γ-amino butyric acid (GABA) and β-glucan, which are antioxidants and reduce blood cholesterol levels (Saleh et al., 2013; Sharma et al., 2016; Vinoth et al., 2017). Barnyard millet is recommended as one of the best foods for type II diabetics due to its low carbohydrate content among millets. In the Indian Himalayan region, barnyard millet has long been used as a substitute for rice. Dehulled millet grains are prepared and consumed similarly to rice. In Uttarakhand, barnyard millet porridge (also known as madirakikheer) is a famous sweet meal. Barnyard millet is used in classic South Indian dishes including idli, dosa, and chakli. This fast-growing fodder crop generates a lot of fodder, livestock feed, and produces extremely palatable feed and can be used to make hay or silage (Sood et al., 2015).

8.7.1 BIOFORTIFICATION

The practice of genetically adding extra nutrients to important food crops is known as biofortification. It differs from fortification in that it uses agronomy, breeding, and biotechnology in agriculture to improve the nutrients of crops at the source. The agronomic approach to improving the mineral content of plants involves adding the relevant nutrient as an inorganic compound to the fertilizer. Two current methods for biofortification involve the discovery of genomic regions/potential genes for high nutritional content by the tagging/identification of important genes or the mapping of quantitative trait loci (QTL) followed by their introgression into common varieties. Genetic modification technology is the introduction of genes from novel sources for desirable target traits, and it has the advantages of having unlimited access to the genes of interest, targeted expression in target tissues, rapid and direct application through introduction into popular varieties, and stacking of different genes (Neeraja et al., 2017).

Induced mutagenesis has also been successfully used in minor millets to generate cultivars and expand the genetic basis. The majority of barnyard millet genotypes are non-glutinous (non-waxy), giving food items a poor texture. Consumers in Asian countries, on the other hand, prefer a sticky texture such as rice. As a result, developing waxy barnyard millet genotypes with a sticky texture would be extremely beneficial. Foliar spray is a simple approach to providing essential nutrients to plants. Barnyard millet might benefit from the use of foliar spray to improve crop development and productivity (Kumar et al., 2020). The grain Zn content can be genetically enhanced by modulating metal transporters (Guerinot, 2000).

8.7.2 NUTRITIONAL QUALITY ENHANCEMENT

In comparison to other major grains and millet, barnyard millet has a higher protein content (11.2–12.7%). Barnyard millet has a comparable amount of minerals, ash, fat, and amino acids to other cereals and millet, but the amount of iron in the grain is substantially higher (Renganathan et al., 2020). The grain of barnyard millet contains about 65% carbohydrates, with a large portion of those being dietary fiber and non-starchy polysaccharides. This helps prevent constipation, lower blood cholesterol levels, and delay the absorption of glucose into the system after meals. When compared to other millets, barnyard millet has been proven to be the most effective in lowering blood glucose and cholesterol levels. The presence of phytates, phenols, and tannins in millets limits their nutritional value, and this can be reduced by pretreatment or processing of grains. Around 50% of the iron in the diet could be bonded to tannins and phytic acid. Tannin content is reduced through processing methods such as soaking, roasting, boiling, germination, fermentation, and malting. The bioavailability of iron and zinc is boosted via a processing technology known as 'popping,' which increases the bioavailability of iron and zinc from 12.1 to 16.4 ppm and 15.3 to 22.8 ppm without changing the whole grain content (Neeraja et al., 2017). When compared to other millets, barnyard millet is shown to have an extremely low glycemic index (Sood et al., 2020). Barnyard millet husks can be used to make ethanol by undergoing various pretreatments and fermentation processes.

8.7.3 GRAIN PROTEIN ENHANCEMENT

The protein content of the millet can be significantly affected by germination. The free amino acid gamma-aminobutyric acid (GABA) content in the barnyard millet grain can be significantly increased by germination. GABA may aid in preventing cancer cell proliferation while simultaneously lowering blood pressure. Except for a severe lysine shortage, this millet's amino acid composition is nutritionally satisfactory. The protein content of barnyard millet grain can be enhanced by fermentation (Sharma et al., 2016).

8.7.4 STARCH CONTENT ENHANCEMENT

Barnyard millet starch, which is high in amylose, is now being considered an antioxidant packaging material in the biodegradable film industry. The application of borage seed oil to barnyard millet starch increases the elongation range of the starch while lowering its tensile strength, water permeability, and moisture content, making it ideal for biofilm development. These biofilms have been proven to be resistant to a variety of microorganisms and to inhibit light transparency and the generation of free radicals in the food industry (Cao et al., 2017).

8.7.5 SEED SHELF-LIFE ENHANCEMENT

The crop's fast maturity allows it to dodge drought. Under ideal weather conditions, barnyard millet can produce ripe grains 45 days after sowing. September–October and February–March are the best months for seed production. For quality and effective seed setting, pollination should not coincide with rainfall. Certified and foundation seeds should have a minimum of 97% physical purity, 75% germination potential, and 12% moisture content. To achieve a safe moisture level of 12%, the cleaned seeds should be sun-dried under the right conditions; seeds can be kept for a year. The crop's seeds are nutrient-dense, and major fungal infections do not affect barnyard millet (ICAR-IIMR, 2021).

8.7.6 POST-HARVEST SEED PROCESSING TRAIT ENHANCEMENT

Processing the barnyard millet crops is extremely difficult due to their small seed size. In barnyard millet, grain moisture is the most crucial factor since it influences not only storage but also the setup of processing equipment. The moisture content of barnyard millet grain has a significant impact on the quality as well as the milling and polishing time. To make barnyard millet grain appropriate for human consumption, it must first be dehulled. The technique requires repeated pounding in a mortar, which takes time and effort. Manual processing entails a great deal of work, which contributes to a decrease in millet consumption. To ease the drudgery, the Vivek Millet Thresher 1 was invented, a machine that can thresh 40–60 kg of barnyard millet grains per hour. Due to its affordable price, compact size, light weight, and dual functionality (threshing and dehusking), it is suitable for marginal farmers in hilly and tribal areas. Another similar low-cost, easy-to-use mill has been built by Indian and Canadian researchers, and it is 98% efficient in dehusking with only 2% broken grains. In an hour, this machine can generate roughly 2 kilograms of clean seed. Large farms can also benefit from higher capacity machines/mills that can process 100 kg/h. These larger machines were created solely for dehusking. The Central Institute of Agricultural Engineering (CIAE, ICAR), Bhopal, India, invented one such machine that can dehusk and grind 100 kilograms in an hour at 10–12% moisture content. It includes a one-horsepower single-phase electric motor and can process up to one kilogram of grains (Sood et al., 2015). Winnowing and aspiration are the other methods for seed cleaning. During winnowing, sand and other light contaminants are separated to the front of the basket from where they are thrown off. The method is tedious and time-consuming, taking up to an hour to clean a single 100 kg bag (Ahmad et al., 2021).

8.8 PROSO MILLET

Proso millet is grown in India, China, Russia, and some areas of North America and Eastern Europe. In different parts of the world, proso millet is known as broomcorn millet, common millet, hog millet, Russian millet, and so on. Proso is a shallow-rooted plant with a wide range of morphology, ranging from 25 cm to 230 cm in length. There are several tillers, each with an open panicle and drooping branches. Panicle branches range in length from 4–5 cm to 15 cm. Individual seeds range in size from 5 to 9 milligrams. Its inflorescence is a slender panicle that can grow up to 45 cm long and can be open or compact. The caryopsis is usually white, oval, and smooth in appearance. In comparison to other minor millets, proso millet has a higher protein content. Its grain also contains significantly more of the essential amino acids leucine, isoleucine, and methionine. Because of its high vitamin and mineral content, proso millet is a superior source of nutrition compared with other minor millets. Proso millet grains have a nutritional profile that is comparable to or superior to that of everyday cereals. Proso millet grains are also used as bird feed (Rao et al., 2016).

8.8.1 NUTRITIONAL QUALITY ENHANCEMENT

Proso millet has a higher nutritional value when compared to common grains since it has a higher concentration of minerals and dietary fiber. Iron (Fe), calcium (Ca), potassium (K), phosphorus (P), zinc (Zn), magnesium (Mg), vitamin B-complex, niacin, and folic acid are all found in abundance in proso millet. Except for lysine, the limiting amino acid, proso millet provides much larger amounts of essential amino acids. Proso millet, on the other hand, has a nearly 51% higher essential amino acid index than wheat. Furthermore, proso millet-based products have a lower glycemic response than staple cereal-based products (Das et al., 2019; Gowda et al., 2022).

One drawback of millets is that they contain more anti-nutritional compounds than wheat or rice. Polyphenols, tannins (0.61%), phytates (0.48%), trypsin inhibitors, and oxalates are all found in finger millet, and they may interfere with micronutrient absorption and protein digestion. Anti-nutritional factors such as metal chelation and enzyme inhibition reduce the bioavailability of nutrients, particularly minerals and proteins. Anti-nutritional factors, such as polyphenolic compounds, have been described as nutraceuticals in recent years due to their antioxidant characteristics. The majority of anti-nutrient secondary metabolites can induce severely harmful biological reactions, whereas some are employed in nutrition and pharmacologically active medications. Pretreatment or processing processes for food grains, such as debranding, soaking, germination, fermentation, and autoclaving, meet the need to remove anti-nutrients. These techniques improve the bioavailability of a few minerals, such as Ca, Fe, and Zn, as well as the absorption of proteins (Birania et al., 2020; Rao et al., 2017).

8.8.2 GRAIN PROTEIN ENHANCEMENT

Proso millet is high in protein, with an average protein content of 11–13%. Millet grains can undergo biochemical modifications during germination or malting that produce malt with enhanced nutritional content that can be used in a range of conventional recipes. The amount of free amino acids increased during the germination of proso millet grains.

8.8.3 SUGAR CONTENT ENHANCEMENT AND BIOFUEL APPLICATION

The total sugar content in proso millet grains can be increased by germination. Proso millet is used for ethanol production. The waxy starch in the grain is fermented with yeast, resulting in ethanol production. Dried distillers grains with soluble (DDGS) are a valuable byproduct of ethanol manufacturing. Proso millet lines yielded higher DDGS and had more protein, which can be used for animal feed (Santra and Rose, 2013).

8.8.4 STARCH CONTENT ENHANCEMENT

Humans' primary nutritional energy source, starch, has a wide range of uses in food and other sectors. Starch can be treated to a variety of processing conditions and additions during food preparation. The use of flour in food products depends on the starch's structural, functional, morphological, rheological, and physicochemical characteristics. Amylose content is greater in proso millet. Based on the amylose and amylopectin concentration of the starch, proso millet starch is divided into low amylose content waxy and high amylose content non-waxy starch. The ostensible amylose content of several proso millet cultivars ranged from 0.75 to 28.3%, as per the report of Zhang et al. (2021). Native starch has a wide range of structure and function depending on the plant source, and its direct food and non-food applications are limited.

On the other hand, the capability to modify starch makes it one of the most important and sought-after components in both food and non-food uses. The three most common methods of modifying starch are physical modification (such as heat treatment and non-thermal treatment), enzymatic modification (such as acid treatment and oxidation), and chemical modification (esterification, etherification, and cross-linking). Chemical modification of proso millet starch entails integrating a new chemical moiety or functional group into the starch structure, which alters the starch structure and, as a result, impacts the starch's functionality, such as water absorption, swelling, solubility, and gelling property. Substitution and elongation reactions are involved in these chemical processes. Physical modification of starch entails structural and morphological changes in the presence of physical elements such as temperature, pressure, moisture, electromagnetic radiation, milling, and so on. The physicochemical and functional properties of starch are directly influenced by changes in its structure and shape. Thermal and non-thermal changes are the two types of physical modifications. Hydrothermal therapy for physical alterations comprises heat moisture treatment (HMT), autoclaving, dry heat treatment, and microwave treatment, and non-thermal treatment includes ultra-high-pressure treatment. The starch characteristics of proso millet are primarily responsible for its eating quality and cooking parameters. The starch is separated into waxy and non-waxy categories based on the amylose level. In comparison to waxy starch, proso millet's non-waxy starch has a wide range of food applications. The high resistant starch content and slowly digesting starch of proso millets, especially non-waxy types, are other key reasons to consume them. The fluctuation in amylose concentration and structural integrity of proso millet starch (PMS) is responsible for its pasting and thermal behavior. PMS, like the other millet starches, had A-type crystallinity. Proso millet starch is shown to have some promising functional features that could make it a viable alternative to commercially available starch in food and non-food applications (Bangar et al., 2021).

8.8.5 SEED SHELF-LIFE ENHANCEMENT

Certified and foundation seeds should have a minimum physical purity of 97%, with a minimum germination potential of 75% and moisture content of 12%. The inert matter should not exceed 2.0% of the total weight of the product. June-July and February-March are the best months for seed production. For quality and effective seed setting, pollination should not coincide with rain. Grain quality characteristics such as seeds without melanopathy (blackening caused by a seed disease) are crucial. In general, a dry atmosphere with clear skies and high temperatures aids in the production of high-quality seeds.

8.8.6 POST-HARVEST SEED PROCESSING TRAIT ENHANCEMENT

Proso millet ripens inconsistently in the panicle, as seeds in the upper portions are ripe while those in the lower are still green. These lower seeds are harvested and swathed until the green has gone from the panicle. Once the earheads have reached physiological maturity, they are harvested. In most cases, the crop is ready to harvest 65 to 75 days after sowing. When two-thirds of the seeds are ready, the crop should be picked. The collected earheads are threshed by hand or by bullocks trampling. Winnowing is used to clean the threshed grains further. To achieve a safe moisture level of 12%, the cleaned seeds should be sun-dried. Under the right conditions, seeds can be preserved for up to 13 months. Seed cleaning involves two techniques: winnowing and aspiration. The winnowing process involves approximately 2–3 kg of threshed millet grains being placed on a flat reed or raffia woven basket and winnowed by up-and-down strokes, causing the grains to fly through the air and settle back onto the basket. Sand and other light pollutants are separated to the front of the basket and thrown off during this process. The operation is arduous and time-consuming, taking up to an hour to clean a single 100 kg bag (Ahmad et al., 2021).

8.9 FOXTAIL MILLET

Foxtail millet, also called Italian millet, Thenai, or Kangni, is thought to be a Chinese native and is one of the world's oldest cultivated crops. This millet ranks second in total world millet output, and as a result, it continues to play an important role in global agriculture. Foxtail millet feeds millions of people, mostly on poor or marginal soils, especially in Southern Europe and temperate, subtropical, and tropical Asia. Its grain is utilized for human and poultry consumption, as well as cage birds. Foxtail gruel with jaggery is beneficial for postpartum weakness, liver illness, and urinary infections. For septic boils, foxtail millet flour poultice is highly effective. The anti-hyperglycemic properties of foxtail millet aqueous extracts are excellent. The nutritional value of foxtail millet is high, notably in terms of starch, protein, vitamins, and minerals. The digestible section of foxtail millet grains is about 79% due to the coarse nature of the grain, while the remaining undigestible half of the grain contains significant levels of fiber as well as various anti-nutritional components.

Like most millets, foxtail millet is high in crude fiber, which aids in digestion and induces bowel movement, resulting in a laxative effect that is useful to a healthy gut (Sharma and Niranjan, 2018).

8.9.1 NUTRITIONAL QUALITY ENHANCEMENT

Some phytochemicals found in foxtail millet have been shown to have potent anti-nutrient properties, lowering bioavailability and quality. Milling, heating, extrusion, germination, and fermentation are some of the processing processes that can be used to reduce the anti-nutritional effect and increase the bioavailability of micronutrients in foxtail millet. Popping or roasting is a high-temperature, short-time procedure used to enlarge millet grains. The weight/volume ratio and the roasting/popping qualities of foxtail millet were shown to be highly correlated. The nutritional properties of popped grains were discovered to be better, with an increase in crude protein content, a decrease in total fat content, and improved starch and protein digestibility due to the separation of starch granules from the protein matrix during popping, making it more receptive to enzymatic digestion. Roasting also led to the removal of volatile aroma compounds and odor. Fermentation is a quick, simple, and cost-effective way to boost the nutritional value of foxtail millet while also improving its sensory and functional qualities. The availability of bioactive substances such as total phenolics, antioxidants, total flavonoids, dietary fiber, proteins, and minerals was increased and modified by the germination of foxtail millet seeds, whereas anti-nutritional factors were reduced (Rao et al., 2016).

8.9.2 GRAIN PROTEIN ENHANCEMENT

Foxtail millet protein is easily digested and has a high net utilization rate in the body. The employment of enzymes to hydrolyze (enzymatic hydrolysis) foxtail millet protein is a helpful way to produce modified proteins with required functional properties like solubility, emulsification, gelation, water- and fat-holding capacities, and foaming ability (Kamara et al., 2020).

8.9.3 STARCH CONTENT ENHANCEMENT

Starch is one of the most important carbohydrate constituents in foxtail millet, accounting for around 61% of the dry content. It is made up of two primary molecular components, amylose and amylopectin, which are linked to the production and quality of foxtail millet products in a weight ratio of 25:75. The granules of foxtail millet starch are usually tiny or large polygonal (particularly pentagonal) in shape, with round edges and occasional surface indentations generated by the protein bodies. The structure of the starch granules is very similar to that of maize. Among all other millets, the starch granules of foxtail millet are the biggest. Cooking foxtail millet raises the amylose content while decreasing the resistant starch concentration, giving the starch a high degree of syneresis and gel consistency. Gelatinization parameters, molecular structure, amylose and amylopectin levels, gel-forming properties, and digestibility and absorption are all starch attributes that determine the quality of a foxtail millet-based product. These varied qualities can be purposefully controlled using a variety of physical, chemical, and biological methods to produce a certain type of starch available in sufficient quantities for consumption (Dey and Sit, 2017).

8.10 KODO MILLET

Kodo millet, also known as koda millet or kodra millet, is a member of the Poaceae family. Kodo millet, sometimes called cow grass, rice grass, ditch millet, native Paspalum, or Indian crown grass, is a tropical African plant. Varagu, haraka, and arakalu are some of the other names for this cereal. It was domesticated over 3000 years ago in India. It can be found in humid tropics and subtropics all over the world. It is an annual millet that takes longer to mature, grows to a height of 30 to 90 cm, and has between 10 and 48 basal tillers. The inflorescence is tiny, measuring 2 to 12 cm in length. It is self-pollinating, and the florets are usually closed throughout flowering. De-branding is tough since the grain is enclosed in a hard husk. The crop is drought-resistant, resilient, and can be cultivated in a variety of soil conditions. Kodo millet is easy to digest, has a lot of lecithin, and is beneficial for boosting the nervous system's strength. B vitamins, including niacin, B6, and folic acid, as well as minerals like calcium, iron, potassium, magnesium, and zinc, are abundant in kodo millet (Rao et al., 2017). It does not contain gluten and is ideal for gluten-intolerant individuals. Regular consumption of kodo millet is extremely advantageous for postmenopausal women with cardiovascular disease symptoms such as elevated blood pressure and cholesterol levels (Bunkar, 2021). Kodo grain has notoriety for being harmful to both cattle and humans. The fungus is responsible for the poisoning of the grain. Nausea, vomiting, confusion, depression, drunkenness, and unconsciousness are all symptoms of kodo poisoning in humans. The toxic chemical cyclopiazonic acid, which causes 'kodua poisoning' when found in kodo millet, is thought to be produced by *Aspergillus flavus, Aspergillus tamarii*, and *Phomopsis paspalli* (Deepika et al., 2022).

8.10.1 NUTRITIONAL QUALITY ENHANCEMENT

Polyphenols, antioxidants, tannins, phosphorus, and phytic acids are all abundant in kodo millets. The anti-nutrients present in kodo millet bind to micronutrients including iron, calcium, and zinc, lowering their solubility and bioavailability. Tannins also inhibit protein and carbohydrate use by creating complexes, leading to decreased growth, feeding efficiency, metabolizable energy, and amino acid bioavailability. Traditional methods of reducing anti-nutrients such as decortication, soaking, germination, and fermentation boost amino acid and mineral bioavailability and improve protein and starch digestibility. Both phytate and total phosphorus can be removed by dehulling to the extent of 40 to 50% (Dey et al., 2022).

8.11 LITTLE MILLET

The millet known as little millet is fibrous and is divided into two types: nana and robusta. The millet seeds are nutrient-dense and tolerant to drought and waterlogging.

It was domesticated in India and is similar to proso millet in size. Little millet consumption can lower total and LDL cholesterol while increasing HDL cholesterol. Because small millet has a low GI, it may be useful to diabetics. Nana plants range in height from 60 to 170 cm. The inflorescence is upright, open, and densely branching, measuring 14 to 15 cm in length. When these branches reach maturity, they may droop. The robusta race produces plants that are 120 to 190 cm in height. The inflorescence measures 20 to 45 cm in length, is open or compact, and is densely branched, smaller in stature, and contains fewer panicles and seeds (Rao et al., 2017). Antioxidants such as phenolics, tannins, and phytates are abundant in kodo millets. This aids in the reduction of blood glucose, cholesterol, blood plasma lipid peroxides, and tail tendon collagen glycation (Dey et al., 2022).

8.12 CONCLUSION

Minor millets are small-seeded grains that are an abundant source of salubrious components that benefit human health. These are climate-resilient crops remarkably tolerant to various agro-climatic conditions such as drought, low soil fertility, infrequent rainfall, and waterlogging. Minor millets comprise finger millet, little millet, kodo millet, proso millet, barnyard millet, and foxtail millets; among these, finger millet takes up roughly 60% of the area and contributes 70% to the total production of minor millets. Ultrasonication, plant breeding, fermentation, and germination technologies can all be utilized to improve the nutritional value of minor millets. Therefore, the biofortification of millets offers a tried-and-true strategy to alleviate micronutrient deficiencies, ultimately addressing the population's nutritional insecurity. The absence of suitable processing technologies to create value-added products is one of the factors preventing the diversified use of minor millets; therefore, it's critical to redouble efforts to improve millet farming systems and develop innovative approaches to enhance the crop and promote sustainable and healthful living.

KEYWORDS

- **antioxidants**
- **biofortification**
- **dried distillers grains**
- **kodo millet**
- **lactic acid bacteria**
- **LDL cholesterol**
- **minor millet**
- **nutritional quality**

REFERENCES

Ahmad, T., & Jeena, A. S., & Tripathi, A. (2021). Seed conditioning and seed enhancement in millets. *National Seed Association of India.* https://www.researchgate.net/publication/350754168_Seed_Conditioning_and_Seed_Enhancement_in_MILLETS (accessed on 7th November 2025).

Babu, A. S., Mohan, R. J., & Parimalavalli, R. (2019). Effect of single and dual-modifications on stability and structural characteristics of foxtail millet starch. *Food Chemistry, 271*, 457–465.

Bangar, S. P., Ashogbon, A. O., Dhull, S. B., Thirumdas, R., Kumar, M., Hasan, M., Chaudhary, V., & Pathem, S. (2021). Proso-millet starch: Properties, functionality, and applications. *International Journal of Biological Macromolecules, 190*, 960–968. https://doi.org/10.1016/j.ijbiomac.2021.09.064 (accessed on 7th November 2025).

Banshidhar, P. S., Singh, A., Jaiswal, P., Singh, M. K., Meena, K. R., & Singh, S. K. (2023). The potentialities of omics resources for millet improvement. *Functional & Integrative Genomics, 23*(3), 210. https://doi.org/10.1007/s10142-023-01149-2 (accessed on 7th November 2025).

Birania, S., Rohilla, P., & Kumar, R., & Kumar, N. (2020). Post harvest processing of millets: A review on value added products. *International Journal of Chemical Studies, 8*(1), 1824–1829. https://doi.org/10.22271/chemi.2020.v8.i1aa.8528 (accessed on 7th November 2025).

Bunkar, D. S. (2021). Nutritional, functional role of kodo millet and its processing: A review. *International Journal of Current Microbiology and Applied Sciences, 10*(01), 1972–1985.

Cao, T. L., Yang, S. Y., & Song, K. B. (2017). Characterization of barnyard millet starch films containing borage seed oil. *Coatings, 7*(11), 183. https://doi.org/10.3390/coatings7110183 (accessed on 7th November 2025).

Das, S., Khound, R., Santra, M., & Santra, D. K. (2019). Beyond bird feed: Proso millet for human health and environment. *Agriculture, 9*(3), 64. https://doi.org/10.3390/agriculture9030064 (accessed on 7th November 2025).

Deepika, C., Hariprasanna, K., Das, I. K., Jacob, J., Ronanki, S., Ratnavathi, C. V., Bellundagi, A., Sooganna, D., & Tonapi, V. A. (2022). 'Kodo poisoning': Cause, science and management. *Journal of Food Science and Technology, 59*(7), 2517–2526. https://doi.org/10.1007/s13197-021-05141-1 (accessed on 7th November 2025).

Dey, A., & Sit, N. (2017). Modification of foxtail millet starch by combining physical, chemical and enzymatic methods. *International Journal of Biological Macromolecules, 95*, 314–320.

Dey, S., Saxena, A., Kumar, Y., Maity, T., & Tarafdar, A. (2022). Understanding the antinutritional factors and bioactive compounds of kodo millet (*Paspalum scrobiculatum*) and little millet (*Panicum sumatrense*). *Journal of Food Quality*, 2016.

Estivi, L., Brandolini, A., Condezo-Hoyos, L., & Hidalgo, A. (2022). Impact of low-frequency ultrasound technology on physical, chemical and technological properties of cereals and pseudocereals. *Ultrasonics Sonochemistry, 86*. https://doi.org/10.1016/j.ultsonch.2022.106044 (accessed on 7th November 2025).

Gautam, N., Garg, S., & Yadav, S. (2021). Underutilized finger millet crop for starch extraction, characterization, and utilization in the development of flexible thin film. *Journal of Food Science and Technology, 58*(11), 4411–4419. https://doi.org/10.1007/s13197-020-04926-0 (accessed on 7th November 2025).

Gowda, N. N., Siliveru, K., Prasad, P. V., Bhatt, Y., Netravati, B. P., & Gurikar, C. (2022). Modern processing of Indian millets: A perspective on changes in nutritional properties. *Foods, 11*(4), 499. https://doi.org/10.3390/foods11040499 (accessed on 7th November 2025).

Guerinot, M. L. (2000). The ZIP family of metal transporters. *Biochemical et Biophysical Acta (BBA)–Bio Membranes, 1465*(1, 2), 190–198. https://doi.org/10.1016/s0005-2736(00)00138-3 (accessed on 7th November 2025).

Kamara, M. T., Amadou, I., Tarawalie, F., & Huiming, Z. (2020). Effect of enzymatic hydrolysis on the functional properties of foxtail millet (*Setaria italica* L.) proteins. *International Journal of Food Science and Technology, 45*(6), 1175–1183.

Kaur, B., Singh, A., Suri, S., Usman, M., & Dutta, D. (2023). Minor millets: A review on nutritional composition, starch extraction/modification, product formulation, and health benefits. *Journal of the Science of Food and Agriculture*. https://doi.org/10.1002/jsfa.12493 (accessed on 7th November 2025).

Kaur, J., Salve, R. V., Mathew, A., Anil, A., Zutso, V., & Kaur, N. (2023). Shelf-life extension of millets: A comprehensive overview of various preservation techniques. *The Pharma Innovation Journal, 12*(6), 1258–1264.

Kulkarni, D. B., Sakhale, B. K., & Giri, N. A. (2018). A potential review on millet grain processing. *International Journal of Nutritional Sciences, 3*(1), 1–8. Available at: www.austinpublishinggroup.com (accessed on 7th November 2025).

Kumar, N. N., Lakshmi, S., & Sathya, S. (2020). Effect of pulse sprout spray as a foliar nutrition to enhance seed yield and quality in barnyard millet (*Echinochloa frumentacea* L.). *Journal of Applied and Natural Science, 12*(3), 394–400. https://doi.org/10.31018/jans.v12i3.2333 (accessed on 7th November 2025).

Longvah, T. R., Ananthan, K., & Bhaskarachary, K. V. (2017). *Indian Food Composition Table*. NIN, ICMR.

Mobolade, A. J., Bunindro, N., Sahoo, D., & Rajashekar, Y. (2019). Traditional methods of food grains preservation and storage in Nigeria and India. *Annals of Agricultural Sciences, 64*(2), 196–205. https://doi.org/10.1016/j.aoas.2019.12.003 (accessed on 7th November 2025).

N/A. (1995). *Sorghum and Millets in Human Nutrition*. Food and Nutrition Series, No. 27. FAO: Rome, Italy. https://www.fao.org/3/t0818e/T0818E02.htm#Grains%20and%20their%20structure\pdf (accessed on 7th November 2025).

N/A. (1998). *Dietary Guidelines for Indians–A Manual*. Hyderabad: NIN. http://www.sahajasamrudha.org/ (accessed on 7th November 2025).

N/A. (2021). *Annual Report 2021*. Rajendranagar, Hyderabad 500 030, India, 152 pp.

Neeraja, C. N., Babu, V. R., Ram, S., Hossain, F., Hariprasanna, K., Rajpurohit, B. S., Prabhakar, T., Longvah, T., Prasad, K. S., Sandhu, J. S., & Datta, S. K. (2017). Biofortification in cereals: Progress and prospects. *Current Science*, 1050–1057. https://doi.org/10.18520/cs/v113/i06/1050-1057 (accessed on 7th November 2025).

Pampangouda, P., Munishamanna, K. B., & Gurumurthy, H. (2015). Effect of *Saccharomyces boulardii* and *Lactobacillus acidophilus* fermentation on little millet (*Panicum sumatrense*). *Journal of Applied and Natural Science, 7*(1), 260–264.

Petrova, P., & Petrov, K. (2020). Lactic acid fermentation of cereals and pseudocereals: Ancient nutritional biotechnologies with modern applications. *Nutrients, 12*(4), 1–26.

Puranik, S., Kam, J., Sahu, P. P., Yadav, R., Srivastava, R. K., Ojulong, H., & Yadav, R. (2017). Harnessing finger millet to combat calcium deficiency in humans: Challenges and prospects. *Frontiers in Plant Science, 8*, 1–16.

Ramadas, S., Vellaichamy, S., Ramasundaram, P., Kumar, A., & Singh, S. (2020). Biofortification for enhancing nutritional outcomes and policy imperatives. In *Wheat and barley grain biofortification* (pp. 309–327). Woodhead Publishing.

Ramashia, S. E., Anyasi, T. A., Gwata, E. T., Meddows-Taylor, S., Israel, A., & Jideani, O. (2019). Processing, nutritional composition and health benefits of finger millet in Sub-Saharan Africa. *Food Science and Technology, 39*, 253–266.

Rao, B. D., Malleshi, N. G., Annor, G. A., & Patil, J. V. (2017). *Millets Value Chain for Nutritional Security: A Replicable Success Model from India*. Indian Institute of Millets Research (IIMR): Hyderabad, India. https://doi.org/10.1079/9781780648309.0024 (accessed on 7th November 2025).

Rathore, T., Singh, R., Kamble, D. B., Upadhyay, A., & Thangalakshmi, S. (2019). Review on finger millet: Processing and value addition. *The Pharma Innovation Journal, 8*(4), 283–291.

Renganathan, V. G., Vanniarajan, C., Karthikeyan, A., & Ramalingam, J. (2020). Barnyard millet for food and nutritional security: Current status and future research direction. *Frontiers in Genetics, 11*, 500. https://doi.org/10.3389/fgene.2020.00500 (accessed on 7th November 2025).

Saleh, A. S., Zhang, Q., Chen, J., & Shen, Q. (2013). Millet grains: Nutritional quality, processing, and potential health benefits. *Comprehensive Reviews in Food Science and Food Safety, 12*(3), 281–295. https://doi.org/10.1111/1541-4337.12012 (accessed on 7th November 2025).

Santra, D. K., & Rose, D. (2013). Alternative uses of proso millet. *University of Nebraska-Lincoln, NebGuide G, 2218*, 3–6.

Sharma, D., Jamra, G., Singh, U. M., Sood, S., & Kumar, A. (2017). Calcium biofortification: Three-pronged molecular approaches for dissecting complex trait of calcium nutrition in finger millet (*Eleusine coracana*) for devising strategies of enrichment of food crops. *Frontiers in Plant Science, 7*, 1–18.

Sharma, N., & Niranjan, K. (2018). Foxtail millet: Properties, processing, health benefits, and uses. *Food Reviews International, 34*(4), 329–363. https://doi.org/10.1080/87559129.2017.1290103 (accessed on 7th November 2025).

Sharma, R., Garg, P., Kumar, P., Bhatia, S. K., & Kulshrestha, S. (2020). Microbial fermentation and its role in quality improvement of fermented foods. *Fermentation, 6*(4), 1–20.

Sharma, S., Saxena, D. C., & Riar, C. S. (2016). Analyzing the effect of germination on phenolics, dietary fibers, minerals and γ-amino butyric acid contents of barnyard millet (*Echinochloa frumentaceae*). *Food Bioscience, 13*, 60–68. https://doi.org/10.1016/j.fbio.2015.12.007 (accessed on 7th November 2025).

Sharma, S., Saxena, D. C., & Riar, C. S. (2016). Isolation of functional components β-glucan and γ-amino butyric acid from raw and germinated barnyard millet (*Echinochloa frumentaceae*) and their characterization. *Plant Foods for Human Nutrition, 71*, 231–238. https://doi.org/10.1007/s11130-016-0545-6 (accessed on 7th November 2025).

Sood, S., Joshi, D. C., & Pattanayak, A. (2020). Breeding advancements in barnyard millet. In *Accelerated Plant Breeding, Volume 1: Cereal Crops* (pp. 391–409). https://doi.org/10.1007/978-3-030-41866-3_15 (accessed on 7th November 2025).

Sood, S., Khulbe, R. K., Gupta, A. K., Agrawal, P. K., & Upadhyaya, H. D., & Bhatt, J. C. (2015). Barnyard millet–a potential food and feed crop of future. *Plant Breeding, 134*(2), 135–147. https://doi.org/10.1111/pbr.12243 (accessed on 7th November 2025).

Srivastava, B., & Reddy, P. B. (2023). Recent explorations on millets: The Nutri-cereals. *International Multidisciplinary Research Foundation*, India.

Thamburaj, S., Palanivel, R., Ramakrishnan, P., & Rajagopal, V. (2023). Intervention of fermentation in preservation and nutritional enhancement of common millets. In *Postharvest Management of Fresh Produce* (pp. 217–240).

Vagdevi, A., Trupthi, S., Pandey, M., Reddy, N. B., & Singh, M. (2023). Nutritional qualities, processing and health benefits of finger millet (*Eleusine coracana* L.). *The Pharma Innovation Journal, 12*(1), 1291–1297.

Vidhyalakshmi, R., & Meera, M. S. (2023). Dry heat and ultrasonication treatment of pearl millet flour: Effect on thermal, structural, and *in-vitro* digestibility properties of starch. *Journal of Food Measurement and Characterization*, 1–11.

Vinoth, A., & Ravindhran, R. (2017). Biofortification in millets: A sustainable approach for nutritional security. *Frontiers in Plant Science, 8*, 29. https://doi.org/10.3389/fpls.2017.00029 (accessed on 7th November 2025).

Yadav, S., Mishra, S., & Pradhan, R. C. (2021). Ultrasound-assisted hydration of finger millet (*Eleusine coracana*) and its effects on starch isolates and antinutrients. *Ultrasonics Sonochemistry, 73*. https://doi.org/10.1016/j.ultsonch.2021.105542 (accessed on 7th November 2025).

Zhang, T., Li, K., Ding, X., Sui, Z., Yang, Q. Q., Shah, N. P., Liu, G., & Corke, H. (2021). Starch properties of high and low amylose proso millet (*Panicum miliaceum* L.) genotypes are differentially affected by varying salt and pH. *Food Chemistry, 337*, 127784. https://doi.org/10.1016/j.foodchem.2020.127784 (accessed on 7th November 2025).

PART III

Improved Cultivation Practices

CHAPTER 9

Improved Cultivation Practices of Sorghum

V. VISHA KUMARI,[1,*] SWARNA RONANKI,[2] PURABI BANERJEE,[1] K. A. GOPINATH,[1] G. VENKATESH,[1] and V. K. SINGH[1]

[1]*ICAR–Central Research Institute for Dryland Agriculture, Hyderabad, Telangana, India*

[2]*ICAR–Indian Institute of Millets Research, Hyderabad, Telangana, India*

*Corresponding author

9.1 INTRODUCTION

Sorghum [*Sorghum bicolor* (L.) Moench], known as jowar, is an important crop in semi-arid regions. It is the primary food source for over 750 million people residing in tropical semi-arid regions of Asia, Africa, and Latin America (Hassan, 2023). Due to its high photosynthetic efficiency and productivity, it is considered a star crop (Khalifa and Eltahir, 2023; Tari et al., 2013). It is believed to have origins at the Egyptian-Sudanese border or in Northern Africa around 5000 to 8000 years ago. The crop is extensively grown in over 100 nations worldwide (Hao et al., 2021). The crop is also known for its great tolerance and broad adaptation to numerous biotic and abiotic stressors (Machado and Paulsen, 2001; Tari et al., 2013). It is a multipurpose crop that can provide food, fodder, fiber, and fuel. In addition to being utilized as food for humans, sorghum is also used as animal feed, fencing, and brooms Doggett, 1988; House, 1985; Rooney and Waniska, 2000). Sorghum stems and leaves are frequently used to feed animals. All categories of livestock feed obtained from it in the forms of green fodder, dry fodder, hay, and silage are substantially nutritious. Sorghum is also gaining importance in livestock due to its nutritional value. Sorghum is also used for the production of grain alcohol, ethanol, starch, beverages, flours, adhesives, and paper (Ajeigbe et al., 2019). Besides, it can be used as a raw material in different industries in various countries worldwide. The crop is more heat and drought-tolerant, making it especially desirable for arid and semi-arid regions (Hossain et al., 2022). Maharashtra, Karnataka, Andhra Pradesh, Madhya Pradesh, Gujarat, Rajasthan, Uttar Pradesh, and Tamil Nadu are the major sorghum-growing states in India. Despite sorghum's numerous uses and significant potential to enhance national food security and sustainable livelihoods in dryland areas, its cultivation is declining. This decline is due to factors such as the limited adoption of improved varieties, shifting consumer food preferences, and low productivity. Therefore, it is essential to concentrate on boosting production through improved agronomic practices.

Millets in Sustainable Agriculture: Genetic Improvement, Nutritional Quality, Cultivation, and Integrated Pest–Disease Management. Roshni Vijayan, A. John Joel, R. Ambika Rajendran, and Adithya P. Balakrishnan (Eds.)

DOI: 10.1201/9781779642332-12

9.1.1 *CLASSIFICATION AND ORIGIN*

The systematics, history, and evolution of sorghum have been extensively documented by numerous authors, including de Wet and Harlan (1971). Dahlberg (2000) utilized an integrated classification approach to describe the variations within cultivated sorghums, providing a comprehensive overview of current classifications. Sorghum, belonging to the genus Sorghum (Clayton and Renvoize, 1986), was categorized by De Wet (1978) to include Sorghum bicolor, as well as the rhizomatous species *S. halepense* and *S. propinquum*, covering all annual cultivated, wild, and weedy sorghums. Mann et al. (1983) traced the origin and domestication of sorghum to northeastern Africa around 5000 years ago. After originating in Africa, sorghum spread globally, reaching the Indo-Pak subcontinent early on, China in the 13th century, and the United States in the 17th century (Bean and McCollum, 2006; Undersander et al., 1990). In India, sorghum is predominantly cultivated in Peninsular and Central regions, with major producers including Maharashtra, Andhra Pradesh, Karnataka, Madhya Pradesh, Gujarat, Tamil Nadu, Rajasthan, and Uttar Pradesh due to their low rainfall.

9.2 CLIMATIC REQUIREMENTS

Sorghum is a C4 plant and adapts better to harsh environments compared to other crops. Temperature, day length, water, and soil fertility are the most important factors affecting sorghum growth and development.

9.2.1 *TEMPERATURE*

Warm-weather crops like sorghum thrive in high temperatures, which are essential for optimal germination and growth. The minimum temperature for germination ranges from 7 to 10°C, with the ideal planting time being when the soil temperature at a depth of 10 cm is 15°C or higher and has sufficient moisture. Post-germination, temperatures significantly impact growth and development. Optimal growth occurs at temperatures between 27 and 30°C, though temperatures as low as 21°C do not drastically affect growth and yield. Exceptionally high temperatures, however, can hinder production. Increased day and night temperatures delay flower initiation and the development of flower primordia. If exposed to temperatures below 18°C, plants with 4–6 mature leaves will produce lateral branches. However, apical dominance inhibits lateral shoot development in plants at or beyond the eight-leaf stage. Sorghum is susceptible to damage from below-freezing temperatures and can die. Young plants (one to three weeks old) can survive temperatures as low as –5°C.

9.2.2 *DAY LENGTH*

Sorghum is a short-day plant that requires longer nights and shorter days to reach the reproductive stage. The optimal photoperiod for flower growth is between 10 and 11 hours. Photoperiods longer than 11 to 12 hours encourage vegetative growth. Short-season sorghum varieties are typically less sensitive to photoperiod changes than tropical varieties. The plant is most sensitive to photoperiod during flower initiation.

9.3 WATER REQUIREMENT

Maintaining optimal soil moisture at 20–25 days after sowing (DAS) aids in developing strong secondary roots. There are four critical irrigation stages: flower primordial initiation (25–30 DAS), flag leaf stage (50–55 DAS), flowering (60–70 DAS), and grain filling (80–90 DAS).

9.4 SOIL REQUIREMENT

A wide range of soil is suitable for sorghum cultivation, ranging from heavy clay to light-textured sandy soils, with pH varying from 5.5 to 8.5. It performs best on deep, fertile, well-drained, loamy soils with good water-holding capacity and a pH between 6 and 7.5. Sorghum does not grow well on shallow and sandy soils with a high clay content; however, it performs better than maize in those areas. Additionally, sorghum can tolerate alkaline salts, aluminum toxicity, and short periods of waterlogging better than other cereals (Ajeigbe et al., 2020).

9.4.1 LAND PREPARATION

Sorghum requires fine-grained, well-drained soil with adequate moisture for proper germination, root growth, and seedling establishment. The field should be ploughed once or twice, and then harrowing needs to be done to ensure a weed-free seedbed. Chiseling is recommended at the start of each season in soils with a subsoil hardpan. FYM at 12.5 metric tons/ha or composted coir pith, together with 2 kg of Azospirillum (10 packets/ha), can be applied, and the manures should be incorporated into the soil with a country plow. Furthermore, well-decomposed poultry manure at 5 metric tons/ha can be applied to improve soil physical properties and crop productivity.

- **Sowing Time and Sowing Depth:** Sorghum can be grown in almost all seasons of the year. The best time for sowing the kharif crop is from the last week of June to the first week of July, depending on the onset of the monsoon. The rabi crop is sown from October to November, and the summer crop from January to February. While planting can take place from May to July during the kharif season, depending on the amount of rainfall, it should be completed between September 25 and October 15 during the rabi season (Ratnavathi et al., 2011). The optimum sowing depth depends on the type of cultivar, growing season, soil moisture, and soil texture. Sorghum seeds should be planted 2–5 cm deep. Seeds should be planted deeper in sandy, dry soils than in heavy, wet soils (Ajeigbe et al., 2020).
- **Planting Geometry:** Optimum spacing depends on the method of planting, season, and soil-moisture status (Mubarik et al., 2022). In drought-prone areas, wider spacing increases grain yield because it enables sorghum plants to use limited water more efficiently. As a result, grain sorghum is often planted with a 45 cm row spacing during the kharif season; however, a 60 cm row spacing can be used during the rabi season (Madhusudhana et al., 2007). The ideal

intra-row spacing for grain sorghum is primarily based on soil-water availability and varies from 7 to 30 cm globally. Wider inter-row spacing enables plant roots to absorb a larger volume of water from the soil during critical growth periods, thereby increasing grain yield under dryland conditions. However, narrower intra-row spacings that increase resource-use efficiency and crop–weed competition should be preferred to achieve high grain-yield potential under wet conditions.

- **Optimum Seed Rate:** The optimal seed rate for sorghum depends on factors like the growing season, seed test weight, germination percentage, weather conditions, and the variety's duration. The required plant population per unit area dictates the seed rate, which is influenced by inter-row and intra-row spacing and seed germination rates. For irrigated and transplanted sorghum, the recommended seed rate is 7.5 kg/ha. Direct sown sorghum needs 10 kg/ha, while rainfed direct sown sorghum requires a higher seed rate of 15 kg/ha.

9.5 SEED TREATMENT

Seed treatment prior to sowing is necessary to protect the crop from soil-borne disease. The following materials may be used for treating sorghum seeds:

- 300-mesh sulfur powder at 4 g, followed by *Azotobacter* at 25 g/kg of seed
- Carbendazim, Captan, or Thiram at 2 g/kg of seed

9.6 NUTRIENT MANAGEMENT

Sorghum extracts a large amount of nutrients from the soil with its well-developed fibrous root system. Fertilizers should be applied at the optimum rate and at the optimum time, considering the soil test results and the expected yield average. Excessive application of fertilizers results in environmental pollution, NOx emissions, and soil structure degradation, while insufficient fertilization results in reduced crop yields and quality (Nazli et al., 2014).

9.6.1 NITROGEN (N)

Sorghum is generally grown in less fertile soil, which drives the need for efficient use of resources, including N (van Oosterom et al., 2001). Nitrogen (N) is one of the major nutrients limiting sorghum growth; an adequate supply of this nutrient is essential for sustainable sorghum production (Ostmeyer et al., 2022). N plays a vital role in numerous plant functions, including the constituents of protoplasm, enzymes, chlorophyll, and the synthesis of protein and sucrose (Dovale et al., 2012). Insufficient nitrogen nutrition inhibits vegetative growth, lowers protein content, delays flowering, and causes partial grain filling, thus affecting sorghum biomass and grain yields. Older leaves typically exhibit deficiency symptoms as they turn pale green with pale yellow chlorosis and pale brown necrosis (Sawyer, 2004). Excess N nutrition results in increased costs, soil health deterioration, nitrous oxide (NOx) emission, and groundwater pollution (Kabange et al., 2022). Extensive research has demonstrated the impact of available nitrogen (N) on the final grain yield of sorghum. Key yield determinants are grain number and weight. The number of kernels is set early in the growth cycle (during panicle initiation, gametogenesis, and anthesis), while kernel weight is influenced by post-flowering processes (Maman et al., 2004). Mahama et al. (2004) found that sorghum grain yield increased with higher N application rates (0, 45, and 90 kg N ha^{-1}), mainly due to an increase in grain number. Thus, optimum N application is essential for increasing the yields of sorghum. The amount of N to be applied varies with the cultivar, organic matter in the soil, and irrigation facilities. The optimal N fertilization rates required to produce sustainable yield from sorghum vary between 40 and 80 kg ha^{-1} (Table 9.1). In rainfed conditions, total N is applied as basal due to the unreliability of the occurrence of rainfall at later growth stages. In irrigated conditions, half of the fertilizer N is broadcasted at planting, and then the remaining half is top-dressed in the form of urea, ammonium nitrate, or ammonium sulfate.

AICRP on sorghum has developed zone-wise recommendations for the crop (https://aicrp.icar.gov.in/). Recommended application of fertilizers (Kugedera et al., 2018), application of FYM (Tamboli et al., 2013), use of *Azospirillum (*Bhuriya et al., 2015), application of vermicompost (Kalibhavi et al., 2001), etc., have shown a positive result in improving the yield. All India Coordinated Research on *Sorghum* (AICRP-S) reported that withholding the fertilizers and weed management caused the highest yield reduction.

9.6.2 PHOSPHORUS (P)

Phosphorus is the second most important nutrient for the growth of sorghum, after nitrogen. It plays a crucial role in respiration, photosynthesis, cell enlargement, cell division, and root development (Asif et al., 2022). P nutrition improves the quality and yield of sorghum as it increases the stalk and stem strength, flowering, and uniform and early crop maturity. Deficiency of P results in poor seedling vigor, root development, dark green younger leaves with reddish-purple margins, brown and irregular stains on older leaves, and delayed maturity (Mattos et al., 2020). The amount of P fertilizer to be applied to sorghum varies with soil and organic matter content (Asif et al., 2022). Studies show that the optimum P fertilization rate for sorghum varied between 20 and 40 kg ha^{-1} (Table 9.1). The entire dose of phosphorous will be applied as basal at the time of planting (Bhat et al., 2019).

9.6.3 POTASSIUM (K)

In addition to nitrogen and phosphorus, potassium (K) plays several key roles in plant growth and is a crucial nutrient for sorghum crop production. It plays an essential role in various metabolic processes like osmotic regulation, enzyme activation, stomatal conductance, nutrient translocation, cell expansion, photosynthesis, root growth, and protein synthesis (Hasanuzzaman et al., 2018). K is important for strengthening plant stalks and thus aids in imparting resistance to bacterial and fungal attacks as well as lodging. K fertilization increases plant resistance to drought via regulation of osmotic pressure, maintenance of turgor potential, stimulation of root growth, reduction of transpiration, and enhancement of photosynthetic rate under water deficit conditions (Hasanuzzaman et al., 2018). It also protects the crop from salinity stress by reducing oxidative damage. Yellow or necrotic leaf margins that start at the lower leaves and move up to the top leaves are critical symptoms of K deficiency in sorghum. K is usually applied at a rate of 10–40 kg/ha (Table 9.1), depending on the available potassium concentration in the soil, either as broadcast or band placement before planting (Bhat et al., 2019).

TABLE 9.1 Recommended Dose of Fertilizers in Sorghum

Season/Soil	Dose
Kharif Sorghum	80:40:40 Kg NPK ha^{-1}
Rabi Rainfed (shallow to medium soil)	40:20:00 Kg NPK ha^{-1}
Rabi Rainfed (deep soil)	60:30:00 Kg NPK ha^{-1}
Irrigated	80:40:40 Kg NPK ha^{-1}

The package and practice for sorghum, as given by ICAR-IIMR, is given in Table 9.2.

TABLE 9.2 Package and Practices of Sorghum

Practices	Management
	Time of Sowing
Kharif	Jun-Jul
Rabi	Sep-Oct
Summer	Jan-Feb
Seed rate (per ac)	3–4 kg
Spacing	45 cm × 15 cm
Seed Treatment	Imidacloprid (70 WS) 3 g/kg seed
	Nutrient input (per acre)
Organic (FYM)	4 tons
Inorganic (*Kharif*) (N:P=32:16)	36 kg DAP +56 kg Urea or 100 kg SSP + 70 kg Urea
Inorganic (*Rabi*) (N:P=16:8)	20 kg DAP + 30 kg Urea or 50 kg SSP + 36 kg Urea
Weed MGMT (Pre-emergence)	Atrazine 0.5–1.0 kg a.i./ha
Intercropping	Sorghum + Pigeonpea; Sorghum + Cowpea
Crop Rotation	Sorghum-chickpea/Safflower /Mustard

Source: ICAR-IIMR.

9.7 NUTRIENT BIOFORTIFICATION

Biofortifying sorghum is a viable strategy to combat micronutrient deficiencies and alleviate malnutrition among poor rural communities, given the high consumption of sorghum in major regions of India. The significant variability in grain micronutrient content suggests the potential to develop superior varieties with enhanced nutritional quality. Micronutrients such as zinc and iron are particularly promising for this purpose.

9.8 WATER MANAGEMENT

Water stress can significantly impact sorghum production, with yield loss depending on the crop stage affected. This stress influences the plant's physicochemical traits, affecting growth and development from germination to the reproductive and grain-filling stages, leading to reduced grain yield and quality (Queiroz et al., 2019). Monitoring soil moisture and timely irrigation are crucial for maintaining yield potential. Sorghum's water requirements vary by season, soil, nutrition, and environmental conditions. In the Kharif season (July), 1 to 3 irrigations may suffice, while summer crops might need 6 to 7 due to higher temperatures. If only one irrigation is possible, it should be applied just before the booting stage (40–50 days). Various methods, such as bund formation, tied ridging, and *in-situ* mulching with dhaincha during Kharif, can enhance soil moisture content.

9.9 WEEDS

Weeds compete with sorghum plants for nutrients, moisture, and sunlight, mostly during the initial development stages, which may result in yield losses of up to 70% (Mishra, 2015). Some of these weed species serve as alternate hosts of pests and diseases, thus leading to a reduction in crop yields and grain quality (Nyoni et al., 2022). Weed control should be started immediately after the harvest of the previous crop. The efficiency of a weed control method largely depends on environmental conditions, cultivar, method of weed control (cultural or chemical), and intensity and biology of weeds (Kaczmarek et al., 2012). Suppressing weeds during the first eight weeks after sowing is crucial, as the crop competes intensively with weeds for nutrients and water during this period. Applying a pre-emergence spray of atrazine at 1.0 kg a.i/ha one day after sowing, but before weed emergence, is effective. Additionally, mechanical weeding twice up to 35–40 days of crop age helps control weed growth. In many sorghum-growing areas of Maharashtra, Karnataka, and Andhra Pradesh, striga, also known as witchweed (*Striga litura* L.), is a serious problem. Depending on the severity of its infestation, yield can be reduced by 15–100%. Groundnut, cotton, soybean, pigeon pea, and other trap crops are used in crop rotation to destroy the seeds and reduce losses. If the *Striga* menace is severe, spraying 2, 4-D sodium salt after emergence at 1 kg/ha is recommended between 20–60 DAS in addition to the above herbicides. Table 9.3 presents the recommended herbicides in sorghum.

TABLE 9.3 Herbicides Recommended for Sorghum

Herbicide	Dose (Kg/Ha)	Time of Application	Weeds Controlled	Remarks
Atrazine	0.75–1.0	Pre-emergence/early post-emergence	Broad-spectrum weed control. Some grasses are tolerant	For sole crop only. Did not control *Acrachne racemosa, Brachiaria reptans* and *Commelina benghalensis*
Pendimethalin	0.75–1.0	Pre-emergence	Effective control of grasses	Suitable for intercropping, higher doses may cause phytotoxicity
Alachlor	1.5–2.0	Pre-emergence	Effective control of grasses	Suitable for intercropping
Metolachlor	1.0–1.5	Pre-emergence	Effective control of grasses	Suitable for intercropping
2,4-D	0.50–0.75	Post-emergence	Effective against broad-leaved weeds	For sole crop only (4-6 DAS).
Atrazine + pendimethalin	0.75+0.75	Pre-emergence	Broad-spectrum weed control	For sole crop only
Atrazine + alachlor	0.75+0.75	Pre-emergence	Broad-spectrum weed control	For sole crop only
Atrazine + metolachlor	0.75+0.75	Pre-emergence	Broad-spectrum weed control	For sole crop only

Source: Mishra (2015).

9.10 DISEASES

Sorghum is susceptible to a wide range of diseases, including bacterial, viral, and fungal infections. Among fungal pathogens, significant ones such as grain mold, ergot, smut, anthracnose, downy mildew, rust, and charcoal rot can infect various parts of the plant, from roots and stems to leaves, peduncles, and grains, leading to substantial losses in yield and quality. Diseases like zonate leaf spot, anthracnose, sooty stripes, leaf blight, and rust are particularly economically damaging, reducing both yield and the quality of forage. Specific diseases like maize stripe virus (a viral infection), sugary disease (ergot, caused by a fungus), and head mold (grain mold) become significant under specific environmental conditions. Management strategies primarily focus on developing resistant sorghum varieties through breeding, with limited reliance on agrochemicals for disease control.

9.10.1 MANAGEMENT OPTIONS

Farmers that lack resources produce sorghum in marginal soil with minimal to no input. Because of this, the production system is less at risk from high input consumption, or it is, by default, an organically managed agro-system with limited use of synthetic plant growth regulators, fertilizers, or pesticides. The most cost-effective disease management strategy for such a system is for farmers to use host plant resistance. The use of bioagents or their metabolites can be beneficial to boost the host's natural resistance. Some soil-borne diseases affecting sorghum still lack adequate host resistance. Biocontrol agents have proven effective in managing diseases like sorghum charcoal rot. Specifically, strains of *Trichoderma* and *Pseudomonas* are beneficial against foot rot and sheath rot. Practices such as deep summer plowing, crop-residue removal, and crop rotation with nonhost plants contribute to reducing the spread of soil-borne diseases such as downy mildew, smut, charcoal rot, and certain fungal and bacterial leaf diseases by lowering inoculum levels.

9.11 PEST

Globally, sorghum is vulnerable to infestation by a diverse array of insect species, numbering at least 150. These insects target various developmental stages of sorghum plants, leading to substantial economic losses. In India, for instance, insect pests cause approximately 32% of sorghum crop losses during the rainy season (Borad and Mittal, 1983) and 26% during the post-rainy season (Delaware, 2012). Key insect pests affecting sorghum worldwide include the sorghum shoot fly, spotted stemborer, Oriental armyworm, shoot bug, sugarcane aphid, sorghum midge, head bugs, and head caterpillars. These pests pose significant challenges to sorghum cultivation by damaging different parts of the plant and reducing overall yield and quality.

9.11.1 MANAGEMENT OPTIONS

The timing of sorghum sowing plays a crucial role in mitigating insect damage. Typically, sorghum is planted with the onset of the first good monsoon rains. To minimize production losses from pests like shoot fly, midge, and head bugs, it is effective to synchronize the sowing of sorghum cultivars within similar maturity groups across large areas within a short timeframe. Managing pests such as shoot fly, stem borer, midge, sugarcane aphid, shoot bug, and other panicle pests involves practices like eliminating volunteer and alternative host plants. Intercropping sorghum with pigeonpea, cotton, soybean, cowpea, safflower, and other leguminous crops can also help reduce pest pressure. Specifically, intercropping with leguminous crops has been shown to reduce shoot fly damage. Additionally, treating sorghum seeds with Imidacloprid (0.165 mg/kg of seed) or Thiamethoxam (Cruiser) 70 WS at 3 g/kg of seeds enhances plant stand and seedling vigor while reducing damage caused by shoot fly and sucking pests. Insecticides should only be used when necessary, based on established economic thresholds, to effectively manage pest populations and minimize crop damage.

9.12 SPECIAL PRACTICES FOR SORGHUM

9.12.1 TRANSPLANTING

The transplanting of young sorghum seedlings is practiced in some places with assured irrigation facilities. A 300 m^2 area is required for raising the nursery, which is sufficient for transplanting in 1 ha.

9.12.2 SEED HARDENING

Pre-sowing seed hardening helps to enhance drought tolerance, especially during long dry spells. Treating the seeds by soaking in 1.0% $CaCl_2$ is found to be highly beneficial. Sorghum seeds are also pre-soaked in a 2% solution of potassium hydrogen phosphate (KH_2PO_4) for effective drought tolerance.

9.12.3 RATOONING

Though ratooning is a common practice in multi-cut forage sorghum varieties, it is also followed in grain sorghum. Ratooning is suitable in places with good irrigation facilities.

9.12.3.1 SORGHUM RATOON MANAGEMENT

To develop sorghum ratoon, the main crop is harvested by leaving 15 cm stubbles. Then, the first two formed sprouts are removed from the main crop, and the later ones are permitted to grow. As a whole, two tillers are grown per hill. The emerging weeds are removed immediately after the harvest of the main crop. Hoeing and weeding are practiced twice on the 15th and 30th days after cutting. Nitrogen is applied at 100 kg/ha in two split doses on the 15th and 45th days after cutting. About 50 kg P_2O_5/ha is also applied along with the application of a second split of N. Irrigation is provided immediately after cutting the main crop. In fact, irrigation should not be delayed for more than a single day after cutting. The second irrigation is applied 3 days after cutting. The ratoon crop can be subsequently irrigated once every 7–10 days, which should be continued up to 70–80 days after ratooning. The pest and disease management of ratoon crops should be taken into major concern. The crop can be harvested when the grains begin to yellow. Altogether, the duration of the ratoon crop remains 15 days less than that of the main crop.

9.12.3.2 SORGHUM CULTIVATION IN RICE FALLOWS

The area under Kharif sorghum is decreasing rapidly due to various production constraints. This emerging situation demands a search for potential niches for its cultivation in non-traditional areas. In rice fallows, sorghum cultivation has been found to have considerable potential with higher yields under labor and input-intensive cropping systems (Chapke and Tonapi, 2016). The profit margin of the farmers can be further increased by proper mechanization and the introduction of a standard package of practices. Keeping in view the yield benefits, the innovative knowledge of the farmers should be validated on their fields with extensive transfer of viable technologies to standardize location-specific production technologies along with sustaining productivity and soil health in the long run.

9.13 DROUGHT MITIGATION IN SORGHUM THROUGH CONTINGENCY PLANNING

9.13.1 SORGHUM IN KHARIF

9.13.1.1 EARLY ONSET

Timely sowing along with the onset is highly recommended.

9.13.1.2 TIMELY ONSET OF MONSOON FOLLOWED BY A LONG GAP

To maintain moisture, frequent inter-cultivation activities, including hoeing, weeding, and mulching, are recommended. Reduce the plant population by 2/3 by thinning it under conditions of extreme moisture stress so that moisture can be used effectively. A low plant population ensures that the crop will grow normally. Secondly, under such circumstances, sporadic armyworm outbreaks are anticipated. Spreading toxic bait in the field by combining 250 ml of Metacystox with 4 kg of rice bran and 8 liters of water could control both the moths and the larvae.

9.13.1.3 DELAYED ONSET

In case of a delay in the monsoon by 2–3 weeks, short-duration cultivars such as CSH 6 can be an alternative. The seed rate needs to be increased almost 1.5 times, and shoot fly attacks need to be prevented by the application of Carbofuran @ 20 kg/ha or Phorate 3 g granules.

- **Prolonged Monsoon**

These circumstances at the seed growth stage are likely to result in a severe infestation of grain mold. It is advised to harvest the crop when it reaches physiological maturity to prevent grain mold damage or to spray a fungicide containing 0.3% Dithane M-45 3 times at 10-day intervals beginning at 50% flowering.

9.13.1.4 FAILURE OF MONSOON

If the monsoon fails after the sowing of the crop, the population can be thinned to half. Alternative crops like castor, bajra, and horse gram may be considered in case of a long delay of the monsoon.

9.13.2 RABI SORGHUM

9.13.2.1 DELAYED TERMINATION OF MONSOON

The crop should be protected from shoot fly infestation if resistant varieties are not planted. In order to ensure grain production under severe drought, thin out alternate rows so that available moisture can support the retained plants.

9.14 HARVESTING, THRESHING AND YIELD OF SORGHUM

Sorghum can be harvested 90–140 days after planting, depending on the cultivar and the climatic conditions (Badigannavar et al., 2017). Grain sorghum should be harvested when the moisture level reaches 12–14% to ensure safe and long-term storage. However, harvesting may begin at 20–25% grain moisture if drying facilities are available, and drying may continue until the grain reaches 12–14% moisture. Pre-harvest desiccants, such as glyphosate and diquat, are applied to grains to speed up the drying process and enable an earlier harvest since they prevent the crop from using additional water (Ballagh et al., 2021). Crop desiccants are applied at physiological maturity when the grain moisture content is 25–35% (Nyoni et al., 2022).

9.15 POSTHARVEST MANAGEMENT

Sorghum, or jowar, is considered a healthy food, especially for diabetic people. The grains can be consumed by cooking and can be made into flour and chapatis. The grain is also used to produce industrial alcohol and solvents. Value-added products from sorghum flour, like cookies, muffins, sorghum flakes, puffed sorghum, and cake, are gaining importance in the country. After harvest, the crop is threshed, and the grain is separated from the husk. Once the grains are separated, they can be processed in three different ways: dry milling, wet milling, and fermentation. In the dry milling process, water is added to the clean grain to soften the endosperm, which is milled by conventional roller mills. Another milling process uses clean grain, wetted for hardly 2–3 minutes and immediately milled in a rice huller. This type of milling gives us products that contain more protein. Sorghum grain, as malt, is used for infant food as well as in the pharmaceutical industry.

9.16 CONCLUSION

Sorghum indeed plays a crucial role in agriculture, particularly in rainfed regions like India, where it demonstrates remarkable adaptability to diverse climatic conditions and water scarcity. Despite its resilience, declining cultivation areas are attributed to factors like fluctuating market prices and changing consumer preferences. Beyond its traditional role as a food crop, sorghum holds promise for various alternative uses. Novel and processed foods, as well as its potential as animal feed for dairy cattle, underscore its versatility. Moreover, sorghum can be utilized in industrial applications such as starch production, beverage manufacturing, and ethanol production, contributing to economic diversification. Improving sorghum yield and sustainability involves the adoption of new varieties, enhanced agronomic practices, and effective pest and disease management strategies. These advancements are pivotal in ensuring continued productivity and expanding its utility across different sectors. Its adaptability and the potential for diverse applications make it a valuable crop deserving of continued research and development efforts.

KEYWORDS

- ***Azospirillum***
- **drought mitigation**
- **kharif season**
- **maize stripe virus**
- **nutrient deficiency**
- **ratooning**
- **seed treatment**
- **zonate leaf spot**

REFERENCES

Ajeigbe, H. A., Angarawai, I. I., Aba, D. A., Akinseye, F. M., Inuwa, A. H., Kunihya, A., & Opara, G. O. (2019). *Handbook on Improved Agronomic Practices of Sorghum Production in Nigeria* (p. 28).

Ajeigbe, H. A., Angarawai, I. I., Akinseye, F., Inuwa, A. H., Abdulazeez, T., & Vabi, M. B. (2020). *Handbook on Improved Agronomic Practices for Sorghum Production in North East Nigeria.*

Asif, M., Adnan, M., Safdar, M. E., Akhtar, N., Khalofah, A., & Alzuaibr, F. M. (2022). Integrated use of phosphorus and growth stimulant (Actibion) improves yield and quality of forage sorghum (*Sorghum bicolor* L.). *Journal of King Saud University–Science, 34*(7). https://doi.org/10.1016/j.jksus.2022.102118 (accessed on 7th November 2025).

Badigannavar, A., Kumar, A., Girish, G., & Ganapathi, T. R. (2017). Characterization of post-rainy season grown indigenous and exotic germplasm lines of sorghum for morphological and yield traits. *Plant Breeding and Biotechnology, 5*(2), 106–114.

Ballagh, A., Kezar, S., Harris, S. C., & Lofton, J. (2021). Grain sorghum desiccation: Impacts on plant and grain dry-down. *Agrosystems, Geosciences & Environment, 4*(4).

Bean, B., & McCollum, T. (2006). *Summary of Six Years of Forage Sorghum Variety Trials*. Pub. SCS-2006-04. Texas Cooperative Extension and Texas Agricultural Experiment Station: College Station, TX, USA.

Bhat, B. V., Ratnavathi, C. V., Dayakar Rao, B., Chapke, R., Swarna, R., Padmaja, P. G., Shyam Prasad, G., Das, I. K., Aruna, C., Venkateswarlu, R., Sooganna, S. S., Rao, A., Singode, H. S., Talwar, K., & Babu, V. A. (2019). *Manual on Good Agricultural Practices in Millets* (Vol. 53). Rajendranagar, India.

Bhuriya, K. P., Mistry, G. J., & Kumar Prajapathi, D. (2015). Effect of integrated nutrient management on growth and yield of forage sorghum (*Sorghum bicolor* L. Moench) during summer season. *Trends in Biosciences, 8*(17), 475–4758.

Borad, P. K., & Mittal, V. P. (1983). Assessment of losses caused by pest complex of sorghum hybrid, CSH-5. In K. Rao & B. H. Murthy (Eds.), *Crop Losses Due to Insect Pests* (pp. 271–278). Entomological Society of India: Rajendranagar, Hyderabad.

Chapke, R. R., & Tonapi, V. A. (2016). *Best Practices for Sorghum Cultivation and Importance of Value-Addition.* Training Manual (Vol. 17). Hyderabad, India.

Clayton, W. D., & Renvoize, S. A. (1986). *Genera graminum: Grasses of the World. Kew Bulletin, Additional Series XIII*, 338–345.

Dahlberg, J. A. (2000). Classification and characterization of sorghum. In W. A. Smith & R. A. Frederiksen (Eds.), *Sorghum: Origin, History, Technology, And Production* (pp. 99–130). John Wiley & Sons Inc: New York, Delaware.

De Wet, J. M. J. (1978). Systematics and evolution of *Sorghum* sect. *Sorghum* (Gramineae). *American Journal of Botany, 65*, 477–484.

De Wet, J. M. J., & Harlan, J. R. (1971). The origin and domestication of *Sorghum bicolor*. *Economic Botany, 25*, 128–135.

Delaware, D. G., Ambilwade, P. P., & Kamble, R. J. (2012). Evaluation of integrated pest management components for the control of sorghum shoot pests in Rabi season. *Indian Journal of Entomology, 74*, 58–61.

Doggett, H. (1988). *Sorghum*. Longman Scientific & Technical: Essex, England.

Dovale, J. C., Delima, R. O., & Fritsche-Neto, R. (2012). Breeding for nitrogen use efficiency. In R. Fritsche-Neto & A. Borem (Eds.), *Plant Breeding for Abiotic Stress Tolerance* (pp. 1–20). Springer: Berlin.

Hao, H., Li, Z., Leng, C., Lu, C., Luo, H., Liu, Y., Wu, X., Liu, Z., Shang, L., & Jing, H. C. (2021). Sorghum breeding in the genomic era: Opportunities and challenges. *Theoretical and Applied Genetics, 134*(7), 1899–1924. https://doi.org/10.1007/s00122-021-03789-z (accessed on 7th November 2025).

Harlan, J. R. (1975). *Crops and Man*. American Society of Agronomy: Madison, Wisconsin.

Hasanuzzaman, M., Bhuyan, M. B., Nahar, K., Hossain, M. S., Mahmud, J. A., Hossen, M. S., Masud, A. A. C., Moumita, & Fujita, M. (2018). Potassium: A vital regulator of plant responses and tolerance to abiotic stresses. *Agronomy, 8*(3), 31.

Hassan, S. M. (2013). Nutritional, functional and bioactive properties of sorghum (*Sorghum bicolor* L. Moench) with its future outlooks. *Open Journal of Nutrition and Food Sciences, 5*, 1030.

Hossain, M. S., Islam, M. N., Rahman, M. M., Mostofa, M. G., & Khan, M. A. R. (2022). Sorghum: A prospective crop for climatic vulnerability, food and nutritional security. *Journal of Agriculture and Food Research, 8*(100300), 100300. https://doi.org/10.1016/j.jafr.2022.100300 (accessed on 7th November 2025).

House, L. R. (1985). *A Guide to Sorghum Breeding* (2nd ed.). ICRISAT: Patancheru, A.P., India.

Kabange, N. R., Lee, S. M., Shin, D., Lee, J. Y., Kwon, Y., Kang, J. W., Cha, J. K., Park, H., Alibu, S., & Lee, J. H. (2022). Multiple facets of nitrogen: From atmospheric gas to indispensable agricultural input. *Life, 12*(8).

Kaczmarek, S., Matysiak, K., & Kierzek, R. (2012). Weed control efficacy and selectivity of preemergence herbicides in *Sorghum vulgare* Pers. cultivation. *Progress in Plant Protection, 52*, 47–50.

Kalibhavi, C. M., Kachapur, M. D., & Patil, R. H. (2001). Performance of Rabi sorghum under integrated nutrient management system. *Indian Journal of Dryland Agricultural Research and Development, 16*(1), 45–50.

Khalifa, M., & Eltahir, E. A. B. (2023). Assessment of global sorghum production, tolerance, and climate risk. *Frontiers in Sustainable Food Systems, 7*.

Kugedera, A. T., Kokerai, L. K., & Chimbwanda, F. (2018). Effects of *in-situ* rain water harvesting and integrated nutrient management options on sorghum production. *Global Scientific Journals, 6*(12).

Machado, S., & Paulsen, G. M. (2001). Combined effects of drought and high temperature on water relations of wheat and sorghum. *Plant and Soil, 233*, 179–187.

Madhusudhana, R., Padmaja, P. O., Satish, K., Mohan, S. M., Srinivas, G., & Reddy, R. N. (2007). Heterosis for shoot fly (*Atherigona soccata* Rondani) resistance in sorghum. *Indian Journal of Genetics and Plant Breeding, 67*(3), 299–300.

Mahama, G. Y., Prasad, P. V. V., Mengel, D. B., & Tesso, T. T. (2014). Influence of nitrogen fertilizer on growth and yield of grain sorghum hybrids and inbred lines. *Agronomy Journal, 106*(5), 1623–1630. https://doi.org/10.2134/agronj14.0092 (accessed on 7th November 2025).

Maman, N., Mason, S. C., Lyon, D. J., & Dhungana, P. (2004). Yield components of pearl millet and grain sorghum across environments in the Central Great Plains. *Crop Science, 44*(6), 2138–2145. https://doi.org/10.2135/cropsci2004.2138 (accessed on 7th November 2025).

Mann, J. A., Kimber, C. T., & Miller, F. R. (1983). *The Origin and Early Cultivation of Sorghums in Africa*. Texas Agricultural Experiment Station Bulletin 1454.

Mattos, B. B., Marriel, I. E., De Sousa, S. M., Lana, U. D. P., Schaffert, R. E., Gomes, E. A., & Oliveira-Paiva, C. A. (n.d.). Sorghum genotypes response to inoculation with phosphate solubilizing bacteria. *Revista Brasileira de Milho e Sorgo*, 1177.

Mishra, J. S. (2015). Weed management in millets: Retrospect and prospects. *Indian Journal of Weed Science, 47*(3), 246–253.

Mubarik, M. K., Hussain, K., Abbas, G., Altaf, M. T., Baloch, F. S., & Ahmad, S. (2022). Productivity of sorghum (*Sorghum bicolor* L.) at diverse irrigation regimes and sowing dates in semi-arid and arid environment. *Turkish Journal of Agriculture and Forestry, 46*(1), 1–18.

Nazli, R. I., Kuşvuran, A., Inal, I., Demirbaş, A., & Tansi, V. (2014). Effects of different organic materials on forage yield and quality of silage maize (*Zea mays* L.). *Turkish Journal of Agriculture and Forestry, 38*(1), 23–31.

Nyoni, N., Dube, M., Bhebhe, S., Sibanda, B., Maphosa, M., & Bombom, A. (2022). *Sorghum Production Manual: A Guide to Agronomic Practices for Sorghum Production in Zimbabwe*. Zimbabwe.

Ostmeyer, T. J., Bahuguna, R. N., Kirkham, M. B., Bean, S., & Jagadish, S. V. (2022). Enhancing sorghum yield through efficient use of nitrogen—Challenges and opportunities. *Frontiers in Plant Science, 13*. https://doi.org/10.3389/fpls.2022.987654 (accessed on 7th November 2025).

Queiroz, M. S., Oliveira, C. E. S., Steiner, F., Zuffo, A. M., Zoz, T., Vendruscolo, E. P., Silva, M. V., Mello, B. F. F. R., Cabral, R. C., & Menis, F. T. (2019). Drought stresses on seed germination and early growth of maize and sorghum. *Journal of Agricultural Science, 11*(2), 310. https://doi.org/10.5539/jas.v11n2p310 (accessed on 7th November 2025).

Ratnavathi, C. V., Chakravarthy, S. K., Komala, V. V., Chavan, U. D., & Patil, J. V. (2011). Sweet sorghum as feedstock for biofuel production: A review. *Sugar Tech, 13*(4), 399–407.

Rooney, L. W., & Waniska, R. D. (2000). Sorghum food and industrial utilization. In C. W. Smith & R. Frederiksen (Eds.), *Sorghum: Origin, History, Technology, and Production* (pp. 689–729). John Wiley & Sons Inc: New York.

Sawyer, J. (2004). *Nutrient Deficiencies and Application Injuries in Field Crops*. Iowa State University.

Tamboli, B. D., Bagwan, I. R., Pawar, A. B., Bhakare, B. D., Shelke, S. R., & Kadam, J. R. (2013). Integrated nutrient management for sustaining Rabi sorghum yield economics and soil fertility on Inceptisol under dryland condition. *Indian Journal of Dryland Agricultural Research and Development* (1).

Tari, I., Laskay, G., Takács, Z., & Poór, P. (2013). Response of sorghum to abiotic stresses: A review. *Journal of Agronomy and Crop Science, 199*(4), 264–274. https://doi.org/10.1111/jac.12017 (accessed on 7th November 2025).

Undersander, D. J., Smith, L. H., Kaminski, A. R., Kelling, K. A., & Doll, J. D. (1990). *Forage Sorghum* (No. 20). http://corn.agronomy.wisc.edu/Crops/SorghumForag (accessed on 7th November 2025).

Van Oosterom, E. J., Carberry, P. S., & Muchow, R. C. (2001). Critical and minimum N contents for development and growth of grain sorghum. *Field Crops Research, 70*, 55–73.

CHAPTER 10

Improved Cultivation Practices of Pearl Millet

N. ANANDA,[1,*] M. R. UMESH,[2] T. V. KRISHNA,[3] and CHARAN BABU ANKELA[4]

[1]*ICAR-AICRP on Linseed, Main Agricultural Research Station, University of Agricultural Sciences, Raichur, Karnataka, India*

[2]*ICAR-AICRP on Sunflower, Main Agricultural Research Station, University of Agricultural Sciences, Raichur, Karnataka, India*

[3]*ICAR-AICRP on Kharif Pulses, Zonal Agricultural Research Station, University of Agricultural Sciences, GKVK, Bangalore, Karnataka, India*

[4]*ICAR-Central Research Institute for Dryland Agriculture, Santhoshnagar, Hyderabad, Telangana, India*

*Corresponding author

10.1 INTRODUCTION

Pearl millet [*Pennisetum glaucum* (L.) R. Br. emend. Stunz] is also known as spiked millet, candle millet, cattail millet, bulrush millet, *bajra* in Hindi, Bengali, Oriya, Punjabi, and Urdu, *bajree* in Rajasthan, Gujarati, Marathi, *sajje* in Kannada, Kambu in Tamil, and *sajja* in Telugu.

Pearl millet is the second most important millet in India, after sorghum, both in terms of area and production, and the fifth most important food crop in the world. Especially in arid regions of Africa and Asia, pearl millet (bajra) holds a crucial place. In addition to its use as a food grain for human consumption, it is also utilized as green fodder in India, Pakistan, Russia, Australia, and the USA. Bajra grain has rich nutritional quality, containing 10.5 to 14.5% protein with higher levels of amino acids, 5% fat, 67% carbohydrates, and 2.7% minerals. It is rich in carotene, riboflavin, and niacin but deficient in lysine, like other cereals (Rajendra Prasad, 2012). The grain of pearl millet is superior in nutritive value to sorghum grain but inferior in feeding value. It is consumed in various forms: cooked like rice or chapattis/rotis prepared with flour in India, parboiled balls in Western Africa, and porridge or fermented beverages in Namibia. It provides staple food for the poor in relatively dry areas of the country in a short period (Reddy and Reddy, 2017), as it grows well even under adverse conditions. Thus, it is often considered as poor man's food.

Millets in Sustainable Agriculture: Genetic Improvement, Nutritional Quality, Cultivation, and Integrated Pest–Disease Management. Roshni Vijayan, A. John Joel, R. Ambika Rajendran, and Adithya P. Balakrishnan (Eds.)

DOI: 10.1201/9781779642332-13

- **Forage/Feed:** Bajra grain is used as feed for poultry (Myers, 2002) and as green fodder or dry karbi for cattle. It can be used to make silage and hay, and it can also be consumed directly as green chop or through grazing (Newman et al., 2010). Preserved hay or silage is extremely useful in dry regions (Ramesh et al., 2006). Compared to other warm-season millets, such as brown top, Japanese millet, and proso millet, bajra is the most preferred choice for forage.

  Pearl millet is a short-duration crop and one of the most drought-tolerant cereals grown in arid and semi-arid regions of the world (Bhagavatula et al., 2013; Lee et al., 2012). It thrives well under conditions with rainfall as low as 16–26 inches per year and does not grow well at altitudes above 6500–9000 feet (Hannaway and Larson, 2004), and it cannot withstand standing water. Due to its efficient use of soil moisture and higher heat tolerance compared to sorghum and maize, pearl millet is highly suitable for semi-arid climates (Shah et al., 2012). As a short-duration, drought-tolerant crop, it serves as a better alternative to grain sorghum (Plett et al., 1991) and fits well into various cropping systems, particularly in regions with low rainfall and relatively short growing periods (Mukund Joshi, 2018). Furthermore, pearl millet can also tolerate extreme salinity, poor soil fertility, and high temperatures.

10.2 ORIGIN AND HISTORY

Wide morphological diversification of pearl millet was found in the Sahel region of Western Africa. It was domesticated in the Sahel region sometime during 2500 BC, and it must have reached India around 2000 BC. Since 1500 BC, pearl millet was started in South Africa. During 1850 AD, it reportedly reached the USA, and in the late 1960s, it reached Brazil.

10.3 AREA AND DISTRIBUTION

Since prehistoric times, pearl millet has been extensively cultivated in Africa and Asia. It is grown as a principal crop, replacing sorghum in sandy soils and drier regions in Africa. It is an important millet crop in Asian countries such as India, Pakistan, and China, among others. In the world, pearl millet production is 30 Mt from an area of 27 M ha. Major pearl millet-growing countries include India, China, Myanmar, Nepal, Nigeria, Pakistan, Sudan, Egypt, and Arabia. India is leading the world in pearl millet production (10.86 Mt) with an area and productivity of 7.65 M ha and 1420 kg/ha, respectively (FAI, 2023). Although it is a crop of tropical regions, it is also grown in the warmer months of the sub-tropics. In India, it is mainly grown in Rajasthan, followed by Maharashtra, Uttar Pradesh, and Gujarat, among other states, including Madhya Pradesh, Karnataka, and Haryana. North Haryana (2372 kg/ha) and Madhya Pradesh (2256 kg/ha) are leading in productivity, while high-spread states like Rajasthan and Maharashtra are known to have low productivity (955–1049 kg/ha), as shown in Table 10.1.

TABLE 10.1 Area, Production, and Productivity of Pearl Millet in Important States of India During 2020–21

State	Area (000' ha)	Production (000'tons)	Productivity (kg/ha)
North Haryana	569.2	1350.1	2372
Uttar Pradesh	907.0	2014.4	2221
Karnataka	222.0	275.5	1241
West Gujarat	460.3	1008.9	2192
Madhya Pradesh	327.0	737.7	2256
Maharashtra	687.5	565.6	955
Rajasthan	4348.4	4561.5	1049
All India	7652.1	10863.2	1420

Source: FAI, 2023, Fertilizer Statistics 2022–23, Fertilizer Association of India, New Delhi.

India has been divided into four ecological zones for pearl millet cultivation.

1. **Ecological Zone I:** Adequate to abundant rainfall and light soils (Punjab, Uttar Pradesh, Delhi, Haryana, and adjacent Madhya Pradesh).
2. **Ecological Zone II:** Limited rains but heavy to loamy soils (Gujarat, Maharashtra, and Madhya Pradesh).
3. **Ecological Zone III:** Low rainfall and light sandy and sandy loam soils (Rajasthan, Karnataka, and North-Central Andhra Pradesh).
4. **Ecological Zone IV:** Limited but well-distributed rainfall (Tamil Nadu, Coastal Andhra Pradesh).

10.4 BOTANICAL DESCRIPTION

Pennisetum typhoidea is synonymous with *P. americanum, P. glaucum,* and *P. typhoideum*. It belongs to the family Gramineae (Poaceae), subfamily Panicoideae, and genus *Pennisetum*. Out of the 32 species described by Stapf (1934), only two species – *Pennisetum glaucum* (pearl millet) and *Pennisetum purpureum* (elephant grass) – are known outside the African continent, and both have a common ancestor. Pearl millet is an annual herb and a tall plant, growing up to 3 m in height with a profuse tillering habit. It is a short-day plant that thrives in warm weather. The following are the parts of the bajra plant:

10.4.1 ROOTS

Primary roots that develop from the seed are thin and fine, often growing up to 30 cm in length. From the base of the lowermost nodes, secondary adventitious roots begin to grow at the 2–3 leaf stage. Additionally, anchor roots (prop roots) may develop from the second or third nodes above the ground.

10.4.2 STEM

The stem is solid and usually single, often with a tillering habit. Each stem has swollen nodes with a characteristic groove above each node, housing an axillary bud. The bud at the terminal node develops into a spike. The internodal length increases from the base upwards. Lower internodes are completely covered by the leaf sheath immediately above them, while the upper internodes are only partially covered.

10.4.3 LEAVES

The leaves are lanceolate, long, alternate, auricled, and pubescent, with short ligules. Each plant typically has 8–14 leaves. The leaf margins have saw-like teeth, and the leaf sheath is thicker than the blade, covering the internodes immediately below. Stomata are present on the dorsiventral surfaces of the leaves.

10.4.4 SPIKE (INFLORESCENCE)

The inflorescence consists of 800–3000 cylindrical, densely packed spikelets, which often occur in pairs. Each spikelet has 2 glumes and 2 florets, with the lower floret being male and the upper floret being hermaphroditic. The upper floret consists of a thin palea, a thicker lemma, 3 stamens, and a characteristic penicillate anther. Two to three days after the spike emerges, the styles protrude and become viable, remaining receptive for 2–3 days. New spikelets begin to emerge as the styles dry up. Pearl millet exhibits a protogynous condition, which facilitates cross-pollination.

10.4.5 GRAIN (CARYOPSIS)

The grain, or caryopsis, is oval, tapering at one end, with a maximum width of 3–4 mm. The color varies from whitish-yellow to gray or dull blue, with a reddish embryo. Germination, tillering, boot leaf stage, milky grain, and dough stages are the crucial growth stages in the life cycle.

10.5 SOIL AND CLIMATIC REQUIREMENTS

Pearl millet can be grown under a range of climatic conditions. It thrives in warm, dry climates and can withstand drought to a greater extent. Even under adverse conditions such as arid regions with drought stress and higher temperatures, it yields reasonably well. It is predominantly grown in arid and semi-arid environments characterized by hot and dry weather (Khairwal et al., 2007). The crop is suitable for areas with annual rainfall ranging from 400 to 750 mm and cannot grow in saturated soils, especially in low-lying areas with waterlogging conditions. Moist weather is necessary during the vegetative growth phase, but rainfall during flowering can be detrimental, as it washes off the pollen and leads to poor seed setting. The crop performs best under conditions of light showers followed by bright sunshine.

Commonly, pearl millet is grown in regions where sorghum cannot be cultivated due to high temperatures and low rainfall. In North India, it is grown as a kharif crop, while in Tamil Nadu, Karnataka, and Punjab, it is cultivated as a summer crop under irrigated conditions. Pearl millet seeds can germinate across a wide range of temperatures (15 to 40°C); however, the optimal range for growth is 20 to 28°C, with the ideal temperature for high photosynthesis being 30 to 35°C. Lower temperatures can delay the grain-filling stage. Generally, longer day lengths (beyond 12 hours) result in higher yields. Under higher humidity during the grain development stage, the risk of ergot disease increases. Pearl millet has a high rate of photosynthesis, with a maximum rate observed of 277 $mg/cm^2/s$ (McPherson and Statyer, 1973).

10.5.1 CLIMATE RESILIENCE AND PEARL MILLET

Pearl millet is known for its climate resilience, as it grows under a range of adverse climate conditions such as high temperatures, low moisture levels, and poor soil fertility. It performs well in regions with very low annual rainfall (200–500 mm) and can survive under higher temperatures (>30°C). Among the hardiest of crops, pearl millet is often grown under conditions that are too hot and dry for most other crops, signifying its crucial role in food security in the scenario of climate change.

Temperature is a key climatic factor influencing the growth and development of pearl millet. In both India and Africa, soil temperatures often exceed 45°C, occasionally going as high as 60°C. In India, approximately 25% of pearl millet is cultivated in the arid A1 zone, which includes parts of Rajasthan, Gujarat, and Haryana, where annual rainfall is less than 400 mm. This region is prone to extended dry spells, droughts, and heat waves, with soil temperatures frequently crossing 45°C. Cultivars suitable for this arid zone are characterized by early maturity (~75 days), profuse tillering, and higher grain and fodder yields (Jinu Jacob et al., 2023). Pearl millet is suitable as a summer crop in northern Gujarat, eastern Uttar Pradesh, and parts of Rajasthan, as it can withstand heat stress during the reproductive stage (up to 42°C).

10.5.2 SOIL REQUIREMENT

Pearl millet is adaptable to a wide range of soils, including sandy soils in Rajasthan, sandy loams in Punjab, heavy clay soils in Maharashtra and Karnataka, and red sandy soils in Tamil Nadu. It performs best in well-drained soils with optimum moisture but is sensitive to waterlogging and acidic conditions. It can also grow in soils with low fertility and tolerates slight salinity. In India, pearl millet is successfully cultivated in black cotton soils, poor sandy soils, and alluvial soils.

10.5.2.1 THE RESILIENCE OF PEARL MILLET TO LOW MOISTURE AND POOR SOIL FERTILITY

Pearl millet has a deep root system, which helps it withstand water scarcity. Through the extension of its root length and maintaining a high leaf water status, which supports effective nutrient acquisition, pearl millet can tolerate drought. The plant's ability to thrive in low-nutrient soils with minimal chemical fertilizer requirements is attributed to its high photosynthetic efficiency, dry matter production, and productivity. As a C_4 plant, pearl millet is efficient in atmospheric CO_2 fixation and has higher water use efficiency (WUE). Additionally, the substantial deposition of cuticular wax on its leaves reduces transpiration water losses and reflects radiation, facilitating heat and drought tolerance. Its ability to yield reasonable economic returns even in poor and marginal soils under adverse conditions makes pearl millet a suitable crop for arid and semi-arid regions worldwide.

10.5.2.2 PEARL MILLET: A PREFERRED CROP IN SALINITY-AFFECTED ARID TRACTS

Soil salinity is becoming a major problem for crop production, and it is increasing due to both natural and human activities. Salinity is more common in semi-arid and arid regions where there is less rainfall and higher evaporation. Pearl millet is known for its ability to withstand salinity stress effectively, and its genes have been used to develop salinity-tolerant rice and groundnut plants. It is also recommended as an alternative crop for areas affected by salinity. Since major pests and diseases attacking pearl millet crops are less common, it requires fewer pesticides and fungicides, making it a sustainable crop with minimal impact on the environment (Jinu Jacob et al., 2023).

10.5.3 SEASONS FOR CULTIVATION

The time of sowing is crucial for the performance of any crop. Pearl millet can be grown in all three seasons, depending on day length and temperature. In Tamil Nadu, Karnataka, and Andhra Pradesh, it is sown during the rabi season (September–October). In other states, pearl millet is primarily grown as a kharif crop during June-July. As a summer crop, it is sown in January-February in Punjab, Gujarat, Tamil Nadu, Karnataka, and Andhra Pradesh (Rajendra Prasad, 2012).

10.6 PREPARATION OF LAND

As the pearl millet seeds are small, they require a fine, clod-free seedbed. The ideal land preparation involves deep ploughing (at least 15 cm) with a soil-inverting plough, followed by 2–3 passes with a harrow. It is important to prevent rainwater stagnation during levelling. Under moisture-limited conditions, proper tillage before and after sowing is crucial for rainwater conservation. In such regions, summer ploughing each year and deep ploughing every 3–4 years are recommended. Deep ploughing helps control perennial weeds, releases nutrients from the subsoil, and retains moisture (Malik et al., 1973).

Apply the recommended quantity of FYM or compost 2–3 weeks before sowing. Optimum moisture content in the seedbed at the time of sowing results in good germination. Subramanian et al. (1973) found that incorporating organic materials like rice husk or FYM at 25 tons/ha to a depth of 20–45 cm, 8 weeks before sowing, improved soil water holding capacity and increased grain yield. A flat seedbed can store sufficient moisture, resulting in significantly higher grain yields and enhancing NPK uptake compared to ridge and furrow beds (Ali and Prasad, 1974).

10.7 SOWING OF PEARL MILLET

In the kharif season, pearl millet is sown with the onset of the monsoon, during the first fortnight of July in the northern and central regions of the country. As a rabi crop in Tamil Nadu, it is sown in the first fortnight of October. If the plant population is sparse, gap filling has to be carried out by transplanting seedlings 2–3 weeks after sowing. In the Marathwada region of Maharashtra, it is recommended to dry sow before the first monsoon rains. Summer pearl millet is sown between the 4th and 5th standard meteorological weeks (SMW), which corresponds to the last week of January through the first week of February, to achieve higher yields.

10.7.1 METHOD OF SOWING

Pearl millet can be sown using three different methods: (i) on a flat surface, (ii) ridge and furrow method, and (iii) broad-bed and furrow method. Seeds should be placed at a depth of 2.5 to 3 cm. Sowing can also be done using the plough sole method or an improved six-tine seed drill. Thinning is essential, as high plant density can reduce yields.

10.7.1.1 TRANSPLANTING OF BAJRA

For delayed sowing, transplanting is recommended in pearl millet. In intensive irrigated cropping systems, transplanting can save 3–4 weeks of crop duration. Transplanted pearl millet performs comparably to directly seeded crops in terms of productivity. To transplant pearl millet, a nursery of 500 m^2/ha should be prepared in advance, with seedlings sown 3–4 weeks before transplantation. In August, these seedlings are transplanted into the main field, with spacing of 50 cm x 20 cm (plant density of 175,000 to 200,000 plants/ha). It is ideal for transplantation on a rainy day, and irrigation is necessary if there is no rainfall (Rajendra Prasad, 2012).

10.7.2 SEED TREATMENT

Seed treatment is crucial for disease management and enhancing seed performance. Seed treatment with bio-pesticides like *Trichoderma harzianum* @ 4 g/kg or thiram 75% dust @ 3 g/kg is recommended to control soil-borne diseases. Smut can be controlled by applying 300-mesh sulphur powder @ 4 g/kg of seeds. For ergot-affected seeds, soaking them in a 10% salt solution helps remove the affected seeds. To manage downy mildew, treat seeds with metalaxyl @ 6 g/kg. Seeds are treated with *Azospirillum* (570 g) and *Phospho bacterium* (375 g) to improve the availability of nitrogen and phosphorus.

10.7.3 SEED RATE, SPACING AND PLANT POPULATION

The recommended seed rate for pearl millet is 3 kg/ha. In the arid western plains of Rajasthan, Haryana, and Kutch in Gujarat (A1 Zone), pearl millet should be planted in rows 60 cm apart, with a plant population of 100,000–125,000 plants/ha. In areas receiving more than 450 mm of rainfall (Zones A and B), the crop can be sown at a spacing of 45 × 10–15 cm, maintaining a plant population of 175,000–200,000 plants/ha.

To maintain optimum plant stand, use a seed rate of 3 to 4 kg/ha. Under normal conditions, a plant population of 180,000 plants/ha (72,000 plants/acre) is recommended. For irrigated or high-management systems on productive soils, a population of 225,000 plants/ha (100,000 plants per acre) is ideal. On extremely sandy or drought-prone soils, a lower population of about 90,000 plants/ha (40,000 plants per acre) is recommended.

10.8 VARIETIES AND HYBRIDS

Pearl millet cultivation mainly involved traditional and improved varieties with limited yield potential until the 1970s. After the identification of cytoplasmic male sterility, which is essential due to the crop's specific flowering habit and cross-pollinated nature, hybrids became popular in pearl millet. Today, hybrid pearl millet is grown extensively, while traditional varieties are still preferred in drought-prone regions (Anon., 2022). Below is a list of the latest hybrids and varieties of pearl millet.

Region/State	Crop Season	Hybrid	Variety
Rajasthan	*Kharif*	BHB-1202 (Bikaner Hybrid Bajra-1202), RHB 223 (MH 1998), HHB 299 (MH 2076), AHB 1200, MP 535 (Pusa Composite 701), HHB 67, MPMH-21, MPMH 17, KBH 108, GHB 905, 86M89, 86M86, 86M86, Kaveri Super Boss, Bio 448, RHB-173	MBC 2, PC 443, JBV3, PC 383, ICMV 221, Raj 171
	Summer	Nandi 72, Nandi 70, 86M64	-
	Kharif – arid parts	HHB 234, Bio 70, HHB-226, RHB-177	CZP 9802
Gujarat	*Kharif*	RHB 223 (MH 1998), HHB 299 (MH 2076), AHB 1200, MP 535 (Pusa Composite 7-1), MP 7872, MP 7792, HHB 272 (MH1837), HHB 67, MPMH 21, MPMH 17, KBH 108, GHB 905, 86M89, Kaveri Super Boss, Bio 448, 86M86, 86M66, RHB -173	MBC 2, PC 443, JBV3, PC 383, ICMV 221, Raj 171
	Summer	Nandi 70, Nandi 72, 86M64	-
	Kharif – arid parts	HHB 234, Bio 70, HHB-226, RHB-177	CZP 9802
Haryana	*Kharif*	RHB 223 (MH 1998), HHB 299 (MH 2076), AHB 1200, MP 535 (Pusa Composite 7-1), MP 7872, MP 7792, HHB 272 (MH1837), HHB 67, MPMH 21, MPMH 17, KBH 108, GHB 905, 86M89, Kaveri Super Boss, Bio 448, 86M86, 86M66, RHB -173	MBC 2, PC 443,
	Kharif – arid parts	HHB 234, Bio 70, HHB-226, RHB-177	CZP 9802
Punjab	*Kharif*	HHB 299 (MH 2076), AHB 1200, MP 535 (Pusa Composite 701) MP 7872, MP 7792, PHB 2884, KBH 108, GHB 905, 86M89, MPMH 17, Kaveri Super Boss, Bio 448, 86M86, 86M66, RHB-173	PCB 164, ICMV 221, Raj 171

(Continued)

Region/State	Crop Season	Hybrid	Variety
Delhi	*Kharif*	Pusa 1201 (MH 1849), HHB 299 (MH 2076), AHB 1200, MP 535 (Pusa Composite 701) MP 7872, MP 7792KBH 108, GHB 905, 86M89, MPMH 17, Kaveri Super Boss, Bio 448, 86M86, 86M66, RHB-173	JBV 3, PC 383, ICMV 221, Raj 171
Uttar Pradesh	*Kharif*	MP 535 (Pusa Composite 701) MP 7872, MP 7792, KBH 108, GHB 905, 86M89, MPMH 17, Kaveri Super Boss, Bio 448, 86M86, 86M66, RHB-173	JBV 3, PC 383, ICMV 221, Raj 171
Madhya Pradesh	*Kharif*	MP 535 (Pusa Composite 701) MP 7872, MP 7792, KBH 108, GHB 905, 86M89, MPMH 17, Kaveri Super Boss, Bio 448, 86M86, 86M66, RHB-173	JBV 4, JBV 3, PC 383, ICMV 221, Raj 171
Maharashtra	*Kharif*	HHB 299 (MH 2076), AHB 1200, CO 10, CO 9, kaveri Super Boss, Pratap, Shine, MP 7792, 96M86, 86M53, PAC 909.	ABPC 4-3, PC 612, Parbhani Sampada, Samrudhi, ICMV 221, Raj 171, ICMV 155
	Summer	Nandi 72, Nandi 70, 86M64	-
Tamil Nadu	*Kharif*	HHB 299 (MH 2076), AHB 1200, CO 10, Co9, Kaveri super Boss, Pratap, Shine, MP 7792, 86M64, 86M53	AbP4-3, PC 612, Parbhani Sampada, Samrudhi, ICMV 221, RAJ 171, ICMV 155
	Summer	Nandi 72, Nandi 70, 86M64	-
Andhra Pradesh	*Kharif*	AHB 1200, Kaveri Super Boss, Pratap, Shine, MP 7792, 86M86, PAC 909, 86M64, 86M53	PC 612, Samrudhi, ICMV 221, Raj 171, ICMV 155, Ananta
Karnataka	*Kharif*	Kaveri Super Boss, Pratap, Shine, MP 7792, PAC 909, 86M64, 86M53	PC 612, Samrudhi, ICMV 221, Raj 171, ICMV 155

Source: Chapkae *et al.*, 2020.

10.9 NUTRIENT MANAGEMENT

In India, pearl millet is majorly grown in rainfed areas, characterized by poor soil fertility and erratic rainfall patterns, leading to lower yields compared to potential yield (Suzuki et al., 2017). Inherently low organic carbon and poor management practices to replenish soil fertility are the major reasons for lower soil fertility. Under such circumstances, synthetic fertilizers have been promoted to achieve higher yields. The application of synthetic fertilizers alongside organic resources could be an effective strategy for boosting soil fertility and crop yields using minimal resources (Thumar et al., 2016). In addition to nutrient losses, a ton of pearl millet grain can remove 26.6 kg N, 9.4 kg P, and 32.6 kg K from the soil (Khairwal et al., 2007).

It is recommended to apply FYM or compost at 6 t/ha before sowing. In arid regions, 40 kg N and 20 kg P_2O_5 per hectare, and in semi-arid regions, 60 kg N and 30 kg P_2O_5 per hectare are recommended for both sole crops and intercropping. At the time of seedbed preparation, only half of the recommended nitrogen can be applied in light soils to avoid unnecessary leaching losses due to heavy rains, and the remaining half can be applied as side dressing at 25 DAS. In heavy soils, where leaching is minimal, such as black soils, all the nitrogen may be applied during seedbed preparation. As pearl millet seeds are sensitive to fertilizer burn, care should be taken not to apply fertilizer in the furrow with the seed or very near the seed in the row after sowing. It should be applied as a side dressing.

The use of bio-fertilizer Phosphate Solubilizing Bacteria (*Azospirillum* and PSB) can economize the N and P fertilizer application. In zinc-deficient soils of the pearl millet growing area of the country, the application of $ZnSO_4$ at 10 kg/ha is recommended. To correct the zinc deficiency in standing crops, a spray of 0.2% $ZnSO_4$ at the tillering to the pre-flowering stage is recommended. Under prolonged dry spells, skip the top dressing of N and spray 2% urea. Under excessive rain situations during the vegetative phase, an additional dose of N at 20 kg/ha should be given.

Using organic manures in combination with chemical fertilizers has a more beneficial effect on microbial biomass and, consequently, on soil health. Integrated nutrient management helps in both maintaining productivity and enhancing soil microbial load (Reddy, 2006). Farmers often struggle to use costly fertilizers due to economic constraints and irregular rainfall in dryland areas. Moreover, the exclusive use of chemical fertilizers has led to several issues, including micronutrient deficiencies, nutritional imbalances in soil and plant systems, pest infestations, environmental degradation, and soil health deterioration. Therefore, in dryland areas, combining inorganic and organic fertilizers may be more advantageous than using either type alone.

Abdullahi et al. (2014) investigated the effects of combining biofertilizers with poultry manure or cow dung and reported that the combination of biofertilizers with poultry manure resulted in improved plant growth, including plant height, number of tillers per plant, and both shoot and root biomass. This improvement was attributed to the increased availability of energy sources for microbes from organic materials, which in turn enhanced biological activity, nitrogen availability, phosphorus mobilization, and growth-promoting hormones. Similarly, Choudhary et al. (2014) observed that increased doses of vermicompost and the recommended dose of fertilizers (RDF) led to greater plant height and tiller numbers per meter of row in pearl millet. This was due to the essential plant nutrients provided by vermicompost, as well as its positive impact on soil physical and biological properties. Narolia et al. (2009) found that increasing levels of vermicompost from 0 to 2 t/ha progressively improved yield attributes in pearl millet and also reported that drilling 2 t/ha of vermicompost was more effective than incorporation. Integrated nutrient management showed significant potential for addressing the growing nutrient needs of pearl millet while maintaining high productivity and enhancing soil health. A balanced approach using organic, inorganic, and biofertilizers is crucial for maintaining a healthy soil environment and providing energy for soil microbial activity. Thus, integrated nutrient management practices are highly beneficial and should be optimized, with adjustments of either 25% or 50% on an equivalent basis.

10.10 WATER MANAGEMENT

The lower productivity of pearl millet is largely due to its cultivation in dryland areas and inadequate irrigation management practices. Therefore, careful use of irrigation water assumes a key role in enhancing both the quality and productivity of this resilient crop. Typically, two irrigations are sufficient, with the most critical stages being after germination, during tillering, at ear head emergence (heading), and during grain filling. Pearl millet does not tolerate waterlogging, so effective drainage is essential. Moisture conservation practices such as deep ploughing, broad bed furrows, and application of manures can help in the efficient use of rainwater in rainfed conditions. During the kharif season, the irrigation requirement generally ranges from 150–200 mm across most regions. Pearl millet can tolerate up to 75% depletion of available soil water in the 0–30 cm depth. The criteria for scheduling irrigation vary depending on specific conditions. Although several methods are available, the choice of the best method should be based on experience. Among different approaches, the climatological method, that is, the IW/CPE method, has been found to be the most effective. This method considers all weather parameters and their relative importance in the soil-water-plant continuum (Parveen Kumar and Amit Kumar, 2021).

Patel et al. (2010) reported significantly higher values for growth, yield attributes, grain yield, and fodder yield of pearl millet when irrigation was scheduled at a 1.1 IW: CPE ratio. The grain yield increased by 18.2% and 4.6% compared to the 0.7 and 0.9 IW: CPE ratios, respectively. Sonawane et al. (2010) found that the application of 40 mm of irrigation water resulted in significantly higher grain yield, fodder yield, and water use efficiency compared to a 60 mm irrigation depth. Although the 1.0 IW/CPE ratio reported the highest grain yield, water use efficiency was significantly higher at 0.5 IW/CPE, which was comparable to irrigation applied at critical growth stages. To achieve higher yields and better water use efficiency in summer pearl millet, the application of 40 mm of irrigation water at critical growth stages is most effective.

10.11 INTERCULTIVATION AND WEED MANAGEMENT

Weeds pose a significant challenge to increasing pearl millet productivity, particularly during the rainy season. Sharma and Jain (2003) reported a loss in grain yield of up to 40% due to weed competition in pearl millet. Since pearl millet is predominantly grown during the warm, rainy season, it is vulnerable to various types of weeds that can hamper its growth. Effective weed management is crucial for enhancing pearl millet productivity, as weeds compete with the crop for nutrients, water, light, and space. They reduce both crop yield and quality, particularly during the early growth stages (Leela Rani et al., 2018).

Weeds result in lower grain and straw yields of kharif pearl millet. Due to its initial slow growth, pearl millet is a relatively poor competitor with weeds during its initial development stages (Yaduraju et al., 2016). Practices such as planting in wider rows to facilitate inter-row cultivation or using ditch furrow irrigation can increase weed density and growth. The slow development of the crop canopy provides minimal shading for weeds between rows until midseason, by which time many weeds are well established. With the adoption of improved agricultural technologies, efficient weed management becomes even more critical; otherwise, weeds may benefit from the costly inputs rather than the crops.

Weeds compete with crops for nutrients, soil moisture, sunlight, and space, leading to reduced yields, lower grain quality, and increased production costs (see Table 10.2). The extent of these losses depends on factors such as crop cultivars, the type and intensity of weed infestation, spacing, duration of weed presence, environmental conditions, and management practices. According to Ram et al. (2005), weeds can deplete nutrients from pearl millet by up to 61.8 kg N, 5.6 kg P, and 57.6 kg K per hectare.

TABLE 10.2 Losses Due to Weeds in Pearl Millet

Reduction in Grain Yield (%)	References
40	Sharma and Jain (2003)
55	Banga et al. (2000)
16–94	Balyan et al. (1993)
31–46	Kaushik and Gautam (1984)
35–90	Umrani et al. (1980)

10.11.1 MAJOR WEEDS OF PEARL MILLET IN INDIA

Most commonly observed weeds in pearl millet field are *Ageratum conyzoides, Alhagi camelorum, Ageratum conyzoides, Amaranthus viridis, A. spinosus, Borreria articularis, Commelina benghalensis, Celosia argentea, Corchorus olitorius, Corchorus aestuans, Canabis sativa, Chrozophora rottleri, Cleome viscose, Convolvulus arvensis, Croton bonplandianum, Digera arvensis, Euphorbia geniculata, E. hirta, Eclipta alba, Fumaria parviflora, Ipomoea purpurea, Leucas aspera, Oxalis latiifolia, Phyllanthus niruri, Striga asiatica, Sonchus arvensis, Trianthema portulacastrum, Tridax procumbens* among the broad leaved weeds; *Brachiaria ramose, Brachiaria eruciformis, Brachiaria ramose, Cynodon dactylon, Dactyloctenium aegyptium, Digitaria sanguinalis, Digitaria marginata, Digitaria ciliaris, Echinochloa colona E. crusgalli, Eleusine aegypticum, Ischenedespaire, Paspalum paspaloides, Panicum dichotomiflorum, Panicum repens, Saccharum spontaneum and Sorghum halepense* among grasses and *Cyperus rotundus, C. esculentus, C. iria, Eragrostis major, Fimbristylis diphylla* among sedges (Charul Chaudhary et al., 2018).

10.11.1.1 CONTROL MEASURES

1. **Mechanical Weeding**: Manual and mechanical weeding are the most commonly used methods for weed control in millets. Hand weeding or inter-row cultivation can provide effective weed management. However, during the rainy season, the limited number of clear days often leads to delays in intercultural operations, allowing weeds to overtake the crops and significantly reduce yields. Additionally, rising labor wages and the occasional scarcity of adequate labor further complicate manual weed control over large areas and at the critical stages. The lowest weed density and highest weed control efficiency are achieved by performing inter-cultivation followed by hand weeding at 20 and 40 DAS.

2. **Cultural Operations:** Cultivation of mung bean, groundnut, cowpea, soybean, and similar crops as intercrops with sorghum or pearl millet can effectively suppress weeds. Planting intercrops in widely spaced rows not only reduces weed intensity but also provides additional yield. Choudhary et al. (2012) concluded that intercropping pearl millet with green gram (2:2) was superior with higher yield, net return and Land Equivalent Ratio (LER) compared to growing pearl millet alone

3. **Chemical Control:** Due to unfavorable weather conditions and shortage of labour, manual weeding and inter-cultivation are often delayed. In such cases, chemical weed control plays a crucial role in the weed management strategy for pearl millet (see Table 10.3). Herbicide application is particularly beneficial when labor is scarce or expensive. Appropriate use of recommended herbicides can result in completely weed-free environment for the crop, while manual or mechanical weeding can only be performed after weeds have emerged. The decision to use herbicides should be based on their relative effectiveness compared to manual and mechanical weeding methods. However, herbicide application requires some expertise to avoid potential risks, as improper use can lead to significant crop loss.

TABLE 10.3 Herbicides Recommended for Pearl Millet

Herbicide	Time of Application	Dose (kg/ha)	Type of Weeds Controlled	Notes	References
Atrazine	Pre-emergence/ early post-emergence	0.50	Broad spectrum weeds control. Some grasses are tolerant	For sole crop only	Munde et al. (2012)
Pendimethalin	Pre-emergence	0.75–1	Effective control of grasses	Suitable for intercropping	Ram et al. (2005)
Oxadiazon	Pre-emergence	1	Annual grasses and broadleaved weeds	For sole crop only	Ram et al. (2005)
2,4-D (ethyl ester)	Post-emergence	0.50–0.75	Effective against broad-leaved weeds	For sole crop only. Apply at 4–6 weeks after planting. Good as sequential application to pre-emergence herbicides	Singh and Singh (2010)
Saflufenacil	Pre-emergence	0.05	Broad spectrum weed control	For sole crop only	Reddy et al. (2014)

4. **IWM with Herbicides as a Component:** Research studies were carried out on Integrated Weed Management (IWM) to use herbicides as part of a broader weed management strategy rather than relying solely on them. This approach includes the application of herbicides along with crop rotations, cropping systems, tillage, mulching, and hand weeding. Legumes, as part of cropping systems, offer several benefits: they provide nutritious food and feed, enhance soil fertility through biological nitrogen fixation, reduce pests and disease incidence, add organic matter, conserve resources, and allow for more flexible weed control. Pearl millet, with its allelopathic properties, shows potential as a component of IWM for reducing weed

populations. Analysis of various IWM strategies suggests that the most common and effective approach is integrating hand weeding with herbicide application. The critical period for weed competition is 35 days after sowing (DAS). In line-sown crops with a minimum inter-row spacing of 45–60 cm, most weeds can be effectively controlled through inter-cultivation. Performing two hoeing and weeding operations at 15 and 30 DAS is generally sufficient for effective weed management. Chemical control measures include the pre-emergence application of atrazine 50 WP at 0.5 kg a.i./ha, followed by one hand weeding, which has proven effective. Additionally, proper weed management helps conserve soil moisture (Abdelrahman et al., 2003). It is important to ensure that there are no clods at the time of spraying and that soil moisture is adequate. Excessive spraying beyond the recommended rate can hinder proper germination and reduce germination percentages.

10.12 PEARL MILLET-BASED CROPPING SYSTEMS IN KHARIF SEASON

Instead of growing the same cultivars over the years, using different cultivars through rotation can be adopted to avoid downy mildew disease problems (see Table 10.4). Hybrids and open-pollinated varieties (OPVs) should be used in alternate years/seasons. It is suggested not to grow the same hybrid or open-pollinated variety continuously on the same piece of land.

TABLE 10.4 Various Crop Rotation Systems Followed Across Different States in India

States	Crop Rotation Order
Rajasthan	Pearl millet + cluster bean/cowpea/green gram/moth bean/sesame
Haryana	Pearl millet + green gram/sesame/cluster bean/cowpea
Gujarat	Pearl millet + green gram/sesame/cowpea
Uttar Pradesh	Pearl millet + green gram/sesame/cowpea
Madhya Pradesh	Pearl millet + Black gram/soybean/Pigeon pea/cowpea
Delhi	Pearl millet + Pigeon pea/groundnut/castor
Punjab	Pearl millet + Chickpea/fodder sorghum/wheat
Maharashtra	Pearl millet + moth bean/pigeon pea/soybean/black gram, green gram/ cowpea/sunflower
Karnataka	Pearl millet + Pigeon pea/green gram/sunflower/soybean
Tamil Nadu	Pearl millet + Pigeon pea/green gram/sunflower/soybean/cowpea
Andhra Pradesh and Telangana	Pearl millet + Pigeon pea/green gram/sunflower/soybean/groundnut

1. **Mixed Systems**: Bajra with sorghum, pigeon pea, green gram, cowpea, field bean, horse gram and sesame.
2. **Intercropping Systems:** Pigeon pea, groundnut or cowpea are cultivated as replacement series in pearl millet.
3. **Sequence Systems:** Bajra is rotated with rice, cotton, sorghum, groundnut, or wheat.

10.13 HARVESTING

The ideal time to harvest pearl millet is when the plants reach physiological maturity, which is indicated by the presence of a black spot at the bottom of the grain in the hilar region. At this stage, the leaves turn yellow and the plants have nearly dried-up appearance. The grains become hard and firm. Typically, the harvesting process involves cutting the ear heads first, followed by the stalks. The stalks (straw) are usually cut after a week, allowed to dry, and then stacked. Grain with a moisture content of 14% or lower is considered dry. For long-term storage (> 6 months), the grain moisture content should be less than 10%.

10.14 THRESHING, CLEANING, DRYING, AND STORING

After the ear heads are dried, they can be threshed using a mechanical thresher, or alternatively, by spreading them out and using a stone roller or cattle to thresh the grain. Once dried, the grain should be stored in gunny bags to ensure proper preservation.

10.15 YIELD

Under irrigated conditions, a grain yield of 30–35 q/ha and a straw yield of 80–100 q/ha can be expected. Under rainfed conditions, the expected grain yield is 25–28 q/ha, with a straw yield of 40–50 q/ha.

10.16 NON-CASH INPUTS

Sow the crop in summer during January and kharif from June 15th to July 15th

Apply a recommended dose of FYM/compost, 2–3 weeks before sowing the crops.

The most critical stages of irrigation are after germination, tillering, earhead emergence (Heading), grain forming stage and grain filling.

10.17 ANTI-NUTRITIONAL FACTORS

Bajra grains contain polyphenols (tannins) and phytic acid, similar to many other cereals. Tannins can impair protein digestion, while phytic acid may reduce the bioavailability of essential minerals. Additionally, bajra can contain mycotoxins such as Fusarium toxins, which are carcinogenic. Regular consumption of bajra as a staple food may also be associated with the risk of goiter, a condition caused by abnormal thyroid hormone secretion. This is linked to the presence of vitexin and 8-glycosylapigenin in bajra grains, which inhibit thyroid peroxidase activity and can contribute to thyroid-related abnormalities (El Tinay, 2005).

10.18 CONCLUSION

Pearl millet is an important crop in rainfed areas due to its drought-escaping characteristics, allowing it to thrive with rainfall ranging from 50 to 1300 mm. Beyond its use as a grain, pearl millet is also utilized as green fodder. The grain is rich in minerals, protein, essential amino acids, and fat. India is the largest producer of pearl millet, with Rajasthan, Gujarat, Haryana, Uttar Pradesh, Maharashtra, and Karnataka being the leading producing states.

Pearl millet is a tall, warm-weather, short-day annual plant belonging to the Gramineae (Poaceae) family. Its growth stages include germination, seedling emergence, seedling establishment, tillering, final leaf stage, booting, panicle emergence, half bloom stage, fertilization, seed set, milking, soft dough stage, hard dough stage, and physiological maturity.

For optimal growth, pearl millet requires fine seedbed free from clods due to its small seed size. It is recommended to apply 6 t/ha of FYM or compost 2–3 weeks before sowing. Recommended fertilizer dose is 40 kg N and 20 kg P_2O_5 per hectare in arid regions, and 60 kg N and 30 kg P_2O_5 per hectare in semi-arid regions, for both sole crop and intercrop systems. The use of biofertilizers such as phosphate-solubilizing bacteria (*Azospirillum* and PSB) can reduce the requirement of external application of nitrogen and phosphorus fertilizers.

Although pearl millet is primarily a rainfed crop, it requires irrigation during the summer, especially at critical stages such as germination, tillering, ear head emergence (heading), and grain filling. The critical period for weed competition is 35 days. With a minimum inter-row spacing of 45–60 cm, most weeds can be controlled through intercultivation. Promising and profitable intercrop mixtures in various agroclimatic regions include pearl millet with cowpea, pigeon pea, groundnut, mung bean, and sesame.

KEYWORDS

- **intercultivation**
- **Kharif**
- **Pearl millet**
- ***Pennisetum glaucum***
- ***Pennisetum typhoidea***
- **seed**
- **weed management**

REFERENCES

Abdelrahman, S. M., Hagir, B. E., Wisal, H. I., Elfadil, E. B., Abdullahi, S., Prem, C. S., & Rajbir, S. (2003). *Modern Techniques of Raising Field Crops* (2nd ed., pp. 55–83). Oxford and IBH: New Delhi.

Abdullahi, R., Sheriff, H. H., & Buba, A. (2014). Effect of bio-fertilizer and organic manure on growth and nutrients content of pearl millet. *ARPN Journal of Agricultural and Biological Science, 9*(10), 351–355.

Ali, M., & Prasad, R. (1974). Effect of mulches and type of seed bed on pearl millet under semi-arid conditions. *Experimental Agriculture, 10*, 263–272.

Balyan, R. S., Kumar, S., Malik, R. K., & Panwar, R. S. (1993). Post emergence efficacy of atrazine in controlling weeds in pearl millet. *Indian Journal of Weed Science, 25*, 7–11.

Banga, R. S., Yadav, A., Malik, R. K., Pahwa, S. K., & Malik, R. S. (2000). Evaluation of tank mixture of acetochlor and atrazine or 2,4-D Na against weeds in pearl millet. *Indian Journal of Weed Science, 32*, 194–198.

Bhagavatula, S., Parthasarathy, R. P., Basavaraj, G., & Nagaraj, N. (2013). *Sorghum and Millet Economies in Asia–Facts, Trends and Outlook* (p. 80). International Crops Research Institute for the Semi-Arid Tropics, Patancheru, Andhra Pradesh, India.

Chapke, R.R., Shyam Prasad, G., Das, I.K., Hariprasanna, K., Singode, A., Kanthisri, B.S. and Tonapi, V.A. 2020. Latest millet production and processing technologies. Booklet, ICAR-Indian Institute of Millets Research, Hyderabad 500 030, India: 82p. (ISBN: 81-89335-90-X).

Chaudhary, C., Dahiya, S., Rani, S., & Pandey, S. (2018). Review and outlook of weed management in pearl millet. *International Journal of Chemical Studies, 6*(2), 2346–2350.

Choudhary, R., Dodia, I. N., Choudhary, R., & Golada, S. L. (2012). Effect of pearl millet-based pulses intercropping in rainfed conditions. *International Journal of Forestry & Crop Improvement, 3*(2), 112–115.

Choudhary, R., Yadav, L. R., & Parihar, S. (2014). Studies on the interactive effect between vermicompost and fertility levels on pearl millet (*Pennisetum glaucum* L.). *Annals of Agricultural and Biological Research, 19*(3), 430–433.

El-Tinay, H. (2005). Antinutritional factors content and minerals availability of pearl millet (*Pennisetum glaucum*) as influenced by domestic processing methods and cultivar. *Journal of Food Technology, 3*, 397–403.

Hannaway, D. B., & Larson, C. (2004). *Forage fact Sheet: Pearl Millet (Pennisetum americanum)* (p. 83). Oregon State University, Corvallis, Oregon.

Jacob, J., Reddy, P. S., Visarada, K. B. R. S., Malathi, V. M., & Venkateswarlu, P. (2023). Pearl millet, a benefitting crop for the changing climate. *Indian Farming, 73*(1), 33–34.

Joshi, M. (2018). *Textbook of Field Crops*. PHI Learning Limited: New Delhi.

Kaushik, S. K., & Gautam, R. C. (1984). Weed control studies in pearl millet under rainfed condition. *Indian Journal of Agronomy, 29*, 31–36.

Khairwal, I. S., Rai, K. N., Diwakar, B., Sharma, Y. K., Rajpurohit, B. S., Nirwan, B., & Bhattacharjee, R. (2007). *Pearl Millet: Crop Management and Seed Production Manual* (p. 104). International Crops Research Institute for the Semi-Arid Tropics, Patancheru, Andhra Pradesh, India.

Kumar, P., & Kumar, A. (2021). Water management for improving pearl millet production under irrigated environment: A review. *Agricultural Reviews, 42*(2), 225–229.

Lee, D., Hanna, W., Buntin, G. D., Dozier, W., Timper, P., & Wilson, J. P. (2012). *Pearl Millet for Grain.* Bulletin #B 1216. College of Agricultural and Environmental Sciences, University of Georgia Cooperative Extension.

Malik, A. S., Kumar, V., & Mebane, M. K. (1973). Dryland research in North West India. Effect of tillage on soil moisture storage and yield of pearl millet. *Agronomy Journal, 65*(12–14), 1973.

McPherson, H. G., & Slatyer, R. D. (1973). Mechanisms regulating photosynthesis in *Pennisetum typhoides. Australian Journal of Biological Sciences, 26*, 329–339.

Munde, S. D., Aghav, V. D., Pagar, R. D., & Patel, J. C. (2012). Effect of herbicides on weeds and yield of rainy season pearl millet (*Pennisetum glaucum* (L.) R. Br. emend. and Stuntz). *Crop Research, 44*(3), 288–291.

Myers, R. L. (2002). *Alternative Crop Guide: Pearl Millet.* Jefferson Institute: Washington, D.C. http://www.jeffersoninst.org/ (accessed on 7th November 2025).

N/A. (2022). *Improved Production Techniques of Millets*. ITC Private Limited.

N/A. (2023). *Fertilizer Statistics*. New Delhi.

Narolia, R. S., Poonia, B. L., & Yadav, R. S. (2009). Effect of vermicompost and inorganic fertilizers on productivity of pearl millet (*Pennisetum glaucum*). *Indian Journal of Agricultural Sciences, 79*(7), 506–509.

Newman, Y., Jennings, E., Vendramini, J., & Blount, A. (2010). *Pearl Millet (Pennisetum glaucum): Overview and Management*. University of Florida IFAS Extension. Publication #SS-AGR-337.

Patel, P. M., Patel, J. J., Patel, G. G., & Gediya, K. M. (2010). Influence of irrigation schedules, mulches and antitranspirant on growth and yield of summer transplanted pearl millet (*Pennisetum glaucum* L.). *International Journal of Forestry & Crop Improvement, 1*(2), 60–63.

Plett, S., Nelson, L. A., & Clegg, M. D. (1991). Alternative crop prospects in Western Nebraska. *Journal of Production Agriculture, 4*, 162–166.

Prasad, R. (2012). *Textbook of Field Crops Production: Food Grains Vol-I* (p. 184). Indian Council of Agricultural Research: New Delhi.

Ram, B., Chaudhary, G. R., Jat, A. S., & Jat, M. L. (2005). Effect of integrated weed management and intercropping systems on growth and yield of pearl millet. *Indian Journal of Agronomy, 50*(3), 254–258.

Ramesh, S., Santhi, P., & Ponnuswamy, K. (2006). Photosynthetic attributes and grain yield of pearl millet (*Pennisetum glaucum* L. R. Br.) as influenced by the application of composted coir pith under rainfed conditions. *Acta Agronomica Hungarica, 54*(1), 83–92.

Rani, P. L., Sreenivas, G., Bhatt, P. S., & Jayashankar, P. (2018). *Basic Concepts of Weed Management*. Indian Council of Agricultural Research, New Delhi.

Reddy, S. R. (2006). *Agronomy of Field Crops*. Kalyani Publishers: Ludhiana.

Reddy, S. S., Stahlman, P. W., Geier, P. W., Charvat, L. D., Wilson, R. G., & Moechnig, M. J. (2014). Tolerance of foxtail, proso and pearl millets to saflufenacil. *Crop Protection, 57*, 57–62.

Reddy, Y., & Reddy, S. (2017). *Principles of Agronomy*. Kalyani Publishers: Ludhiana.

Shah, A. H., Rahman, H., Shah, S. M. A., Rahman, S., & Noor, M. (2012). Characterization of pearl millet germplasm for various morphological and fodder yield parameters. *Journal of Botany, 44*, 273–279.

Sharma, O. L., & Jain, N. K. (2003). Integrated weed management in pearl millet (*Pennisetum glaucum*). *Indian Journal of Weed Science, 35*, 34, 35.

Singh, R., & Singh, H. R. (2010). Effects of method of sowing and herbicides on the yield of pearl millet (*Pennisetum glaucum*). *Progressive Agriculture, 10*(1), 111–113.

Sonawane, P. D., Wadile, S. C., Girase, P. P., Chitodkar, S. S., & Sonawane, D. A. (2010). Response of summer pearl millet (*Pennisetum glaucum* L.) to depth and time of irrigation scheduling. *International Journal of Agricultural Sciences, 6*(1), 283–285.

Stapf, O. (1934). *Flora of Tropical Africa Dyer Graminac* (Vol. 9, pp. 954–1090).

Subramanian, S., Loganathan, S., Ravi Kumar, V., & Krishnamoorthy, K. K. (1973). Effect of tillage and organic amendments on the physical properties of soil and yield of bajra. *Madras Agricultural Journal, 62*, 100–109.

Suzuki, K., Matsunaga, R., Hayashi, K., Matsumoto, N., Tobita, S., & Bationo, A. (2017). Effects of long-term application of mineral and organic fertilizers on dynamics of nitrogen pools in the sandy soil of the Sahel region, Niger. *Agriculture, Ecosystems & Environment, 24*(2), 76–88.

Thumar, C. M., Dudhat, M. S., Chaudhari, N. N., Hadiya, N. J., & Ahir, N. B. (2016). Growth, yield attributes, yield and economics of summer pearl millet (*Pennisetum glaucum* L.) as influenced by integrated nutrient management. *International Journal of Agricultural Sciences, 59*(8), 3344–3346.

Umrani, M. K., Bhoi, P. G., & Patil, N. B. (1980). Effect of weed competition on growth and yield of pearl millet. *Journal of Maharashtra Agricultural University, 5*, 56–57.

Yaduraju, N. T., Sharma, A. R., & Das, T. K. (2016). *Weed Science and Management*. Indian Society of Weed Science, Jabalpur and Indian Society of Agronomy, New Delhi.

CHAPTER 11

Improved Cultivation Practices of Finger Millet

U. TRIVENI,* N. ANURADHA, T. S. S. K. PATRO, and Y. SANDHYA RANI

Agricultural Research Station, Acharya N. G. Ranga Agricultural University, Vizianagaram, Andhra Pradesh, India

*Corresponding author

11.1 INTRODUCTION

Finger millet (*Eleusine coracana* (L) Gaertn) occupies a prime position among the small millets in India. It stands unique among the cereals with its high nutritional content. It is rich in calcium, dietary fiber, minerals, proteins, essential amino acids, phenolics, and phytates. The abundance of these phytochemicals makes finger millet a "powerhouse of health-benefiting nutrients." Finger millet cultivation is mainly concentrated in tropical, sub-tropical, and slightly temperate regions. Finger millet is the primary energy source next to cereals, particularly in the semi-arid tropics and drought-prone regions of Asia and Africa. Due to its nutritional significance and affordability to the weaker sections of the population, it is now viewed as a latent solution for malnutrition worldwide. Due to its climate resilience, wider adaptability to different soils, and low input requirements, finger millet is being considered a "climate-smart crop" or "future crop for changing climatic conditions and global warming."

11.2 AREA AND DISTRIBUTION

Global production of millet is about 97.75 million tons from 78.43 million hectares, with a productivity of 1442 kg/ha (FAOSTAT database, http://www.fao.org/faostat/en/). India is the leading producer of millet and contributed to 15% of the global millet production. Current millet production in India is around 15.53 million tons from 12.45 million hectares. Bajra, jowar, and finger millet together constituted 98% of the total millet production. India, Ethiopia, Nepal, Uganda, and Malawi are the top five finger millet-producing countries in the world. In India, finger millet is being grown in a 1.01 million hectare area with 1.67 million tons of production and 1.75 t/ha productivity (Final Estimates-2021–22, Dept. of Economics and Statistics, DAC&FW, GoI, New Delhi). Karnataka, Tamil Nadu, Andhra Pradesh, Maharashtra, Odisha, Uttarakhand, and Gujarat are major finger millet-growing states in India.

Millets in Sustainable Agriculture: Genetic Improvement, Nutritional Quality, Cultivation, and Integrated Pest–Disease Management. Roshni Vijayan, A. John Joel, R. Ambika Rajendran, and Adithya P. Balakrishnan (Eds.)

DOI: 10.1201/9781779642332-14

Realizing the nutritional significance, finger millet is now a highly demanded crop, and hence there is a need to enhance its productivity. This calls for the adoption of the following improved agronomic practices for sustaining and improving finger millet productivity.

11.3 LAND PREPARATION AND FARM MACHINERY PRACTICES

Finger millet requires a fine tilth due to its smaller seed size. Land preparation may vary depending on the method of establishment, water availability, and prevailing weather conditions. Initially, deep ploughing with a disc or moldboard plough must be done during the summer months to expose weed seeds and hidden insect pests to higher temperatures. With the onset of the monsoon, 2–3 ploughings with a cultivator or country plough followed by leveling with disc harrows will be done to achieve a fine tilth. Proper leveling of the field is necessary to drain out excess water.

In lowlands with plenty of water available, finger millet will initially be grown in nursery beds, followed by transplanting in the main field. Initially, 2–3 baskets of well-decomposed farmyard manure (FYM) along with 1.0 kg super phosphate, 0.5 kg muriate of potash, 0.5 kg ammonium phosphate, and 750 g zinc sulfate are to be applied in the nursery bed as a basal dose. Broadcast the seeds uniformly on the bed and cover the seeds with well-decomposed FYM or soil. Watering of the bed is to be done regularly until the seeds germinate. Finger millet seeds require 4–5 days for germination. Top dress with urea 500 g per bed when the seedlings are 12–14 days old. Seedlings of 21–25 days old are ideal for transplanting.

Mechanized tractor-drawn or animal-drawn multi-row seed drills are available nowadays to sow the finger millet seed in lines. Ferti-cum-seed drills are more advantageous because they reduce the cost of cultivation and improve fertilizer use efficiency. Line sowing and/or planting allows other mechanized intercultural operations, which in turn reduces the dependence on human labor and improves crop productivity and profitability.

Weeds are a major problem, especially in direct-seeded finger millet, contributing to yield losses of up to 30–70%. Thorough land preparation is required to suppress weed growth. Mechanized cycle weeders or wheel hoes can be used as weeding tools, which not only remove the weeds but also incorporate them into the soil. A study conducted on mechanized weeding with cycle weeders reported savings of 87% of labor, 87% of the total cost of cultivation, and 25% of the time compared to manual weeding (Ali et al., 2017).

11.4 SOWING WINDOW (EARLY, MEDIUM AND LATE SOWING)

Successful crop production depends on the availability of primary growth factors like rainfall, temperature, light, nutrition, and other factors in a given location. For rainfed crops, the length of the crop growing period will be decided based on the total rainfall, normal rainfall distribution pattern, temperature, potential evapotranspiration, etc. The onset of the monsoon is most important, as it determines the sowing window of crops (Raes et al., 2004). An optimum sowing window is particularly important to achieve good crop harvests, as it ensures optimum soil moisture availability throughout the crop growth stages. Hence, the identification of a safe growing period in a region based on the moisture availability index (MAI) at different probability levels would be an important aspect of smart agricultural planning under rainfed conditions (Sattar et al., 2018). The selection of the optimum sowing window not only enhances crop growth and yield but also reduces pest and disease occurrence in the crop. In addition to climatic and soil factors, the sowing window also varies with the location, crop, variety, duration of the crop, pest and disease occurrence, cropping pattern of the region, etc.

More than 90% of the finger millet crop is being grown as rainfed on marginal and sub-marginal soils. With the provision of irrigation facilities, it can be grown throughout the year. In order to determine the optimum sowing window, many experiments have been conducted across the world. In India, the rainfed early season crop is sown in May or early June, while the kharif season main crop is sown from June to July. If rains are not received on time, sowing time may be delayed up to the end of August. Finger millet is sown during rabi and summer in Karnataka, Tamil Nadu, and Andhra Pradesh from September to October and January, respectively. The reduction in yield-contributing characters of finger millet was noticed with the delayed planting (Pandiselvi et al., 2010).

Further, sowing windows also influence pest and disease outbreaks in the crops. Several studies have proven that blast-causing fungi prefer low temperatures (<20°C) with high humidity, heavy rainfall, and low light for outbreaks (Bhatt and Chauhan, 1985). June 1st sowing, coupled with average minimum and maximum temperatures of 21°C and 29°C, respectively, along with RH of 70–81%, has registered the lowest incidence of leaf, neck, and finger blast and the highest grain yield compared to later dates of sowing extending up to August 20th in the Jagdalpur region of Chhattisgarh (Netam et al., 2014). Senthil et al. also reported the lowest incidence of leaf, neck, and finger blast with July 1st sowing compared to later sowing dates (July 16th, August 1st, August 15th, September 1st, and September 16th) in Coimbatore, Tamil Nadu (2012).

11.5 CHEMICAL AND MICROBIAL SEED PRIMING

Seed priming is a pre-sowing seed treatment that allows controlled hydration of seeds to imbibe water and go through the first stage of germination but does not allow radicle protrusion through the seed coat (McDonald, 2000). Rapid and uniform seed germination is critical for proper crop establishment, efficient use of production resources, and in turn, for attaining higher productivity. Seed priming is a low-cost technology that helps to improve the speed of germination, seedling vigor, and good stand establishment, promotes uniform crop maturity, enhances biotic and abiotic stress tolerance, and increases growth and yield attributes, which ultimately results in increased crop yields. Various kinds of seed priming techniques have been developed, and those will be broadly categorized into chemical and microbial seed priming methods.

11.5.1 CHEMICAL SEED PRIMING

Chemical seed priming involves seed treatment with different chemical solutions, viz., hydrogen peroxide, sodium nitroprusside, selenium, chitosan, fungicides, macro and micronutrients, antioxidants, hormones, etc., with the objective of establishing a healthy crop stand to increase stress tolerance, yield, nutritional qualities, and other characters that fetch a premium price in the market. Chemical priming is further categorized into osmo priming, nutrient seed priming, and hormonal priming.

11.5.1.1 OSMOPRIMING

Osmopriming or halopriming involves the soaking of seeds in an osmotic solution with low water potential instead of pure water. Due to the low water potential of osmotic solutions, water enters the seed slowly, which allows gradual seed imbibition and activation of early phases of germination but prevents radicle protrusion (Di Girolamo and Barbanti, 2012). Usually, the water potential of the priming agent varies from −1.0 down to −2.0 MPa (Kubala et al., 2015). Different compounds used in the osmopriming procedure include polyethylene glycol (PEG), mannitol, sorbitol, glycerol, and inorganic salts such as NaCl, KCl, KNO_3, K_3PO_4, KH_2PO_4, $MgSO_4$, and $CaCl_2$ (Yacoubi et al., 2013). Osmopriming allows the crops to grow under soil salinity conditions. Significant improvements in yield-attributing characters, seed yield, and seed quality parameters were obtained when finger millet seeds were primed with KH_2PO_4 (Patil et al., 2018; Thakur et al., 2016).

11.5.1.2 NUTRIENT SEED PRIMING

'Nutrient seed priming' is a technique in which seeds are soaked in macro or micronutrient solutions and subsequently dried back to their initial moisture content for storage. It is an efficient approach to increase seed nutrient contents (biofortification) along with general priming effects. However, the concentrations of these nutrients need to be optimized to avoid nutrient toxicity and seed losses that can impair germination (Rehman et al., 2013).

Seed priming with nano ZnO in finger millet produced grain yields that were found to be on par with seed priming + foliar application of nano ZnO both under normal and stress conditions, which signifies the importance of seed priming (Shankar, 2017). Cu-chitosan nanoparticles applied through seed priming and foliar spray to finger millet showed induction of defense enzymes such as chitinase, chitosanase, β-1,3 glucanase, peroxidase, polyphenol oxidase, and protease, which enhanced resistance to the blast fungus *Pyricularia grisea* (Sathiyabama and Manikandan, 2018).

11.5.1.3 HORMONAL PRIMING

It involves seed imbibition in the presence of plant growth regulators like abscisic acid (ABA), ascorbate, auxins, gibberellins, kinetin, ethylene, polyamines, and salicylic acid (SA). Hormonal seed priming techniques can be used to improve seed germination of dormant seeds or under abiotic stress conditions.

11.5.2 MICROBIAL SEED PRIMING

Microbial seed priming is the seed hydration in the presence of living bacterial inoculum in the medium. Plant growth-promoting rhizobacteria and biocontrol agents can be added to priming media to improve seed germination and vigor, uniform crop stand, plant growth, and yield attributes, including biotic and abiotic stress tolerance. Commonly used plant growth-promoting rhizobacteria include *Azotobacter, Pseudomonas, Azospirillum, Bacillus*, and *Agrobacterium* to improve growth and yield, to reduce seed-borne and soil-borne diseases, and to enhance abiotic stress tolerance, whereas *Trichoderma* and *Pseudomonas* improve biotic stress resistance (Mahmood et al., 2016).

In organic agriculture, microbial seed priming is considered an efficient method to save the crop from seed and soil-borne pathogens and to increase crop growth and yield. Seed priming with *Pseudomonas* isolates JUPW 121 and JUPC 113 significantly suppresses the blast incidence caused by *Magnaporthe grisea* through increased production of phenolic acid content in finger millet. The primed finger millet plants also showed a significant increase in root and shoot lengths as compared to the control (Patil and Kumudini, 2019). Finger millet primed with *Pseudomonas aeruginosa* isolates produced more cross-linking proteins and callose deposition, which play a significant role in preventing blast pathogen entry into plants. Further, priming with *Pseudomonas aeruginosa* isolates enhances the level of plant hormones like salicylic acid (SA) and jasmonic acid, which are known to play a major role in the plant immune system (Patil et al., 2020). Seed bio-priming with *Trichoderma asperellum* Th-14 + FYM pre-colonized by Th-14 was found significantly superior in reducing foot rot disease caused by *Sclerotium rolfsii* and also exhibited maximum growth, yield, and yield-contributing characters in finger millet (Rawat et al., 2019).

11.6 SPACING EFFECT ON YIELD

Finger millet is mainly grown by resource-poor small and marginal farmers as a rainfed crop under poor crop management practices, which lead to low production and productivity. However, due to its nutritional superiority, finger millet is being recognized as a low-priced alternative functional food to alleviate malnutrition in rural households. Certain changes in crop management, without incurring additional monetary inputs, result in a tremendous increase in economic yield. Crop geometry is one such important agronomic management practice to achieve higher production by better utilization of moisture and nutrients from the soil (root spread) and above ground (plant canopy) by harvesting maximum possible solar radiation and, in turn, better photosynthate formation (Uphoff et al., 2011).

Broadcasting of dry seed and random transplanting with 25–35-day-old seedlings are the most common practices adopted to establish the finger millet crop. These practices have resulted in the uneven distribution of plants, leading to a reduction in growth and yield. The optimum plant population of finger millet per hectare ranges between 3.5 to 4.5 lakhs/ha. In order to maintain optimum plant population, row-to-row spacing of 22.5 cm to 30 cm and plant-to-plant spacing of 7.5 cm to 10 cm should be followed. Line sowing with tractor-drawn or animal-drawn seed drills with a seed rate of 10 kg/ha is a good practice that can allow mechanization. Sowing with ferti-cum seed drill is more advantageous as it reduces the cost of fertilizer application as well as improves fertilizer use efficiency. Seedling age is one of the important aspects that can decide the number of tillers per plant. For short to medium-duration finger millet varieties, 18–25 days old seedlings, and for long-duration varieties, 30 days old seedlings were found to be optimum for transplanting.

Higher yield attributes, grain yield, straw yield, harvest index (HI), net return, and benefit-to-cost (B:C) ratio were recorded when finger millet was planted at a spacing of 30 x 10 cm with two hand weeding's at 15 and 30 days after transplanting (DAT) (Kumar et al., 2019). Dry matter production of finger millet was higher at a wider spacing of 25 cm × 10 cm over the closer spacing of 25 cm x 6 cm (Roy et al., 2002). A study conducted with three spacings (30 × 10 cm, 25 × 25 cm, and 30 × 30 cm) in combination with three graded levels of NPK (100%, 125%, and 150% RDF, respectively) revealed that adaptation of 25 × 25 cm + 150% RDF showed significantly higher growth characters and yield attributes of finger millet (Aravinth et al., 2019). An experiment conducted in northwestern Ethiopia with four seed rates (10, 15, 20, and 25 kg/ha) and four-row spacings (20, 30, 40, and 50 cm) revealed that the highest grain yield was recorded with the lowest seed rate (10 kg/ha) and 30 cm row spacing (Bitew and Asargew, 2014).

11.6.1 SYSTEM OF FINGER MILLET INTENSIFICATION (SFMI)

The system of finger millet intensification (SFMI) method is nothing but the application of concepts and methods of the System of Rice Intensification to finger millet with the aim of increasing productivity. It comprises the transplanting of 12–15 day-old seedlings with a mass of soil attached to the roots at a spacing of 25 cm x 25 cm (20 cm x 20 cm in less fertile soil) in a square pattern. The first weeding will be done 20 days after transplanting with a roller weeder, and the second weeding is done manually by hoeing. Dry matter accumulation, productive tillers per plant, and grain yield of finger millet were higher with 30 cm x 30 cm spacing compared to 22.5 cm × 10 cm spacing on sandy loam soils (Kalaraju et al., 2011). In an experiment conducted in the mid to high hills of Nepal on the effect of establishment methods and spacings, finger millet planted at 25 cm x 25 cm spacing with 15-day-old seedlings produced superior growth traits and showed 82.43% and 24.66% higher yield than direct sowing (20 cm x 10 cm) and conventional transplanting (10 cm x 10 cm) with 30-day-old seedlings, respectively (Bhatta et al., 2017). Transplanting of 13-day-old seedlings at 22.5 cm x 22.5 cm spacing with a single seedling per hill recorded significantly higher values of growth and yield-contributing characters, grain yield, and straw yield compared to transplanting 2–3 seedlings per hill on a flatbed at 30 cm x 10 cm (Kumar, 2018).

"Guli Vidhana" is another method evolved by farmers in the Haveri district of Karnataka during the 1980s. This method also has more similarities with the System of Rice Intensification (SRI) by having a square planting of 25 cm to 60 cm based on soil fertility. Two special animal-drawn inter-cultivation implements, viz. Yade kunte (weeder) and Koradu (hollow wooden log), will be used in this method. Yade kunte (weeder) is used for removing weeds as well as aerating the soil. Koradu (hollow wooden log) is to be pulled across the field several times between 15 to 45 days after transplanting, to bend over the young plants without breaking them off in order to enhance tiller production (SRI International Network and Resource Centre).

11.7 FERTILIZER, MICRO-NUTRIENTS AND BIO-FERTILIZER APPLICATION

Finger millet is one of the highly nutritious crops; however, most finger millet cultivation is confined to marginal and poor soils under rainfed conditions with lesser inputs. The inherent resource-poor conditions, coupled with poor nutrient management practices, result in lower yield and grain nutritional qualities. Finger millet also has great fodder value in dryland areas. Stover of finger millet is considered good feed in India and Nepal. Finger millet stover yield is about 2–3 t/ha under rainfed conditions and may reach up to 6–10 t/ha under irrigated conditions (Bhat et al., 2023). Continuous harvest of grain and fodder year after year without adding nutrients makes the soil unviable and can no longer support sustainable crop production. Hence, optimum nutrient management practices must be adopted to maintain the sustainable finger millet yield potential and grain nutritional values. Nutrient requirements of the finger millet crop may vary with the soil type, and hence, site-specific nutrient management will be adopted. India, being the largest producer of finger millet, has specific spacing and fertilizer recommendations for each state. *See* Table 11.1.

TABLE 11.1 Recommended Spacing and NPK Fertilizer Doses for Finger Millet in Different States

State	Spacing (cm)	Fertilizer (N: P: K kg ha-1)	
		Rainfed	Irrigated
Andhra Pradesh	22.5x10.0 cm	40:20:20	60:30:30
Bihar	22.5x10.0 cm	40:20:20	40:20:20
Jharkhand	22.5x10.0 cm	40:20:20	40:20:20
Gujrat	30.0x7.5 cm	40:20:10	-
Himachal Pradesh	22.5x10.0 cm	40:20:0	-
Karnataka	22.5–30 x 10.0 cm	50:40:25	100:50:50
Maharashtra	22.5x10.0 cm	25:20:20	50:25:00
Chhattisgarh	25x10.0 cm	60:30:20	-
Madhya Pradesh	25x10.0 cm	40:40:00	-
Orissa	22.5x10.0 cm	40:40:20	60:20:20
Tamil Nadu	22.5x15.0 cm	40:40:20	90:45:45
Uttaranchal	22.5x15.0 cm	60:30:20	-

Source: Reprinted from ICAR-Indian Institute of Millets Research (IIMR), Hyderabad. https://millets.res.in/technologies/1 Recommended_Package_of_Practices-Finger_Millet.pdf

Recently developed high-yielding finger millet varieties respond very well to fertilizer applications. In order to improve soil health and nutrient use efficiency, 7.5–10 t/ha of FYM is to be applied one month before sowing. Entire phosphorus and potassium fertilizers are to be applied in the last ploughing at the time of sowing. Nitrogen is to be applied in two equal split doses, one at the time of sowing and the other at 30–35 days after sowing. Apart from macronutrients, adequate quantities of secondary and micronutrient elements are required for optimum growth and development. Continuous withdrawal of micronutrients through crop harvests without sufficient replenishment may develop micronutrient deficiencies in soil and harvested produce (grain and fodder). Hence, soil application of micronutrients once in three crop seasons is one of the important ways to correct the micronutrient deficiencies. Seed coating, seed priming, seedling root dip, and foliar spray are the other methods of micronutrient application. Foliar application is the most effective and fastest way of correcting micronutrient deficiencies in finger millet crops. In the field experiment conducted in Andhra Pradesh, 150% RDF + $ZnSO_4$ 0.5% foliar spray + $FeSO_4$ 0.2% foliar spray resulted in a 38.7% and 5.0% increase in grain yield of finger millet over 100% RDF and 150% RDF alone (Rani et al., 2017). Seed treatment of finger millet with ZnO @ 10 mL/kg seed followed by root dipping of seedlings in 0.5% $ZnSO_4$ resulted in a 32.0% and 30.1% increase in grain yield and straw yield, respectively, compared with no Zn application (Dholariya et al., 2020). Significantly higher grain yield (3,498 kg ha^{-1}) and straw yield (4,640 kg ha^{-1}) of finger millet were recorded with foliar application of boron @ 0.5 kg ha^{-1} compared with the control (3,083 kg ha^{-1} and 4,118 kg ha^{-1}, respectively) (Shankar et al., 2017). Imbalanced fertilizer use without knowing the soil fertility status or crop nutrient requirement is leading to adverse effects on soil and crop productivity. Fertilizer prescriptions based on Soil Test Crop Response's targeted yield approach proved to be good for enhancing soil health and sustaining crop productivity. The targeted yield approach considers the nutrient demand of crops for a targeted yield goal and relative contributions from soil and fertilizer sources under a given set of farming situations. Biofertilizers are one of the important components in integrated nutrient management. Biofertilizers mobilize or solubilize plant nutrients under favorable conditions. Hence, the integration of biofertilizers with chemical fertilizers saves a certain quantity of chemical fertilizers and, in turn, reduces environmental pollution. Treating finger millet seeds with *Azospirillum brasilense* (N-fixing bacterium) and *Aspergillus awamori* (P-solubilizing fungus) @ 25 g kg^{-1} seed was found to be beneficial compared with treatment without biofertilizers (Prabhakar et al., 2017). Application of 100% RDP (40 kg ha^{-1}) + PSB @ 750 mL ha^{-1} + VAM @ 12.5 kg ha^{-1} produced higher yield-attributing characteristics, grain yield, stover yield, and grain quality parameters (protein and carbohydrate content) in finger millet compared with the control (Kejia et al., 2019). Nanotechnology is a promising area yet to be exploited in agriculture. Nano-fertilizer usage is one of the frontier technologies that are in a nascent stage in India. However, nano fertilizers are more advantageous over conventional fertilizers; they are highly reactive, have more surface area and sorption capacity, can penetrate through the epidermis, control release and target distribution, and thus improve nutrient use efficiency. The use efficiency of conventional fertilizers is very low. Studies have proven that the use efficiencies of applied nitrogen are <50%, phosphorus <10%, and potassium 40–50% under tropical conditions. In the case of micronutrient fertilizers, use efficiency is less than 2%. Application of conventional fertilizers in bulk quantities to the soil would lead to environmental pollution and loss of biodiversity. Unlike conventional fertilizers, nano fertilizers will be applied in very small quantities, deliver the nutrients at the right time and in the right place, and hence will not contribute to environmental pollution. Nano fertilizers can be applied through various methods like seed priming, seed coating, seedling root dip, soil application, and foliar spraying. Seed priming with nano-scale nutrients is an efficient process that can change seed metabolism and signaling pathways, affecting not only germination and seedling establishment but also the entire plant life cycle.

Foliar application of the nano fertilizers is the most efficient method, which not only increases crop growth and yield but also improves crop quality.

11.8 IRRIGATION, DRIP IRRIGATION, AND MULCHING

11.8.1 MICROIRRIGATION

Finger millet is being grown under rainfed and irrigated conditions. Finger millet sown during the kharif season does not require irrigation. However, if dry spells occur for a prolonged period during sensitive crop growing stages like tillering and flowering, life-saving irrigations need to be provided. In rainfed areas, proper soil moisture conservation measures like deep summer ploughing, ploughing across the slope, formation of dead furrows at 4.0 m intervals, and formation of ridges and furrows are to be followed for sustainable crop production. Finger millet does not tolerate waterlogging conditions; hence, draining out excess water is essential. Under irrigated conditions, finger millet can be grown throughout the year. In light soils, the crop must be irrigated at 6–8 day intervals, and in heavy soils, the crop needs irrigation at 12–15 day intervals.

Flooding is the most common method of irrigation, having a water use efficiency of only 35–40%. Efficient utilization of water resources becomes crucial due to dwindling water resources. Microirrigation, particularly drip irrigation, has emerged as an appropriate water-saving technique in water-scarce regions, undulated, sandy, and hilly regions, where finger millet is being grown as a major crop. With drip irrigation and fertigation, crop productivity exhibits a threefold increase with a water use efficiency of 80–90%. A study on drip irrigation in millets by Kalpana and Fanish in the state of Tamil Nadu revealed that through the adoption of drip irrigation in finger millet, an additional 20% of the area could be brought under millet cultivation with the saved water (2014). They also stated that 44.0% of additional production of finger millet was through the adoption of drip fertigation. Microirrigation combined with fertigation is one of the best management options to increase the productivity of irrigated crops as it matches crop nutrient demand with nutrient supply and increases nutrient and water use efficiencies. However, in order to reduce the installation and maintenance costs of the drip irrigation system, finger millet can also be grown in sequence with micro-irrigation-supported high-value crops like vegetables, cotton, groundnut, etc. In view of increased awareness about the nutritional significance of millets and increased demand for millet-based foods and value-added products in present times, it may be favorable for the adoption of micro-irrigation in millets.

11.8.2 MULCHING

Soil moisture conservation measures are most important as finger millet is being grown as a rainfed crop. Mulching is one of the main soil moisture conservation measures adopted in drylands. Mulching restricts evaporation and loss of moisture from the soil, reduces weed growth, and thereby increases soil moisture availability and crop yields. Mulching also regulates and moderates soil temperature, improves root growth, and enhances soil organic carbon. Sugarcane trash mulching at 10 cm thickness in finger millet not only enhanced the consumptive use of water and water-use efficiency but also produced higher grain yields (Nagarajan and Wahab, 2001). Application of horse gram residue mulch enhanced grain and straw yields of finger millet through increased soil infiltration rate, water holding capacity, and soil microbial as well as enzymatic activities (Hatti et al., 2018).

11.9 MECHANIZATION IN HARVESTING AND POST-HARVEST MANAGEMENT

Harvesting at physiological maturity will improve the quality of the harvested produce. Manual harvesting with sickles is the most common practice, which is highly time-consuming and involves greater drudgery. Traditional harvesting with sickles requires nearly 25 man-days to cover one hectare of area. However, due to a shortage of human labor, timely harvest becomes difficult and ultimately affects the productivity and quality of finger millet grain.

In order to save labor, time, and drudgery, mechanized harvesting equipment, such as combine millet harvesters, millet reapers, portable millet harvesting machines, etc., are available nowadays in the market. Combine millet harvesters are suitable for harvesting a larger area of finger millet within a short time. With small modifications in RPM and size of the mesh, paddy combine harvesters can be used for harvesting finger millet. Combine harvesters cover one hectare of area within a short time. The straw cut by the harvesters will be used to feed the cattle. Considering the drawbacks, combine harvesters are not suitable for smaller holdings and hill slopes. As they leave larger stubbles in the fields, lesser straw yields will be produced. Soil compaction may also occur due to their heavy machinery.

In India, finger millet is cultivated in smaller isolated patches by resource-poor small and marginal farmers. In these situations, mechanical reapers are found to be the best implements for harvesting as they reduce drudgery, dependence on labor, and save time. Unlike combine harvesters, reapers cut the whole plant at the bottom. The whole plants, along with earheads, will be heaped for 2–3 days followed by drying and threshing for separation of grains. The usage of mechanized reapers reduces the time taken to harvest 1 hectare of finger millet area by 72%, the total labor requirement by 92%, and the total cost of cultivation by 81% respectively, as compared to the conventional method of manual harvesting (Ali et al., 2017). In hilly areas, finger millet is being cultivated on slopes and contour benches in very small patches in subsistence farming. Combine harvesters or mechanized reapers are not suitable under these conditions. Particularly for hilly and undulated topographies, specially fabricated portable millet harvesters need to be developed. Ghimire et al. (2019) of Nepal designed and fabricated such a machine, which cuts the gearheads and straw separately at a time and was found to be most effective as compared to manual harvesting. It is a semi-automated machine operated by using both manual and engine power. It will cover one hectare of area in 6.9 man-days. Lodging of the crop due to heavy rain or wind is another problem that causes heavy loss in grain recovery and minimizes the scope for machine harvesting. Dwarf varieties of finger millet with lodging resistance at the time of maturity must be developed to encourage mechanical harvesting. In the event of continuous rainfall during maturity, non-dormant finger millet seeds germinate on the earheads themselves. This condition calls for evolving varieties with seed dormancy.

Threshing is the first step in post-harvest processing. Beating with sticks and trampling with bullock-drawn rollers are the common practices being adopted for the threshing of finger millet. However, in recent times, tractors and mechanical threshers have also been used by some sections of farmers. In hilly areas, where farmers cultivate millets in smaller isolated patches, small threshers are highly suitable. ICAR-Vivekananda Parvatiya Krishi Anusandhan Sansthan, Almora, Uttarakhand developed one millet thresher for threshing smaller quantities of earheads. Similarly, Odisha University of Agriculture and Technology (OUAT) also developed a finger millet thresher cum pearler, which can be used for threshing, pearling, and cleaning. The output of this thresher was found to be 90 kg/hr with 93.5% threshing efficiency and 92.4% cleaning efficiency. Grading with a destoner cum grader is an important step after winnowing to remove the leftover stones and other impurities from the grain. This instrument separates the grains based on their weight. Depending on its usage, the grain can be packed and marketed. Pulverizer machines are available for preparing flour from finger millet grain.

11.10 CONCLUSION

Being a "powerhouse of health-benefiting nutrients," finger millet is regarded as an efficient food source to reduce malnutrition among the poor in Africa and Asia. However, more than 90% of the finger millet crop is grown under rainfed conditions, and hence its production and productivity are subjected to the prevailing weather conditions. Moreover, the majority of the cropped area is poorly managed due to the low economic status of the farmers in the finger millet-growing areas. In order to increase finger millet production to meet the growing demand across the world, improved crop management practices need to be adopted. Choosing the right variety, using quality seed, pre-sowing seed treatment with fungicides and insecticides, and seed priming with biofertilizers would facilitate uniform germination and healthy seedling growth. Line sowing with mechanized seed drills, timely weed control, and proper irrigation management must be practiced. *In-situ* soil and moisture conservation measures like deep ploughing, green manuring, application of tank silt, contour cultivation, the opening of conservation furrows, and intercropping need to be practiced. In order to encourage sustainable finger millet production in dryland areas, integrated nutrient management practices involving the application of FYM, green manures, biofertilizers, crop rotation with legumes, etc., should be included. Mechanization in sowing, harvesting, threshing, cleaning, and post-harvest processing must be adopted to ensure timely operations and reduce the cost of cultivation and the drudgery involved in millet cultivation.

Currently, at the global level, the area under finger millet is in a declining trend, and hence focus must be on bringing more area under finger millet cultivation by introducing it in fallow lands, wastelands, and non-traditional areas by diversifying the existing monocropping systems. Adequate availability of quality seed is important to increase the finger millet area and production. The establishment of millet seed hubs and linking them with national and state seed corporations need to be done for large-scale multiplication and supply of quality seed. Integrated nutrient management (INM) practices in millet cultivation should be given priority for the maintenance of soil health. The usage of nano fertilizers is one of the novel technologies adopted in recent times to increase fertilizer use efficiency and to reduce the drudgery, cost of cultivation, and environmental pollution. Adequate research needs to be carried out to determine the possibility of integrating nano fertilizers into INM practices. There is much demand for organically grown food crops. By default, millets are organically grown with available organic manures in tribal areas. However, adequate research must be done on the efficient use of organic manures, biofertilizers, green manures, recycling of crop residues, etc., to acquire a premium price for organically grown produce. More emphasis should be placed on developing conservation agriculture-based strategies aiming for moisture conservation, maintenance of soil quality, and carbon sequestration. Seed-to-seed mechanization, starting from sowing, inter-cultivation, harvesting, threshing, and post-harvest processing, particularly suitable for small and marginal farm holdings, should be developed to produce quality produce, overcome the shortage of human labor, reduce the drudgery, and increase returns. More attention needs to be given to the transfer of technology in finger millet-growing areas by involving research organizations, Department of Agriculture functionaries, other extension agencies, and non-governmental organizations. A greater number of Farmer Producer Organizations (FPOs) must be encouraged to take up production, processing, and value addition of finger millet on a cluster basis in order to get a remunerative price and also to increase the bargaining power of farmers.

KEYWORDS

- **biofortification**
- **contour cultivation**
- **finger millet**
- **foliar spraying**
- **nano-fertilizer**
- **nutrient management**
- **organic agriculture**
- **pulverizer machines**
- **seed priming**

REFERENCES

Ali, M. A., & Nagraj, K. H. (2017). Role of mechanization in effective management of time and labor in ragi (*Eleusine coracana* L.) cultivation. *Agricultural Engineering Today, 41*(2), 16–20.

Aravinth, K. V., Kumar, N. S., Joseph, M., & Paramasivan, M. (2019). Effect of plant geometry and graded NPK levels on growth and yield of transplanted finger millet (*Eleusine coracana* L.). *International Journal of Agriculture Sciences, 11*(10), 8491–8493.

Bhat, B. V., Singode, A., & Balakrishna, D. (2023). Millet crops as source of fodder. *Indian Farming, 73*(01), 19–21.

Bhatt, J. C., & Chauhan, V. S. (1985). Epidemiological studies on neck blast of rice in U. P. hills. *Indian Phytopathology, 38*, 126–130.

Bhatta, L. R., Subedi, R., Joshi, P., & Gurung, S. B. (2017). Effect of crop establishment methods and varieties on tillering habit, growth rate and yield of finger millet. *Agricultural Research & Technology, 11*(5), 1–3.

Bitew, Y., & Asargew, F. (2014). Determination of seed rate and inter row spacing for finger millet production (*Eleusine coracana* Gaertn.) in North Western Ethiopia. *International Journal of Research & Review, 1*(4), 1–8.

Dholariya, H. P., Zinzala, V. J., Patel, J. V., & Patel, V. M. (2020). Zinc nutrition in finger millet [*Eleusine coracana* (L.) Gaertn.] for better nutritional security. *International Journal of Current Microbiology and Applied Science. Special Issue, 11*, 1082–1086.

Ganiger, P. C., Boraiah, B., Bhat, S., Nandini, C., Tippeswamy, V., & Manjunath, H. A. (2017). Improved production technologies for finger millet. Project Coordinating Unit ICAR-AICRP on Small Millets GKVK, 10–12.

Ghimire, S. K., Awal, R., Paneru, P., Bokati, P., & Wasti, S. (2019). Design, fabrication and testing of finger-millet harvesting machine. *Journal of the Institute of Engineering, 15*(1), 71–76. https://doi.org/10.3126/jie.v15i1.27707 (accessed on 7th November 2025).

Girolamo, D., & Barbanti, G. (2012). Treatment conditions and biochemical processes influencing seed priming effectiveness. *Italian Journal of Agronomy, 7*, 8–18.

Hatti, V., Ramachandrappa, B. K., Sathish, A., & Thimmegowda, M. N. (2018). Soil properties and productivity of rainfed finger millet under conservation tillage and nutrient management in Eastern Dry Zone of Karnataka. *Journal of Environmental Biology, 39*, 612–624.

Kalaraju, K., Kumar, N. D., Nagaraja, N., & Ningappa, K. B. (2011). Effect of methods of planting on growth and yield of finger millet genotypes under organic farming. *Research on Crops, 10*, 20–24.

Kalpana, R., & Fanish, A. (2014). Micro irrigation and fertigation to maize and millets–A review. *Agricultural Review, 35*(2), 103. https://doi.org/10.5958/0976-0741.2014.00087.7 (accessed on 7th November 2025).

Kejiya, P., Vajantha, B., Naidu, M. V. S., & Nagavani, A. V. (2019). Effect of phosphatic fertilizer and biofertilizers on yield and quality of finger millet (*Eleusine coracana* L.). *International Journal of Current Microbiology and Applied Sciences, 8*(07), 846–852. https://doi.org/10.20546/ijcmas.2019.807.101 (accessed on 7th November 2025).

Kubala, S., Garnczarska, M., Wojtyla, Ł., Clippe, A., Kosmala, A., Żmieńko, A., Lutts, S., & Quinet, M. (2015). Deciphering priming-induced improvement of rapeseed (*Brassica napus* L.) germination through an integrated transcriptomic and proteomic approach. *Plant Science, 231*, 94–113. https://doi.org/10.1016/j.plantsci.2014.11.008 (accessed on 7th November 2025).

Kumar, A. P., Parasuraman, P., Sivagamy, K., & Sivakumar, B. (2019). Growth, yield and economics of irrigated finger millet as influenced by system of finger millet intensification (SFI) practices in North Eastern Zone of Tamil Nadu. *Journal of Pharmacognosy and Phytochemistry, 8*(3), 660–663.

Kumar, C. (2018). *Performance of Finger Millet (Eleusine coracana L. Gaertn.) Under Different Crop Establishment Methods* (M.Sc. thesis, Dept. of Agronomy, Ranchi, Jharkhand).

Mahmood, A., Turgay, O. C., Farooq, M., & Hayat, R. (2016). Seed biopriming with plant growth promoting rhizobacteria: A review. *FEMS Microbiology Ecology, 92*(8), fiw112. https://doi.org/10.1093/femsec/fiw112 (accessed on 7th November 2025).

McDonald, M. B. (2000). Seed priming. In M. Black & J. D. Bewley (Eds.), *Seed Technology and Its Biological Basis*. Sheffield Academic Press.

N/A. (n.d.). Cornell. http://sri.ciifad.cornell (accessed on 7th November 2025).

N/A. (n.d.). Cornell.edu. http://sri.ciifad.cornell.edu/aboutsri/othercrops/fingermillet/index.html (accessed on 7th November 2025).

Nagarajan, R., & Wahab, K. (2001). Influence of irrigation and mulching practices on water use, yield components and yield of finger millet (*Eleusine coracana*). *Indian Journal of Agronomy, 46*(4), 702–706.

Netam, R. S., Tiwari, R. K. S., Bahadur, A. N., Singh, D. P., & Patel, D. P. (2014). Effect of sowing dates and meteorological factors on the development of blast disease in finger millet crop. *American International Journal of Research in Formal*, 14–101.

Pandiselvi, T., Narayanan, A. L., & Karthikeyan, R. (2010). Evaluation of optimum time of sowing of finger millet (*Eleusine coracana* G.) varieties in Karaikal region. *International Journal of Agricultural Sciences, 6*(1), 94–96.

Patil, P. P., Shinde, A. K., Gadhave, P. M., Chavan, A. P., & Mahadkar, U. V. (2018). Effect of sowing methods, nutrient management and seed priming on seed yield and yield attributes of finger millet (*Eleusine coracana* G.). *Advanced Agricultural Research, 2*, 12–17.

Patil, S. V., & Kumudini, B. S. (2019). Seed priming induced blast disease resistance in finger millet plants through phenylpropanoid metabolic pathway. *Physiological and Molecular Plant Pathology, 108*, 101428. https://doi.org/10.1016/j.pmpp.2019.101428 (accessed on 7th November 2025).

Patil, S. V., Kumudini, B. S., Pushpalatha, H. G., De Britto, S., Ito, S. I., Sudheer, S., Singh, D. P., Gupta, V. K., & Jogaiah, S. (2020). Synchronized regulation of disease resistance in primed finger millet plants against the blast disease. *Biotechnology Reports (Amsterdam), 27*, e00484. https://doi.org/10.1016/j.btre.2020.e00484 (accessed on 7th November 2025).

Raes, D., Sithole, A., Makarau, A., & Milford, J. (2004). Evaluation of first planting dates recommended by criteria currently used in Zimbabwe. *Agricultural and Forest Meteorology, 125*(3–4), 177–185. https://doi.org/10.1016/j.agrformet.2004.05.001 (accessed on 7th November 2025).

Rani, Y., Triveni, U., Patro, T. S. S. K., & Anuradha, N. (2017). Effect of nutrient management on yield and quality of finger millet (*Eleusine coracana* (L.) Gaertn). *International Journal of Chemical Studies, 5*(6), 1211–1216.

Rawat, L., Tiwari, A., Bisht, T. S., Nautiyal, A., & Naithani, D. C. (2019). Enhancement of growth, yield and yield contributing traits with a particular reference by using *Trichoderma* and *Pseudomonas* through seed bio-priming technique and value added FYM in finger millet (*Eleusine coracana* L.) under field conditions. *Journal of Pharmacognosy and Phytochemistry, 8*(4), 3333–3337.

Rehman, A., Farooq, M., Cheema, Z. A., & Wahid, A. (2013). Role of boron in leaf elongation and tillering dynamics in fine-grain aromatic rice. *Journal of Plant Nutrition, 36*, 42–54.

Roy, N. R., Chakraborty, T., Sounda, G., & Maitra, S. (2002). Growth and yield attributes of finger millet as influenced by plant population and different levels of nitrogen and phosphorus. *Indian Agriculturist, 46*(1), 65–71.

Sathiyabama, M., & Manikandan, A. (2018). Application of copper-chitosan nanoparticles stimulate growth and induce resistance in finger millet (*Eleusine coracana* Gaertn.) plants against blast disease. *Journal of Agricultural and Food Chemistry, 66*(8), 1784–1790. https://doi.org/10.1021/acs.jafc.7b05921 (accessed on 7th November 2025).

Sattar, A., Khan, S. A., Banerjee, S., & Nanda, M. K. (2019). Assessing sowing window and water availability of rainfed crops in eastern Indian state of Bihar for climate smart agricultural production. *Theoretical and Applied Climatology, 137*(3–4), 2321–2334. https://doi.org/10.1007/s00704-018-2741-9 (accessed on 7th November 2025).

Senthil, R., Manikandan, R., Ramanathan, A., & Raguchander, T. (2012). Influence of dates of sowing on blast disease incidence in finger millet. *Madras Agricultural Journal, 99*(1–3), 99–101.

Shankar, M. A., Thimmegowda, M. N., Bhavitha, N. C., & Manjunatha, B. N. (2017). Comparative efficiency of soil and foliar application of boron on growth and yield of finger millet (*Eleusine coracana*). *Mysore Journal of Agricultural Sciences, 51*(2), 430–435.

Thakur, A. K., Kumar, P., Salam, P., Patel, R. K., & Netam, C. R. (2016). Effect of different sowing methods, nutrient management and seed priming on growth, yield attributing characters, yield and economics of finger millet (*Eleusine coracana* L.) at Bastar Plateau. *Journal of Pure and Applied Microbiology, 10*(1), 407–415.

Viswanath, S., & Channamma, L. (1988). Survey and surveillance of rabi diseases and their management in Karnataka. *National Seminar on Pest Surveillance for Integrated Pest Management*.

Yacoubi, R., Job, C., Belghazi, M., Chaibi, W., & Job, D. (2013). Proteomic analysis of the enhancement of seed vigour in osmoprimed alfalfa seeds germinated under salinity stress. *Seed Science Research, 23*(2), 99–110. https://doi.org/10.1017/s0960258513000093 (accessed on 7th November 2025).

CHAPTER 12

Improved Cultivation Practices of Minor Millets

V. S. SUSHA,[1,*] ATUL JAYAPAL,[1] LISHA JOSE KAPPEN,[1] ANCY G. MARTIN,[1] and SHARIFULLAH SHARIFI[2]

[1]*Department of Agronomy, College of Agriculture, Kerala Agricultural University, Vellayani, Kerala, India*

[2]*Afghanistan National Agricultural Sciences and Technology University (ANASTU), Kandahar, Afghanistan*

*Corresponding author

12.1 FOXTAIL MILLET

Foxtail millet (*Setaria italica*), known by various names such as Italian millet, Chinese millet, or German millet, is an important minor millet with a short growing season, high photosynthetic efficacy, tolerance to salinity, disease and pest resistance, as well as nutritional and therapeutic properties (Abedin et al., 2022). It's widely grown in the arid and semi-arid regions of the world, like Asia, South Africa, South America, and North America. Foxtail millet is considered the most digestible and non-allergic grain, with 10–12% protein, 4.7–6.6% fat, 60.6–64.5% carbohydrates, and 2.29–2.7% lysine. It also contains various phytochemicals such as phenols, ferulic acid, chlorogenic acid, p-coumaric acid, flavonoids, carotenoids, tocopherol, and tocotrienol. Furthermore, foxtail millet has been reported to possess beneficial properties, including anti-inflammatory, antioxidant, anticancerous, anti-hypolipidemic, and anti-hyperglycemic effects. These properties make foxtail millet potentially valuable from a pharmacological perspective, contributing to its significance in the realm of human nutrition and health (Kalsi and Bhasin, 2023).

Known as one of the oldest crops cultivated by humans, it has also been used as bird feed in African regions. This particular crop has the capability of being harvested for green fodder or hay once it reaches a stage of approximately 70–75 days, depending on the specificity of the variety. It undergoes self-pollination and can produce seeds within a timeframe of 85–95 days, specifically in the case of grain seed production. It possesses a significant level of adaptability to abiotic stresses, particularly concerning drought conditions and poor soil quality. Foxtail millet exhibits a diverse range of seed colors, including white, yellow, orange, brown, red, black, and grey (Verma et al., 2020). Foxtail millet is regarded as a promising crop in the organic production system of the North Eastern Himalayan Region (NEHR) of India. Among various millets evaluated under the organic production

Millets in Sustainable Agriculture: Genetic Improvement, Nutritional Quality, Cultivation, and Integrated Pest–Disease Management.
Roshni Vijayan, A. John Joel, R. Ambika Rajendran, and Adithya P. Balakrishnan (Eds.)

DOI: 10.1201/9781779642332-15

system, Chla, Chlb, Chla/b ratios, and carotenoids were significantly higher for foxtail millet cv. SiA 3088 (Layek et al., 2023).

12.1.1 LAND PREPARATION AND FARM MACHINERY

The land should be tilled three to four times using a moldboard plough, followed by the application of well-decomposed compost and neem cake at the rate of 40 quintals/acre and 50 kg/acre, respectively (Somashekhar and Loganandhan, 2020). The first ploughing is done before the onset of the monsoon. With the onset of the monsoons, the field is usually harrowed or ploughed with a local plough twice in North India or with blade harrows in South India (Chapke et al., 2020).

12.1.2 SOWING WINDOW

Sowing time varies from region to region. In Tamil Nadu, kharif irrigated crop is cultivated from the very beginning of June to the end of July, the rabi crop in August-September, and the summer irrigated crop in January, whereas in the states of AP and Telangana, foxtail millet is sown in the first fortnight of July. July-August is the preferred season in Karnataka, whereas in Maharashtra, sowing is performed in the second and third week of July. In the plains of Uttar Pradesh and Bihar, it is the middle of June (Chapke et al., 2020). Srikanya et al. (2020) observed that the crop sown on 30th August performed significantly well regarding the growth parameters and yield when compared to the crop sown in September. This may be due to the reception of a prolonged photoperiod, which would have resulted in greater accumulation of assimilates in the early sown crop. Delay in the sowing of foxtail millet beyond July drastically reduced the yields and monetary returns at Nandyal (Sukanya et al., 2022). Mubeena et al. (2019) observed that sowing of foxtail millet in the second fortnight of June resulted in the highest uptake of nitrogen (112.12 kg ha^{-1}), phosphorus (10.82 kg ha^{-1}), and potassium (60.66 kg ha^{-1}). Also, the highest grain yield (2,368 kg ha^{-1}), stover yield (8,916 kg ha^{-1}), net returns (24,978 ha^{-1}), and benefit–cost ratio (2.46) were recorded with sowing in the month of June. Lokesh et al. (2023) studied the effect of different sowing windows and row spacings on the yield of foxtail millet during the rabi season.

12.1.3 SEED PRIMING

Foxtail millet seed can be treated with Ridomil @ 2 g/kg and Carbendazim @ 2 g/kg for the management of seed-borne diseases (Chapke et al., 2020). Gangadharayya et al. (2019) observed that growth and yield attributes in foxtail millet significantly improved on seed priming with *Azospirillum sp.* @ 20% + *Pseudomonas fluorescens* @ 20% + Phosphobacter @ 20% + $ZnSO_4$ @ 0.1% + Boron @ 0.1%. Soumya et al. (2021) reported that finger millet seeds bio-primed with 1% *Trichoderma harzianum* recorded significantly superior germination percent (86.99%), seedling length (12.30 cm), seedling dry weight (36.83 mg), test weight (6.23 g), and seedling vigor index I and II (1072.76 and 3204.56). This may be due to a quick extensive network of hyphae from *Trichoderma*, which can help in the nourishment of seeds through nutrient uptake and moisture conservation near seeds. Somasekhar et al. (2020) observed that microbial seed treatment with *Azospirillum* (N-fixing) and PSM (P-solubilizing) @ 25 g/kg of seed ensured better nutrient availability to the plant. Nasibi et al. (2022) conducted a study on seed priming with selenium nanoparticles and plant growth-promoting rhizobacteria, which improved seedling development of foxtail millet (*Setaria italica*) under salinity stress. The results of the study on growth, biochemical, and physiological parameters revealed that soaking millet seeds in a solution of selenium nanoparticles can increase the plant's ability to tolerate salinity stress by increasing the production of photosynthetic pigments, compatible solutes, and reduced sodium uptake, which resulted in reducing oxidative stress, thereby protecting the cell membrane. A more pronounced effect of this treatment was found in the presence of PGPR bacteria (such as *Bacillus cereus* and *Pseudomonas fluorescens*). Therefore, it was suggested that these two biostimulators can be used together for greater effects.

12.1.4 SPACING EFFECT

The Indian Institute of Millet Research, Hyderabad, has recommended the spacing in foxtail millet as 25–30 cm between rows, 8–10 cm between plants, and a sowing depth of 2–3 cm. Foxtail millet grown with 20 cm × 10 cm spacing exhibited significantly higher grain yield and straw yield (2227 kg ha^{-1} and 4349 kg ha^{-1}, respectively) and higher quality parameters like protein content (10.08%) and fiber content (6.33%). Using varying spacing, the harvest index was reported as non-significant. Additionally, it was noted that the vertical expansion of plants with higher growth and dry matter production was mostly responsible for the rise in stover yields with closer spacing (Nandini and Sridhara, 2019). Lokesh et al. (2023) studied the effect of different sowing windows and line spacings on the yield of foxtail millet during the Rabi season. Compared to closer spacings (30 cm and 45 cm), line spacing of 60 cm showed higher yield attributes. The 30 cm line spacing yielded more grain and straw per hectare, even though the 60 cm line spacing had higher yield parameters. This indicated that while individual plant performance may have been better with wider spacing, the total yield per hectare was better with the narrower 30 cm line spacing.

12.1.5 MULCHING

Argaman and Stavi (2023) evaluated the benefits of straw mulching and polyacrylamide techniques, making use of on-site rainfall simulation to determine the efficiency of polyacrylamide and straw-mulch cover in controlling runoff from foxtail millet (*Setaria italica*) fields under three seasonal conditions, viz., high-intensity rainfall (characteristic of autumn, on bare soil surfaces), moderate-intensity rainfall (characteristic of winter, following the tillering phase of the crop), and high-intensity rainfall (characteristic of spring, following the flowering phase). The most significant finding was obtained for spring simulation, wherein straw-mulch cover yielded no runoff, whereas the polyacrylamide and control treatments yielded similar runoff rates and runoff ratios.

12.1.6 INTERCROPPING

Foxtail millet is a suitable crop for intercropping. Intercropping has several benefits, including increased resource usage, decreased populations of detrimental biotic agents, improved soil health and resource conservation, increased system production, and system sustainability. Experimental results revealed that intercropping of green gram or black gram with foxtail millet at 3:3 ratios is beneficial. In Andhra Pradesh, intercropping systems with foxtail millet + groundnut (2:1) and foxtail millet + cotton (5:1) are very common, whereas in the Rayalseema region of Andhra Pradesh, intercropping of foxtail millet and pigeon pea (5:1) is preferred. More gross and net incomes and benefit-cost ratios were obtained with the sowing of foxtail millet + pigeon pea (5:1) with sowing during the first fortnight of August (Maitra et al., 2020).

12.1.7 NUTRIENT MANAGEMENT

Fertilizer recommendations in finger millet vary from state to state, but the general recommendation as per the Indian Institute of Millet Research, Hyderabad, is the application of 5 t/ha FYM, 40 kg N, 20 kg P2O5, and 20 kg K_2O/ha. The entire dose of phosphorus and half the dose of nitrogen should be applied basally, and the remaining half of nitrogen should be applied 30 days after sowing. In a field experiment conducted at Tribhuvan University, Nepal, Ojha et al. (2018) observed that the balanced application of nutrients (6 t FYM ha^{-1} + 60:30:20 kg NPK ha^{-1}) gave higher grain yield along with high monetary returns. The higher grain yield could be attributed to the favorable effect of more effective tillers, increased panicle length, more grains per panicle, and more test weight. Raviraja et al. (2020) reported that integrated nutrient management involving chemical fertilizers, organic manures, and biofertilizers enhanced the yield, protein content, and provided better net returns when compared with the conventional application of chemical fertilizers on the black soils of ARS Raichur. The recommended dose of fertilizer (30:15:15 kg $N:P_2O_5:K_2O$ ha^{-1}) + 2.5 t ha^{-1} vermicompost + biofertilizer + foliar spray of 19:19:19 @ 1% at 15, 30, and 45 DAS was found to be the most effective nutrient dosage for foxtail millet. A study conducted at TNAU, Coimbatore, on the effect of different foliar treatments in combination with the recommended dose of fertilizers on the yield and growth parameters of foxtail millet found that foliar treatment of 0.6% TNAU maize maxim + RDF resulted in higher plant height (132.38 cm), number of tillers (2.37), leaf area index (1.37), dry matter production (5937.74 kg/ha), productive tillers per plant (1.95), panicle length (23.59 cm), panicle weight (15.29 g), grain yield (2130.20 kg/ha), and straw yield (4108.14 kg/ha). Based on an AICRP trial conducted at Nandyal on foxtail millets during 2014–2016, it was found that soil application of $ZnSO_4$ @ 25 kg/ha + foliar application of $ZnSO_4$ @ 0.5% was more effective concerning yield and productivity. The impact of varying fertilizer ratios of different nitrogen, phosphorus, and potassium fertilizers on the agronomic and qualitative characteristics of foxtail millet during the jointing, heading, and filling stages was studied by Xing et al. (2023). The investigation demonstrated that an application of 160 kg of nitrogen (N), 90 kg of phosphorus (P), and 150 kg of potassium (K) per hectare ($N_{160}P_{90}K_{150}$) performed the best in terms of various trait indicators for the variety JG21.

12.1.8 IRRIGATION MANAGEMENT

Foxtail millet is a rainfed crop, and it does not require irrigation; only minimal irrigation is required during the kharif season. However, if a dry spell prevails for a longer period, 1–2 irrigations should be given. The summer crop requires 2–5 irrigations, depending on soil type and climatic conditions (Chapke et al., 2020). Heidari et al. (2012) found that alternate furrow irrigation resulted in an increased percentage of water-soluble carbohydrates and decreased crude protein percentage compared to conventional furrow irrigation in arid and semi-arid regions of Iran. Seghatoleslami et al. (2018) observed that deficit irrigation in foxtail millet resulted in the highest number of seeds per ear because of the longer duration between stem elongation and ear emergence. The study conducted by Meng et al. (2020) indicated that using a moderate planting density with ridge and furrow rainwater harvesting (RFRH) for foxtail millet—specifically, three rows within a 60 cm wide furrow and a plant spacing of 7.5 cm—was successful in enhancing yield and crop water productivity in semi-arid areas of China.

12.1.9 WEED MANAGEMENT

Ravali et al. (2021) documented the most significant growth traits, such as plant height (cm), number of tillers (No. m^{-2}), total dry matter production (kg ha^{-1}), and yield parameters, including the number of panicles m^{-2}, panicle length, panicle weight, total grains per panicle, filled grains per panicle, and yield of foxtail millet, when subjected to hand weeding at 15 and 30 DAS. Among the herbicide-applied treatments, the combination of herbicide and cultural practice, that is, intercultivation at 15 DAS fb ethoxysulfuron @ 18.75 g a.i. ha^{-1} as a postemergence application at 30 DAS, recorded the highest plant height, number of tillers, total dry matter accumulation, and grain yield.

12.1.10 HARVESTING

Foxtail millet reaches maturity within a span of 80–100 days. The crop is gathered once the ear heads have dried, either through the method of severing the entire plant with a sickle or through the detachment of the individual ears. Typically, the crop is harvested during the Kharif season, which spans from September to October, as well as the Rabi season, taking place between January and February. The grain yield ranges from 15 to 18 metric quintals per hectare, while the straw yield falls within the range of 20 to 40 metric quintals per hectare (Chapke et al., 2020). The effect of harvesting time on the quantitative and qualitative forage yield of lines and cultivars of foxtail millet (*Setaria italica* L.) in the Birjand Region of Iran was evaluated (Azarinasrabad et al., 2023). The results revealed that delaying the harvest of foxtail millet can have a positive impact on both the yield and yield components of forage, as well as forage protein yield.

12.2 KODO MILLET

Kodo millet (*Paspalum scrobiculatum* L.), also known as cow grass and ditch millet, is mainly cultivated in India, China, Russia, Japan, and Africa. Native to the Indian subcontinent, kodo millet has been cultivated for ages and is referenced in the Brihad Sanhita. Kodo millet is recognized for having the greatest drought resilience among all minor millets and is reported to yield well within a short growing duration of 80–135 days. Kodo millet is rich in amino acids like lysine, isoleucine, valine, threonine, and sulfur-containing amino acids but is deficient in tryptophan (Bunkar et al., 2021). Kodo millet is a suitable dietary option for individuals who exhibit gluten intolerance due to its absence of gluten content. The presence of a heightened quantity of lecithin within this millet variety renders it highly conducive to efficient digestion, thereby promoting the optimal functioning of the neurological system. Furthermore, postmenopausal women displaying symptoms of cardiovascular afflictions, such as elevated blood pressure and excessive cholesterol, can greatly benefit from the regular incorporation of kodo millet into their diets. This particular millet variant harbors a greater abundance of antioxidants, which effectively shield against oxidative stress and serve to maintain stable glucose levels among patients afflicted with type 2 diabetes (Singh et al., 2023). Kodo millet contains a higher level of lecithin, which is essential for the health of the neurological system, making it easy to digest (Sonkar et al., 2023).

Kodo millet is known as the famine grain in many parts of the country, as the grains possess excellent storage properties and can be stored for several years without pest damage, even under ordinary storage conditions. This crop stands out for its impressive drought resistance and is primarily grown under rainfed conditions, though in small areas receiving irrigation (Samundeswari et al., 2023). Studies have shown that the growth and yield of kodo millet can be significantly improved through improved agricultural practices.

12.2.1 LAND PREPARATION AND FARM MACHINERY

Land preparation involves one deep ploughing and two light ploughing followed by harrowing. (E-Krishi, UAS Bangalore)

12.2.2 SEED PRIMING

Seeds are treated with *Azospirullum brasilense* (nitrogen-fixing bacterium) and *Aspergillus awamori* (phosphate solubilizing fungus) @ 25 g/kg (Chapke et al., 2020). Seeds are also treated with 2 g Thiram or Carbendazim to provide disease resistance (CPG, 2012). Studies conducted by Sridevi and Manonmani (2016) on the impact of seed priming on physiological traits of Kodo millet and barnyard millet found that the application of *Pseudomonas fluorescens* at 20% for 6 hours was the best treatment for improving the physiological seed quality parameters, such as speed of germination, germination rate, shoot-root length, dry matter production, vigor index, and seed metabolic efficiency. The use of seed enhancement treatments, such as priming with beneficial microorganisms like *Pseudomonas fluorescens,* represents an environmentally acceptable alternative to traditional fungicide seed treatments. A study on the standardization of different pre-sowing seed treatments on growth, yield, and yield-attributing traits of Kodo millet indicated that a magnetic field of 200 mT gave the best results in field and lab conditions regarding field emergence and germination rate (98.5%). Significantly higher yield and morphological trait values, as well as seed quality measures like shoot length, root length, seedling length, vigor index I and II, and fresh and dry weight of seedlings, were obtained with a 400 mT magnetic field (Kavitha et al., 2022).

12.2.3 SOWING WINDOW

Ungata et al. (2020), in a field trial at ARS Vizianagaram, observed that Kodo millet, when sown in the first fortnight of July, gave maximum growth and yield attributes compared to other sowing dates. For seed saving for Kodo millet, the single-row millet planter-cum-fertilizer drill was found to be more effective than the traditional method of line sowing and broadcasting (Nandede et al., 2018).

12.2.4 SPACING EFFECT

A spacing of 22.5 cm between rows and 5 to 7.5 cm between plants, with a seed rate of 15 kg/ha (600–800 thousand plants per ha), is found to be optimum for Kodo millet. Observations at the Paspalum Improvement Center at Dindori, Madhya Pradesh, found that a crop sown at 22.5 cm but unthinned within the row and a broadcast field yielded comparably, assuring that the crop responds to a higher population. Average seed spacing (ASS), seed damage (SD), seed rate (SR), and coefficient of uniformity (Uc) were considered for identifying the suitable size of the slot in the single-row millet planter-cum-fertilizer drill for metering Kodo millet. The slot size of 3×3 mm was found suitable for Kodo millet as it delivered optimum ASS, SD, SR, and Uc values of 5.21 cm, 3.14%, 2.92 kg/ha, and 0.88, respectively (Nandede et al., 2018).

12.2.5 NUTRIENT MANAGEMENT

The application of organic manures is preferred in Kodo millet as it helps in the retention of water besides supplying nutrients. FYM should be applied a month before sowing at the rate of 5-–10 t/ha. Fertilizers of 40 kg N, 20 kg P_2O_5, and 20 kg K_2O per hectare are applied in the furrow at the time of sowing. In high rainfall areas of Madhya Pradesh and Chhattisgarh, N should be applied in two splits, that is, half at sowing and the remaining half at 35–40 days after sowing (Chapke et al., 2020). Singh et al. (2023) observed that with an increase in the level of nitrogen supply, the number of grains per panicle and test weight of Kodo millet increased significantly, maximizing at a tested dose of 40 kg N/ha. Another study conducted by Choudhari et al. (2018) on integrated nutrient management for growth, yield, and economics of Kodo millet (*Paspalum scrobiculatum* L.) indicated that the application of 100% NPK recommendation (40 kg N, 20 kg P_2O_5, and 10 kg K_2O per hectare) was the most beneficial for the cultivation of Kodo millet var. JK-13 in the rainfed conditions of the Kymore plateau in Madhya Pradesh. The application of the full recommendation of NPK fertilizer resulted in maximum panicles (6.56/plant), grain count (237.5/panicle), panicle length (14.86 cm), grain weight (1.24 g/panicle), and test weight (6.36 g), with a grain yield of 27.55 q/ha. Hemalatha et al. (2021) conducted a study on the effect of different nitrogen management practices on the growth and yield attributes of Kodo millet. The experiment with 10 treatments of different combinations of nitrogen sources, such as inorganic fertilizers, farm yard manure, and poultry manure, revealed that the growth parameters and yield attributes were found to be best with the application of 100% RDN through fertilizer. Among the organic sources tried, 75% RDN through fertilizer + 25% RDN through poultry manure and 75% RDN through fertilizer + 25% RDN through FYM proved to be the most promising integrated nitrogen management practice. Soil application of zinc sulfate, combined with foliar application of ferrous sulfate, has been found to enhance growth and yield parameters (Ram et al., 2021). The integration of all the inputs, 50% NP + 100% K + *Azotobacter* + PSB + FYM, proved best and significantly increased the Kodo yield (Dwivedi et al., 2016).

12.2.6 IRRIGATION MANAGEMENT

During dry periods, irrigations are required every 4–7 days depending on the severity of the drought and type of soil. In the case of a limited irrigation facility first irrigation can be given at 25–30 DAS and second irrigation at 40–45 DAS. Drain out the excess rainwater from the field during heavy and continuous rains.

12.2.7 INTERCROPPING

Among different intercropping systems in Kodo millet, pigeon pea intercropped with Kodo millet followed by black gram exhibited the highest grain and straw yield. The grain equivalent yield of Kodo millet was computed to be in the range of 19.5–54.8 qt/ha (Mahajan, 2016).

12.2.8 WEED MANAGEMENT

Among the various weed management approaches, hand weeding twice on 20 and 40 DAT resulted in a lower weed intensity percentage on 30 and 60 DAT, as well as a higher weed control efficiency and treatment efficiency index. This was comparable to the application of the post-emergence herbicide bispyribac sodium @ 20 g a.i. ha^{-1} on 20 DAT. Among the herbicides, the post-emergence herbicide bispyribac sodium @ 20 g a.i. ha^{-1} on 20 DAT had the lowest weed index, followed by Pretilachlor @ 0.5 kg a.i./ha applied pre-emergence on 3 DAT. The application of the post emergence herbicide bispyribac sodium @ 20 g a.i. ha^{-1} on 20 DAT has immense capacity to control all types of weeds in transplanted Kodo millet (Jawahar et al., 2020).

12.2.9 HARVESTING

The impact of eight harvest stages at four-day intervals on seed hardness, moisture, test weight, yield, germination, field emergence, electrical conductivity, seedling vigor, and α-amylase activity was studied on different Kodo millet varieties. Based on significant differences in seed yield, vigor, and storability traits across harvest stages, the appropriate stage of physiological maturity was identified at 33 DAA (Days After Anthesis) in JK36 and 28 DAA in TNAU86; and harvest maturity at 37 DAA in JK36 and 32 DAA in TNAU86 during the rainy season. However, during the post-rainy season, the stages at 32 DAA in JK36 and 26 DAA in TNAU86 were ideal for both seed physiological maturity as well as harvest maturity. Additionally, the second-order polynomial regression analysis predicted maximum seed yield, germination, vigor, and storability more or less close to these stages (Kannababu et al., 2023).

12.3 BARNYARD MILLET

From ancient times, the Indian barnyard millet (*Echinochloa frumentacea*) and Japanese barnyard millet (*Echinochloa utilis*) have been cultivated in temperate regions of the world. It is cultivated extensively in Asia and Africa, serving as a valuable source of both human food and livestock feed. It thrives well even in challenging environmental conditions such as drought, salinity, extreme cold, and heat, while also being adaptable to various soil types (Renganathan et al., 2023). Apart from India, China, Japan, Pakistan, Africa, and Nepal also contribute to the production of this millet crop. Being the fourth most produced minor millet in the world, IIMR (2018) credits India as its biggest producer in terms of area (0.146 m ha^{-1}) and production (0.147 mt). In India, the major production is from the hilly tracts of Uttarakhand, Tamil Nadu, Andhra Pradesh, Karnataka, and Gujarat (Chandra et al., 2021; Gomashe, 2017; Kumar et al., 2007). Barnyard millet is one of the fastest-growing millets and can be harvested in 70–100 days (depending on the variety), making it an ideal alternate crop at times if the major crop fails (Gupta et al., 2009).

The crop is known for its better yield and high nutritional value (Prabha et al., 2010) compared to major cereal crops and can be raised successfully with minimum inputs. The millet is a source of highly digestible proteins and carbohydrates, rich in minerals like iron and zinc (Saleh et al., 2013), making it an ideal food for humans engaged in sedentary activities (Sujatha et al., 2013). Barnyard millet possesses anti-nutrients like phytates, polyphenols, and tannins, which reduce the bioavailability of zinc and iron by chelation (Renganathan and Vanniyarajan, 2018). It contains several bioactive phytochemicals, including polyphenolic compounds such as Formononetin, Luteolin, Apigenin, Catechin, Kaempferol, and 3,7-Dimethylquercetin (a flavonoid), and Isorhamnetin. Like other millets, barnyard millet is nutritionally rich compared to traditional cereals, but its utilization remains limited. However, this lesser-known millet can be incorporated into a variety of products, including baked goods, nutritious snacks, and germinated complementary foods (Dutta et al., 2023). In India, barnyard millet is primarily cultivated in two distinct agroecological regions. One is in the northern part of the country, specifically in the lower and mid-hills of the Himalayan region, and the other is in the southern part, mainly within the Deccan plateau region (Sood et al., 2016). Barnyard millet is known for its rapid growth compared to other millet varieties and is particularly well-suited to fragile and hilly mountain ecosystems (Gururani et al., 2021). In India, it is grown in most of the southern and central states, with significant cultivation taking place in Odisha, Maharashtra, Gujarat, Madhya Pradesh, Tamil Nadu, Bihar, and the hilly regions of Uttar Pradesh (Chandra et al., 2021). Barnyard millet is a potential crop suitable for the bio-fortification of micronutrients, as its grains are naturally rich in essential micronutrients like iron (Fe) and zinc (Zn) (Renganathan et al., 2020).

12.3.1 VARIETIES

Even though an important millet crop, very little improvement work is being done on barnyard millet. The varieties developed are selections made from the local germplasm of the Indian barnyard millet (Table 12.1).

TABLE 12.1 Barnyard Millet – Released Varieties at National Level in India

Sl. No.	Variety	Institute Developed	Maturity (Days)	Average Yield (kg ha-1)
1.	K - 1	TNAU	90–110	1000
2.	Kudiraivali K– 2	TNAU	90	2300–3000
3.	VL Madira – 8	VPKAS, Almora	90–120	1800
4.	Anurag	ARS	75	1700–2000
5.	Gujarat Bank - 1	AICSIP, GAU, Surat	85–90	2000
6.	Rau – 3	IDRO, RAU, Dholi	80–50	1800
7.	VL Madira – 21	VPKAS, Almora	85–100	2500–2600
8.	VL Madira – 29	VPKAS, Almora	88	2500
9.	Chandan	-	75–85	1200–1600
10.	Kanchan	CSAUAT, Kanpur	85–90	1500–1800
11.	VL Madira – 172	VPKAS, Almora	90–95	2100–2300
12.	Sushrutha	RAU, Pusa and UAS, Dharwad	90–95	2000–2200
13.	VL Madira – 181	VPKAS, Almora	90–95	1600–1700
14.	PRJ – 1	GBPUA&T, Uttaranchal	118	2500
15.	Praptap sanwa – 1	MPUA&T, Udaipur	85–91	1500–1700
16.	VL Madira – 207	VPKAS, Almora	85–100	1800–1900
17.	Co (KV) 2	TNAU	-	2100
18.	Co 1	TNAU	105–110	1600

Source: Adapted from Gomashe (2017).

12.3.2 SEED RATE

Lee et al. (2021) studied the effect of planting density and planting methods on the dry matter yield of barnyard millet and found that the highest dry matter yield was obtained with an increased seed density of 30 kg ha^{-1} under both broadcasting and seed drilling. In broadcasting, with the increase in seed density, the crop suffered lodging. Taking into account the lodging effect, seed price, and labor cost, the seed rate adopted can be limited to 15–20 kg ha^{-1}.

12.3.3 SEED TREATMENT

Macro and micronutrients are vital for plant growth and development. Even though a marginal crop, barnyard millet responds well to macro and micronutrients. Works on barnyard seed fortification with micronutrients are meager. Sujatha et al. (2013) studied the effect of seed fortification in barnyard millet and observed that seedling quality and germination were improved due to seed fortification with IAA @ 100 ppm or KH_2PO_4 @ 1%, which can be used as a pre-sowing treatment for barnyard millet. Seed quality, seed rate, and uniformity of seedling emergence can be enhanced with seed priming treatment (Balaji and Narayana, 2019). Gajalakshmi et al. (2022) studied the effect of germination of barnyard millet by seed priming with various zinc sources and observed that seed priming with 0.5% of $ZnSO_4$ (12 hours) improved the seed germination and seedling growth parameters like shoot length, root length, dry weight, germination percentage, speed of germination, and vigor index. Suruthi et al. (2019), in a study conducted to evaluate the effect of different seed priming methods viz., hydro priming (H_2O), hormo priming (GA3 10 ppm), organic priming (vermi wash 10%), botanical priming (Calotropis leaf extract 15%), halo priming (KH_2PO_4 1%), and chemo priming (succinic acid 100 ppm), observed that hormo priming (GA3 10 ppm) recorded higher values in germination (100%), shoot length (10.85 cm), and vigor index (2353) compared to other priming treatment methods. Iswariya et al. (2019) reported that barnyard millet seeds soaked in an equal volume of liquid microbial culture *Azophos* + *Pseudomonas fluorescens* @ 20% for 8 hours exhibited higher seedling vigor. This may be due to phytohormones or enhanced nutrient mobilization from the seed. Gajalakshmi et al. (2022) found that seed priming with 0.50% $ZnSO_4$ enhanced germination percentage (99%), germination index (3.91), and seedling growth parameters such as shoot length (14.7 cm), root length (16.5 cm), seedling dry weight (0.44 g), vigor index I (3099), and II (44.6).

12.3.4 SOWING AND SPACING

Since barnyard millet is grown as a marginal rainfed crop, the seeds are often broadcast to reduce labor costs. An appropriate planting method with balanced nutrition is essential for better productivity (Michaelraj and Shanmugham, 2013). Sowing seeds at optimum spacing is more likely to ensure uniform germination (Maitra et al., 2020), and the competition among the plants can be reduced, thus producing a better yield (Krishnaveni and Avudaithai, 2020). Achieving a satisfactory yield is contingent on effective stand establishment, which in turn relies on the sowing of high-quality seeds at the right spacing to ensure uniform germination (Maitra et al., 2020). Optimal crop geometry plays a crucial role in minimizing competition among plants, and the appropriate plant population during harvest significantly influences yield outcomes (Anandha and Avudaithai, 2020). To maximize barnyard millet yields, it's important to maintain an optimum plant density and provide a balanced fertilization regimen, including the three primary nutrients: nitrogen, phosphorus, and potassium (Maitra et al., 2020).

12.3.5 NUTRIENT MANAGEMENT

Barnyard millet is usually grown in marginal soils under rainfed conditions. The improved varieties respond well when provided with better nutrition. Maitra et al. (2020) have reported that a balanced fertilizer application of NPK produced a higher yield in barnyard millet. Farmyard manure (FYM) @ 5 t ha^{-1} along with 40:20:0 kg NPK ha^{-1} is recommended for the soils of Bihar and Uttar Pradesh (Prabhakar et al., 2017). However, the application of potassium is found to improve grain filling, grain weight, straw strength and even account for disease resistance. Janani et al. (2022) observed that providing a spacing of 22.5 cm x 10 cm along with the application of NPK @ 40, 20, and 10 kg ha^{-1} helped to obtain a higher yield in barnyard millet. The split application of 150% of RDF (99:49.5:33), nitrogen, and potassium 25% as basal and the remaining 75% in equal split at tillering and flowering, along with 1% PPFM application at 30 and 50 days after sowing, increased the grain yield and straw yield of barnyard millet under a rainfed ecosystem (Balaji et al., 2019). The length of the millet increased significantly with nitrogen fertilization during the harvest period, but the difference was insignificant during the milk stage. Similarly, barnyard millet dry matter yield increased significantly with an increase in nitrogen fertilization levels, but there was no significant difference in dry matter yield among nitrogen fertilization levels during the heading stage (Lee et al., 2021).

12.3.6 INTERCROPPING

The effect of intercropping other crops with barnyard millet is yet to be fully realized. In rainfed and dryland conditions where barnyard millet is usually raised, due to unpredictable rainfall and drought, their pure stand is risky. Crop diversification by intercropping millets with legumes ensures a more stable yield. Even though the grain yield of barnyard millet was reported to be higher in pure stands compared to crop mixtures (Singh and Arya, 1999), a combination tried with soybean (seed mixture of 90:10%) recorded a higher LER of 1.45 and barnyard equivalent yield. Moreover, Chapke et al. (2018) have also recommended the intercropping of barnyard millet with rice bean in a 4:1 ratio for higher yield in Uttaranchal.

12.3.7 WATER MANAGEMENT

Barnyard millet is rarely irrigated since it is raised as a rainfed crop. Life-saving irrigation should be ensured at the tillering stage (25–30 days after sowing) and panicle initiation stage (45–50 days after sowing), which are considered the critical stages for irrigation (Trivedi et al., 2017). The crop does not perform well under excess moisture conditions; hence, proper drainage facilities should be ensured.

12.3.8 WEED MANAGEMENT

Kumar et al. (2015) and Prithvi et al. (2015) observed a reduction in the grain yield of millets due to weed competition. Shamina et al. (2019) assessed the effect of different weed management practices to suppress the weed population and enhance the growth of barnyard millet and found that adjusting the spacing and weed management practices had significant results in suppressing weeds. Barnyard millet raised with a spacing of 25 cm x 10 cm with hand weeding twice (20 and 40 days after sowing) produced yields on par with the weed-free check. In general, for barnyard millet, two hoeing sessions (on the 15th day and the 40th day after sowing) check the weed population under the threshold level.

12.4 PROSO MILLET

Proso millet (*Panicum miliaceum* L.) is an underutilized minor millet. The grains of Proso millet contain a high protein content of 12.5% and offer superior nutritional value in terms of vitamins, minerals, essential amino acids, fats, and dietary fiber when compared to major cereals like rice, maize, and wheat (Amadou et al., 2013). Habiyaremye et al. (2017) have reported Proso millet as a staple food source in specific regions of Asia and Africa. Farmers cultivate Proso millet for subsistence and income using native landraces to meet their dietary needs (Ghimire et al., 2018). Proso millet has a short growing period of approximately 70 days and minimal water requirements for growth and yield. It thrives in adverse conditions and produces reasonable yields in areas where crops like maize and rice struggle or produce poor yields (Liu et al., 2015). Its adaptability to various soils and climates positions it as a promising crop for the future, especially in the context of climate change (Gregory et al., 2019). Proso millet fits well in crop rotations for weed, disease, and pest management (Heyduck et al., 2008).

12.4.1 SOWING AND SOWING TIME

Gavit et al. (2017) studied the effect of establishment techniques and sowing time on yield and yield attributes of Proso millet (*Panicum miliaceum* L.) and opined that sowing Proso millet in the 25th meteorological week resulted in the maximum and significantly greater grain yield and straw yield compared to other sowing times. Additionally, transplanting Proso millet yielded the highest grain and straw yield when compared to other establishment techniques. Dimple et al. (2019) investigated the productivity and nutrient uptake in Proso millet as influenced by time of sowing and found that among various sowing times, the crop sown in the second fortnight of June resulted in higher grain yield, straw yield, and nutrient uptake at the time of harvest. Aiswarya et al. (2022) also studied the effect of weather parameters on Proso millet (*Panicum miliaceum* L.) cultivars under different sowing windows and observed that Proso millet sown in the first fortnight of June (var. GPUP 8) utilized more growing degree days, photothermal units, and exhibited higher heat use efficiency compared to sowing in the second fortnight of June and the first fortnight of July. Al-Hassoun et al. (2023) examined the effect of sowing rates and potassium foliar feeding on the growth and yield of *Panicum miliaceum* L. and observed that different sowing rates had a notable impact on various aspects of the crop's vegetative growth. Notably, the treatment involving a sowing rate of 10 kg per hectare outperformed other rates in terms of plant height, leaf area, tiller number, and most of the yield and its components, such as the number of grains and individual plant yield.

12.4.2 SEED PRIMING

Balaji et al. (2019) investigated the effect of osmo-priming on seed quality in certain minor millets and found that when Proso millet cv. CO 4 seeds were osmo-primed with 2% mannitol (for 6 hours), they exhibited superior seed quality, viz. higher speed of emergence, greater germination percentages, longer seedling length, increased dry matter production, and a higher vigor index. Among the different seed priming treatments, Sridevi et al. (2020) found that 20% *Pseudomonas fluorescens* could produce significantly higher growth parameters such as field emergence, plant height, chlorophyll content, net assimilation rate, dry matter efficiency, earliness in days to booting, days to first flowering, and days to 50% flowering. In comparison to other priming techniques, the treatment including 20% *Pseudomonas fluorescens* had higher yield characteristics such as panicle weight, panicle to seed recovery percentage, per plant seed yield, seed yield per plot, and 1000 seed weight (Table 12.2).

12.4.3 VARIETIES

TABLE 12.2 Recently Released Varieties of Proso Millet in India (2005 to 2018)

Sl. No.	Name of Variety	Pedigree	Institute Where Developed	Year of Release	Maturity (days)	Avg. Yield (q/ha)	Area of Adaptation	Special Features
1	TNAU 145	PV 1454 x TNAU 96	TNAU, Coimbatore	2007	70–72	18–20	Tamil Nadu	High-yielding, superior grain quality for value addition.
2	CO(PV) 5(TNAU 143)	PV 1403 x GPUP 21	TNAU, Coimbatore	2007	70–75	23–25	National	High yield, profuse tillering, drought Tolerant
3	TNAU 151	TNAU 96 x PV 1673	TNAU, Coimbatore	2008	72–75	18–20	National	Bold grains, tolerant to shoot fly
4	TNAU 164	TNAU 137 x CO 4	TNAU, Coimbatore	2009	70–75	18–20	National	Nonlodging, tolerant to shoot fly and Rust
5	Pratap Cheena-1 (PR-18)	Pureline Selection	MPUA &T, Udaipur	2006	65–70	15–17	National	Early duration, dual purpose.
6	PRC 1	Selection from GPMS 519	Ranichauri, GBPUA&T, Pantnagar	2008	70–75	10–12	Uttarakhand hills	Resistant to leaf blight
7	TNAU 202	PV 1453 x GPUP 16	TNAU, Coimbatore	2011	70–75	18–20	National	Profuse tillering and bold grains
8	TNPm 230	TNAU 164 x IPM-19	TNAU, Coimbatore	2017	70–75	21–23	National	Short duration, drought tolerant variety
9	DHP-2769	Selection from IPM-2769	ARS, Hanuman Matti, UAS, Dharwad	2018	70–72	20–22	Agro-climatic zone-3 and Zone 8 of Karnataka state	Suitable for contingency planting

Source: Adapted from Bhat et al. (2019).

12.4.4 NUTRIENT MANAGEMENT

Kakad et al., 2021 experimented on the effect of different nutrient combinations on the growth and yield of Proso millet. Five nutrient combinations;100% recommended dose of fertilizer given completely via chemical fertilizer(N1), 75% RDF+25% FYM, 50% RDF + 50% FYM, 25% RDF+ 25% RDF,100% FYM. The recommended dose of fertilizer was 80 kg ha^{-1} nitrogen and 40 kg ha^{-1} phosphorus. He observed that when 100% recommended dose of fertilizer was applied chemically in the form of urea (nitrogen source) and single super phosphate (phosphorus source) recorded significantly higher mean plant height hill^{-1}, number of functional leaves hill^{-1}, number of functional tillers hill^{-1} and plant dry matter accumulation hill^{-1} than the other treatments. Even when the economic aspect was considered 100% RDF through chemical fertilizer gave a B: C ratio of 1.27 whereas it was 1.02 for the treatment; 75% RDF with 25% FYM.

12.4.5 IRRIGATION

Proso millet can be grown as a dryland crop with no additional irrigation. Proso millet is only irrigated when it is planted as a replacement crop. Proso millet can be cultivated as a catch crop when other crops fail due to aberrant weather conditions (Habiyaremye et al., 2017). The Kharif season cultivations do not necessitate any irrigation. Nevertheless, if a prolonged period of dry spell persists, it would be prudent to provide one to two irrigations during the tillering stage to enhance the yield. The initial irrigation ought to be administered approximately 25 to 30 days after sowing, followed by a subsequent one at around 40 to 45 days after sowing. Conversely, the summer harvest necessitates two to four irrigations contingent upon the composition of the soil and the prevailing climatic conditions. In times of excessive precipitation, it is advisable to drain out any superfluous water from the field (IIMR, 2018).

12.4.6 HARVESTING

Due to its susceptibility to shattering when fully ripe, it is a rather precarious endeavor to allow Proso grain to fully mature and dry while still in the standing position. The initial variety of Proso, known as Even, a minor gust of wind can result in considerable shattering. Consequently, it is advisable to engage in swathing when the main head's uppermost section contains mature seeds. The identification of a mature seed is possible when the center of the seed is starch-like in consistency, as the Proso seed develops from the seed coat towards the center. Threshing can then be postponed until the grain's moisture content falls below 13%. The attainment of a moisture level of 13% can be a challenge in certain years, as high humidity, frequent dew, and lower temperatures in September are prevalent. Properly setting up the combine is essential for the proper harvest. The cylinder speed should be approximately 20% slower than that used for wheat harvest, or around 850 rpm. The primary issue encountered when marketing Proso is the removal of the hulls, leaving only the yellow-colored inner berry behind. A useful guideline to follow is to retain as many of the loose outer glumes as there are hulls that have been removed (Baltensperger et al., 1914).

The seed heads of Proso millet tend to break apart when they are dry, thus it is advisable to carry out the harvesting process when most of the seed head has lost its green hue, while the stem and leaves may still retain their greenness. The most optimal method for harvesting, which minimizes the risk of seed loss, involves cutting and binding. For the storage of the seed, it is recommended to keep it in an environment with low humidity, specifically below 13%, and away from direct exposure to sunlight (Tran, 2015).

12.5 LITTLE MILLET

Little millet (*Panicum sumatrense* Roth) is a resilient minor cereal crop from the Poaceae family and is regarded as native to India. It is predominantly grown in India and Nepal, with India being the world's largest producer, yielding approximately 0.12 million tons annually. In India, it is cultivated extensively in states such as Karnataka, Andhra Pradesh, Tamil Nadu, Odisha, Bihar, Maharashtra, and Madhya Pradesh (Meena et al., 2021). Belonging to the C4 plant family, little millet is considered physiologically efficient due to its ability to produce nutritious grains and fodder within a short growing season, particularly in dry and hot conditions. The crop is highly drought-tolerant and nutritionally on par with other cultivated cereals. The nutritional value of little millet is better than that of many other cereals. Research conducted by Selvi et al. (2015), Divyashree et al. (2018), and Dey et al. (2022) highlights little millet's excellent nutritional profile, which includes 67.0 g of carbohydrates, 7.7 g of proteins, 4.7 g of fats, 12.2 g of dietary fiber, 1.5 g of mineral matter, 17.0 g of calcium, 220.0 mg of phosphorus, and 6.0 mg of iron. It is comparatively rich in protein, fiber, and iron, and it is gluten-free. Therefore, it can be a beneficial dietary option for individuals with nutrient deficiencies and diabetes (Ambati and Sucharita, 2019). Little millet is also known for its extended storability under normal conditions, making it a valuable famine reserve crop. Little millet is now gaining popularity among farmers due to its resilience under moisture-stress conditions and flexibility in sowing timing. Since the cultivation of little millet is confined to marginal and sub-marginal lands with inadequate nutrition and poor agricultural practices, the productivity of little millet is limited.

12.5.1 VARIETIES

Little millet is mainly cultivated in Karnataka, Madhya Pradesh, Andhra Pradesh, Odisha, Tamil Nadu, Gujarat, Chhattisgarh, and Maharashtra (Maitra and Shankar, 2019) (Table 12.3).

TABLE 12.3 Little Millet Varieties Introduced in India

Sl. No.	State	Varieties
1.	Orissa	OLM 203, OLM 208, OLM 217, DHLM 36-3, DHLM 14-1
2.	Madhya Pradesh	JK 4, JK 8, JK 36, JK 137, BL 6, DHLM 36-3
3.	Andhra Pradesh	OLM 203, JK 8, BL 6, DHLM 36-3
4.	Tamil Nadu	Paiyur 2, TNAU 63, CO 3, CO 4, K1, OLM 203, OLM 20, BL 6, DHLM 36-3, DHLM 14-1
5.	Chhattisgarh	JK 8, BL 6, JK137, BL 4, JK 36, DHLM 36-3
6.	Karnataka	OLM 203, JK 8, BL 6, DHLM 36-3, DHLM 14-1
7.	Gujarat	GV 2, GV 1, OLM 203, JK 8, BL 6, DHLM 36-3, DHLM 14-1
8.	Maharashtra	Phule Ekadashi, JK 8, OL 203, BL 6, DHLM 36-3, DHLM 14-1

Source: Adapted from Chapke et al. (2018).

12.5.2 SEASON

Little millet can be planted during the Kharif (June–September) and Rabi (September–December) seasons. Crop establishment, vital for maintaining optimum plant population, has a lot to do with the efficient utilization of all available resources for sustained crop growth and yield realization in little millet.

12.5.3 SEED PRIMING

Seed priming with 2% KH2PO4 for 8 hours or 20% liquid *Pseudomonas fluorescens* was found effective in little millet (Patel et al., 2020).

12.5.4 SEED RATE AND SOWING

Farmers usually go for broadcasting to save time and labor. If broadcasted, the seed rate to be adopted is 12 kg ha^{-1}. In the case of line sowing, 8 kg ha^{-1} can be adopted. For effective crop establishment, the method of sowing is important. Farmers usually broadcast little millet seeds to save labor and time. According to Himasree et al. (2021), when 20-day-old little millet seedlings were transplanted at a spacing of 30 cm x 10 cm, the growth parameters of the little millet were found to be less than when they were broadcasted. Studies have shown that early sowing in June gave higher grain as well as straw yield compared to later sowings. When the crop was sown in the first fortnight of June, at a plant density of 890,000 plants per ha, a higher yield compared with either lower populations or very high populations could be obtained. Planting later than the first fortnight of June resulted in lower yields. The optimum plant population at which maximum yield was obtained was 890,000 plants per ha. In the southern agro-climatic zone of Andhra Pradesh, Sarala et al. (2020) studied the effects of fertilizer levels and spacing on the growth and yield of little millet. Transplanting seedlings of little millet at a spacing of 25 cm × 25 cm resulted in the highest grain yield (13.14 q ha^{-1})

12.5.5 MANURING AND FERTILIZER APPLICATION

The potential yield of little millet is not realized due to its cultivation in marginal soils where it is given little to no nutrition. Balanced nutrient management is the key to achieving higher biomass and also to maintaining soil fertility. Among the macronutrients, nitrogen is very important for little millet as its application promotes rapid vegetative growth. Himasree et al. (2021) have reported that the application of N @ 40 kg ha^{-1} produced higher growth parameters like plant height, tillers m^{-2}, and leaf area. Including a legume crop in the cropping system helps to improve the yield and also reduces the input requirement for the succeeding crop. Umadevi et al. (2019) investigated the impact of phosphorus and organic manures on cowpea and their impact on subsequent little millet. They discovered that the addition of the legume increased little millet yield. The yield of little millet was also significantly improved when nitrogen was applied at 20 kg ha^{-1} to little millet after raising cowpea as a preceding crop. Divyashree et al. (2018) tested different levels of NPK fertilizers on little millet and found that application of 30:20:10 kg NPK ha^{-1} produced taller plants with a greater number of leaves, more tillers, high leaf area, high grain, and straw yields.

Biofertilizers are formulations that contain live microorganisms when applied to seeds, plants, or soil, improve crop growth. Depending on the formulation used, the application of biofertilizer fixes the nitrogen or solubilizes the insoluble phosphorus in soil thereby making it available to the crop. Pagare et al. (2022) studied the response of liquid biofertilizers on growth, yield, and economics of little millet and observed that seed treatment with liquid bio-fertilizer (5 ml kg^{-1} seed) followed by soil application @ 3liter ha^{-1} along with

100% recommended dose of fertilizer had produced significantly higher grain yield and straw yield. In developed and developing countries, the food habits of people are rapidly changing and consumers are becoming more health conscious. As a result, the market for organic goods is constantly expanding. Little millet responds well to organic nutrition. Rani et al. (2022) studied the yield and quality of little millet as influenced by organic manures and inorganic fertilizers. It was observed that long-term application of organic manures like FYM, neem cake, and *in situ* green manuring restores soil health. The grain yield and straw yield may not increase in a short period, but the soil quality and crop yield can be improved over time. Kottegoda et al. (2011) have reported the increased fertilizer use efficiency in crops when nano fertilizers are used. The use of nano fertilizers is said to improve soil fertility and is less harmful to the environment, as only a low quantity is used compared to commercial fertilizers. The effect of nano fertilizers on the growth, yield, and economics of little millet (*Panicum sumatrense* Roth) under rainfed conditions was studied by Rajput et al. (2022). It was found that seed treatment with nano fertilizer and application of 75% recommended dose of fertilizer, along with foliar spray of nano fertilizer at 7–10 days before flowering, could improve the growth and yield parameters in little millet.

In an AICRP field trial conducted at Dindori from 2003–2005 in little millet, it was observed that the best crop response was obtained with an integrated nutrient supply of 7.5 t FYM/ha along with recommended NPK (20:20:10 kg/ha). The second-best practice was a little millet-legume two-year rotation with recommended organic manures. In a field study conducted at a wetland farm in Tamil Nadu Agricultural University, Coimbatore, India, to evaluate the effect of sowing dates and nutrient levels on the yield of little millet cultivated under rice fallow, it was observed that sowing 15 days after rice harvest with 75% RDF produced a yield that was on par with the yield produced with 100% RDF (44:22:0 NPK kg ha^{-1}) sown 15 days after harvest of rice (Sapthagiri et al., 2020). In a field study conducted at ARS, Vizianagaram, India it was observed that the highest grain and straw yields were recorded with the application of 100% RDF(20:20:20 NPK kg ha^{-1}) + neem cake @ 1 t ha^{-1} (Rani et al., 2017). Sarala et al., 2020 assessed the effect of spacing and fertilizer levels on the growth and yield of little millet in Andhra Pradesh and observed that, among the fertilizer levels, 40:40:20 kg of NPK ha^{-1} produced the highest grain yield (12.66 q ha^{-1}) of little millet in sandy loam soils.

12.5.6 INTERCROPPING

The basic concept of intercropping is raising two or more different crops on the same piece of land together so that all the crops can exploit the environment better and ensure yield stability. Sharmili and Manoharan (2018) studied the effect of intercropping on little millet and found that improvements in plant height, dry matter production, productive tillers per plant, grains per panicle, and little millet grain equivalent were found to be greater than in sole cropping. Keeerthanapriya et al. (2019) also studied little millet-based intercropping and found that all the intercrops tried dominated over the sole crop of little millet. Little millet intercropped with black gram in a 4:1 ratio produced a higher grain and straw yield for little millet and was found to be on par with little millet intercropped with black gram, green gram, and cowpea.

12.5.7 *WATER MANAGEMENT*

Little millet is the least water-demanding crop. However, lifesaving irrigation may be given at the critical stages of growth. Two irrigations at 25–30 days after sowing and 45–50 days after sowing are recommended for little millet (Chapke et al., 2018). Proper drainage must be ensured to prevent waterlogging.

12.5.8 *WEED MANAGEMENT*

Both grassy and broadleaved weeds are usually found in fields where little millet is grown. Inter-cultivating the field twice, followed by one hand weeding in line-sown crops and two hand weeding's in the broadcasted field, is sufficient for effective weed control (Maitra and Shankar, 2019). In the case of broadleaved weeds, post-emergent application of 2, 4-D @ 1 kg ai/ha at 20–25 days after sowing is found effective. Chapke et al. (2018) have reported the pre-emergence spray of Isoproturon @ 1 kg ai/ha for effective weed control.

12.5.9 *PEST AND DISEASE*

Being a minor millet crop, the literature available on the pests of little millet is meager. However, grain smut, rust, blight, and udbatta are the important fungal diseases occurring at different plant growth stages of little millet (Chauhan, 2014; Jain et al., 1997; Pall et al., 1980). Grain smut is ovaricolous, and the symptoms appear at the grain formation stage. The ovary of the affected plants gets converted to smuts. Some of the grains that grow late still have a greenish tint. Kumar et al. (2017) have reported that seed treatment with carbendazim @ 2g kg^{-1} seeds managed the disease and produced the highest yield.

12.5.10 *POST-HARVEST MANAGEMENT IN MINOR MILLETS*

Post-harvest, for long-term storage, millets are subjected to various pre treatments such as drying, threshing, winnowing, and other post-harvest processing methods. Numerous storage methods have been examined to determine the most suitable way to store minor millets. These processes play a pivotal role in the entire food chain of millet production, from harvest to consumption.

12.6 BARNYARD MILLET

Barnyard millet grains are to be dehulled first to make them suitable for human consumption (Lohani et al., 2012). Since the grains are tightly encased in their lemma and palea, traditional dehulling methods involve labor-intensive and time-consuming pounding using a mortar. The manual labor involved in this process is very high, and the smaller size of the seeds adds to the difficulty in processing. To alleviate this labor-intensive task, Singh et al. (2003) developed the Vivek Millet Thresher 1. This machine can thresh 40–60 kg of barnyard millet grains per hour. It is particularly suitable for small-scale farmers in hilly and tribal regions due to its affordability, compact size, lightweight design, and ability to perform both threshing and dehusking tasks.

12.7 FOXTAIL MILLET

Processing foxtail millet is also difficult due to its smaller size. Ushakumari et al. (2004) compared the properties of flaked grains and popped grains in foxtail millet. A rice milling machine was used to decorticate (remove the outer layer of grains) foxtail millet. The decorticated grains were then further processed by flaking, extrusion, and roller drying, while native grains were used for roasting or popping. The results indicated that the popped grains had the best nutritional composition due to the intact bran layer during roasting. Because of structural alterations in the starch granules, it was discovered that the functional characteristics of both popped and flaked grains were comparable in terms of swelling power, solubility, and oil and water absorption capacities.

Foxtail millet has been identified to contain certain phytochemicals that possess anti-nutritional properties. These antinutrients have the potential to reduce the absorption and nutritional quality of nutrients in the grain (Devisetti et al., 2014). To reduce their effects and enhance the bioavailability of micronutrients in foxtail millet, various processing techniques can be employed, including milling, cooking, extrusion, germination, and fermentation (Nazni and Shobhana, 2016). Numerous techniques have been applied to process foxtail millet to improve its nutritional, sensory, and economic properties. Pawar and Machewad (2016) observed that even simple steps like dehulling, soaking, and cooking could reduce the anti-nutrient content by about 50% and increase the *in vitro* protein digestibility by 32.71%. Processed products based on foxtail millet can be successfully prepared using these procedures.

12.8 KODO MILLET

The grains of Kodo millet are not only nutritious but also possess antioxidant properties and various medicinal benefits. 'Kodo poisoning,' a condition caused by fungal contamination and the generation of cyclopiazonic acid, a toxic fungal secondary metabolite, frequently prevents them from being consumed. To minimize this 'kodo poisoning,' it is essential to implement appropriate post-harvest grain management practices prior to storage. Hell et al. (2008) observed that field crops should be dried quickly after harvest to reduce their moisture levels to between 10% and 13%. Drying, storage, and processing are crucial aspects of post-harvest control where contamination can be reduced (Choudhary and Kumari, 2010). In Kodo millet, drying is observed as a vital step in ensuring high-quality, fungi-free grains (Villers, 2014). During storage, the moisture, temperature, and relative humidity must be controlled for efficient storage. A grain moisture content of 12% or less and relative humidity of 70% or less will inhibit the establishment of storage molds and the generation of their toxins (Kabak, 2009; Neme and Mohammed, 2017). Maintaining a water activity level of less than 0.7 will also discourage fungal growth and mycotoxin production in the grains (Manna and Kim, 2017).

12.9 LITTLE MILLET

Processing little millet is difficult as it comes with husk and bran covering the edible endosperm. Traditionally, these husk and bran layers are manually removed through labor-intensive hand pounding, as suitable mechanical dehusking or polishing units for little millet processing are not widely available. A comparative study of post-harvest processing methods for little millet (*Panicum miliare* L.) was conducted by Ravindra et al. (2018), highlighting the necessity of pretreatment to eliminate the husk. A provender mill was found to efficiently mill not only little millet but also other cereals and millets in a cost-effective manner. To minimize milling losses and maximize returns, it is recommended to parboil and dry little millet before milling. This process enhances the overall efficiency and quality of little millet processing.

Post-harvest management of minor millets is of paramount importance to ensure the quality, safety, and availability of these valuable crops. Minor millets, such as foxtail millet, barnyard millet, Kodo millet, and little millet, offer numerous nutritional benefits and can contribute to food security. However, their potential is often hindered by various challenges, including fungal contamination, antinutrient properties, and limited consumer awareness. Effective post-harvest management practices are essential to address these challenges and maximize the utility of minor millets. These practices include proper drying, storage, processing, and the reduction of antinutrient levels. Additionally, processing techniques can enhance the nutritional quality of millet-based products and reduce anti-nutrients. Furthermore, raising awareness and promoting the consumption of minor millets among consumers is essential. Efforts should be made to overcome hurdles like 'kodo poisoning' through proper education and dissemination of safe handling practices. The effective post-harvest management of minor millets is a multifaceted approach that involves both technical and awareness-building aspects. By implementing these measures, we can unlock the full potential of these nutritious grains, promoting food security and healthier diets.

12.10 CONCLUSION

As we are celebrating the year 2023 as the "International Year of Millets," minor millets, considered the warehouse of nutrients, are gaining popularity. In the era of nutritional awareness and a health-conscious lifestyle, these nutri-cereals are to be scientifically cultivated not just as a contingent crop but as a potential staple crop. Minor millets are easy to cultivate and are well-suited to low rainfall areas due to their higher water and nutrient use efficiency, and they will excellently fit into climate-resilient agriculture. Low input demand, stable yield, drought tolerance, and high water use efficiency give minor millets the green signal to be popularized for future climate-smart agriculture. The prospect of minor millets as a nutrient-rich food, fodder, and fuel would further help gain popularity among farmers and consumers, allowing them to cultivate these climate-smart, nutrient-rich crops. Designing good agricultural practices, developing stable high-yielding varieties, and establishing reliable market channels would bring more area of millets under cultivation.

KEYWORDS

- **climate-resilient crops**
- **intercropping**
- **minor millets**
- **nutrient management**
- **post-harvest management**
- **seed priming**

REFERENCES

Abedin, M. J., Abdullah, A. B. T., Satter, M. A., & Farzana, T. (2022). Physical, functional, nutritional, and antioxidant properties of foxtail millet in Bangladesh. *Helion, 8*(10). https://doi.org/ (accessed on 7th November 2025).

Aiswarya, G., Rani, P. S., Umamahesh, V., & Sarala, N. (2022). Effect of weather parameters on proso millet (*Panicum miliaceum* L.) cultivars under different sowing windows. *Andhra Pradesh Journal of Agricultural Sciences, 8*(2), 94–99.

Al-Hassoun, O. S. M., & Mohsen, H. A. H. (2023). Effect of sowing rates and potassium foliar feeding on some growth and yield of millet (*Panicum miliaceum* L.). *IOP Conference Series: Earth and Environmental Science, 1225*(1), 012071. https://doi.org/10.1088/1755-1315/1225/1/012071 (accessed on 7th November 2025).

Amadou, I., Gounga, M. E., & Le, G. W. (2013). Millets: Nutritional composition, some health benefits and processing—A review. *Emirates Journal of Food and Agriculture, 25*(7), 501–508.

Ambati, K., & Sucharitha, K. V. (2019). Millets—Review on nutritional profiles and health benefits. *International Journal of Recent Scientific Research, 10*(7), 33943–33948.

Anandha Krishnaveni, S., & Avudaithai, S. (2020). Effect of crop geometry and nutrient management in barnyard millet under sodic soil condition. *International Journal of Current Microbiology and Applied Sciences, 9*(06), 276–280.

Argaman, E., & Stavi, I. (2023). Runoff mitigation in croplands: Evaluating the benefits of straw mulching and polyacrylamide techniques. *Agronomy, 13*, 1935.

Ashoka, P., & Priya, P. (2020). Application of nutrients through soil and foliar methods on growth and yield of foxtail millet (*Setaria italic* L.). *Research Journal of Agricultural Sciences, 12*, 86–89.

Azarinasrabad, A., Javadi, H., & Ataei, R. (2023). Effect of harvest time on quantitative and qualitative forage yield of lines and cultivars of foxtail millet (*Setaria italica* L.) in Birjand region. *Crop Science Research in Arid Regions.*

Balaji, D. S., & Narayana, G. S. (2019). Effect of various bio priming seed enhancement treatments on seed quality in certain minor millets. *Plant Archives, 19*(1), 1727–1732.

Balaji, D. S., & Narayanan, G. S. (2019). Effect of osmo priming seed treatments on seed quality in certain minor millets. *Plant Archives, 19*(1), 508–514.

Balaji, G. A., Sakthivel, S., Chinnusamy, C., & Muthuramu, S. (2019). Effect of nutrient management and PPFM application on growth and productivity of barnyard millet under rain-fed ecosystem. *Journal of Pharmacognosy and Phytochemistry, 8*(3), 3031–3033.

Baltensperger, D., Lyon, D., Anderson, R., Holman, T., Stymiest, C., Shanahan, J., Nelson, L., Deboer, K., Hein, G., & Krall, J. (1914). *Producing and Marketing Proso Millet in the High Plains*. USDA. https://www.ars.usda.gov/ARSUserFiles/30100500/Publications%2019991990/41%20Producting%20and%20marketing%20proso%20millet.pdf (accessed on 7th November 2025).

Bhat, S., Nandini, C., Srinathareddy, S., & Jayarame, G. (2019). Proso millet (*Panicum miliaceum* L.) a climate resilient crop for food and nutritional security: A review. *Environment Conservation Journal, 20*(3), 113–124.

Bunkar, D. S., Goyal, S. K., Meena, K. K., & Kamalvanshi, V. (2021). Nutritional, functional role of kodo millet and its processing: A review. *International Journal of Current Microbiology and Applied Sciences, 10*(1), 1972–1985.

Chandra, A. K., Chandora, R., Sood, S., & Malhotra, N. (2021). Global production, demand, and supply. In *Millets and Pseudo Cereals* (pp. 7–18). Woodhead Publishing.

Chapke, R. R., Prabhakar, Shyamprasad, G., Das, I. K., & Tonapi, V. A. (2018). *Improved Millet Production Technologies and Their Impact*. Technical Bulletin, ICAR-Indian Institute of Millets Research, Hyderabad 500 030, India, 84 p.

Chapke, R. R., Prasad, S. G., Das, I. K., Hariprasanna, K., Singode, A., Kanthi, S. B. S., & Tonapi, V. A. (2020). *Latest Millet Production and Processing Technologies*. Booklet, ICAR-Indian Institute of Millets Research, Hyderabad, India, 82 p.

Chauhan, S. S. (2014). *Studies on Banded Leaf and Sheath Blight of Little Millet (Panicum sumatrense) Caused by Rhizoctonia solani Kuhn* (M.Sc. Agri. thesis, J.N.K.V.V., Jabalpur, M.P., India), 95 p.

Choudhari, M. K., Tiwari, R. K., Mishra, R. M., & Namdeo, K. N. (2018). Integrated nutrient management on growth, yield, and economics of kodo millet (*Paspalum scrobiculatum* L.). *Annals of Plant and Soil Research, 20*(4), 405–408.

Choudhary, A. K., & Kumari, P. (2010). Management of mycotoxin contamination in preharvest and post-harvest crops: Present status and prospects. *Journal of Phytology, 46*(6), 1435–1437.

CPG, T. (2012). *Crop Production Guide*. Tamil Nadu Agricultural University, Coimbatore, Tamil Nadu.

Devisetti, R., Yadahally, S. N., & Bhattacharya, S. (2014). Nutrients and antinutrients in foxtail and proso millet milled fractions: Evaluation of their flour functionality. *LWT–Food Science and Technology, 59*, 889–895.

Dey, S., Saxena, A., Kumar, Y., Maity, T., & Tarafdar, A. (2022). Understanding the antinutritional factors and bioactive compounds of kodo millet (*Paspalum scrobiculatum*) and little millet (*Panicum sumatrense*), 1–19 p. https://doi.org/10.1155/2022/1578448 (accessed on 7th November 2025).

Dimple, K., Nagamani, C., Chandrika, V., Kumar, A. N., & Sagar, G. K. (2020). Productivity, nutrient uptake, and post-harvest soil nutrient status as influenced by times of sowing and nitrogen levels in proso millet. *Andhra Pradesh Journal of Agricultural Sciences, 6*(3), 115–119.

Divya, S., & Maurya, B. M. (2013). Response of kodo millet (*Paspalum scrobiculatum*) to varying levels of nitrogen under rainfed conditions. *International Journal of Science and Research, 2*(8), 10–11.

Divyashree, U., Kumar, M. D., Sridhara, S., & Naik, T. B. (2018). Effect of different levels of fertilizers on growth and yield of little millet. *International Journal of Farm Sciences, 8*(2), 104–108.

Divyashree, U., Kumar, M. D., Sridhara, S., & Naik, T. B. (2018). Effect of different levels of fertilizers on growth and yield of little millet (*Panicum sumatrense* Roth ex Roem and Schult). *International Journal of Farm Sciences, 8*(2), 104–108.

Dong, B., Liu, M., Jiang, J., Shi, C., Wang, X., Qiao, Y., Liu, Y., Zhao, Z., & Si, F. (2014). Growth, grain yield, and water use efficiency of rain-fed spring hybrid millet (*Setaria italica*) in plastic-mulched and unmulched fields. *Agricultural Water Management, 143*, 93–101.

Dutta, A., Joshi, D., Suri, S., Trivedi, N., Acharya, R., Malik, R., Bose, S., & Dobhal, A. (2023). Bioactive and techno-functional properties of barnyard millet (*Echinochloa frumentacea*). In *Nutri-Cereals* (pp. 213–231). CRC Press.

Dwivedi, B. S., Rawat, A. K., Dixit, B. K., & Thakur, R. K. (2016). Effect of inputs integration on yield, uptake, and economics of kodo millet (*Paspalum scrobiculatum* L.). *Economic Affairs, 61*(3), 519.

Gajalakshmi, R., Chitdeshwari, T., Maragatham, S., & Ravikesavan, R. (2022). Seed priming with different levels and sources of zinc on the seed germination and seedling growth of barnyard millet (*Echinochloa frumentacea*). *Journal of Applied and Natural Science, 14*(3), 876–884.

Gangadharayya, S. H., Prashant, S. M., Shakuntala, N. M., Kurnalliker, V. K., Yogesh, L. N., & Krishnamurthy, D. (2019). Effect of seed priming with chemicals, micronutrients, and bio-inoculants on physiological attributes of resultant seed in foxtail millet (*Setaria italica* L.). *International Journal of Chemical Studies, 7*(6), 46–49.

Gavit, H. D., Rajemahadik, V. A., Bahure, G. K., Jadhav, M. S., Thorat, T. N., & Kasture, M. C. (2017). Effect of establishment techniques and sowing time on yield and yield attributes of proso millet (*Panicum miliaceum* L.). *International Journal of Current Microbiology and Applied Sciences, 6*(5), 1523–1528.

Ghimire, K. H., Joshi, B. K., Dhakal, R., & Sthapit, B. R. (2018). Diversity in proso millet (*Panicum miliaceum* L.) landraces collected from the Himalayan mountains of Nepal. *Genetic Resources and Crop Evolution, 65*(2), 503–512.

Gomashe, S. S. (2017). Barnyard millet: Present status and future thrust areas. In J. V. Patil (Ed.), *Millets and Sorghum: Biology and Genetic Improvement* (pp. 184–198). https://doi.org/10.1002/9781119130765.ch7 (accessed on 7th November 2025).

Gregory, P. J., Mayes, S., Hui, C. H., Jahanshiri, E., Julkifle, A., & Kuppusamy, G. (2019). Crops for the Future (CFF): An overview of research efforts in the adoption of underutilized species. *Plantae, 250*(3), 979–988.

Gupta, A., Mahajan, V., Kumar, M., & Gupta, H. (2009). Biodiversity in the barnyard millet (*Echinochloa frumentacea* Link. Poaceae) germplasm in India. *Genetic Resources and Crop Evolution, 56*, 883–889. https://doi.org/10.1007/s10722-009-9462-y (accessed on 7th November 2025).

Gururani, K., Sood, S., Kumar, A., Joshi, D. C., Pandey, D., & Sharma, A. R. (2021). Mainstreaming Barahnaja cultivation for food and nutritional security in the Himalayan region. *Biodiversity and Conservation, 30*, 551–574.

Habiyaremye, C., Matanguihan, J. B., D'Alpoim Guedes, J., Ganjyal, G. M., Whiteman, M. R., & Kidwell, K. K. (2017). Proso millet (*Panicum miliaceum* L.) and its potential for cultivation in the Pacific Northwest, U.S.: A review. *Frontiers in Plant Science.* https://www.frontiersin.org/articles/10.3389/fpls.2016.01961 (accessed on 7th November 2025).

Hariprasanna, K. (2017). Kodo millet, *Paspalum scrobiculatum* L. In *Millets and Sorghum: Biology and Genetic Improvement* (pp. 199–225).

Hell, K., Fandohan, P., Bandyopadhyay, R., Kiewnick, S., Sikora, R., & Cotty, P. J. (2008). Pre- and postharvest management of aflatoxin in maize: An African perspective. In J. F. Leslie, R. Bandyopadhyay, & A. Visconti (Eds.), *Mycotoxins: Detection Methods, Management, Public Health, And Agricultural Trade* (pp. 219–229). CAB International.

Hemalatha, S., Reddy, G. P., Chandrika, V., Naidu, M., & Sagar, G. K. (2021). Growth and yield attributes of kodo millet (*Paspalum scrobiculatum* L.) as influenced by integrated nitrogen management. *Journal of Agricultural Sciences, 7*(3), 134–138.

Heyduck, R. F., Baltensperger, D. D., Nelson, L. A., & Graybosch, R. A. (2008). Yield and agronomic traits of waxy proso in the Central Great Plains. *Crop Science, 48*(2), 741–748.

Himasree, B., Hemalatha, S., Sumathi, V., Sudhakar, P., Nagamadhuri, K. V., & Karunasagar, G. (2021). Growth of little millet as influenced by varied agronomic interventions. *The Pharma Journal, 10*(11), 336–339.

Islam, M. R., Rahman, M. A., Anik, A. R., Biswas, J. K., Akhtar, S., Uddin, M. N., & Ahmed, S. (2023). The role of minor cereals in food and nutrition security in Bangladesh: Constraints to sustainable production. *Food Security, 15*, 1151–1160.

Iswariya, S., Sujatha, K., & Subhashini, R. (2019). Enhancement of seedling vigor through bio-priming for barnyard millet var. MDU 1. *International Journal of Current Microbiology and Applied Sciences, 8*(1), 2254.

Janani, N., Singh, D. R., Ramesh, C., Vendan, T. R., & Selvam, S. (2022). Effect of row spacing and potassium management on the growth and yield attributes of barnyard millet (*Echinochloa frumentacea* L.) under irrigated conditions. *The Pharma Innovation Journal, 11*(7), 1944–1948.

Jawahar, S., Ramesh, S., Kumar, S. R., Kalaiyarasan, C., Arivukkarasu, K., & Suseendran, K. (2020). Effect of weed management practices on weed indices in transplanted kodo millet. *Plant Archives, 20*, 3196–3198.

Kabak, B. (2009). The fate of mycotoxins during thermal food processing. *Journal of the Science of Food and Agriculture, 89*(4), 549–554.

Kakad, C. B., Chavan, A. P., Varnekar, K. D., & Game, V. N. (2021). Effect of seedlings age and nutrient combinations on growth and yield of proso millet (*Panicum miliaceum* L.) in lateritic soils of Western Ghats. *Indian Journal of Agronomy, 66*(2), 246–250.

Kalsi, R., & Bhasin, J. K. (2023). Nutritional exploration of foxtail millet (*Setaria italica*) in addressing food security and its utilization trends in the food system. *eFood, 4*(5), 111. https://doi.org/10.1002/efd2.111 (accessed on 7th November 2025).

Kannababu, N., Deepika, C., Venkateswarlu, R., Hariprasanna, K., & Das, I. K. (2023). Physiological maturity of seeds in kodo millet (*Paspalum scrobiculatum* L.) cultivars across seasons. *Plant Physiology Reports*, 1–12.

Kavitha, M., Rai, P. K., Haritha, B., Kumar, R. K., Kumar, A. H., & Khandka, S. (2022). Standardization of different pre-sowing seed treatments on growth, yield, and yield-attributing traits of kodo millet (*Paspalum scrobiculatum*). *Indian Journal of Agricultural Research, 56*(2), 162–169.

Keerthanapriya, S., Hemalatha, M., Joseph, M., & Prabina, B. J. (2019). Assessment of competitiveness and yield advantages of the little millet-based intercropping system under rainfed conditions. *International Journal of Chemical Studies, 7*(3), 4121–4124.

Kottegoda, N., Munaweera, I., Madusanka, N., & Karunaratne, V. A. (2011). Green slow-release fertilizer composition based on urea-modified hydroxyapatite nanoparticles encapsulated wood. *Current Science, 101*(1), 73–78.

Krishnaveni, S. A., & Avudaithai, S. (2020). Effect of crop geometry and nutrient management in barnyard millet under sodic soil condition. *International Journal of Current Microbiology and Applied Sciences, 9*(6), 276–280.

Kulkarni, L. R., Naik, R. K., & Katarki, P. A. (1992). Chemical composition of minor millet. *Karnataka Journal of Agricultural Sciences, 5*(3), 255–256.

Kumar, A., Jain, A. K., & Lal, N. (2017). Efficacy of *Trichoderma* and agrochemical against grain smut of little millet (*Panicum sumatrense*). *Trends in Biosciences, 10*(6), 1422–1425.

Kumar, J., Kumar, B., & Yadav, V. K. (2007). Small millets research at GB Pant University. *G. B. Pant University of Agriculture and Technology*, Pantnagar, Uttarakhand, India, 3–17.

Kumar, K., Thakur, L., Anjani, K., & Singh, S. K. (2023). Biochemical characterization of grain iron and zinc content in little millet genotypes. *Journal of Pharmaceutical Innovation, 12*(5), 4399–4402.

Kumar, M. K., Shekara, B. G., Yamuna, B. G., & Sunil, C. M. (2015). Crop weed competition for nutrients by weed and drill sown finger millet (*Eleusine coracana* L. Gaertn.). *National Academy of Agricultural Sciences, 33*(3), 2049–2054.

Layek, J., Rangappa, K., Das, A., Ansari, M. A., Choudhary, S., Rajbonshi, N., Patra, S., Kumar, A., Mishra, V. K., Ravisankar, N., Kumar, S., Hazarika, S., Dutta, S. K., Babu, S., Tahasildar, M., & Shettigar, N. (2023). Evaluation of millets for physio-chemical and root morphological traits suitable for resilient farming and nutritional security in Eastern Himalayas. *Frontiers in Nutrition, 10*, 1198023. https://doi.org/10.3389/fnut.2023.1198023 (accessed on 7th November 2025).

Lee, S., Jeon, S., Seo, S., Kim, J., Seong, P., & Baek, Y. C. (2021). Comparative analysis of the energy values of oat hay, tall fescue, annual ryegrass, and barnyard millet using *in vivo* digestibility results and predicted values. *Korean Journal of Agricultural Science, 48*(4), 831–842.

Liu, M., Qiao, Z., Zhang, S., Wang, Y., & Lu, P. (2015). Response of broomcorn millet (*Panicum miliaceum* L.) genotypes from semiarid regions of China to salt stress. *Crop Journal, 3*(1), 57–66.

Lohani, U. C., Pandey, J. P., & Shahi, N. C. (2012). Effect of degree of polishing on milling characteristics and proximate compositions of barnyard millet (*Echinochloa frumentacea*). *Food and Bioprocess Technology, 5*(3), 1113–1119. https://doi.org/10.1007/s11947-011-0518-6 (accessed on 7th November 2025).

Lokesh, T., Choudhary, A. A., & Mairan, N. R. (2023). Effect of different sowing windows and line spacings on yield and economics of foxtail millet during rabi season. *Journal of Pharmaceutical Innovation, 12*(2), 1121–1123.

Mahajan, G. (2016). Effect of Kodo millet (*Paspalum scrobiculatum*) based intercropping system on yield and economics of Kodo millet under rainfed conditions. *New Agriculturist, 27*(1), 121–124.

Maitra, S., & Shankar, T. (2019). Agronomic management in little millet (*Panicum sumatrense* L.) for enhancement of productivity and sustainability. *International Journal of Bioresource Science, 6*(2), 91–96.

Maitra, S., Panda, P., Panda, S. K., Behera, D., Shankar, T., & Nanda, S. P. (2020). Relevance of barnyard millet (*Echinochloa frumentacea* L.) cultivation and agronomic management for production sustainability. *International Journal of Bioinformatics and Biological Sciences, 8*, 27–32.

Maitra, S., Pine, S., Shankar, T., Pal, A., & Pramanick, B. (2020). Agronomic management of foxtail millet (*Setaria italica* L.) in India for production sustainability: A review. *International Journal of Bioresource Science, 7*(1), 11–16.

Maitra, S., Praharaj, S., Hossain, A., Patro, T. S. S. K., Pramanick, B., Shankar, T., Pudake, R. N., Gitari, H. I., Palai, J. B., Sairam, M., & Sagar, L. (2022). Small millets: The next-generation smart crops in the modern era of climate change. In *Omics of Climate Resilient Small Millets* (p. 25). Springer Nature, Switzerland. Available: http://dx.doi.org/10.1007/978-981-19-3907-5_1 (accessed on 7th November 2025).

Manna, M., & Kim, K. D. (2017). Influence of temperature and water activity on deleterious fungi and mycotoxin production during grain storage. *Microbiology, 45*(4), 240–254.

Meena, R. P., Joshi, D. C., Bisht, J. K., & Kant, L. (2021). Global scenario of millet cultivation. In A. Kumar, M. K. Tripathi, D. Joshi, & V. Kumar (Eds.), *Millets and Millet Technology* (pp. 33–50). Springer, Singapore.

Meng, X., Lian, Y., Liu, Q., Zhang, P., Jia, Z., & Han, Q. (2020). Optimizing the planting density under the ridge and furrow rainwater harvesting system to improve crop water productivity for foxtail millet in semiarid areas. *Agricultural Water Management, 238*(1), 106220.

Michaelraj, P. S. J., & Shanmugam, A. (2013). A study on millets-based cultivation and consumption in India. *International Journal of Marketing, Financial Services and Management Research, 2*(4), 49–58.

Mubeena, P., Halepyati, A. S., & Chittapur, B. M. (2019). Effect of date of sowing and nutrient management on nutrient uptake and yield of foxtail millet (*Setaria italica* L.). *International Journal of Bio-Resource and Stress Management, 10*(1), 92–95.

Muthamilarasan, M., & Prasad, M. (2021). Small millets for enduring food security amidst pandemics. *Trends in Plant Science, 26*(1), 33–40. https://doi.org/10.1016/j.tplants.2020.08.008 (accessed on 7th November 2025).

N/A. (2018). *Annual Report 2017–18*. Hyderabad: IIMR.

Nandede, B. M., Chandel, N. S., Kumar, T. S., & Dhimate, A. S. (2018). Development of manually operated single-row millet planter cum fertilizer drill. *Indian Journal of Dryland Agricultural Research and Development, 33*(1), 01–06.

Nandini, K. M., & Sridhara, S. (2019). Response of growth, yield and quality parameters of foxtail millet genotypes to different planting densities. *International Journal of Current Microbiology and Applied Sciences, 8*(2), 1765–1773.

Nasibi, F., Aminian, F., Mohammadinejad, G., & Hassanshahian, M. (2022). Seed priming with selenium nanoparticles and plant growth-promoting rhizobacteria improve seedling development of foxtail millet (*Setaria italica*) under salinity stress. *Research Square*. https://doi.org/10.21203/rs.3.rs-1809244/v1 (accessed on 7th November 2025).

Nazni, P., & Shobana, D. R. (2016). Effect of processing on the characteristic changes in barnyard and foxtail millet. *Journal of Food Processing & Technology, 7*(3), 566–574.

Neme, K., & Mohammed, A. (2017). Mycotoxin occurrence in grains and the role of postharvest management as mitigation strategies. *Food Control, 78*, 412–425.

Ojha, E., Adhikari, B., & Katuwal, Y. (2018). Nutrient management trial on foxtail millet at Sundarbazar, Lamjung. *Journal of the Institute of Agriculture and Animal Science, 35*(1), 89–94.

Pagare, A., Thakur, A. K., Salam, P. K., Chandrakar, T., & Singh, D. P. (2022). Response of liquid bio-fertilizers on growth, yield attributing characters, and economics of little millet (*Panicum sumatrense* Roth) under Marhan (upland) situations. *The Pharma Innovation Journal, 11*(8), 1445–1449.

Pall, B. S., Jain, A. C., & Singh, S. P. (1980). Diseases of lesser millets. *J.N.K.V.V.*, Jabalpur (M.P.), 62–69.

Patel, V., Sengar, S. S., Singh, R. K., Singh, A. K., Kumar, R., & Singh, O. (2020). Primary nutrient content and its uptake in little millet (*Panicum sumatrense*) as influenced by different nutrient management and seed priming. *Journal of Pharmacognosy and Phytochemistry, 9*(4), 1228–1233.

Pavithra, K. S., Senthil, A., Babu Rajendra Prasad, V., Ravikesavan, R., & Djanaguiraman, M. (2020). Variations in photosynthesis-associated traits and grain yield of minor millets. *Plant Physiology Reports, 25*(3), 418–425.

Pawar, V. D., & Machewad, G. M. (2006). Processing of foxtail millet for improved nutrient availability. *Journal of Food Processing and Preservation, 30*, 269–279.

Prabha, D., Negi, Y. K., & Khanna, V. K. (2010). Morphological and isozyme diversity in the accessions of two cultivated species of barnyard millet. *Nature and Science, 8*, 71–76.

Prabhakar, P. C., Ganiger, B. B., Bhat, S., Nandini, C., Tippeswamy, V. K., & Manjunath, H. A. (2017). Improved production technology for barnyard millet. *Project Co-ordinating Unit, ICAR-AICRP on Small Millets*, GKVK, Bengaluru, 16.

Pradhan, A., Thakur, A., Patel, S., & Mishra, N. (2011). Effect of different nitrogen levels on Kodo and finger millet under rainfed conditions. *Research Journal of Agricultural Sciences, 2*, 136–138.

Prashant, S. M., Macha, S. I., Kurnalliker, V. K., Yogeesh, L. N., & Ravikumar, A. (2021). Influence of seed bio priming for enhancing seed quality in finger millet (*Eleusine coracana* L. Garten.). *Journal of Pharmacognosy and Phytochemistry, 10*(1), 102–104.

Prithvi, K. B., Rao, A. S., & Srinivasulu, K. (2015). Weed management in transplanted ragi. *Indian Journal of Weed Science, 47*(2), 214–215.

Rajput, J. S., Thakur, A. K., Nag, N. K., Chandrakar, T., & Singh, D. P. (2022). Effect of nano fertilizers on growth, yield, and economics of little millet (*Panicum sumatrense* Roth) under rainfed conditions. *The Pharma Innovation Journal, 11*(7), 153–156.

Ram, M., Meena, R. C., Parewa, H. P., & Meena, D. (2021). Productivity and profitability of pearl millet as affected by zinc and iron application in arid and semi-arid regions. *International Journal of Agriculture, Environment and Biotechnology, 14*, 375–380.

Rani, P., Charu, S., & Kumar, K. (2023). Growth trend of pearl millet and its impact on Indian economy: Performance analysis. *International Journal of Advanced Multidisciplinary Research Studies, 3*(3), 162–166.

Rani, Y. S., Jamuna, P., Rao, P. J., Triveni, U., Patro, T. S. S. K., & Anuradha, N. (2020). Yield and quality of little millet (*Panicum sumatrense*) as influenced by organic manures and inorganic fertilizers. *Journal of Pharmacognosy and Phytochemistry, 9*(5), 595–598.

Rani, Y. S., Triveni, U., & Patro, T. S. S. K. (2017). Integrated nutrient management for enhancing the soil health, yield, and quality of little millet (*Panicum sumatrense*). *International Journal of Bio-Resource and Stress Management, 8*(1), 26–32.

Ravali, P. S., Prasad, P. V. N., Venkateswarlu, B., & Rao, C. S. (2021). Growth and yield of foxtail millet as influenced by different weed management practices. *The Pharma Innovation Journal, 10*(7), 870–872.

Ravindra, U., Vijayakumari, J., Sharan, S., Raghuprasad, K. P., & Kavaloor, R. (2008). A comparative study of postharvest processing methods for little millet (*Panicum miliare* L.). *Tropical Agricultural Research, 20*, 115–122.

Raviraja, S. R., Balanagoudar, M. V., Ravi, S. N., Bhat, S. N., & Yadahalli, G. S. (2020). Effect of integrated nutrient management practice on growth and yield of foxtail millet (*Setaria italica* L.) in black soil. *International Journal of Current Microbiology and Applied Sciences, 9*(3), 2101–2108.

Renganathan, V. G., & Vanniarajan, C. (2018). Exploring the barnyard millet (*Echinochloa frumentacea* Roxb. Link) segregating population for isolation of high-yielding iron and zinc content genotype. *International Journal of Current Microbiology and Applied Sciences, 7*(4), 3611–3621. https://doi.org/10.20546/ijcmas.2018.704.407 (accessed on 7th November 2025).

Renganathan, V. G., Renuka, R., Vanniarajan, C., Raveendran, M., & Elangovan, A. (2023). Selection and validation of reliable reference genes for quantitative real time PCR in barnyard millet (*Echinochloa* spp.) under varied abiotic stress conditions. *Scientific Reports, 13*, 15573–15586.

Renganathan, V. G., Vanniarajan, C., Karthikeyan, A., & Ramalingam, J. (2020). Barnyard millet for food and nutritional security: Current status and future research direction. *Frontiers in Genetics, 11*, 500–521.

Sachin, S., Asewar, B. V., & Syed Shireen, J. R. (2023). Response of different little millet (*Panicum sumatrense* L.) varieties to fertilizer levels. *Journal of Pharmaceutical Innovation, 12*(1), 1590–1592.

Saleh, A., Zhang, Q., Chen, J., & Shen, Q. (2013). Millet grains: Nutritional quality, processing, and potential health benefits. *Comprehensive Reviews in Food Science and Food Safety, 12*, 281–295. https://doi.org/10.1111/1541-4337.12012 (accessed on 7th November 2025).

Sapthagiri, S., Krishnan, R., Chinnamuthu, C. R., Chandrasekhar, C. N., & Geethalakshmi, V. (2020). Performance of little millet under rice fallow condition. *Madras Agricultural Journal, 107*(1), 1.

Sarala, N., Kumar, M. H., Madhavilatha, L., Vajantha, B., & Hemalatha, T. (2020). Effect of spacing and fertilizer levels on growth and yield of little millet. *Journal of Agricultural Science, 6*(2), 71–74.

Seghatoleslami, M. J., Kafi, M., & Majidi, E. (2008). Effect of deficit irrigation on yield, WUE, and some morphological and phenological traits of three millet species. *Pakistan Journal of Botany, 40*(4), 1555–1560.

Selvi, M. V., Nirmalakumari, A., & Senthil, N. (2015). Genetic diversity for zinc, calcium, and iron contents of selected little millet genotypes. *Journal of Nutrition and Food Sciences, 5*, 417. https://doi.org/10.4172/2155-9600.1000417 (accessed on 7th November 2025).

Shamina, C., Annadurai, K., Hemalatha, M., & Suresh, S. (2019). Effect of spacing and weed management practices on barnyard millet (*Echinochloa frumentacea*) under rainfed conditions. *International Journal of Current Microbiology and Applied Sciences, 8*(6), 330–337.

Sharmili, K., & Manoharan, S. (2018). Studies on intercropping in rainfed little millet (*Panicum sumatrense*). *International Journal of Current Microbiology and Applied Sciences, 7*(2), 323–327.

Singh, K. P., Kundu, S., & Gupta, H. S. (2003). Development of higher capacity thresher for ragi/kodo. In *Recent Trend in Millet Processing and Utilization* (pp. 109–116). CCSHAU, Hissar, India.

Singh, N., Kumari, P., Gulati, J., & Bassi, N. (2023). Nutrient-rich Kodo millet, importance and value addition: An overview. *Journal of Pharmaceutical Innovation, 12*(6), 3713–3719.

Singh, R. V., & Arya, M. P. S. (1999). Effect of seed ratio on barnyard millet (*Echinochloa frumentacea*) based mixed cropping system. *Indian Journal of Agronomy, 44*, 51–55.

Sonhkarri, R., Jeyapandiyan, N., Anitha, M., Kalaiarasi, J. P., Poonguzhali, R. S., Jayapradha, C., Rathikannu, S., & Kumar, K. U. (2023). Impact of different levels of iron fertilizer on growth and yield physiology of Kodo millet under rainfed conditions–An overview. *Journal of Applied Biology and Biotechnology, 11*(2), 33–40.

Sonkar, S., Jaddu, S., Pradhan, R. C., Dwivedi, M., Seth, D., Goksen, G., Sarangi, P. K., & Lorenzo, J. M. (2023). Effect of atmospheric cold plasma (pin type) on hydration and structure properties of Kodo-millet starch. *LWT, 182*, 114889–114901.

Sood, S., Khulbe, R. K., & Kant, L. (2016). Optimal yield-related attributes for high grain yield using ontogeny-based sequential path analysis in barnyard millet (*Echinochloa* spp.). *Journal of Agricultural Science and Technology, 18*(7), 1933–1944.

Sridevi, R., & Manonmani, V. (2016). Seed priming effect on physiological traits of Kodo millet and barnyard millet. *International Journal of Agricultural Science Research, 6*(4), 187–194.

Sridevi, R., Manonmani, V., & Vijayakumar, A. (2020). Effect of seed priming in Proso millet (*Panicum miliaceum*) and barnyard millet (*Echinochloa frumentacea*) under field conditions. *Research on Crops, 21*(2), 390–395.

Srikanya, B., Revathi, P., Reddy, M. M., & Chandrashaker, K. (2020). Effect of sowing dates on growth and yield of foxtail millet (*Setaria italica* L.) varieties. *International Journal of Current Microbiology and Applied Sciences, 9*(4), 3243–3251.

Sujatha, K., Selvarani, K., Vijayalakshmi, V., Vanniarajan, C., & Sivasubramaniam, K. (2013). Seed fortification studies in barnyard millet (*Echinochloa frumentacea*) cv. CO1. *IOSR Journal of Agriculture and Veterinary Science, 5*(4), 22–24.

Sukanya, T. S., Prabhakar, K., Krishne Gowda, K. T., Swarna, R., Hariprasanna, K., & Tonapi, V. A. (2022). Good agronomic practices for higher yield in small millets. *ICAR - All India Coordinated Research Project on Small Millets, ICAR-Indian Institute of Millets Research*, Hyderabad.

Suruthi, S., Sujatha, K., & Menaka, C. (2019). Evaluation of different seed priming methods for improvement of seedling growth of barnyard millet (*Echinochloa frumentacea* L.) cv. MDU1. *Electronic Journal of Plant Breeding, 10*(2), 400–405.

Tran, G. (2015). Proso millet (*Panicum miliaceum*). *Feedipedia, INRA, CIRAD and FAO, 16*, 16. Available: https://www.feedipedia.org/node/722 (accessed on 7th November 2025).

Trivedi, A. K., Arya, L., Verma, S. K., Tyagi, R. K., & Hemantaranjan, A. (2017). Evaluation of barnyard millet diversity in Central Himalayan region for environmental stress tolerance. *Journal of Agricultural Science, 155*(10), 1497–1507.

Umadevi, G. D., Sumathi, V., Reddy, A. P., Sudhakar, P., & Kumari, K. L. (2019). Effect of organic manures and phosphorus on cowpea and their residual effect on succeeding little millet. *Journal of Pharmacognosy and Phytochemistry, 8*(3), 2236–2239.

Ungata, T., Poiba, J., Narasupalli, A., Patro, T. S. S. K., & Yesarapu, S. R. (2020). Influence of sowing windows on the productivity of small millet crops. *Journal of Pharmacognosy and Phytochemistry, 9*(5), 1291–1294.

Ushakumari, S. R., Latha, S., & Malleshi, N. G. (2004). The functional properties of popped, flaked, extruded, and roller-dried foxtail millet (*Setaria italica*). *International Journal of Food Science and Technology, 39*, 907–915.

Verma, K. C., Joshi, N., Rana, A. S., & Bhatt, D. (2020). Quality parameters and medicinal uses of foxtail millet (*Setaria italica* L.): A review. *Journal of Pharmacognosy and Phytochemistry, 9*(4), 1036–1038.

Vilas, A., Tonapi, B., Bhat, V., Kannababu, N., Elangovan, M., Umakanth, A. V., Kulakarni, R., Tonapi, K. V., Rao, K. V. R., & Rao, T. G. N. (2015). *Millet Seed Technology: Seed Production, Quality Control & Legal Compliance* (p. 114).

Villers, P. (2014). Aflatoxins and safe storage. *Frontiers in Microbiology, 5*, 158–164.

Xing, G., Ma, J., Liu, X., Lei, B., Wang, G., Hou, S., & Han, Y. (2023). Influence of different nitrogen, phosphorus, and potassium fertilizer ratios on the agronomic and quality traits of foxtail millet. *Agronomy, 13*(8), 2005.

PART IV

Integrated Pest and Disease Management

CHAPTER 13

Integrated Pest and Disease Management in Sorghum

MANJUNATH HUBBALLI,[1,*] V. A. ANJALI,[1] R. HARISCHANDRA NAIK,[1] and CHRISTOPHER R. LITTLE[2]

[1]*University of Horticultural Sciences, Bagalkot, Karnataka, India*

[2]*Kansas State University, Manhattan, Kansas, USA*

*Corresponding author

13.1 INTRODUCTION

Sorghum, most frequently known as the "food security crop," is widely consumed as food and feed worldwide. It is also a famous biofuel crop and most often surpasses maize in its feeding value. The African and southeastern countries viz. India, Bangladesh, Bhutan, Maldives, Nepal, Pakistan, and Sri Lanka mainly use it as a food crop, while it is of fodder and feed value in the United States. The crop contributes immensely to food security in African countries. Moreover, it is a good source of vitamin B and potassium. Sorghum also contains certain anti-nutritional compounds such as tannins, cyanogenic glucosides, and phytic acid. Although the United States stands first in the production of sorghum worldwide, more variability in wild and cultivated sorghum is found in northeastern Africa. The crop was probably domesticated in Ethiopia 5000 to 7000 years ago and passed along the Middle East to India about 3000 years back. India occupies fifth place in sorghum production, with a total area of 5.5 million hectares and an annual production of 1.69 million tons (FAOSTAT, 2022).

Sorghum, a "climate-smart crop," is well adapted to various environmental conditions and can withstand most unfavorable environments, giving significant yields. As a C_4 plant with an extensive root system and waxy bloom on leaves, the crop is particularly drought tolerant. It can cease growth and resume growth according to water availability. Apart from this, the crop can also withstand waterlogging and is well adapted to a wide range of soil types. Though it is adopted to semiarid tropics, the crop grows well in temperate conditions. Hence, sorghum is considered a hardy crop withstanding a certain limit of abiotic stresses and thus can be cultivated in extreme environmental conditions where many other crops cannot grow (Chaturvedi et al., 2023).

Millets in Sustainable Agriculture: Genetic Improvement, Nutritional Quality, Cultivation, and Integrated Pest–Disease Management. Roshni Vijayan, A. John Joel, R. Ambika Rajendran, and Adithya P. Balakrishnan (Eds.)

DOI: 10.1201/9781779642332-17

Although the crop exhibits a high potential to withstand many abiotic stresses, its production is highly challenged by biotic stresses mainly caused by pests and diseases. The diseases caused by fungi, bacteria, viruses, and nematodes hamper the production to a greater extent. The diseases which cause an extensive yield reduction in the crop are anthracnose, downy mildew, and ergot whereas the, leaf spot, sooty stripe and diseases occurring on leaf due to bacteria inflict minimum yield loss. Tesso et al. (2012) reported the perennial yield drag in charcoal rot, *Fusarium* rot and grain mold diseases. The reported yield losses due to disease and insect pests are given in Table 13.1.

TABLE 13.1 Yield Losses Associated with Insect Pests/Diseases of Sorghum

Sl. No.	Disease/Pest	Yield Loss Reported (%)	References
		Diseases	
1	Anthracnose	50.00	Djaenuddin et al. (2023) and Koima et al. (2023)
		50.00	Binyam et al. (2023)
		80.00	Tariku (2021)
		67.00	Abreha et al. (2021); Koima et al. (2023)
2	Charcoal rot	77.00	Amrate et al. (2023)
3	Fusarium stalk rot	5.00–50.00	Ketema (2023)
4	Head smut	5.00–80.00	Prom et al. (2021)
5	Turticum blight	50.00	Mittal and Boora (2005)
		74.45	Ogolla et al. (2019)
6	Grey Leaf Spot	67.00	Anitha et al. (2020)
7	Rust	65.00	Bandyopadhyay (2000)
		30.00–50.00	Anitha et al. (2020)
8	Downy Mildew	78.00	Anitha et al. (2020)
		Insects Pests	
Insect Pests Infesting Standing Crop			
1	Maize stem borer	29.00–72.00	Puri and Mote (2003)
2	shoot fly	75.60	Balikai and Bhagwat (2010)
3	Sorghum Midge	50.00–100.00	Knutson and Cronholm (2007)
4	Ear head caterpillar	50.00–70.00	Guo et al. (2011)
5	Sorghum Head Bugs (mirid bug)	86.00	Malgwi and Dunuwel (2011)
6	Sugarcane aphid	46.00–78.00	Van den Berg (2002)
Insect Pests Infesting at Storage			
1	Rice weevil	50.00	Sanon et al. (2018)
2	Tamarind Beetle	50.00–70.00	Jood et al. (1993)
3	Lesser grain borer	50.00–70.00	Jood et al. (1993)

Right from the emergence of seedlings to the formation of grains, the crop is susceptible to many insect pests. Armyworms and insects residing in the soil form the insect group that occurs at early stages. The damage caused by these pests is usually minimal due to their low numbers in a grain sorghum crop, and hence the crop can tolerate the damage. In some cases, due to certain climatic conditions, the population may build up and cause significant damage to the crop. Some of the insect pests of sorghum include shoot fly, shoot bug, ear head bug, stem borer, sorghum midge, plant lice (aphids), ear head webworm, gram caterpillar, slug caterpillar, flea beetle, red hairy caterpillar, semi looper, ash weevils, leaf roller, and wingless grasshopper (Olabimpe et al., 2021). The important pests and diseases are discussed in the following units.

13.2 MAJOR DISEASES IN SORGHUM

The diseases occurring in the crop can be categorized as seed rot, seedling diseases, root and stalk diseases, foliar diseases, and panicle and grain diseases.

13.2.1 SEED ROT AND SEEDLING DISEASES

13.2.1.1 PRE-AND POST-EMERGENT DAMPING-OFF

Both seedborne and soilborne pathogens can cause damping-off. Damping-off diseases are the most serious, indicated by the fact that they result in the failure of the seed to germinate or the seedling to establish.

1. **Etiology:** Soilborne pathogens responsible for pre- and post-emergent damping-off include *Rhizoctonia solani, Pythium spp.*, and *Macrophomina phaseolina. Pythium aphanidermatum, P. arrhenomanes, P. graminicola, P. myriotylum,* and *Pythium ultimum var. ultimum* are some of the *Pythium* spp. associated with damping-off. On the other hand, *Fusarium thapsinum* is a seedborne pathogen infecting sorghum florets and developing kernels, resulting in reduced seed germination and seedling blight (Das et al., 2023).

2. **Symptoms**: Regardless of the mode of pathogen entry, whether seedborne, soilborne, or due to improper storage of seeds, the most common symptom is weakened seedlings. Pre-emergent damping-off can occur due to the inability of the seed to germinate because of the rotting of seeds, which show a brown-to-black appearance. Even if the seeds germinate and emerge, they will be weak and stunted, especially when infected with *Fusarium* spp. The infested field is characterized by uneven or sparse stands (Das et al., 2023).

3. **Management:** The soil should have a good drainage system to remove surplus water. Good quality seeds must be used for sowing. In general, hybrids will have more tolerance compared to varieties; hence, hybrids are advised in disease-prone areas. Seed treatment with Captan or Thiram at 4 g/kg and foliar application of Mancozeb 1.25 kg or Captafol 1 kg/ha are suggested to manage the disease (Lettry Vivien, 2023).

13.2.2 ROOT AND STALK DISEASES

In most cases, many pathogens are responsible for root rot in sorghum, hence it can be referred to as a root rot complex. Soilborne pathogens, namely *Fusarium* spp., *Macrophomina phaseolina, Pythium* spp., and *R. solani,* are associated with sorghum root rots. Necrotrophic and hemi-biotrophic pathogens causing root rot can enter through wounds caused by insect or mechanical damage as well as through apertures in the ground. Though many pathogens infect the roots of sorghum at the same time, the peculiar symptoms are produced only when one pathogen dominates. In severe cases, certain pathogens like *Fusarium* spp. or *M. phaseolina* can develop in the stalk portion, causing stunting of the plants, especially in the apical region, further leading to chlorosis and necrosis of leaves, and ultimately wilting.

13.2.2.1 CHARCOAL ROT

A major constraint in sorghum production, charcoal rot is a soil and seed-borne disease with occurrence mostly severe during the post-rainy season, which is congenial for this disease.

1. **Etiology:** Charcoal root rot/stalk rot is caused by the necrotrophic fungus *Macrophomina phaseolina* (Perfect state: *Sclerotium bataticola*). The fungus is both seed- and soil-borne, and microsclerotia act as a source of inoculum for the pathogen. The hyphae of the fungi develop in the soil and infect the smaller and developing roots (Cloud and Rupe, 1994).
2. **Symptoms:** Initially, discoloration occurs at the basal part of the stalk, which appears as water-soaked lesions that later turn brown or black. This is followed by lodging of plants, as such stalks become soft at the base (Anita et al., 2020). Further, internal disintegration and decomposition at the crown level of pith lodging can be seen, resulting in the loss of structural integrity of the stalk. Also, a visible sign of the disease includes the development of microsclerotia where the stalk is damaged (Tesso et al., 2012). Moreover, stalks infected by the fungi appear ash-gray or black in color (Adeyanju, 2014). This is also accompanied by incomplete/poor development of panicles and poor-quality grains. In many cases, panicle and stalk chlorosis are the common symptoms observed in the aerial region.
3. **Management:** Good soil tillage during the summer months must be done to expose the soil-borne inoculum, and an adequate drainage system should be established in the field. Crop rotation with soybean can be practiced to reduce the inoculum in the soil. *Trichoderma asperellum* formulated in talc should be used for seed treatment at 4 g/kg, or a talc-based formulation of *Pseudomonas fluorescens* at 10 g/kg of seed is advisable for the effective management of the disease. In addition to this, fungicides such as carbendazim or thiram at 2 g per kg of seed can also be applied. The selective soil drenching using carbendazim at 1 g/liter or soil application of *P. fluorescens/T. viride* at 2.5 kg/ha with 50 kg of FYM is useful in disease management.

13.2.2.2 FUSARIUM ROOT AND STALK ROT

Another major soil-borne disease, *Fusarium* root and stalk rot is a severe disease as the pathogen mostly builds up in the crop debris, hence it is more prevalent in areas of continuous cropping of sorghum.

1. **Etiology:** Various species of *Fusarium* spp. are reported to cause the root rot disease, namely *F. thapsinum, F. proliferatum, F. subglutinans*, and *F. andiyazi* (Tesso et al., 2010; Tesso et al., 2012).
2. **Symptoms:** Unlike charcoal rot, this disease develops when there is sufficient moisture in the soil, especially at the early stages of crop development. The visible symptoms include stalk discoloration and a scorched appearance of leaves, often appearing "grayed out or freeze damaged" (Adeyanju et al., 2015; Bandara et al., 2015; Tesso et al., 2012). The disease is often characterized by a shredded vascular system, and the pith area may turn red/pink or purple due to the accumulation of 3'-deoxyanthocyanins and other flavonoid-derived anthocyanin pigments (Adeyanju, 2014). Thus, it finally leads to lodging, and the plant undergoes premature death (Figure 13.1).

FIGURE 13.1 Fusarium stalk rot (left panel; *Fusarium* spp.) and charcoal rot (right panel; *Macrophomina phaseolina*) of sorghum.

As in charcoal rot, the predisposing factors for the development of the disease are denser plants, increased nitrogen, and low potassium in the soil. Furthermore, damage due to hail, feeding by other insects, and other diseases occurring on foliage also aggravates the disease (Mughogho, 1984; Trimboli and Burgess 1983). In addition to this, monocropping increases the population of *Fusarium* in the soil, thereby worsening the disease over time (Idris et al., 2008).

3. **Management:** Sorghum hybrids exhibiting better strength in stalks as well as resistance to leaf diseases must be used for sowing. Wherever possible, rotate crops with non-cereal crops. Maintain optimum plant population and spacing. Use the

correct dosage of nitrogen and potassium fertilizers. Manage diseases occurring on leaves using suitable fungicides wherever necessary.

13.2.2.3 MILO DISEASE

Milo disease, otherwise known as root and crown rot of sorghum, is an important quarantine disease in India and many other countries. The disease devastated sorghum production during the 1920s and 1930s in south-central USA, mainly affecting representatives of the milo race.

1. **Etiology:** The disease is caused by the fungus *Periconia circinata* (Perfect stage: *Aspergillus circinatus*). This fungus grows very slowly under artificial media, whereas it survives in soil for a longer period and tolerates temperatures of 10 to 40°C (Churchill et al., 2001). The pathogen produces a polyketide-peptide host-selective toxin, peritoxin, which incites the disease.
2. **Symptoms:** The disease is seen at almost all stages of growth and development, and the host-selective toxin produced by the fungi causes an efflux of electrolytes, resulting in the formation of lesions on young roots. With the progress of the disease, discoloration of the stele along with rotting of the stem and root begins, and conidiophores of *P. circinata* arise at this place. The stele and root epidermis exhibit conidia and conidiophores. In some sensitive genotypes, seedlings develop limited growth and curled leaves. Infected plants have dark crimson discoloration inside their crowns. The diseased plants might produce grains that are sub-normal (Mayers, 1976).
3. **Management:** If resistant varieties are available, they can be utilized as this is the most preferred method over other operations. The use of resistant cultivars is the best method of managing the disease.

13.2.2.4 POKKAH BOENG

Pokkah Boeng is derived from the Javanese term which means "malformed top." The disease, also known as twisted tops, is endemic to the sorghum-growing parts of India.

1. **Etiology:** The fungus, *Fusarium moniliforme* var. *subglutinans*, with the perfect stage *Gibberella fujikuroi* var. *subglutinans*, is the causal agent for the disease (Frederiksen, 1986). The pathogen can survive on host residue and can be spread in the form of conidia through the air. Some reports suggest the seed-borne nature of the disease (Little and Perumal, 2019).
2. **Symptoms:** The disease distorts the growing point of the plant, causing wrinkling and chlorosis of leaves. In most cases, the new leaves do not unfold properly. If the disease is mild, the symptoms resemble those caused by a virus. In severe cases, infections are seen in the stem portion, leading to the death of plants. Rarely, discoloration running vertically is seen above the basal node. The discolored portion exhibits small transverse cuts in the rinds, known as knife cut symptoms, which facilitate stalk breakage under high wind conditions. A period of dry weather

followed by rain and high humidity promotes the disease and is more common in India after the rainy season.

3. **Management:** The use of host plant resistance is an easy and effective means of managing the disease.

13.2.2.5 ERWINIA STALK ROT

A bacterial disease, more commonly known as stalk and top rot, is a disease of major relevance in tropical and subtropical areas.

1. **Etiology:** The disease is caused by the gram-negative bacteria *Dickeya dadantii* (Syn. *Erwinia chrysanthemi*). The remnants of previous crops present in the soil act as means of survival for the bacteria, and rain and irrigation water aid in dispersal and spread. The pathogen normally prefers to multiply at 30°C with frequent rainfall (Saxena et al., 1991).
2. **Symptoms:** During the initial phase, longitudinal patches appear on the uppermost leaves, followed by premature withering. Water-soaked lesions appear on the basal region of the stalk, later turning into an intense brown color. The stem pith, which is infected, starts rotting, exhibiting a bad odor, leading to wilting of the entire plant. If the infection occurs during the early stage of crop growth, it results in the premature death of the plant, whereas the crop lodges in a large area when infection occurs in the later phase of the crop.
3. **Management:** Improving soil organic carbon through the application of FYM, crop residues, and organic manures helps in managing the disease. Furthermore, correcting the terrain of the land ultimately avoids flooding in the field. Nitrogen application should be optimized to avoid overuse, as followed in maize crops (Kloppers and Tweer, 2009; Kumar et al., 2016). According to Kumar et al. (2016), the application of chlorinated water with 22% and 28% chlorine in coated and uncoated granule form, respectively, between rows would reduce the disease severity in maize crops.

13.2.3 FOLIAR DISEASES

Although foliar diseases do not cause much yield loss in sorghum, micro-epidemics can impact sorghum yield under conducive conditions. Foliar diseases can majorly impact yield if the upper foliage gets damaged during the grain-filling period.

13.2.3.1 ANTHRACNOSE

Anthracnose is the most severe foliar disease that affects grain sorghum quality as well as quantity. Anthracnose attacks all parts of the crop throughout its life, exhibiting typical symptoms on leaves, grains, and also as stalk rot (Koima et al., 2023).

1. **Etiology:** *Colletotrichum sublineola* is the pathogen responsible for the anthracnose of sorghum. The pathogen is variable, and numerous pathotypes are present, which makes its management difficult.
2. **Symptoms**: The initial symptoms appear early, during boot formation on leaves, where elliptical lesions with acervuli appear on leaf blades with clear margins. Under favorable situations, particularly when high humidity and rainfall coincide, the lesion size significantly increases and spreads to the entire leaf surface. In the case of highly susceptible varieties/hybrids, the infection can be seen on the peduncle with depressed lesions having sunken areas. If the conditions favor the development of the disease, large dead tissue can be seen in these infected areas (Frederiksen and Odvody, 2000). The stalk gets infected, forming a marbling appearance on the pith. Also, the infection on the stalk and peduncle blocks the movement of food material to the seed, resulting in poor filling of grain. During severe conditions, the plant may lodge. Infections on seeds are characterized by senescence, with black spots containing numerous acervuli on the seed surface. Humid and rainy weather with exceptionally high humidity, coupled with alternating periods of dry weather, is conducive for disease development. The rains during the vegetative growth of the plant, coupled with dew, help in the movement of conidia to the leaf sheath, which ultimately initiates anthracnose on the stalk (Koima et al., 2023).
3. **Management:** Avoiding early sowing (Tsedaley et al., 2016), planting disease-free seeds, removal of crop residues, alternate hosts, and weeds help in managing the disease (Somda et al., 2007). Seed treatment with Apron Plus, along with foliar application of mancozeb, benomyl, and carbendazim, reduces the incidence (Gwary and Asala, 2006).

13.2.3.2 LEAF BLIGHT

Leaf blight is an important polycyclic disease and is mostly prevalent in cooler and humid areas.

1. **Etiology:** *Exserohilum turcicum* is associated with the leaf blight disease. The chlamydospores, mycelia, or sclerotia residing in the crop residue act as primary sources of inoculum. However, transmission over long distances is achieved through wind and seeds (Casela and Frederiksen, 1993).
2. **Symptoms:** The disease is characterized by lesions of 2 to 5 cm with necrotic areas and cigar-shaped lesions. As the disease spreads, these areas further turn dark in color. Later, on the leaf sheaths and leaf blades, larger patches with dead tissue are seen, resulting in total leaf wilting under favorable conditions. Monocerin is a non-host-specific polyketide toxin produced by *E. turcicum*, responsible for lesion development (Axford et al., 2004). The severity of the disease relies on the pathogenicity of the strain, with highly pathogenic strains resulting in epidemic infections.

3. **Management**: Among the cultural practices, crop rotation with non-cereal crops is the most effective means of managing the disease. In addition, deep summer ploughing also exposes the pathogen inoculum present in the soil to sunlight (Barrera and Frederiksen, 1994). Before fungicide application, the foliar application of *Trichoderma harzianum* at 2×10^8 cfu g^{-1} at 0.4% reduced disease intensity and increased grain yield to 53.60 q ha^{-1} (Wani et al., 2017). Furthermore, Kiran and Patil (2019) reported that the application of hexaconazole at 1 ml per liter was found to reduce the incidence of disease, while at the same time, fodder yield also increased in sweet sorghum.

13.2.3.3 DOWNY MILDEW

Downy mildew is one of the most devastating sorghum diseases and is present in tropical and subtropical areas.

1. **Etiology:** Downy mildew, caused by *Peronosclerospora sorghi*, is both soil-borne and airborne in nature. The seeds and soil that are contaminated aid in the long-distance spread of oospores.
2. **Symptoms:** The infection is systemic, where the young infected plants turn green or yellow with white stripes. Systemic infection leads to sterile plants with no panicle. Alternating parallel stripes appear on older leaves. These leaves will have downy growth appearing on the lower surface, resembling white color. The chlorotic areas die, and the leaves become shredded. As the plant matures, white chlorotic streaks change to brown, the interveinal tissue dies, and oospores develop. The streaks usually turn brown and split into long strips. The lamina disintegrates, resulting in shredding symptoms (Pande et al., 1997). The relative humidity in the atmosphere generally increases, and the disease also increases (Bock et al., 2000; Wang et al., 2000).
3. **Management:** The use of disease-free seeds for sowing, crop rotation with non-host crops, deep tillage of residues, and rouging of infected plants are the best ways to minimize the incidence. Reservoir hosts, if they exist, need to be removed. Seed treatment with metalaxyl can effectively manage the disease (Isakeit et al., 2003).

13.2.3.4 ZONATE LEAF SPOT

In the case of sweet sorghum, zonate leaf spot is one of the major foliar diseases, inflicting considerable reductions in forage quality as well as yield.

1. **Etiology:** The disease is caused by *Gloeocercospora sorghi*, which can survive in soil and seed for years.
2. **Symptoms:** The disease is characterized by large, irregular-shaped lesions that exhibit a bull's-eye shape with red to purple bands. As the disease progresses, dark red to blackish-purple or brown lesions appear on the leaves as well as on the leaf sheaths, which ultimately coalesce together, and the entire area becomes blighted.

Additionally, sporodochia appear on the upper surface of these blotches as slimy, salmon-colored masses. In most cases, seeds are infected, and these seeds turn dark brown or red with black oval-shaped spots (Figure 13.2) (Anitha et al., 2020). Finally, dead tissues become greyish-tan with sclerotia lined up in lines.

3. **Management:** An overhead irrigation system is to be avoided in the field. Crop rotation with non-host crops for more than four years and general field sanitation are good practices to minimize the disease. Carbendazim application also effectively reduces disease incidence (Kharayat and Singh, 2012).

FIGURE 13.2 Zonate leaf spot caused by *Gloeocercospora sorghi*. Note the bullseye appearance of lesions along the leaf blade edge.

13.2.3.5 RUST

Sorghum rust is widespread worldwide and is severe in cool and humid regions. Moreover, this disease predisposes the crop to other major diseases like grain mold, charcoal rot, and *Fusarium* stalk rot.

1. **Etiology:** *Puccinia purpurea*, the obligate fungal pathogen, is responsible for rust in the case of sorghum.

2. **Symptoms:** Although the pathogen affects all stages of the crop, initial symptoms appear on lower leaves as reddish-brown pustules. The rust sori on the leaves coalesce and form a blighted appearance. Apart from leaves, the pustules are also formed on stalks and leaf sheaths. These pustules, which are uredinia, rupture to release urediniospores. This is followed by the development of teliospores, which germinate at a favorable temperature of 10–12°C. High rainfall and humidity increase the severity of the disease. Under severe conditions, the plant may lodge (Wirnas, 2020).

3. **Management:** Use of healthy seeds for sowing. Early sowing usually reduces the incidence. The contaminated crop residues as well as weeds such as yellow wood sorrel (*Oxalis corniculate*) must be destroyed. Seed treatment with contact fungicides reduces the incidence. White et al. (2012) reported that spraying oxycarboxin (1 kg active ingredient per 100 L of water per ha) at 10-day intervals from one month post-sowing to physiological maturity was effective in managing the disease.

13.2.3.6 BACTERIAL LEAF STREAK

A bacterial disease of sorghum common in warm, humid areas, is seed-borne, and the bacterium overwinters in plant debris after the crop period.

1. **Etiology:** The bacterial pathogen, *Xanthomonas vasicola* pv. *holcicola*, is of quarantine importance and the causal agent of the disease. The bacteria are seed-borne, perpetuate in plant debris, and are disseminated by irrigation water and wind.
2. **Symptoms:** The symptoms may appear at the seedling stage to maturity stage and are characterized by water-soaked lesions with tiny bead-like bacterial exudates present on them. As the disease proceeds, these lesions become red, opaque oval spots with tan centers and narrow red margins. Gradually, these streaks coalesce, giving a burnt appearance. As the disease advances, dark, narrow margins form between the reddish-brown streaks, and the bacterial exudates will dry to a white-colored scale (Navi et al., 2002).
3. **Management:** Crop residues must be removed from the field. The use of pathogen-free seed for sowing reduces disease incidence. Crop rotation with non-cereal hosts also helps in reducing disease. The varieties offering resistance are far more important with respect to the management of disease, and they can be combined with cultural means to achieve better results (Janse, 2005).

13.2.3.7 BACTERIAL LEAF STRIPE

1. **Etiology:** *Paraburkholderia andropogonis* (syn. *Burkholderia andropogonis*) causes bacterial leaf stripe, which is seed-borne and survives in soil and plant debris. Although plant debris is the primary source of inoculum, secondary spread occurs through irrigation water and rain splashes. Warm temperatures (25 to 29°C), wet weather, and cloudy, humid days following rain favor disease development (Muriithi and Claflin, 1997).
2. **Symptoms:** The initial symptoms are visible as small linear lesions, which can be seen as purple, tan, or yellow on leaves. The peduncle, branches, seeds, and stalks of the plants are also infected and exhibit these characteristic lesions as the disease progresses. Finally, these lesions coalesce along the leaf width with bacterial exudates on the lower surface of the leaf (Anitha et al., 2020).

3. **Management:** Healthy seeds, crop rotation, general field sanitation, and the removal of infected weeds must be implemented to reduce the incidence. According to Anitha et al. (2020), late sowing helps in managing the disease.

13.2.3.8 BACTERIAL EYESPOT OR LEAF SPOT

1. **Etiology:** *Pseudomonas syringae* pv. *syringae* incites bacterial eyespot or leaf spot. The bacterium survives in plant debris and gets transmitted through wind, rain, and infected seeds (Das et al., 2023).
2. **Symptoms:** Infection can start from the seedling stage, affecting the coleoptiles of seeds, causing stunting and discoloration of seedlings. On leaves, the spots are circular to elliptical, with a straw center and dark margin. As severity increases, these spots coalesce, leading to the death of leaves. The bacterial streaks as well as exudates are not seen in bacterial leaf spots; on the other hand, these are seen in bacterial stripe (Hernandez et al., 1992; Nyvall, 1999).
3. **Management:** Healthy seeds, crop rotation, destroying the crop debris, and the use of disease-resistant varieties/hybrids are advisable for disease management.

13.2.3.9 MAIZE STRIPE VIRUS

In India, during 1990, this disease was reported for the first time (Peterschmitt et al., 1991). The viral isolate that infected sorghum is similar to the strain infecting corn. The disease becomes severe during the post-rainy season, depending on vector (*Peregrinus maidis*) survival, cropping pattern, and host specificity.

1. **Symptoms:** The disease is characterized by continuous chlorotic stripes within the veins of leaves. According to the severity of the disease, the width of the stripes increases and slowly turns to a yellow color. As the infection is systemic, the affected plant appears stunted in growth. If the infection is seen in the early stage of crop establishment, the plant dies without the emergence of an ear head, whereas infection at later stages results in shortened internodes with few or no seed formation (Anitha et al., 2020).
2. **Management:** No suitable management practices have been suggested for this disease.

13.2.4 PANICLE AND GRAIN DISEASES

The most important panicle and grain diseases in sorghum include grain mold, smut, and ergot. An important factor in considering these diseases is that they affect the quality and quantity of sorghum production.

13.2.4.1 GRAIN MOLD AND GRAIN WEATHERING

The disease is also called head mold, and many fungal pathogens are associated with it. These fungi infect the internal tissue of spikelets. If infection occurs after physiological maturity and black layer formation, then the disease is known as grain weathering.

1. **Etiology:** About 40 genera of fungal species, including *Alternaria, Aspergillus, Bipolaris, Curvularia, Exserohilum, Fusarium, Penicillium*, and *Phoma*, are reported to cause the disease. According to Bandyopadhyay et al. (2000) and Prom et al. (2011), *Alternaria alternata, Bipolaris australiensis, Curvularia lunata, Exserohilum turcicum, Fusarium andiyazi, F. proliferatum, F. sacchari, F. verticillioides, F. thapsinum, F. nygamai, F. pseudonygamai*, and *Phoma sorghina* have been identified as major pathogens. However, *Fusarium, Curvularia,* and *Alternaria* spp. are considered the principal grain mold-causing pathogens worldwide.
2. **Symptoms:** Initially, the tips of spikelet glumes, lemma, and palea tissues are infected, which gradually proceeds to the base of these parts. As a result, the size of the grain is significantly reduced. The fungal growth is visible at the hilar end of the seed, which then moves towards the pericarp region. If the infection occurs post-maturity of seeds, it results in a moldy appearance (Figure 13.3). The primary fungus causing infection determines the color of the moldiness (Thakur et al., 2006). The warm climate coupled with high humidity during flowering and grain development stages is conducive for mold development, while dry conditions prevent it. Additionally, grain damage from birds and insects feeding produces holes, and through these, facultative pathogenic fungi enter inside (Leslie et al., 2002). Thus, the feeding behavior of insects is greatly associated with the incidence of grain mold disease (Ratnadass et al., 2003).
3. **Management:** The adjustment of dates of sowing or use of photo-insensitive cultivars of sorghum is a viable strategy (Thakur et al., 2006). Further, harvesting at the physiological maturity stage is also equally important in the management of the discasc.

13.2.4.2 ERGOT

1. **Etiology:** Three species of *Claviceps* cause sorghum ergot: *Claviceps sorghicola, C. africana,* and *C. sorghi. C. africana* is the major problem in North and South America and Africa. In Asia, *C. sorghi* is the associated species (Frederickson et al., 1991; Isakeit et al., 1998).
2. **Symptoms:** The disease is characterized by producing sugary exudates referred to as honeydew. The fungal mycelia infect the ovaries of flowers and convert them into a white fungal mass. The fungal spores present in honeydew germinate to produce secondary spores, rapidly spreading long distances through wind or the seed contaminated with honeydew (Anitha et al., 2020) (Figure 13.4).
3. **Management:** Early sowing helps in managing the disease in Indian conditions (Anitha et al., 2020). Rotating the crop with non-host crops, summer ploughing, and removing the infected panicles from the harvest are other methods suggested. Chemical control is not very cost-effective.

FIGURE 13.3 Examples of individual grain molded and discolored sorghum caryopses with signs of different species of fungi.

13.2.4.3 SORGHUM SMUTS

Sorghum smuts are seed-borne diseases that become severe when seeds are not treated. They are classified as covered long smut, head smut, kernel smut, and loose kernel smut.

1. **Covered Kernel Smut:** *Sporisorium sorghi* incites covered kernel smut, otherwise known as grain smut. The fungi infect the grains in the panicle, converting them into conical sori. These break open and release teliospores when the grains are harvested. These teliospores can also be seed-borne and germinate within the seeds, thus infecting the plant growing point.
2. **Loose Kernel Smut:** One of the most destructive smuts known, loose smut is caused by *Sporisorium cruentum* and is more prevalent in areas of scarce rainfall.

FIGURE 13.4 Sorghum panicle with ergot (*Claviceps africana*) with unfertilized florets exuding honeydew (white arrow) and conidia production (black arrow).

Plants affected by this pathogen exhibit stunted growth, with thin stalks, and their heads appear prior to the appearance in healthy plants (Aggarwal and Mehrotra, 2003). Abundant tiller formation is also observed in infected plants (Williams et al., 1978). Infected panicles appear looser, bushier, and darker green than healthy panicles (Malvick, 2008).

3. **Head Smut:** This type of smut is observed in all sorghum-producing regions, incited by *Sporisorium reilianum,* which is a soil-borne, facultatively biotrophic basidiomycete smut fungus. This pathogen, based on its pathogenic ability to infect 10 different lines of sorghum, is classified into eight physiological races. Although infection starts at the seedling stage, symptoms are more pronounced when the plant begins to flower. The infection is formed in the reproductive tissues of the plant, and the young head is filled with smut galls with a thick white membrane. The membrane gets ruptured, exposing the huge spore mass to air currents (Singh et al., 2020) (Figure 13.5).

4. **Long Smut:** Long smut, incited by *Anthracocystis ehrenbergii* (syn. *Tolyposporium ehrenbergii*), is prevalent in sorghum growing zones of India and Africa. This smut is characterized by the appearance of whitish to dull yellow long sori, which are slightly curved with a relatively thick brown-colored membrane called peridium. The peridium splits at the apex to release a black mass of spores, and vascular bundles of the infected ovary are seen as several dark brown filaments (Singh et al., 2020).

13.2.5 MANAGEMENT

Use disease-free seeds for sowing. Harvest grains only from disease-free panicles. Seed treatment using solar energy is performed in summer through seed soaking and drying under natural shade.

FIGURE 13.5 Head smut caused by *Sporisorium reilianum*. Note the sorus with abundant teliospores covered with a white peridium that has replaced the panicle.

The fungicide Apron Plus 50% application reduced the loose smut disease in sorghum (Akpa and Manzo, 1991; Gwary et al., 2007). Further, for the covered kernel smut, the application of metalaxyl and thiram proved to be better fungicides (Mtisi, 1996).

13.3 MAJOR INSECT PESTS

Significant grain yield losses were incurred due to insect pest damage. Nearly 12–50% of the actual crop produce is lost because of insect pests, of which ear head pests were recognized as key constraints for sorghum in reducing grain yield, apart from shoot fly and stem borer (Olabimpe et al., 2021). More than 150 insect species belonging to different orders have been identified and recorded as pests on sorghum (Guo et al., 2011), of which 31 are potential pests that cause significant damage based on their occurrence in different growing regions in India. Economic important pests are shootfly, *Atherigona soccata* Rond., stem borer, *Chilo partellus* Swin., armyworm, *Mythimna separata* Wlk., midge, *Contarinia sorghicola* Coq., headbang, *Calocoris angustatus* Leth., and head caterpillars, *Heliothis armigera* Hb. are important and considered the major pests in India. The information on different key pests and their practicable control measure which were found effective in India are given below.

13.3.1 SHOOT FLY OF SORGHUM

Atherigona soccata is an important pest belonging to Muscidae family of Diptera affect both wild and farmed sorghum in Asia and Africa (Balikai and Bhagwat, 2010). It is distributed throughout Maharashtra, Andhra Pradesh, Tamil Nadu, and Karnataka in India. It occurs often in a variety of hosts, including maize, ragi, bajra, rice, wheat, and grasses. In India, shoot flies infest the crop from July to October; whereas the late-sown crop during early Kharif and Rabi are severely affected, and crop losses exceed 65–75% for grain and fodder (Balikai and Bhagwat, 2010).

Adults are grey in color, and maggots are destructive. Fly adults are active in the morning and evening, and their entire life cycle is completed in 18 to 25 days. Small seedlings may be killed by feeding, stand density and tiller sizes may be lowered, and crop loss may occasionally occur (Patil and Bagde, 2017; Young, 1981). After hatching, the maggot moves to the leaf's upper surface and crawls within the space between the stem and leaf sheath. The maggot slashes the growth point inside the stem after penetrating the earth, causing the "dead heart" feeling. Side tillers are produced by the infected plant. In contrast to the Kharif season, the attack is more severe in the summer (Figure 13.6).

FIGURE 13.6 Shoot fly (Shootfly, *Atherigona varia soccata*) and its damaging symptoms: a. Egg, b. Maggot, c. Adult, d. Side tillering, and e. Gaps in the field due to infestation.

On the underside of leaves, adult flies singly deposit 20–40 rod-shaped eggs. The eggs hatch in two days, and the freshly emerging pale-yellow maggots travel downward between tender leaves by tunneling into the plant whorl. The 10 to 12-mm-long maggot feeds on the decomposing tissue produced when leaf sheaths are cut at their growth point. The central shoot gets damaged, which leads to withering. This symptom is also called "dead heart," which ultimately results in the death of the central shoot and the development of side tillers, which give the plants a bushy appearance. Pupation can take place in the soil or in plant tissue, and larval development is finished in the decomposing tissue (Patil and Bagde, 2017).

The adult flies are typically white-grey in color, and these flies lay their eggs in a cigar shape on the lower surface of leaves in the morning. The egg is white, cylindrical, and has a slightly flattened distal end. Normally, the incubation period ranges from 2 to 3 days. The maggot is apodous and filthy white. Larvae are around 6 mm long and yellow when mature. There are four larval instars and an 8–10-day larval phase. It pupates at the basal portion of the stem or in the soil. The entire life cycle will be completed in 17 to 21 days. The ETL for this pest is 1 egg/plant or damage of 10% (Saurabh et al., 2023).

13.3.1.1 MANAGEMENT

To reduce shoot fly damage, sow sorghum as soon as the monsoon rains start. Resistant cultivars, viz., Co-1, CSH 15R, Maldandi and Hagari, M35-1, Swati, IS - 18551, 5566, 5285, 5613, ICSV 700, ICSV 705, Phule Yashoda, CSH 7, and CSH 8 can be considered. Use a greater seed rate (12.5 kg/ha) and transplant only healthy seedlings, or raise a nursery and remove any seedlings that have been affected by shoot flies during thinning. Remove and destroy any plants that have dead hearts when the plant is being thinned. Install 12 hanging plastic fishmeal traps per hectare (ha) until the crop is 30 days old. Use thiamethoxam 30 FS @ 10 ml, chlorpyrifos 20 EC @ 4 ml, quinalphos 25 EC @ 4 ml, imidacloprid 48 FS 12 ml, or imidacloprid 70 WS 10 g to treat 1 kg of seeds. Apply either carbofuran 3 G or phorate 10 G @ 2.5 kg a.i./ha in granular form to the furrow at the time of sowing. Dimethoate 30 EC @ 500 ml/ha should be sprayed in the main field (Olabimpe et al., 2021).

13.3.2 STEM BORER

Chilo partellus is a significant pest of maize and sorghum in Asia, belonging to the Crambidae family of Lepidoptera. It was first discovered in 1930 in Malawi, Africa. It then spread to most of the countries in eastern and southern Africa (Kfir, 1997). It is reported as a key insect pest of sorghum in India, Pakistan, Sri Lanka, Indonesia, Iraq, Japan, Uganda, Taiwan, Sudan, Nepal, Bangladesh, and Thailand. Its host plants include finger millet, sorghum, sugarcane, bajra, and rice.

Adult *C. partellus* are nocturnal and have a very short lifespan of only 48–72 hours. On the underside of the leaves, females lay between 200 and 600 scale-like eggs. After 4–5 days, the eggs hatch, and the larvae dig into the stalk and through the leaves to feed. Early infestations and older larvae feeding near the base of the stem might kill the growth point, causing the 2–3 center leaves to dry off (dead heart) (Sharma, 2003; Singh et al., 2011). It takes 14 days for complete larval development, and pupation takes place in the stalk, with adults emerging between 5–12 days afterward (Overholt et al., 1994). Due to its short life cycle and capacity for rapid growth, *C. partellus* has taken over as the primary borer in major sorghum-growing countries.

Due to the overlapping generations of *C. partellus*, the pest can attack the same crop at once. However, facultative diapausing larvae in stems can live for up to 6 weeks before pupating throughout the winter and the dry season. Larvae emerge throughout a 12-week period, from early August to early November, after entering diapause during their final instar stages. Larval diapause is terminated by favorable conditions, including high temperatures (28°C), long days (16 hours), and enough water, and pupation can take place nine days later. A month after the crop is sown, the infestation of stem borer can be observed, and infestation prevails until the emergence of the ear heads (Kfir, 1993).

The common damage symptom is "dead heart," withering of the central shoot. On the stem, close to the nodes, the apparent boreholes are seen. Young larvae feed on fragile, folded leaves by crawling over them, producing the familiar "shot hole" symptoms. Caterpillars may be visible internally in the affected stem portions (Figure 13.7). The adult moth is straw-colored and medium in size. On the underside of leaves close to the midrib, it typically lays about 300 flat oval eggs that resemble scales. There is about a 2 to 5 days incubation period before the eggs hatch. The prothoracic shield of the larva is about 25 mm long and has a brown head and a yellowish-brown body. The larval stage has seven instars and lasts for 28 to 50 days. Pupation takes place within the stem, before emerging as an adult within 7–10 days through the entry holes left by the larvae. The entire life cycle takes 30 to 40 days to complete (Singh et al., 2011).

FIGURE 13.7 Stem borer: (*Chilo partellus*, Pyralidae: *Lepidoptera*) and its damaging symptoms a. Larvae b. Pupae c. Bore hole symptom d. Shot hole symptom e. Dead heart.

13.3.2.1 MANAGEMENT

To kill the larvae that are hibernating in the stubbles, they should be ploughed up and burned throughout the winter. Plant resistant cultivars such as ICSV 700, IS 2205, E 302, and E 303. Dead hearts ought to be removed and either utilized as animal feed or buried in manure dumps. To reduce stem borer damage, sow lablab bean or Dolichos bean as an intercrop in a 4:1 ratio. Set up a light trap until midnight to monitor, attract, and kill stem-borer adult moths. It is important to promote the biocontrol agents *Trichogramma minutum* and *Bracon chinensis*. Apply in the leaf whorls any one insecticide, either phorate 10 G @ 8 kg/ha or carbofuran 3 G @ 17 kg/ha, which is mixed and diluted in 50 kg sand for effective control of this pest (Tadeos and Fano, 2019).

13.3.3 PINK STEM BORER

Sesamia inferens belongs to the Noctuidae family of Lepidoptera. This pest lays round, yellowish-white eggs on leaf sheaths or on the surface of the leaf in batches of 30 to 100 eggs about two to three times in parallel. After hatching, fully grown, smooth, and pinkish larvae measure about 30 mm long. During the early stage of the life cycle, the larva does not eat leaves but instead bores directly into the stem. Pupation takes place in stem galleries, and the life cycle takes 40 to 50 days to complete. The forewings of adult moths are straw in color with a dark brown line around the edge. Four to five generations of the pest are observed in a single year (Kaur et al., 2015).

The fawn-colored adult moth has white hind wings and forewings with dark brown streaks. Between the leaf sheath and the stem of the host plant, the female deposits 150 hemispherical, creamy-white eggs in two or three rows. The egg period lasts for 7 days. The larvae are around 25 mm long, pale yellow with a purple-pink tint, and have a reddish-brown head when fully grown. The larval period is 25 days, although it can be 75 days during the colder months. Within 12 days, the adult emerges from the larval tube in the stem after pupation. It takes 6–7 weeks to pass a generation. The total life cycle lasts 45 to 75 days, and 4–6 generations occur annually. This pest is distributed in India and other major sorghum-growing countries. Besides sorghum, it also infects maize, rice, wheat, sugarcane, bajra, ragi, barley, and guinea grasses. The larvae pierce into the main stem and damage the central shoot. Long-duration millets with late sowing are more vulnerable to borer attack. Natural enemies in Africa and India have been studied and reported, viz., *Goniozus procerae* Risbec and other hymenopterous parasitoids of *A. ignefusalis*, including *Euvipio* sp. and *Syzeuctus* sp., appear to be very active on this pest (Gahukar,1989). Some parasitoids and predators that have been seen in India during the life stages of *Sesamia* spp. and *C. partellus* may have a substantial impact on the decline in the incidence of these borers (Das and Padmaja, 2016).

13.3.3.1 MANAGEMENT

Field sanitation, the gathering and eradication of crop residue and weeds after harvest, and the removal of dead hearts can all help control the borer population (Sagnia, 1983). Avoid tall cultivars due to the difficulty of insecticidal sprays, which is neither practical nor cost-effective. Dwarf millet cultivars might be advised. Release the following parasites, viz., *Telenomus* sp., *Trichogramma chilonis*, *Apanteles flavipes*, *Bracon hebetor*, and *Tetrastichus ayyari* for effective suppression of the stem borers. 20 days after the crop germination, apply chlorpyrifos 20 EC at 1.0 L/ha, carbofuran 3 G at 25 kg/ha, or cartap hydrochloride 4 G at 18.75 kg/ha as a chemical control strategy to reduce the infestation (Jeevanandham et al., 2020).

13.3.4 SHOOT BUG

Peregrinus maidis belongs to the Delphacidae family of Hemiptera. It is reported and distributed in Africa, the West Indies, the Philippines, and India. In India, it is distributed in major sorghum-growing areas. The shoot bug is a frequent pest of sorghum; it sucks sap from the leaf whorls and from the space between the leaf sheath and stems. Infested plants show stunted growth (Dharavath, 2022). When there is a serious infestation, the upper leaves begin to dry out first before progressively spreading to the lower leaves, and the plant may eventually die. The leaves crinkle and look sun-tanned. In addition, they produce honeydew, which favors the development of sooty molds. At the later stage of plant growth, infestation may cause the upper leaves to twist and prevent panicle emergence. Infestation is higher during drought conditions (Balikai et al., 2017).

The adults and nymphs cohabit communally on the inner side of leaf sheaths and in plant whorls. To insert the eggs, the females cut a slit in the upper surface of the midrib. The eggs are then covered with a white waxy substance. In 7 days, a female can lay up to 100 eggs. The eggs are cylindrical, white, long, and taper at the ends. Eggs are incubated for 1 to 7 days. The development of a nymph lasts for 16 days and has seven instars. Both adults and nymphs consume plant sap. The attacked plants become weak, yellow, and stunted. The leaves wilt from the top down. If the attack is severe, panicle production is prevented, and the plants perish. The bug's honeydew secretions encourage the formation of sooty mold on leaves. Due to egg-laying, the midribs of the leaves turn crimson and may eventually dry out. The adult has dark brown to yellowish-brown wings that are transparent. The macropterous female is yellowish-brown, while the male and female brachypterous species are both dark brown. The eggs are in groups of 1–4 that are encased in a white waxy material inside the tissue of the leaf. It may produce 97 eggs per female. In 16 days, the nymphal stage goes through five instars. The entire life cycle lasts between 18 and 31 days (Balikai et al., 2017).

13.3.4.1 MANAGEMENT

Protect egg parasitoids like *Paranagrus optabilis* and *Octetrastichus indicus*, as well as predators like *Coccinella septempunctata* and *Geocoris tricolor*. Mist 500 ml of dimethoate or methyl demeton in 500 L of water (Akshatha et al., 2020).

13.3.5 EAR HEAD BUG

Calocoris angustatus belongs to the Miridae family of Hemiptera. Both nymphs and adults feed on panicles from the time of panicle emergence to the hard dough stage of grain maturation. Adults lay their eggs on panicles from the time of emergence until blossoming. By feeding on the milk stage and soft-dough stages of maturing grain, nymphs and adults compress and distort the grain, reducing production (Srivastava, 1984). Nymphs that have just emerged eat solely on seeds, but the third and fourth instars are the most destructive to both new sorghum grain and foliage (Knutson and Cronholm, 2007). This pest can lower grain yield by up to 86% (Gahukar et al., 1989; Malgwi and Dunuwel, 2011). It is distributed over South India, with host species including grasses, maize, sugarcane, and pearl millet. When the grains are in the milky stage, they suck the juice from them. The grains that have been sucked out shrink, turn black, and become ill-filled or chaffy. Older grain shows distinct feeding punctures that reduce grain quality (Sajwan et al., 2022). The adult male is green, while the female is green with a brown border. Blue cigar-shaped eggs are placed either in the center of the florets or beneath the glumes. Each insect produces 150–200 eggs. There is a seven-day egg period. Nymphs are slim and have a green color. The color of the first instar is orange. The nymphal stage lasts 10 to 14 days. From egg to adult, the life cycle takes less than three weeks. When the panicles do not ripen at the same time, at least two generations of the bug can feed on the same crop (Kumar et al., 2022).

13.3.5.1 MANAGEMENT

Grow resistant cultivars like IS1760, IS 17645, CSM 388, and BBR-1. Dust with carbaryl 10% at 12 kg/ha, or quinalphos 1.5% at 12 kg/ha, synchronizing during the milky stage (Olabimpe et al., 2021; Sajwan et al., 2022).

13.3.6 SORGHUM MIDGE

Contarinia sorghicola belongs to the family Cecidomyiidae of Diptera. The midge is a significant pest in all regions of the world where sorghum is grown (Young and Teetes, 1977). There have been reports of grain loss of 50–100% in major growing areas of the world (Knutson and Cronholm, 2007). Adult flies move from their natural host grasses onto sorghum fields where they mate. The orange-colored female midges lay 30–250 eggs in blossoming spikelets. Within two days, the eggs hatch, and the larvae burrow into the seeds to eat before pupating. While pupal development takes three days, larval development takes 9–11 days. Males live for only a few hours, and females seldom live longer than two days; therefore, mating occurs soon after adults emerge. A generation can be completed in two to three weeks, and a growing season can include 9 to 12 generations (McLeod and Greene, 2004).

Larvae overwinter in cocoons in colder climates. Adults from overwintered larvae emerge in late winter or early spring (McLeod and Greene, 2004; Young and Teetes, 1977;). The sorghum midge is active all year round in warmer areas, with higher activity during rainy seasons than during the dry seasons (Kausalya et al., 1997). Poor grain germination and quality are brought on by larval feeding on growing seed, which hinders normal development. Panicle damage varies by location and season, with the worst infestation occurring from mid-September to early October.

The sorghum midge is typically the insect pest that does the most damage to the crop and, if left unchecked, might destroy it. A tiny, orange, "gnat-like" fly, the adult sorghum midge is about 1/8 inch long and can be seen on the heads of sorghum plants that are in bloom. The seed contains little orange maggots called larvae. The sorghum midge can be found across the world. Each sorghum head may have all its seeds destroyed by larvae because they feed directly on the growing seed. When high midge populations are not controlled, severe damage may result, and yield reduction may be close to 100% (Olabimpe et al., 2021).

13.3.6.1 MANAGEMENT

Early and uniform planting should be the primary management strategy. The sorghum midge can produce many generations on the grain sorghum and increase to harmful levels in fields that bloom late in regions or even within individual fields when the sorghum flowering season is prolonged. Only the flowering stage of grain sorghum is vulnerable to midge attack; thus, postponing successive plantings for one to one and a half months may prevent population growth. Early in the season, the destruction of wild host plants, such as Johnson grass, may impede the growth of the midge population. Some hybrids of grain sorghum have some degree of sorghum midge resistance. It is crucial to check as and when the plants blossom in areas with the potential for large yields and profits. About 25% of the field should start to bloom before scouting begins. Look for the orange flies among the sorghum heads on the field's margins. Carefully walk up to the plant and visually check the head on all sides for adult midges. Additionally, adults can be captured by shaking their heads inside a clear half-gallon plastic bag placed over their heads. Adults can be counted by carefully removing the sack and looking inside. The optimal time for sampling is early in the morning with little wind; 100 heads should be counted with a focus on the field borders inside each field (80 acres or smaller). Until the soft dough stage, more scouting should be done every two days or so. Adults discovered at this point cannot infest the seed. Grain sorghum crop residue should be removed after harvest because larvae overwinter in the seed (Olabimpe et al., 2021). Consider cultivating resistant varieties like DJ 6541, AF 28, ICSV 197, ICSV 745, and ICSV 88032. Conserve the predators *Orius albidipennis* and *Tapinoma indicum* and the larval parasitoids *Apanteles* sp. and *Eupelonespopa.* Give the first application when the ear head has almost 90% erupted, and then apply again in 4 or 5 days malathion 50 EC or carbaryl 50 WP or Quinalphos 1.5 D at 25 kg/ha, or 1.0 L (or) malathion 50 EC or carbaryl 50 W.

13.3.7 PLANT LICE (APHIDS)

Rhopalosiphum maidis and *Melanaphis sacchari* belong to Aphididae of Hemiptera. *M. sacchari* is found all over the world, including in Angola, Brazil, China, Colombia, Ecuador, Egypt, Ethiopia, Haiti, Hawaii, India, Indonesia, Japan, Jamaica, the Middle East, Nigeria, Pakistan, Peru, Philippines, Sudan, Thailand, Trinidad and Tobago, Uganda, and Venezuela. It is an economically significant pest on sorghum in China, Taiwan, Japan, India, South Africa, and Botswana, distributed throughout the entire global sorghum-growing region. Sorghum, maize, sugarcane, and ragi are the host range. Aphid colonies are visible in the middle leaf whorl, on stems, or in panicles. Both nymphs and adults ingest the plant juice (Avellara et al., 2022). Marginal leaf necrosis and yellowish mottling of the leaves are frequently caused by this. An excess of honeydew, on which molds develop, is produced by the aphid. Honeydew may make it difficult to harvest panicles. The maize dwarf mosaic virus is also spread by aphids. Dark bluish-green and rather oval, the aphid is identified as *Rhopalosiphum maidis*. It measures 2 mm in length and has black antennae, cornicles, and legs. There are wingless and winged forms. Without mating, females give birth to young ones, and a generation lasts only a week or so. The adult has dark green legs and a yellow body. The sugarcane aphid, *Melanaphis sacchari*, is yellow to buff in color. At the end of the rainy season or during dry spells, populations spike quickly. The female of the wingless type produces 60–100 nymphs during her 13–20-day gestation. A little bit fewer nymphs are produced by the winged type. During the dry season, the life cycle is completed in 5.5–7.0 days (Oberemok et al., 2023).

13.3.7.1 MANAGEMENT

Spray the base of attacked plants with a contact (or) systemic insecticide like dimethoate 30 EC in 500 L of water. The Directorate of Plant Protection, Quarantine and Storage (CIBRC) recommended fungicides and insecticides along with the dosage and application mode to follow in India (see Table 13.2).

TABLE 13.2 Recommended/Crop-Labelled Fungicides/Insecticides for Managing Pests and Diseases as per CIBRC in India

Sl. No.	Insect Pest/ Disease	Insecticide/Fungicide	Dosage/ha		Application Type
			a.i.(g)	Formulation (g or ml)	
Diseases					
1	Anthracnose	Fluxapyroxad 333 g/l FS	0.33	1.0 ml	Seed treatment
2	Downy mildew	Metalaxyl-M 31.8% ES	0.64	2.00 ml	Foliar spray
		Metalaxyl 35% WS	240	600 ml	Seed treatment
3	Grain smut	Sulphur 80% WP	2.4–3.2	3–4g	Seed treatment
		Sulphur85%DP	2.55–3.40	3–4 g	Seed treatment
4	Loose smut	Thiram75%WS	18.8–22.5	25–30	Seed treatment
5	Seedling blight	Thiram75%WS	18.8–22.5	25–30	Seed treatment
Insect Pests					
1	Shoot fly	Carbofuran 03% CG	1000	33300	Whorl application
		Imidacloprid 48% FS	20	1200	Seed treatment
		Imidacloprid 70% WS	700	1000	Seed treatment
		Oxydemeton-methyl 25% EC	250	1000	Foliar Spray
		Quinalphos 25% EC	375	1500	Foliar Spray
		Thiamethoxam 30% FS	03	10	Seed treatment
2	Stem borer	Carbofuran 03% G	250	8300	Whorl application
		Quinalphos 05% G	750	15000	Foliar Spray
3	Ear head Midges	Dimethoate 30% EC	500	1650	Foliar Spray
		Malathion 05% DP	1000	20000	Foliar Spray
		Malathion 50.00% EC	500	1000	Foliar Spray
		Phosalone 04% DP	400	1143	Foliar Spray
		Quinalphos 01.50% DP	400	26600	Foliar Spray
4	Red spider mite, pink mite, purple mite, scarlet mite	Phenthoate 02% DP	400	20000	Foliar Spray
		Phosalone 35% EC	400	1143	Foliar Spray
		Quinalphos 25% EC	375	1500	Foliar Spray
5	Ear head bug	Quinalphos 01.50% DP	375	25000	Foliar Spray

Moreover, there are several new chemical fungicides and pesticides used specifically in sorghum (see Table 13.3).

TABLE 13.3 Newer Insecticide/Fungicide Molecules Effective Against Sorghum Pests and Diseases

Sl. No.	Disease/Insects Pest	Fungicide/Insecticides	References
		Disease	
1.	Anthracnose	Pyraclostrobin 20% WG, Fluxapyroxad 250 + Pyraclostrobin 250 SC, and azoxystrobin 23SC	Acharya et al. (2019)
		Thiram 42-S	Isha Sharma et al. (2015)
2.	Charcoal rot	Carbendazim 50%WP	AICRP Sorghum (2017)
3.	Head smut	Apron plus 50% DS	Gwary et al. (2007)
4.	Turticum blight	Hexaconazole5%EC 0.1%	Kiran and Patil (2019)
		Dithane M-45 75%WP	Wani et al. (2017)
		Azoxystrobin 18.2% + Difenoconazole 11.4% SC	https://www.syngenta.co.ke/fungicides
5.	Rust	Oxycarboxin 750 WP, Hexaconazole (5%EC) and Difenoconazole 25% EC	Isha Sharma et al. (2015)
6.	Downy mildew	Metalaxyl 35% WS	Anitha et al. (2020)
7.	Ergot	Tebuconazole 250EC	Isha Sharma et al. (2015)
		Insect pests	
1.	Wireworms, Shoot fly, Greenbug, and Sugarcane aphid	Imidacloprid 70% WS	Rodríguezdel-Bosque and Terán (2015)
2.	Wireworm, Fire ants, Shoot fly, Green bug, and Sugarcane aphid	Thiamethoxam 30% FS	Jones et al. (2015); Karibasavaraja et al. (2005); Knutson et al. (2018); Pawar et al. (2015); Rodríguezdel-Bosque and Terán (2015)
3.	White grub, Green bug, Sugarcane aphid,	Clothianidin 50%WDG	Jones et al. (2015); Knutson et al. (2018)
4.	Shoot fly, Sorghum midge, Corn earworm, Fall armyworm, Sorghum webworm	Chlorantraniliprole 18.5% SC	Knutson et al. (2018); Patil and Bagde (2017)
5.	Ear head worm	Flubendiamide 39.35 SC	Poul et al. (2020)
6.	Sorghum midge, Corn earworm, Fall armyworm, Sorghum webworm, Maize stem borer	Spinosad 45% SC	Anderson and Teetes (1997); Knutson et al. (2018)
7.	Ear head worm, Stem borer	Emamectin benzoate 5 SG	Anderson and Teetes (1997); Knutson et al. (2018)
8.	Shoot fly, Stem borer and pink stem borer	Phorate 10 G Carbofuran 3 G	Chouraddi and Mallapur (2017)
9.	Stem borer	Phorate 10 G	Chouraddi and Mallapur (2017)

13.4 CONCLUSION

Sorghum, a crop known for its resilience, is not immune to the detrimental effects of pests and diseases. These can result in significant yield losses of varying extents. The crop is susceptible to insect attacks at different stages of growth, including seedlings, blooming structures, and mature grains. Over 150 insect species have been identified globally as posing significant, minor, or occasional threats to sorghum. The management interventions presented within this chapter, when complemented by concerted crop improvement efforts, may serve as a viable remedy.

KEYWORDS

- **aphids**
- **charcoal rot**
- **fusarium root and stalk rot**
- **grain molds**
- **pre- and post-emergent damping-off**
- **shoot fly**
- **stem borers**

REFERENCES

Abreha, K. B., Ortiz, R., Carlsson, A. S., & Geleta, M. (2021). Understanding the *Sorghum–Colletotrichum sublineola* interactions for enhanced host resistance. *Frontiers in Plant Science, 12*. https://doi.org/10.3389/fpls.2021.641969 (accessed on 7th November 2025).

Acharya, B., O'Quinn, T. N., Everman, W., & Mehl, H. L. (2019). Effectiveness of fungicides and their application timing for the management of sorghum foliar anthracnose in the Mid-Atlantic United States. *Plant Disease, 103*(11), 2804–2811. https://doi.org/10.1094/pdis-10-18-1867-re (accessed on 7th November 2025).

Adeyanju, A. (2014). Genetic study of resistance to charcoal rot and *Fusarium* stalk rot diseases of sorghum (Ph.D. dissertation). Manhattan, KS.

Adeyanju, A., Little, C., Yu, J., & Tesso, T. (2015). Genome-wide association study on resistance to stalk rot diseases in grain sorghum. *G3 (Bethesda), 5*(6), 1165–1175. https://doi.org/10.1534/g3.114.016394 (accessed on 7th November 2025).

Aggarwal, A. (2003). *Fundamentals of Plant Pathology*. Tata McGraw-Hill Publishing Company: New Delhi.

Akpa, A. D., & Manzo, S. K. (1991). Chemical seed treatment for the control of loose smut disease of sorghum *Tests Agrochemicals and Cultivars, 12*, 56–57.

Akshatha, S., Karabhantanal, S. S., & Jagginavar, S. B. (2020). Shoot bug, *Peregrinus maidis* (Ashmead): A potential threat to rabi sorghum. *Journal of Entomology and Zoology Studies, 8*(6), 1387–1391.

Al-Jedabi, A. A. (2009). Biological control of *Fusarium* root-rot of sorghum. *Research Journal of Agriculture and Biological Sciences, 5*(4), 465–473.

Amrate, P. K., Shrivastava, M. K., Bhale, M. S., Agrawal, N., Kumawat, G., Shivakumar, M., & Nataraj, V. (2023). Identification and genetic diversity analysis of high-yielding charcoal rot-resistant soybean genotypes. *Scientific Reports, 13*(1), 89–105.

Anderson, R. M., & Teetes, G. L. (1997). Evaluation of insecticides for suppression of sorghum midge on sorghum, 1996. *Arthropod Management Tests, 22*(1), 305–305. https://doi.org/10.1093/amt/22.1.305 (accessed on 7th November 2025).

Anitha, K., Das, I. K., Holajjer, P., Sivaraj, N., Reddy, C. R., & Balijepalli, S. B. (2020). Sorghum diseases: Diagnosis and management. In *Sorghum in the 21st Century: Food–Fodder–Feed–Fuel for a Rapidly Changing World* (pp. 565–619). Springer Singapore: Singapore.

Avellara, G. S., Mendes, S. M., Marriel, I. E., Menezes, C. B., Parrella, R. A. C., & Santos, D. G. (2022). Resistance of sorghum hybrids to sorghum aphid. *Brazilian Journal of Biology, 82.*

Axford, L. C., & Simpson, C. J. (2004). Synthesis and incorporation of the first polyketide synthase free intermediate in monocerin biosynthesis. *Angewandte Chemie, 116*, 745–748.

Balikai, R. A., & Bhagwat, V. R. (2010). Evaluation of integrated pest management components for the management of shoot fly, shoot bug and aphid in rabi sorghum. *Karnataka Journal of Agricultural Sciences, 22*, 532–534.

Balikai, R. A., Kambrekar, D. N., Natikar, P. K., & Anaji, R. (2017). Bio-ecology and management of shoot bug, *Peregrinus maidis* (Ashmead) on sorghum and maize - A review. *Biochemical and Cellular Archives, 17*(1), 27–40.

Bandara, Y., Perumal, R., & Little, C. (2015). Integrating resistance and tolerance for improved evaluation of sorghum lines against *Fusarium* stalk rot and charcoal rot. *Phytopharmaceutical, 43*, 485–499.

Bandyopadhyay, R. (2000). *Compendium of Sorghum Diseases* (2nd ed.). In R. A. Frederiksen & G. N. Odvody (Eds.), APS: St Paul Manhattan.

Bock, C. H., Jeger, M. J., Mughogho, L. K., Cardwell, K. F., Mtisi, E., Kaula, G., & Mukansabimana, D. (2000). Variability of *Peronosclerospora sorghi* isolates from different geographic locations and hosts in Africa. *Mycological Research, 104*, 61–68.

Bowling, R. D., Brewer, M. J., Kerns, D. L., Gordy, J., Seiter, N., Elliott, N. E., Buntin, G. D., Way, M. O., Royer, T. A., Biles, S., & Maxson, E. (2016). Sugarcane aphid (Hemiptera: Aphididae): A new pest on sorghum in North America. *Journal of Integrated Pest Management, 7*(1), 12. https://doi.org/10.1093/jipm/pmw011 (accessed on 7th November 2025).

Casela, C. R. (1993). Survival of *Colletotrichum graminicola* and sclerotia in sorghum stalk residues. *Plant Disease, 77.*

Chaturvedi, P., Govindaraj, M., Sehgal, D., & Weckwerth, W. (2023). Sorghum and pearl millet as climate resilient crops for food and nutrition security, Volume II. *Frontiers in Plant Science, 14*, 117–123.

Chouraddi, M., & Mallapur, C. P. (2017). Impacts of newer molecules of insecticides on management of maize stem borers. *Journal of Entomology and Zoology Studies, 5*(4), 1424–1428.

Churchill, A., Dunkle, L. D., Silbert, W., & Kennedy, K. J., & Macko, V. (2001). Differential synthesis of peritoxins and precursors by pathogenic strains of the fungus *Periconia circinata. Applied and Environmental Microbiology, 67*, 5721–5728.

Das, I. K., & Padmaja, P. G. (2016). *Biotic Stress Resistance in Millets*. Academic Press: Cambridge, MA, USA.

Das, I. K., Chattannavar, S., Gholve, V. M., & Singh, Y. (2017). *Annual Progress Report of All India Coordinated Research Project on Sorghum*. https://www.researchgate.net/publication/361231243_Annual_Progress_Report_of_AICRP-Sorghum_Pathology_2021-22 (accessed on 7th November 2025).

Das, I., Palanna, K. B., & Rajesh, G. (2023). Disease management for improved millet production. *Indian Farming, 73*(01), 67–70.

Dharavath, S., & Karabhantanal, S. (2022). Biology of sorghum shoot bug *Peregrinus maidis* as influenced by varieties. *Indian Journal of Entomology*, 1–3. https://doi.org/10.55446/ije.2021.148 (accessed on 7th November 2025).

Djaenuddin, N., Lestari, E. G., Nur, A., & Nadir, F. A. (2023). The performance of sorghum mutant lines resulting from gamma-ray mutation on main diseases. In *IOP Conference Series: Earth and Environmental Science* (Vol. 1192). IOP Publishing.

Esele, J. P. (2008). Sorghum and pearl millet diseases in the Horn of Africa. In *Sorghum and Millets Diseases* (pp. 383–387). Iowa State Press: Ames, Iowa, USA.

Food and Agriculture Organization. (2021). *World Food and Agriculture: Statistical Yearbook 2020*. Food & Agriculture Organization of the United Nations (FAO): Rome, Italy.

Frederickson, D. E. (1991). *Claviceps africana* sp.: The distinctive ergot pathogen of sorghum in Africa. *Mycological Research, 95*, 1101–1107.

Frederiksen, R. A. (1986). *Compendium of Sorghum Diseases* (Vol. 82). American Phytopathological Society: St. Paul, MN.

Gahukar, R. T. (1989). Insect pests of millets and their management: A review. *Tropical Pest Management, 35*(4), 382–391. https://doi.org/10.1080/09670878909371411 (accessed on 7th November 2025).

Guo, C., Cui, W., Feng, X., Zhao, J., & Lu, G. (2011). Sorghum insect problems and management. *Journal of Integrative Plant Biology, 53*, 178–192.

Gwary, D. M., & Gwary, S. D. A. O. (2007). Management of sorghum smuts and anthracnose using cultivar selection and seed dressing fungicides in Maiduguri, Nigeria. *International Journal of Agriculture and Biology, 6*, 1560–1580.

Hernandez, J., Garrido, M. J., Lopez, O., & Trujillo, G. E. (1992). The bacterial leaf spot of sorghum caused by *Pseudomonas syringae* pv. *syringae* in Venezuela. *Fitopatología Venezolana, 5*, 39–42.

Idris, H. A., Labuschagne, N., & Korsten, L. (2008). Suppression of *Pythium ultimum* root rot of sorghum by rhizobacterial isolates from Ethiopia and South Africa. *Biological Control, 45*, 72–84.

Isakeit, T., & Odvody, R. A. (1998). First report of sorghum ergot caused by *Claviceps africana* in the United States. *Plant Disease, 82*.

Isakeit, T., Odvody, G., Jahn, R., & Deconini, L. (2003). *Peronosclerospora sorghi* resistant to metalaxyl treatment of sorghum seed in Texas. *Phytopathology, 9*, 32–39.

Janse, J. D. (2005). Disease and symptoms caused by plant pathogenic bacteria. In *Phyto Bacteriology: Principles and Practice* (pp. 85–118). CABI Publishing: UK.

Jeevanandham, N., Ramiah, N., Chockalingam, V., & Jegadeesan, R. (2020). An overview of the bionomics, host plant resistance and molecular perspectives of *Sesamia inferens* Walker in cereals and millets. *Agronomy (Basel), 10*(11), 1705. https://doi.org/10.3390/agronomy10111705 (accessed on 7th November 2025).

Jones, N., Brown, S., Williams, S., Emfinger, K., & Kerns, D. (2015). Efficacy of neonicotinoid seed treatments against sugarcane aphid in grain sorghum, 2014: Table 1. *Arthropod Management Tests, 40*(1), F12. https://doi.org/10.1093/amt/tsv139 (accessed on 7th November 2025).

Jood, S., Kapoor, A. C., & Singh, R. (1993). Biological evaluation of protein quality of sorghum as affected by insect infestation. *Plant Foods for Human Nutrition, 43*(2), 105–114. https://doi.org/10.1007/bf01087915 (accessed on 7th November 2025).

Karibasavaraja, L. R., Balikai, R. A., & Deshpande, V. P. (2005). Thiamethoxam 70 WS, a new promising seed dress for the suppression of sorghum shootfly. *Annals of Plant Protection Sciences, 13*, 85–87.

Kaur, J., Kumar, P., Singh, J., Suby, S. B., & Bajya, D. R. (2015). Egg laying pattern of *Sesamia inferens* on maize (*Zea mays*). *Indian Journal of Agricultural Sciences, 85*, 109–113.

Kausalya, K. G., Nwanze, K. F., Reddy, Y. V. R., & Nwilene, F. E. (1997). A simple head cage technique for monitoring sorghum midge (Diptera: Cecidomyiidae). *International Journal of Pest Management, 43*(1), 35–38. https://doi.org/10.1080/096708797228951 (accessed on 7th November 2025).

Ketema, K. N. (2023). Genetic analysis of resistance to *Fusarium* stalk rot (*Fusarium thapsinum*) in sorghum (*Sorghum bicolor* (L.) Moench) (Doctoral dissertation). USA.

Kfir, R. (1993). Diapause termination in the spotted stem borer, *Chilo partellus* (Lepidoptera: Pyralidae) in the laboratory. *Annals of Applied Biology, 123*, 1–7.

Kfir, R. (1997). Competitive displacement of *Busseola fusca* (Lepidoptera: Noctuidae) by *Chilo partellus* (Lepidoptera: Pyralidae). *Annals of the Entomological Society of America, 90*(5), 619–624. https://doi.org/10.1093/aesa/90.5.619 (accessed on 7th November 2025).

Khambalkar, P., Bharose, A., Gawande, N. D., & Manape, T. K. (2014). Identification, characterization and molecular validation of pokkah boeng disease causal organism (*Fusarium moniliforme*) of *Sorghum bicolor* in Maharashtra. *Trends in Biosciences, 7*(17), 2578–2582.

Kharayat, B. S., & Singh, Y. (2013). Unusual occurrence of *Erwinia* stalk rot of sorghum in the Tarai region of Uttarakhand. *International Journal of Agricultural Sciences, 9*(2), 809–813.

Kharayat, B., & Singh, Y. (2012). Evaluation of efficacy of fungicides against zonate leaf spot of sorghum caused by *Gloeocercospora sorghi*. *International Journal of Plant Research, 25*, 136–142.

Kiran, B. M., & Patil, P. V. (2019). Integrated management of leaf blight of sweet sorghum caused by *Exserohilum turcicum* (Pass.) Leonard and Suggs. *Indian Phytopathology, 72*(1), 63–69. https://doi.org/10.1007/s42360-018-0086-z (accessed on 7th November 2025).

Kloppers, A., & Tweer, S. (2009). Maize diseases: *Erwinia* stalk rot. *Erwinia Stalk Rot Fact Sheet*, Version 1. PANNAR Seed (Pty) Ltd: Greytown.

Knutson, A. E., & Cronholm, G. (2007). Economic injury levels for sorghum midge, *Stenodiplosis sorghicola*, and corn earworm, *Helicoverpa zea*, feeding on panicles of sorghum, *Sorghum bicolor*. *Southwestern Entomologist, 32*, 75–85.

Knutson, A., Bynum, E., Kerns, D., Porter, P., Biles, S., & Reed, B. (2018). Managing insect and mite pests of Texas sorghum. *ENTO-085*. Texas A&M Agrilife Extension Service.

Koima, I. N., Kilalo, D. C., Orek, C. O., Wagacha, J. M., & Nyaboga, E. N. (2023). Identification and characterization of *Colletotrichum* species causing sorghum anthracnose in Kenya and screening of sorghum germplasm for resistance to anthracnose. *Journal of Fungi (Basel), 9*(1), 100. https://doi.org/10.3390/jof9010100 (accessed on 7th November 2025).

Kumar, A., Hunjan, M. S., Kaur, H., Singh, P. P., & Kaur, R. (2016). Evaluation of management of bacterial stalk rot of maize (*Dickeya zeae*) using bioagents and chemical agents. *Journal of Applied and Natural Science, 8*, 1146–1151.

Kumar, D., Rolania, K., Sharma, B. L., Kumar, L., & Saini, V. K. (2022). Adaptive dynamics of earhead bugs in different genotypes of sorghum under abiotic factors. *Journal of Agriculture and Environment*, 35–42. https://doi.org/10.53911/jae.2022.13103 (accessed on 7th November 2025).

Kumar, S., Kumar, A., Mishra, I., Madan, M., & Bajpeyi, S. (n.d.). Integrated pest management of sorghum shoot fly (*Atherigona soccata* Rondani). *Agri Articles (e-Magazine for Agricultural Articles)*.

Lettry, V. (2020). PMSP for sorghum in North Carolina, South Carolina, and Virginia. *National IPM Database*.

Lipps, P. E., & Nyvall, R. F. (1990). Field crop diseases handbook. *Mycologic, 82*(5), 663. https://doi.org/10.2307/3760061 (accessed on 7th November 2025).

Little, C. R., Perumal, R., Tesso, T., Bandara, Y. M. A., & Noor, A. (2023). Charcoal rot: A major root and stalk disease of sorghum. Elsevier Publication, Radarweg 29, 1043 NX Amsterdam, The Netherlands.

Malgwi, A. M., & Dunuwel, D. M. (2011). Damage and control of panicle insect pests of sorghum (*Sorghum bicolor* (L.) Moench) in North Eastern Nigeria. *ARPN Journal of Science and Technology, 3*, 195–202.

Malvick, D. K. (2008). Sorghum smuts. Integrated Pest Management. *Reports on Plant Disease*, No. 208.

Mayers, P. E. (1976). The first recordings of milo disease and *Periconia circinata* on sorghums in Australia. *Australasian Plant Pathology, 5*(4), 59. https://doi.org/10.1071/app9760059 (accessed on 7th November 2025).

McLeod, P., & Greene, J. (2004). 5-major insect pests of grain sorghum in Arkansas and their management. *University of Arkansas Extension*. https://www.uaex.edu/publications/PDF/MP297/5_insects.pdf (accessed on 7th November 2025).

Mittal, M., & Boora, K. S. (2005). Molecular tagging of gene conferring leaf blight resistance using microsatellites in sorghum (*Sorghum bicolor* (L.) Moench). *Indian Journal of Experimental Biology, 43*(5), 462–466.

Mtisi, E. (1996). Evaluation of systemic seed dressing for the control of covered kernel smut on sorghum in Zimbabwe. In K. Leouschner & C. S. Manthe (Eds.), *Drought-Tolerant Crops for Southern Africa: Proceedings of the SADC/ICRISAT Regional Sorghum Workshop, Botswana* (pp. 185–188). International Crops Research Institute for the Semi-Arid Tropics, Patancheru.

Muriithi, L. M., & Claflin, L. E. (1997). Genetic variation of grain sorghum germplasm for resistance to *Pseudomonas andropogonis*. *Euphotic, 98*, 129–132.

N/A. (n.d.). Fungicides. Syngenta. https://www.syngenta.co.ke/fungicides (accessed on 7th November 2025).

Navi, S. S., Bandyopadhyay, R., Devi, K. T., & Reddy, D. (2002). Bacterial leaf streak of sorghum - A new report from India. *International Sorghum and Millets Newsletter, 43*, 61–63.

Oberemok, V. V., Gal'chinsky, N. V., Useinov, R. Z., Novikov, I. A., Puzanova, Y. V., Filatov, R. I., Kouakou, N. J., Kouame, K. F., Kra, K. D., & Laikova, K. V. (2023). Four most pathogenic superfamilies of insect pests of suborder Sternorrhyncha: Invisible superplunderers of plant vitality. *Insects, 14*(5). https://doi.org/10.3390/insects14050462 (accessed on 7th November 2025).

Ogolla, F. O., Muraya, M. M., & Onyango, B. O. (2019). Variation in temperature and nutrient source influence the growth of *Exserohilum turcicum* mycelia isolated from sorghum. *Journal of Science and Engineering, 6*, 93–99.

Okosun, O. O., Allen, K. C., Glover, J. P., & Reddy, G. V. P. (2021). Biology, ecology, and management of key sorghum insect pests. *Journal of Integrated Pest Management, 12*(1). https://doi.org/10.1093/jipm/pmaa027 (accessed on 7th November 2025).

Overholt, W. A., Ogedah, K., & Lammers, P. M. (1994). Distribution and sampling of *Chilo partellus* (Lepidoptera: Pyralidae) in maize and sorghum on the Kenya coast. *Bulletin of Entomological Research, 84*(3), 367–378. https://doi.org/10.1017/s0007485300032491 (accessed on 7th November 2025).

Pande, S., Bock, C. H., Bandyopadhyay, R., Narayana, Y. D., Reddy, B., Lenne, J. M., & Jeger, M. J. (1997). Downy mildew of sorghum. *Publication No. 51*. International Crops Research Institute for the Semi-Arid Tropics (ICRISAT), Patancheru.

Patil, S. P., & Bagde, A. S. (2017). Effect of new molecules of insecticides on shoot fly (*Atherigona soccata* Rondani) incidence and grain yield of sorghum. *International Journal of Current Microbiology and Applied Sciences, 6*(8), 2751–2754. https://doi.org/10.20546/ijcmas.2017.609.339 (accessed on 7th November 2025).

Pawar, V. M., Jadhav, G. D., & Kadam, B. S. (1984). Compatibility of Incol 50 SP with different fungicides on sorghum (CS-3541) against shoot fly (*Atherigona soccata* Rondani). *Pesticides, 18*, 9–10.

Perumal, R., Rajendrakumar, P., Maulana, F., Tesso, T., Little, C. R., Ciampitti, I., & Prasad, V. (2016). Genetic changes in sorghum. In *Sorghum: State of the art and future perspectives*. American Society of Agronomy and Crop Science Society of America, Inc.

Peterschmitt, M., Ratna, A. S., Sacks, W. R., Reddy, D. V. R., & Mughogho, L. K. (1991). Occurrence of an isolate of maize stripe virus on sorghum in India. *Annals of Applied Biology, 118*(1), 57–70. https://doi.org/10.1111/j.1744-7348.1991.tb06085.x (accessed on 7th November 2025).

Poul, R. M., Jayewar, N. E., & Bhede, B. V. (2020). Evaluation of novel insecticides against lepidopteron insect pests of sorghum fall armyworm, *Spodoptera frugiperda* and earhead worm, *Helicoverpa armigera*. *Journal of Entomology and Zoology Studies, 8*(6), 1826–1830.

Prom, L. K., Perumal, R., Erattaimuthu, S. R., Erpelding, J. E., Montes, N., Odvody, G. N., Greenwald, C., Jin, Z., Frederiksen, R., & Magill, C. W. (2011). Virulence and molecular genotyping studies of *Sporisorium reilianum* isolates in sorghum. *Plant Disease, 95*(5), 523–529. https://doi.org/10.1094/PDIS-10-10-0720 (accessed on 7th November 2025).

Prom, L. K., Perumal, R., Isakeit, T., Erattaimuthu, S., & Magill, C. (2021). Response of sorghum accessions against newly documented pathotypes 5 and 6 of head smut pathogen, *Sporisorium reilianum*. *American Journal of Plant Sciences, 12*(3), 432–443. https://doi.org/10.4236/ajps.2021.123028 (accessed on 7th November 2025).

Puri, S. N., & Mote, U. N. (2003). Emerging pest problems of India and critical issues in their management. In B. Subrahmanyam, V. V. Ramamurthy, & V. S. Singh (Eds.), *Proceedings of National Symposium on Frontier Areas of Entomological Research* (pp. 13–24). Entomological Society of India: New Delhi.

Ratnadass, A., Marley, P. S., Hamada, M. A., Ajayi, O., Cissé, B., Assamoi, F., Atokple, I. D. K., Beyo, J., Cisse, O., Dakouo, D., Diakite, M., Dossou-Yovo, S., Le Diambo, B., Vopeyande, M. B., Sissoko, I., & Tenkouano, A. (2003). Sorghum head-bugs and grain molds in West and Central Africa: I. Host plant resistance and bug–mold interactions on sorghum grains. *Crop Protection, 22*(6), 837–851. https://doi.org/10.1016/s0261-2194(03)00066-8 (accessed on 7th November 2025).

Rodríguez-del-Bosque, L. A., & Salinas-García, J. R. (2008). Impact of tillage on *Phyllophaga crinita* (Coleoptera: Scarabaeidae) and other soil invertebrates in grain sorghum in Mexico. *Journal of Entomological Science, 43*(1), 117–120. https://doi.org/10.18474/0749-8004-43.1.117 (accessed on 7th November 2025).

Sagnia, S. B. (1983). Possible integrated pest management tools for the effective control of cereal stem-borers in the Gambia. *Insect Science and Its Application, 4*, 217–219.

Sajwan, R., Rawat, L., Joshi, M., Kumar, P., Mamgain, A., & Kharola, G. (2022). Importance, taxonomy, bionomics and damaging symptoms of various insect pests associated with millets. *Vigyan Varta, 3*(10), 50–63.

Sanon, A., Yamkoulga, M., Koussoube, J. C., Waongo, A., & Ouedraogo, I. (2018). Insect pests of post-harvest storage in promising crop sectors in Burkina Faso: Current concerns and prospects for solutions. *Julius Kühn Archiv, 463*, 934–940.

Saxena, S. C., Mughogho, L. K., & Pande, S. (1991). Stalk rot and top rot of sorghum caused by *Erwinia chrysanthemi*. *Indian Journal of Microbiology, 31*, 435–441.

Sharma, H. C. (2003). Evaluation of the effects of plant diseases on the yield and nutritive value of crop residues used for peri-urban dairy production on the Deccan Plateau in India. In S. Pande & P. Parthasarathy Rao (Eds.), *Summary Proceedings of a Workshop*. ICRISAT, Patancheru.

Sharma, I., Bhardwaj, R., & Pati, P. K. (2015). Exogenous application of 28-homobrassinolide modulates the dynamics of salt and pesticides induced stress responses in an elite rice variety Pusa Basmati-1. *Journal of Plant Growth Regulation, 34*(3), 509–518. https://doi.org/10.1007/s00344-015-9486-9 (accessed on 7th November 2025).

Shiferaw, T., & Dargo, F. (2019). Evaluating sorghum stem borer *Chilo partellus* (Swinhoe) (Lepidoptera: Crambidae) management alternatives in Eastern Ethiopia. *Agricultural Research & Technology: Open Access Journal, 20*(4).

Singh, B. U., Rao, K. V., & Sharma, H. C. (2011). Comparison of selection indices to identify sorghum genotypes resistant to the spotted stemborer, *Chilo partellus* (Lepidoptera: Noctuidae). *International Journal of Tropical Insect Science, 31*, 38–51.

Singh, Y., Sharma, D., & Kharayat, B. S. (2020). Major diseases of sorghum and their management. In *Diseases of Field Crops: Diagnosis and Management* (Vol. 1, pp. 153–182). Apple Academic Press.

Tariku, A. (2021). Global impacts of sorghum anthracnose disease on sorghum production and its management option: A review. *Journal of Agricultural Research and Advances, 3*(2), 37–41.

Tesso, T., Ochanda, C., & Claflin, M. R. (2010). *Sorghum bicolor* (L.) Moench: Analysis of host-plant resistance to multiple *Fusarium* species associated with stalk rot disease in sorghum. *Field Crops Research, 118*, 177–182.

Tesso, T., Perumal, C., Adeyanju, G., Prom, L. K., & Magill, C. W. (2012). Sorghum pathology and biotechnology–A fungal disease perspective: Part II. Anthracnose, stalk rot, and downy mildew. *The European Journal of Plant Science and Biotechnology, 6*, 31–44.

Thakur, R. P., Reddy, B., Indira, S., Rao, V. P., Navi, S. S., & Yang, X. B. (2006). Sorghum grain mold. *Information Bulletin No. 72*. International Crops Research Institute for the Semi-Arid Tropics, Patancheru.

Tsedaley, B., Adugna, G., & Lemessa, F. (2016). Distribution and importance of sorghum anthracnose (*Colletotrichum sublineolum*) in Southwestern and Western Ethiopia. *Plant Pathology Journal, 15*(3), 75–85. https://doi.org/10.3923/ppj.2016.75.85 (accessed on 7th November 2025).

Tsedaley, B., Adugna, G., & Lemessa, F. (2023). Evaluation of sorghum genotypes and influence of weather variables on anthracnose (*Colletotrichum sublineolum*) disease development under field conditions at Jimma, Southwestern Ethiopia. *Heliyon, 9*(4), e15297. https://doi.org/10.1016/j.heliyon.2023.e15297 (accessed on 7th November 2025).

Van Den Berg, J., Pretorius, A. J., & Van Liggerenberg, M. (2003). Effect of leaf feeding by *Melanaphis sacchari* (Zehntner) (Homoptera: Aphididae) on sorghum grain quality. *South African Journal of Plant and Soil, 20*, 41–43.

Wang, M. L., Ryley, M., & Meinke, H. (2000). Prediction of sorghum downy mildew risk in Australia using daily weather data. *Australasian Plant Pathology, 29*, 108–119.

Wani, T. A., Ahmad, M., & Anwar, A. (2017). Evaluation of fungicides, bioagents and plant extracts against *Exserohilum turcicum* causing turcicum leaf blight of maize. *International Journal of Current Microbiology and Applied Sciences, 6*(8), 2754–2762. https://doi.org/10.20546/ijcmas.2017.608.329 (accessed on 7th November 2025).

White, J. A., Ryley, M. J., George, D. L., Kong, G. A., & White, S. C. (2012). Yield losses in grain sorghum due to rust infection. *Australasian Plant Pathology, 41*(1), 85–91. https://doi.org/10.1007/s13313-011-0093-3 (accessed on 7th November 2025).

Williams, R. J., & Frederiksen, J. C. (1978). *Sorghum and Pearl Millet Disease Identification Handbook. ICRISAT Information Bulletin*, No. 2.

Wirnas, D. (2020). Genotype by environment interaction affecting leaf rust resistance in sorghum. *SABRAO Journal of Breeding and Genetics, 52*(1), 75–90.

Young, W. R., & Teetes, G. L. (1977). Sorghum entomology. *Annual Review of Entomology, 22*, 193–218.

CHAPTER 14

Integrated Pest and Disease Management in Pearl Millet

P. R. NITHYA[1,*] and PADHIYIL AMRUTHA[2]

[1]*Regional Agricultural Research Station, Kerala Agricultural University, Vellayani, Thiruvananthapuram, Kerala, India*

[2]*School of Agricultural Sciences, Karunya Institute of Technology and Sciences, Coimbatore, Tamil Nadu, India*

*Corresponding author

14.1 INTRODUCTION

Pearl millet (*Pennisetum glaucum* L.R.Br.) is an annual grass, commonly known as "Indian millet" or "bajra." Pearl millet originated in West Africa, with the earliest evidence discovered in Mauritania dating back to 1000 BC. As a crop that is nutrient-rich and climate-resilient, pearl millet is becoming more and more popular. The monophyletic origin of pearl millet and subsequent secondary diversification and migration have resulted in enormous diversity (Pattanashetti et al., 2016). The worldwide yearly millet production is estimated to be 28.4 metric tons (FAO, 2017). It is mostly grown by subsistence and small-scale commercial farmers in the semiarid tropics (Reddy et al., 2021). Approximately 50% of pearl millet is grown in Rajasthan, and Maharashtra, Gujarat, Uttar Pradesh, and Haryana are states that contain about 40% of the total area. It is farmed across 6.93 million hectares, producing an average of 8.61 million tons annually at a productivity of 1,243 kg ha^{-1} (Directorate of Millets Development, 2020). A significant barrier to using this crop's maximum yield potential is the presence of insect pests and diseases. There have been 300 insects discovered feeding on pearl millet in different parts of the world (Sharma et al., 1981). However, there are probably fewer than a dozen pests that cause significant damage, and there are even fewer that do so continuously (Varma, 1980). In the parts of India where pearl millet is grown, downy mildew, leaf blast, ergot, smut, and rust occur virtually often in low to severe forms.

Millets in Sustainable Agriculture: Genetic Improvement, Nutritional Quality, Cultivation, and Integrated Pest–Disease Management.
Roshni Vijayan, A. John Joel, R. Ambika Rajendran, and Adithya P. Balakrishnan (Eds.)

DOI: 10.1201/9781779642332-18

14.2 INSECT PESTS

14.2.1 BORER PESTS

14.2.1.1 SHOOT FLIES

A variety of millets are infested by *Atherigona* shootflies (Gahukar, 1985). *Atherigona approximata* Malloch is known to infect pearl millet in numerous states of India, resulting in yield losses of 12–46% in grain and 57% in dry feed (Natarajan et al., 1973; Sharma et al., 1974). As early as one week following crop emergence, the pest is reported to be attacking the crop. The maggot attacks the plant's central shoot, resulting in a dead heart. Tillering becomes excessive in the later stages of infestation. Additionally, this insect damages the earheads (Sharma et al., 1988).

A temperature range of 24.5 to 29.1°C was reported to have the maximum population in North India on the 37th standard week, and it thereafter began to fall (Ajay et al., 2022). The peak population was observed during the 33rd and 34th standard weeks in arid regions like Rajasthan (Choudhary et al., 2018). The pest relies on wild hosts like *Cynodon dactylon* to live during the off-season (Talati et al., 1978). Pupation lasts for roughly 6 to 11 days and can take place on the plant or in the soil. Three pairs of black dots on the dorsum of the abdomen help to identify adult flies. It takes around three weeks to complete the entire life cycle (Sharma et al., 1988).

Shootflies can be controlled in a variety of ways, including the use of pest-resistant cultivars (MP series-16,19,53,67 and MH series-9, 49,52, 82, 99, 105) (Biradar et al., 2018), increasing the seed rate and subsequently thinning out the damaged seedlings, installing fish meal traps @ 30/Ha, and treating seed with imidacloprid 70WS at 10g kg^{-1} (TNAU, 2016). Application of 5% neem seed kernel extract or 1% NeemAzal, as well as pre-sowing soil applications of granular formulations of phorate 10 G @ 3 kg a.i. ha (Singh et al., 1973) were also successful in controlling this pest.

14.2.1.2 STEM BORERS

One of the most devastating pests wreaking havoc on pearl millet is the stem borer. The two species of stem borers most frequently seen in India are the pink stem borer, *Sesamia inferens* Walker, and the spotted stem borer, *Chilo partellus* Swinhoe (Gahukar et al., 1980). Pests cause dead hearts by boring into the stem, which leads to the formation of numerous tillers with unproductive spikes. Late attacks cause peduncle damage and plant breakage, which contributes to the loss of grains and fodder. When the crop is young, pearl millet seems resistant to borer attack, but as it grows older, it is vulnerable to internode damage (Ahmed et al., 1969). Kishore reported about 23.3 to 36.5% grain losses and 37.5% fodder losses due to shoot fly, and 20–60% yield losses due to stem borer (1966a).

1. **Spotted Stem Borer:** A damaging and widespread species is the spotted stem borer (*C. partellus* Swinhoe). The Gujarat and Delhi regions of India were found to have a higher prevalence than other places where pearl millet is grown. On the underside of the leaves, the female deposits patches of flattened, overlapping,

yellowish oval white eggs. The larvae feed on the developing leaf whorls after hatching and are later borne into the stem. Feeding on the central shoot results in dead heart and plant mortality. The life cycle is completed in 4–6 weeks. *C. partellus* infestation on pearl millet was noted from 35 to 42 standard weeks, with the population peaking on week 37 under Delhi circumstances (Mittal et al., 2006).

2. **Pink Stem Borer:** The pink stem borer attack on pearl millet is not as severe as on sorghum. The pink borer (*S. inferens* Walker) produces batches of 30–100 spherical, yellowish-white eggs laid in two to three parallel rows on the leaf surface or sheaths. When fully grown, the smooth, pinkish larva is about 30 mm long. Early instar larvae do not feed on leaves. Instead, they bore directly into the stem. Pupation takes place in stem galleries, and the life cycle takes 40–50 days to complete. Moths have forewings that are straw in color with a dark brown line around the edge. Under tropical conditions, the pest completes four to five generations in a year.

Field sanitation techniques, such as collecting and burning crop residue and weeds after harvest, as well as digging up dead hearts, can manage the borer population (Jotwani and Butani, 1978; Sagnia, 1983). Late-planted and millet crops with long duration are more vulnerable to borer attack (Harris, 1962). In the Nigerian Sudan Savannah, groundnut pearl millet intercropped in a 1:2 ratio and a 1:1 ratio resulted in lower stem borer infestation (Degri et al., 2014). Pearl millet crops treated with a combination of fipronil + imidacloprid WG at 2.5 g/kg of seed and followed by a spray of *Beauveria bassiana* WP at 0.007%, or treated with clothianidin WDG at 7.5 g/kg of seed and followed by the same *B. bassiana* WP spray, showed reduced infestation of shoot flies and stem borers. The highest benefit-cost ratios were achieved with the fipronil + imidacloprid WG seed treatment and *B. bassiana* WP spray (1:16.19), and the clothianidin WDG seed treatment followed by *B. bassiana* WP spray (1:15.35) (Pateliya et al., 2019).

14.2.2 DEFOLIATORS

There have been numerous reports of hairy caterpillar infestations on millets in India's northern, southern, and western regions, most of the incidents being sporadic. *Amsacta albistriga* Walker, *Estigmene lactinea* Huebner, and *Spilosoma obliqua* Walker are the commonly reported species of hairy caterpillars (Jotwani and Butani, 1978). With the gregarious nature and voracious appetite of the pests, complete defoliation or loss of crop may occur. The larvae are reddish with many blackish hairs. They thrive in the summer via pupa diapause. About a fortnight after the first showers, moths start to appear. Depending on the distribution and amount of rainfall, the pest may complete one or two generations.

14.2.2.1 ARMYWORMS

Mythimna separata Walker, the armyworm, feeds on the vegetation (Balasubramanian et al., 1975). *Spodoptera frugiperda* Smith did not appear to prefer oviposition on different varieties of pearl millet, and resistance seemed to be linked to the absence of leaf pubescence

(Leuck et al., 1968; Leuck et al., 1977). During the rainy season of 1987, outbreaks of the noctuids (*M. separata, M. albistigma Moore,* and *Mocis frugalis* Fab.) were observed in Bangalore and Kolar districts in Karnataka. *S. frugiperda* (J.E. Smith) (Lepidoptera: Noctuidae), an invasive fall armyworm (FAW), has recently been discovered infesting a crop of pearl millet. FAW reports on maize, sorghum, pearl millet, and finger millet were made in several districts of Andhra Pradesh, Karnataka, and Tamil Nadu (Bhavani et al., 2019; ICAR-IIMR, 2019). The yield of grains is seriously endangered by this pest's voracious feeding and long-distance flying habits.

14.2.2.2 GRASSHOPPERS

The gregarious feeding nature of grasshoppers damages the crop by reducing the green matter, thereby affecting the photosynthetic efficiency. Crops of millet are reportedly destroyed by *Hieroglyphus nigrorepletus Bolivar, H. banian Fab., Chrotogonus* spp., and *Colemania spheneroides* Bolivar in India (Gahukar and Jotwani, 1980; Jotwani and Butani, 1978). In the arid areas of western Rajasthan, grasshoppers have occasionally been seen eating pearl millet. The most frequent grasshopper attacking crops in Rajasthan's dry regions at the seedling stage is *Chrotogonus trachypterus*. In Varanasi, Uttar Pradesh, Hieroglyphus nigrorepletus was observed attacking pearl millet, according to Roonwal (1945). In Andhra Pradesh, Karnataka, Maharashtra, and Tamil Nadu, the Deccan grasshopper *Colemania sphenariodes* has been identified as a significant pest of pearl millet (Pruthi, 1969). Besides those mentioned in the written word, several species feed on pearl millet's leaves.

14.2.3 SUCKING PESTS

Sucking pests sap from different plant parts such as leaves, stems, and earheads and act as vectors of some viral diseases. Common sucking pests recorded on pearl millet include *Balclutha* spp. (Khurana and Deshmukh, 1974), pyrilla (*Pyrilla perpusilla* Walker) (Jotwani, 1969), spotted aleyrodid *Neomaskellia bergii* Signoret (Verma, 1980), aphids (*Rhopalosiphum maidis* Fitch), shoot bug (*Peregrinus maidis* Ashmead), and thrips (*Chirothrips mexicanus* Crawford) (Ananthakrishnan and Thirumalai, 1977). The maize aphid *Rhopalosiphum maidis* has been shown to transmit maize streak virus in pearl millet. Nymphs and adults of the shoot bug *P. maidis* (Ashmead) were found gregariously on the peduncle and in compact earheads, leading to chaffy earheads. Gupta and Gupta (1970) reported the mustard-painted bug *Bagrada cruciferarum* as a major pest on early varieties of pearl millet and noted that the bristled earhead types incurred less damage than the non-bristled ones. Since then, reports indicate that it has been a significant pest of spring pearl millet in Punjab (Sandhu, 1975). There have been reports of small pests of pearl millet, such as *Nezara viridula* Linn. and *Dolycoris indicus* Stal., sucking sap from the milky grains (Balikai, 2010). Stink bugs such as *Euschistus servus* (Say) and *Leptoglossus phyllopus* (L.) were reported to attack later-planted pearl millet crops in the southeastern parts of the U.S. (Obeng et al., 2015).

14.2.4 *EARHEAD PESTS*

14.2.4.1 *GRAIN MIDGES*

In Southern India, earhead midges (*Geromyia penniseti*) and gall midges (*Lontinida seminis*) have been reported on pearl millet. *G. penniseti* is a significant commercial pest of pearl millet (Santharam et al., 1976). Larval feeding hinders normal seed development and results in full or partial spike abortion, which can reduce yield by up to 90% (Coutin and Harris, 1968). The adult fly is a tiny, delicate insect with a vivid orange abdomen and two transparent wings. A female typically produces between 30 and 35 eggs, 6–10 eggs per floret, with an incubation period of 3 to 4 days. The maggot is colorless but dark orange when it matures. It has four instars with a duration of 9 to 11 days. The maggot feeds on the growing grains and pupates there. Maggots significantly damage the ovaries, which affects the growth of the seeds and produces chaffy panicles. Pupal cases can be seen attached to the glumes of the damaged spikelet or protruding out from the chaffy grains with exit holes. In a season, midges produce four to five overlapping generations.

14.2.4.2 *HEAD BEETLES*

Millets have been found to be the home to several species of beetles from the *genera Anomala, Cetonia, Cylindrothorax, Mylabris*, etc (Balikai, 2010; Dhaliwal et al., 1974; Jotwani and Butani, 1978). They consume blooms and prevent the growth of grains. In India, beetles like *Epicauta* (*Lytta*) *tenuicollis* Pall., *Mylabris pustulata* Thunb., and *Psalydolytta rouxi* Laporte have long been regarded as major pearl millet pests (Balikai, 2010; Gahukar and Jotwani, 1980; Singh, 1967). Occasionally, scarabaeids can be a serious threat, including *Rhinyptia laeviceps* Arrow (Mishra et al., 1979; Yadava et al., 1973), *R. meridionalis* var. *puncticollis* Arrow (Pal and Sharma, 1973), *Rhinyptia* spp. (Verma, 1979), and *Chiloloba acuta* Wied. (Srivastava et al., 1971) in India. Farmers typically use insecticides by dusting or spraying them. The crop was found to be effectively protected from blister beetles and plant bugs for up to 10 days by being dusted with 20 kg ha of malathion and sprayed with 2 mL per liter of endosulfan. In contrast, no other treatments provided more than 25% protection (Balikai and Guggari, 2006).

14.2.4.3 *HEAD CATERPILLARS*

Helicoverpa armigera Hbn, *Eublemma silicula* Swinh., and *Cryptoblabes gnidiella* Mill. are common caterpillars feeding on pearl millet earheads (Jotwani et al., 1966; Jotwani and Butani, 1978; Patel and Dhagat 1981). On pearl millet, the noctuid *E. silicula* (*Autoba silicula*) has recently become more severe. A severe outbreak was recorded in Delhi, India, during the rainy season of 1995. The larvae consumed developing grains while hiding under a little dome-shaped or longitudinal gallery made of anthers and silk. The most active period for the pest was from late August to early September (Kishore, 1996b). The pearl millet earheads are consumed by the sorghum earhead worm *Cryptoblabes gnidiella* (AICMIP 1970–71).

14.2.5 SOIL DWELLING INSECTS

14.2.5.1 WHITE GRUBS

Pearl millet is extensively attacked by *Holotrichia consanguinea* Blanch throughout central and northwest India (Jotwani and Butani, 1978; Khairwal and Yadav, 2005). The "C"-shaped grubs consume the roots, causing the plant to wither and die. Infected seedlings continue to grow stunted, and seed formation is hindered. Even three or four grubs could attack one plant at once (Lal et al., 1976). Eggs typically hatch in 1–3 weeks, whereas grubs need 8–22 weeks to mature. The plant becomes stunted, yellow, wilted, and dried as a result of the *H. longipennis* Blanch larvae feeding on living roots. In the hilly areas of Uttarakhand, the degree of damage ranged from 2.28% to 65.15%, depending on the height. The mid-hills suffered the most damage (5.67–65.16%), followed by the high-hills (4.96–62.92%). High-hill crops showed less mean damage (Singh et al., 2004). If the climate is suitable, the beetle emerges by November or December after a pupal phase of 1 to 8 weeks. If not, the pest overwinters and becomes active as an adult during May through July of the following year (Jotwani and Butani, 1978). Seed treatment with imidacloprid 600FS at a rate of 10–12 ml/kg was highly effective in decreasing seedling mortality from 28.6% to a mere 2.1% (Choudhary et al., 2018). Incorporation of phorate 10 G, carbofuran 3 G, or quinalphos 5 G with pearl millet seeds at a rate of 12 kg/ha into furrows at the time of sowing effectively reduced crop damage. Endosulfan, 4% dust, applied at a rate of 25 kg/ha is also very effective against white grub. Furrow application of a mixture of chlorpyrifos or quinalphos dust with farmyard manure (FYM) in a 2:3 ratio also showed effectiveness in managing white grubs. When pearl millet is intercropped with pulses like green gram (*Vigna radiata* var. radiata Wilczek), cowpea (*Vigna unguiculata* Walp.), or clusterbean (*Cyamopsis tetragonoloba* L. Taubert), the damage is greatly reduced (Khairwal and Yadav, 2005). It's possible for birds to prey on beetles (Pandey and Yadava, 1974). The use of insecticides in endemic regions and the modification of cultural norms seem appropriate for pest management against white grubs.

14.3 DISEASES OF PEARL MILLET

Even though pearl millet is a crop that is grown in a dry climate, it is infected with numerous illnesses (see Table 14.1). Bacterial and viral infections rarely occur compared to fungus-related illnesses. Downy mildew was ranked as having a relative importance of 45%, striga 32%, smut 9%, ergot 7%, rust 3%, viruses >1%, and other diseases 3% in (King, 1992) ranking of pearl millet illnesses. Both striga and downy mildew are significant in Western Africa and, to a lesser extent, in India.

14.3.1 DOWNY MILDEW

Sclerospora graminicola causing downy mildew (DM) continues to be a significant biotic barrier to the production of pearl millet (*Pennisetum glaucum* (L.) R. Br.) in India due to its frequent occurrence. The disease affects all areas of India and Africa where pearl millet

TABLE 14.1 List of Pearl Millet Diseases

Disease	Causal Organism
Fungal diseases	
Downy mildew	*Sclerospora graminicola*
Blast	*Pyricularia grisea*
Ergot	*Claviceps fusiformis*
Smut	*Moesziomyces penicillariae*
Rust	*Puccinia penisetti* (*P. substriata* var. *indica*)
Top rot	*Fusarium moniliforme*
Bipolaris leaf spot	*Bipolaris setariae*
Cercospora leaf spot	*Cercospora penniseti*
Curvularia leaf spot	*Curvularia penniseti*
Drechslera leaf spot	*Drechslera dematioidea*
Phyllachora leaf spot	*Phyllachora penniseti*
Myrothecium leaf spot	*Myrothecium roridum*
Zonate leaf spot	*Gloeocercospora sorghi*
Southern blight	*Sclerotium rolfsii*
Phyllosticta leaf blight	*Phyllosticta penicillariae*
Rhizoctonia blight	*Rhizoctonia solani; Rhizoctonia zeae*
Exserohilum leaf blight	*Exserohilum rostratum*
Bacterial diseases	
Bacterial spot	*Pseudomonas syringae*
Bacterial leaf streak	*Xanthomonas campestris* pv. *pennamericanum*
Bacterial leaf stripe	*Acidovorax avenae*
Viral diseases	
Black streaked dwarf virus	
Guinea grass mosaic virus	
Maize dwarf mosaic virus	
Maize streak virus	
Panicum mosaic virus	
Satellite panicum mosaic virus	
Wheat streak mosaic virus	

is grown. The main downy mildew-prone regions for pearl millet in India are Gujarat, Rajasthan, and Haryana in the northwest, and Tamil Nadu and Karnataka in the south. The problem of downy mildew was first faced by high-yielding, single-cross hybrids in the late 1960s, and the first epiphytotics on the widely used hybrid HB 3 occurred in 1971 (Nene and Singh, 1976). The limited genetic foundation of the hybrids' parental lines was a major contributing factor in the outbreak. The disease has an impact on grain yield, resulting in a loss of 4.6 million metric tons (Gwary et al., 2009) and posing a negative impact on fodder quality. A severe and widespread infection decreases the shoot's fresh weight as well as the quantity of basal and nodal tillers (Gupta and Singh, 1996). Based on a few localized estimates, the average yearly yield losses in India can reach up to 40%, while losses of between 10% and 50% have been documented in Nigeria.

14.3.1.1 SYMPTOMS

In pearl millet, downy mildew, often known as green ear, typically presents as a systemic disease with few, if any, local lesion signs. Systemic symptoms are visible in the leaves and earhead. The illness progresses as yellowing or chlorosis symptoms move from the lower leaves to the top leaves until all of the leaves are impacted. Pearl millet seedlings grown in soil with downy mildew inoculum start to show symptoms 5–6 days after they emerge. Systemic symptoms can appear from the seedling through the flowering stage. Despite being susceptible to the disease during the whole growth period of the crop, pearl millet is most susceptible during the coleoptile stage, and as the crop ages, this susceptibility decreases.

The disease is typically identified by spore formation on the abaxial leaf surface, chlorosis, stunted growth, and earhead deformity. When a leaf only exhibits symptoms on its lower half, this is referred to as half-leaf symptoms. Numerous spores are produced on the lower surface of the afflicted leaves. Infected chlorotic leaf areas on the lower leaf surface can support a significant amount of white asexual sporulation. Severe damage causes plants to become stunted and may even prevent them from producing earheads. Sometimes, only the earhead of infected plants displays symptoms in the shape of the leafy structure. The term “green ear” disease or “virescence” is also used to describe this stage of downy mildew (Upasana et al., 2010). Under conditions of high relative humidity (>95%) and moderate temperature (20–22°C), abundant white asexual sporulation occurs on the lower leaf surface of an infected chlorotic leaf, resulting in white downy growth.

14.3.1.2 PATHOGEN CHARACTERISTICS

S. graminicola, is a member of the family Sclerosporaceae, order Sclerosporales, phylum Oomycota, and class Oomycetes. The pathogen is homothallic; however, reports of homothalism also generate both sexual and asexual (sporangia, zoospores) spores (oospores) (Michelmore et al., 1982; Pushpavathi et al., 2006). Two kinds of mating have been recognized in S. graminicola. The pathogen derives nutrition from the formation of haustoria and haustorial mother cells. Coenocytic, multinucleate, and abundantly growing hyphae produce a large number of asexual spores on the lower leaf surfaces.

14.3.1.3 ASEXUAL PHASE

At the terminals of the sporangiophore branches are sterigmata, where sporangia are generated. Short, robust, dichotomously branching structures called sporangiophores emerge from systemically infected leaves through stomata. In nature, sporangia are formed between the hours of one and four in the morning. The spores are hyaline, thin-walled, ellipsoid or broadly elliptic, and papillate sporangia with dimensions of 15–22 x 12–21 μm are fully grown (Jouan and Delassus, 1971). Creating zoospores and sporangia indirectly initiates germination. One to twelve zoospores may be present in each sporangium (Shetty, 1987). Zoospores encyst (swim for 30–60 minutes), form a germ tube, and then begin to germinate.

Recent research has shown that sporangial production can take place at night when the temperature is moderate, and humidity is high, that is, between 10 and 30°C, with 20°C being the ideal temperature for production (Singh et al., 1987). Additionally, sporangial production is at its peak between 95 and 100% relative humidity (RH). Similar to this, germ tubes develop at temperatures between 15 and 35°C. At 30°C, zoospores keep their infectiousness for around 4 hours; at lower temperatures, they do so for an extended period (Singh and Gopinath, 1990).

14.3.1.4 SEXUAL PHASE

In *S. graminicola*, antheridia (male) and oogonia (female) start the sexual reproductive process, which is completed by the creation of oospores. Oospores are spherical, thick-walled, brownish-yellow resting spores with a diameter of 32 μm (22–35 μm). Oospores can only be formed when compatible *S. graminicola* mating types are present in the same tissue or when homothallism is active, and only when the host achieves senescence and finishes its development cycle, and at the same time, the pathogen completes its asexual spore cycle and changes to producing antheridia and oogonia, which fuse to produce oospores in large quantities. One of the key characteristics that distinguish this genus is the fusion of the oogonial and oospore walls. The oogonial wall is typically merged with the oospore wall in *Sclerospora* spp., which is a key characteristic of this species. When dealing with several species of *Sclerospora*.

Oospores have been reported to be able to persist for eight to 10 years, and they germinate either directly through germ tubes or indirectly by the release of zoospores, with the frequency of germination being very variable. Oospore germination is best at a temperature of 28±2°C. Oospores that are about a year old or older are more infectious (98%) than fresh or older oospores (Nene and Singh, 1976; Singh and Navi, 2000). SSR, ISSR, RFLP, and RAPD have been used in the pathogen's molecular characterization to explore variability (Sastry et al., 1995; Sharma et al., 2010; Sudisha et al., 2009). For the quick identification and targeted detection of *S. graminicola* pathotype 1, a SCAR primer pair (UOM3-Sg-Path1-F/R) has been created (Sudisha et al., 2009).

14.3.1.5 DISEASE CYCLE

The underground sections of plants become infected by oospores, which are found in soil, primarily when the plants are in the seedling stage. Oospores of *S. graminicola* typically germinate by germ tubes that develop an appressorium, marking the beginning of the infection process. Oospore infection (whether direct by germ tubes or indirect by zoospores) can affect the roots, underground stem bases of older plants, and seedlings' coleorhiza, radicle, and lower coleoptile. When the pathogen infects the developing leaves or inflorescence at the growth point, systemic symptoms emerge. The apical meristem region of the host is invaded by the pathogen, and symptoms manifest when the colonized organs expand and unfold during the process of tissue differentiation. As a result, symptoms start to show up after the crucial colonization and infection processes. Systemically infected leaves that are damp produce a lot of sporangia on the abaxial surface. If environmental conditions are favorable, sporangia are critical for the secondary dissemination of the illness within and between fields. These sporangia exude zoospores that grow, enter the skin or stomata, and subsequently cause infection. Young leaves from an infected plant that is in the coleoptile stage will show systemic signs in 4–7 days. If the conditions are right, diseased leaves will continue to generate sporangia until the tissues senesce or become necrotic.

14.3.1.6 MANAGEMENT STRATEGIES

The most practical and cost-effective approach to preventing downy mildew (*S. graminicola*) on pearl millet is through genetic resistance. Downy mildew resistance genes from wild plants will be helpful in the control of this disease (Singh and Navi, 2000). 26 different types of pearl millet were examined by various Western and Central African nations, where they discovered several sources of resistance types (Dieng et al., 1999). Shishupala et al. identified a protein from a virulent pathotype of *S. graminicola* and used its binding reaction to distinguish between cultivars of pearl millet that were susceptible to the disease and those that were resistant to it (1996). According to anatomical research on the leaves of three cultivars of pearl with varying degrees of susceptibility and resistance to downy mildew, the leaves of the highly resistant cultivar (P-310-17) had significantly thicker cuticular and epidermal layers, a lower stomatal index, and higher epicuticular wax than the leaves of the highly susceptible cultivar (7042 DMS) (Yadav and Thakur, 2001).

The most economically viable strategy for controlling downy mildew is the use of resistant cultivars. The identification of resistance sources and the breeding of resistant cultivars have both advanced significantly.

Lata et al. used Particle-inflow-gun (PIG) technology to create transgenic pearl millet lines that express the pin gene and have high resistance to downy mildew disease (2006). It was found that T1 plants that are downy mildew-resistant displayed tolerance to Basta, suggesting that the pin and bar genes co-segregate, and Southern and Northern analyses of the resistant T1 plants revealed that the pin gene was positive, confirming the pin gene's stable integration, expression, and transmission.

Innate immunity can be induced in plants by using inducers. Deepak et al. reported the use of cerebrosides, glycosphingolipids taken from several plant diseases, identified as resistance-inducing substances against downy mildew disease, which induced resistance on the second day (2003). Amino acid proline and Ind-Ile-Me (1-oxo-indanoyl-L-isoleucine methyl ester) were tested as seed treatments and foliar sprays on pearl millet to see if they could induce resistance against downy mildew disease. It was found that, under greenhouse conditions, 50% protection was achieved, and under field conditions, 60% protection was achieved (Raj et al., 2004; Shantharaj, 2007). Under greenhouse and field circumstances, seed treatment with aqueous sodium nitroprusside (SNP) with or without polyethylene glycol (PEG) priming produced the most host resistance to downy mildew (Manjunatha et al., 2008).

Both systemic and protective fungicides have been employed; however, systemic fungicides are more efficient. For about the first five weeks after sowing, the disease is controlled by seed treatment with the systemic fungicide metalaxyl at 2 g a.i./kg seed. Seeds treated with 6 g Apron 35 WS kg^{-1} seed and 4 g Ridomil MZ L^{-1} reduced disease incidence and increased yield (1,617 kg ha^{-1}) compared with the control (160%) (Pandya et al., 2000). When used as a foliar spray or a seed treatment, metalaxyl formulations successfully managed pearl millet downy mildew. The seed treatment alone was found to be substantially inferior to the seed treatment combined with a single foliar application (Shankara Rao et al., 1987). According to Anaso and Anaso, Apron Star 42WS, a novel metalaxyl formulation for seed dressing, is affordable and long-lasting in northern Nigerian regions where downy mildew is widespread (2010). Chitosan seed priming effectively warded off downy mildew disease in pearl millet plants (Manjunatha et al., 2008).

Raj et al. noticed the use of PGPRs in pearl millet downy mildew management and for promoting plant growth both as fresh suspensions and powdered formulations (2003). Treatments with *B. subtilis* strain GB03 and *B. pumilus* strain SE34 provided 50% and 43% protection. *Trichoderma viride, Trichoderma harzianum, Trichoderma hamatum,* and *Pseudomonas fluorescens* were used as biological control agents to soak pearl millet seeds. The efficacy of *P. fluorescens* and *T. viride* and Mancozeb 64% + metalaxyl 4% and metalaxyl reduced disease incidence by up to 52.40% compared to the control but not on the same level as metalaxyl (Atri et al., 2023; Latake and Kolase, 2007).

Siddaiah et al. evaluated the efficiency of zinc oxide nanoparticles, which are green synthesized, and found them effective in controlling the spore germination of *S. graminicola* by enhancing the activities of polyphenol oxidase, phenylalanine ammonia-lyase, peroxidase, and lipoxygenases (2018). Similarly, Pushpalatha et al. also reported high disease resistance against downy mildew pathogens by using chitosan nanoparticles. The use of endophytic *Streptomyces* spp. in controlling the disease was reported by Thomas et al. (1945).

14.3.2 ERGOT OR SUGARY DISEASE

In India, ergot is a serious disease that affects pearl millet and has quarantine implications caused by *Claviceps fusiformis*. The 1940s saw the first reports of pearl millet ergot coming out of India (Bhide and Hegde, 1957). In 1956, Maharashtra, India, reported having the first case of the disease (Khairwal et al., 2007). Ergot grew severe and occurred virtually yearly in India with the introduction of novel hybrids HB1 and HB3 in the late 1960s (Marley et al., 2002). Currently, many Indian states, including Rajasthan, Maharashtra, Karnataka, Haryana, and Uttar Pradesh, are familiar with the condition. Seed output, seed quality, germination, and seedling emergence are all negatively impacted by ergot infection. In hybrids, grain yield loss has been calculated to reach 58–70% (Thakur et al., 1982). Additionally, other African nations have reported cases of the disease, including Sudan, Ethiopia, Eritrea, Benin, Togo, the Central African Republic, Ghana, Senegal, Tanzania, Cameroon, and Nigeria (Loveless, 1967).

14.3.2.1 SYMPTOMS

Exudates of a sticky liquid (cream to pink mucilaginous droplets) from afflicted florets are the first sign of the disease. Numerous micro- and macro-conidia of the pathogen are present in the honeydew-like discharge. Within 10 to 15 days, these droplets dehydrate, and hard, dark brown to black, seed-sized structures with pointed apexes emerge from the florets in place of the grain. Through mechanical contact, the infection is transferred to neighboring florets. Numerous spikelets that emit honeydew drip down the ear head when the infection is severe (Siddiqui and Khan, 1973). Later, these drops darken and combine to cover a significant portion of the corn. Within 10 to 15 days, these droplets dehydrate and are replaced by sclerotia, which are hard, dark brown to black structures with a pointed apex that emerge from the florets in place of grains. Fungal sclerotium, which is produced by infected florets, ranges in color from light pink to dark brown to black, is spherical or elongated in shape, larger than seeds, and has a pointy apex. These sclerotia fall to the ground during harvest and threshing, mingle with the grain, and act as a key source of inoculum for the following crop.

14.3.2.2 PATHOGEN CHARACTERS

Thakur et al., Mantle et al., Chahal et al., and Tenberge provided the description of *C. fusiformis*. In both honeydew and culture, the fungus produces both macro- and microconidia (1984; 1997; 1985; 1999). Hyaline, fusiform, 12.0–26.4 x 2–4 μm in size, unicellular macroconidia germinate by generating one to three lateral or polar germ tubes. By separating at the tips, these germ tubes form secondary macro- and microconidia. Hyaline, spherical, unicellular microconidia measuring 2.4–10.8 x 1.2–4.8 μm only produce one germ tube during germination. On the ends of germ tubes, macro- and microconidia are both formed; macroconidia are created in chains. Sclerotia, the precursor to sexual spores, come in a variety of sizes, colors, shapes, and densities (Chahal et al., 1985). Sclerotia vary in size (3.6–1 x 1.3–1.8 mm), shape (elongated to round), color (light pink to dark brown to black), and compactness (hard to brittle with cavities). *Pyriform perithecia* are where asci are made. Instead of grains, each perithecium's protruding hyaline, nonseptate, and filliform ascospores. A spherical, dark brown capitulum is present at the head of the 1–16 fleshy stipes that sclerotia produce.

14.3.2.3 DISEASE CYCLE

When *C. fusiformis* ascospores carried by the wind settle on susceptible wild and forage grasses with feather-like stigmas in the spring, the disease begins to spread. Pollen and ascospores can be effectively captured by the stigmas (Tenberge, 1999). Ascospores are the initial (primary) inoculum, and they germinate and infect the ovary in less than 24 hours. A special stroma, which develops in the infected ovary, produces masses of haploid, one-celled conidia that are extruded into a viscous fluid that resembles honeydew. Flying insects and moths and splashes of rain play a significant role in the spread of disease by spreading honeydew to other flowers. Production of honeydew continues until the development of sclerotia begins, and when sclerotia mature within four to five weeks, they take the place of the seeds (Sharma et al., 1984). At the end of the growing season, the sclerotia are harvested along with the grain and remain on the soil as well.

The sclerotia left in the field during harvest or mixed with seed at the time of threshing and seeded with the seed the next season is where the primary disease cycle starts. Sclerotia, which can vary in size and shape, are formed when the honeydew droplets that once contained mycelial mass, conidia, and sugary liquid dry up and turn into hard, dark brown to black structures larger than the seed (Tenberge, 1999). These sclerotia produce numerous ascospores after rain showers, which are then carried by air currents to the stigmas of pearl millet panicles that are in bloom, where they germinate and spread infection at moderate temperatures (20–30°C) and air movement during crop blossoming.

14.3.2.4 MANAGEMENT

Ergot can be managed using disease control techniques such as cultural, chemical, and biological control. Deep ploughing and weeding the area around pearl millet are helpful for lowering the inoculum. By hand or by soaking in a salt solution, a healthy seed can be distinguished from a seed that has been contaminated or combined with ergot. However, the most economical way to control ergot is to utilize resistant cultivars. Ergot incidence can be considerably decreased by using sclerotia-free seeds, disease-resistant pearl millet, early sowing, and prudent fertilizer use (NPK) (Pathak et al., 1984; Thakur and King, 1992; Thakur et al., 1983; Thakur et al., 1998). Mechanical separation of sclerotia can be done by immersing the seeds in 10% NaCl (Thakur et al., 1983).

Intercropping (*Vigna radiata*), long crop rotations, adjusting sowing dates, and frequent deep ploughing can prevent soil-borne inoculum (Thakur et al., 1983). Sharma reported the practice of early crop planting. In soil amendment with urea, diammonium phosphate, and NPK, sclerotia germination was entirely suppressed (2012). The most economical technique for eradicating ergot illness is the use of resistant cultivars (Singh and King, 1991). The spread of the disease can be stopped by removing weeds like *P. antidotale* and *C. ciliaris* from the vicinity of pearl millet fields. *Fusarium semitectum* and *F. sambucinum*, antagonists, may be able to manage the disease in pearl millet (Bhosale et al., 2022). The sclerotia of *C. fusiformis* are prevented from germinating by *T. viride* and *Bacillus subtilis*. Ergot incidence and subsequent dissemination are reduced when panicles are sprayed with fungicides (0.1% Bavistin, 0.2% Tilt, or 0.2% Mancozeb) during the flowering stage. For the prevention of disease, several experts advise 2–3 sprays of ziram, copper oxychloride + zineb, and wettable sulfur at intervals of 5–7 days beginning immediately before ear head emergence. Brefeld reported the use of various bio-agents (*T. viride, T. harzianum, Aspergillus niger, T. koningii, T. longibrachiatum, P. fluorescens*, and *B. subtilis*) against bajra ergot under *in vitro* conditions and reported their efficacy in controlling the disease (1985).

14.3.3 SMUT

Smut is a common disease of pearl millet, which is brought on by *Tolyposporium penicillariae* Bref (Chevalier, 1931). Numerous regions in Africa, including Pakistan, India, the United States, and others, have recorded cases of the disease where it is widespread. Depending on the season, disease incidence in India ranges from 2 to 20% and can directly reduce grain yield by up to 30%. The first report of the disease was from Senegal and India (Ajrekar and

Likhite, 1930; Rachie and Majmudar, 1980). In the 1950s, a survey of India found that the severity of smut in farmers' fields varied from 1 to 30% in several regions of Tamil Nadu, Andhra Pradesh, and Maharashtra (Peregrine and Siddiqui, 1972). Smut disease has been documented by Rothwell, Peregrine, Siddiqui, and Bhatt in Pakistan, Cameroon, Nigeria, the USA, and many other countries (1983; 1972; 1946).

14.3.3.1 SYMPTOMS

The disease starts to show symptoms when the grain starts to set. Ovaries are transformed into structures known as sori in the infected florets (singular sorus). Few florets are infected by the disease, which causes them to develop into big, oval-shaped sacs (sori) that contain black powder (smut spores). The sori initially resemble healthy grains but are larger and greener. In addition, as the sori mature, they turn a dark brown color, become readily broken, and release a vast number of black spore balls. They are typically 3–4 mm long and 2–3 mm wide at the top. A modest scattering of sori amid the grain on panicles, all the way up to the panicles being wholly covered with sori, are signs of infection.

14.3.3.2 PATHOGEN

The fungus was isolated from *P. glaucum* (Peregrine and Siddiqui, 1972; Rachie and Majmudar, 1980; Subba Rao and Thakur, 1983). The teleutospores are found in dense, sphere-like aggregates known as spore balls. Spore balls ranged in shape from circular to nearly polyhedral and were 42–325 x 50–175 μm in size 200 and 1400 teliospores were gathered into balls, and individual teleutospores are angular to spherical, light brown, and measure 7–12 μm in diameter. Teleutospore germination is greatest at 30°C, forming a four-celled promycelium producing lateral and terminal sporidia. Sporidia ranged in length from 8 to 25 μm and were spindle-shaped.

14.3.3.3 DISEASE CYCLE

The pathogen lives in the soil, which is the principal source of primary inoculum in the field, or as teleutospores (resting spores) along with infected seeds. When highly contaminated grains are used for planting, during rain, teleutospores begin to germinate and release a large number of airborne sporidia, which infect the pearl millet crop when it is in flower (Rothwell, 1983). The spores in the air start to grow and generate sporidia, which enter the spikelets and infect the ovary. Chlamydospores carried by the wind cause the secondary spread. In order to create a dikaryotic infection hypha, two sporidia of compatible mating types are thought to be required. When the conditions are right, the spores grow to form a dense mycelial chain. Teleutospores from promycelia and sporidia developed during flowering cause infection of the floral portion (Bhatt, 1946).

14.3.3.4 DISEASE MANAGEMENT

There are various reports of chemical control having varying degrees of success despite the fact that no practical and effective cultural or chemical control mechanisms exist. Chemical control methods are neither practical nor affordable for farms with a major low economic value of the crop. The most efficient and cost-effective method of treating the disease is, of course, to use resistant cultivars. An integrated method was followed to manage bajra rust. Use of tolerant and resistant varieties, crop and field sanitation, seed treatment with vitavax, and spraying with carbendazim or zineb during panicle initiation and deep ploughing to remove teliospores before germination are followed by Subba Rao and Thakur (1983).

14.3.4 RUST

Puccinia substriata var. *indica* Ramachar and Cumm, is the causative agent of pearl millet rust. In Asia, Africa, North America, and South America, pearl millet rust is seen. The disease recently manifested in central Brazil at a harmful percentage (Thakur et al., 1986). Tamil Nadu, Karnataka, and other states in India are particularly prone to the illness, including Gujarat, Maharashtra, and Rajasthan. Depending on the season, disease severity varies from 10% to 45% throughout India. Although it might influence forage crops, it has little impact on grain productivity. Dashora and Kumar noted a 0% to 30% reduction in disease in the USA (2008).

14.3.4.1 SYMPTOMS

Typically, sori (fungal bodies) first appear on the distal half of the leaf before spreading to both surfaces, forming reddish pustules, although there are more on the top layer. The leaves may prematurely dry out if necrotic areas form around clusters of pustules. Leaf tissue from the leaf apex to the base will wilt and turn necrotic as infection severity increases. Uredinia are replaced with sub-epidermal, black, elliptical telia at infection sites that appear late in the season. The leaves become wilted throughout. More rust pustules may develop on stems, and the plants may wilt if disease incidence is higher.

14.3.4.2 PATHOGEN

The pathogen is heteroecious; its life cycle requires two separate host plants. The alternate host, brinjal (*Solanum melongena*), is where the fungus thrives and produces spermagonia, acea, and aeciospores (Thakur et al., 1986). Aeciospores that are airborne infect the leaves and then generate uredinia and telia. Produced in telia, teleutospores infect a different host to complete the life cycle. Typically, pustules (sori) first appear on the distal half of the leaf before spreading to both surfaces.

14.3.4.3 MANAGEMENT

The most cost-effective ways to control rust are to remove alternative hosts and grassy weeds and to employ resistant cultivars. According to reports, rust can be prevented by using three sprays of mancozeb plus dinocap at 10-day intervals beginning at ear emergence. ICML 11 is a germplasm that has just been discovered to be rust-resistant. Subba Rao and Thakur detailed an integrated disease management practice for rust (1983). The practices include timely sowing, following crop rotation with pulses and cereals, using disease-resistant varieties, spraying Mancozeb or hexaconazole starting from the appearance of the disease, removing infected residue and burning it, and keeping the field free from weeds while growing brinjal near bajra.

14.3.5 BLAST

Pearl millet blast, which is caused by *Magnaporthe grisea,* was first noted in 1952 at the Research Farm, Uttar Pradesh (Math et al., 2022). In India, where it was once thought to be a minor disease, the prevalence of pearl millet blast (*Magnaporthe grisea*) has alarmingly grown, primarily on commercial hybrids in numerous areas (de Carvalho et al., 2006). The disease has recently started to cause persistent yield losses. It has become a severe problem for India's grain and pearl millet fodder production in Rajasthan, Gujarat, Maharashtra, and Madhya Pradesh. On hybrids and types, the mean disease severity varies from 10% to 30%, and depending on the season, it may reach 50%. Therefore, blast disease worsens in humid weather, especially in densely packed plant stands. Foliage lesions are 2.5–3.5 x 1.5–2.5 mm and oval or diamond-shaped. The centers of lesions are initially grey and water-soaked, but they soon turn brown. They are frequently encircled by a chlorotic halo that will eventually develop necrotic areas and resemble concentric rings (Wilson et al., 1996). From the seedling to flowering stage, these symptoms emerge on leaves, stems, and boot leaves.

14.3.5.1 SYMPTOMS

The disease causes greyish, wet sores on foliage that dry to a brown color, spread, and turn necrotic, causing extensive chlorosis and premature drying of young leaves. The lesions have a grey center and are elliptical or diamond-shaped. Lesions begin close to the leaf tips, leaf margins, or both and extend down the outside edge. Under highly humid conditions, lesions produce a large amount of sporulation.

14.3.5.2 PATHOGEN

Magnaporthe grisea, the pathogen's sexual stage, is a filamentous, heterothallic ascomycetous haploid fungus that parasitizes more than 50 grass species. Asexual conidia are pyriform, hyaline, primarily 3-celled, and have a small appendage on the base cell. Conidia have dimensions ranging from 17.5 to 30.8 x 5.9 to 8.8 μm (Math et al., 2022). Each stoma produces a cluster of the pathogen's conidiophores. With 2–4 septa, they hardly ever live alone. The conidiophores' basal region is enlarged and tapers toward their lighter apex. Every stoma produces a cluster of conidiophores. They have 2–4 septa and are rarely solitary. Conidiophores have a swollen base that tapers away to a lighter apex.

14.3.5.3 MANAGEMENT

Blast disease prevention and control strategies include early sowing, disease-free seed planting, avoiding excessive nitrogen-based fertilizers, burning agricultural leftovers such as sick straw and stubble, and planting disease-free seeds. Several fungicides, including benomyl, iprobenfos, pyroquilon, felimzone, diclocymet, carpropamid, and metominostrobin, are used to treat blast disease (Wilson et al., 1996). Biocontrol agents such as

Trichoderma harzianum and *Pseudomonas fluorescens*, as well as PGPR strains such as *Bacillus subtilis* and *Bacillus pumilus*, were used to control *Pyricularia blast* disease (Kato, 2001; Thakur et al., 2009). The use of host resistance and various breeding techniques was also a disease management option.

The most efficient methods for decreasing finger millet blast disease are seed treatment and two foliar sprays with *Pseudomonas fluorescens* (Nayaka et al., 2017). According to Krishnamurthy and Gnanamanickam, *P. fluorescens* foliar sprays were successful in reducing blast occurrence (1998). According to Yoshihiro et al. (2003), trifloxystrobin plus tebuconazole was successful in treating blast disease in pearl millet with increased yield. According to Kumar et al., field experiment findings indicated that three sprays of Tebuconazole + Trifloxystrobin or propiconazole were more effective at reducing blast incidence and increasing yields in pearl millet (2011). Tebuconazole plus trifloxystrobin, according to Netam et al., efficiently reduced rice blast and increased yield (2014). Ajay et al. observed that spraying infected plants with *P. fluorescens* at 20 days after sowing (DAS), along with a fungicide spray of Trifloxystrobin + Tebuconazole (0.04%, 35 DAS), and other treatment combinations with spraying Trifloxystrobin + Tebuconazole at 0.04% at 20 DAS and later adding *B. subtilis* at 35 DAS, were effective against blast (2018).

14.4 CONCLUSION

Pearl millet, or bajra, is a widely grown millet in semi-arid tropics and tropics. The major production constraints are pests and diseases affecting the crop at various stages. Early-stage infestations of shoot flies and stem borers can drastically reduce the crop stand as well as yield. Downy mildew is a serious threat to crop cultivation. In the case of arthropod pests, breeding strategies for improved resistance or tolerance to pest attacks can be sought out to enhance the existing production of pearl millet. Cultural practices, coupled with biological and chemical methods, can check the development of pest populations as well as pathogens at levels causing economic loss.

KEYWORDS

- **armyworm**
- **deccan grasshopper**
- **disease management**
- **downy mildew**
- **leaf blast**
- **pest control**
- **pink stem borer**
- **sporangiophore branches**
- **unicellular microconidia**

REFERENCES

Ahmed, S. M., & Young, W. R. (1969). Field observations on the susceptibility of sorghum, maize and bajra to *Chilo zonellus* (Swinhoe). *Indian Journal of Entomology, 31*(3), 235–238.

Ajay, K., Pandya, R. K., & Bhagyashree, S. (2018). Performance of botanicals and fungicides against blast of pearl millet (*Pennisetum glaucum* ANN). *Plant and Soil Research, 20*(3), 258–262.

Ajay, P. M., Parmar, G. M., & Solanki, B. (2022). Seasonal incidence of shoot fly and stem borer of pearl millet in relation to weather parameters. *Pharma Innovation Journal, No. 8*, 240–242.

Ajrekar, S. L., & Likhite, V. N. (n.d.). Observations on *Tolyposporium penicillariae* Bref. (the bajri smut fungus). *Current Science*.

Ananthakrishnan, T. N., & Thirumalai, G. (1977). The grass seed infesting thrips, *Chirothrips mexicanus* Crawford on *Pennisetum typhoides* and its principal alternate host *Chloris barbata*. *Current Science, 46*, 193–194.

Anaso, A. B., & Anaso, C. E. (2010). Pearl millet downy mildew as affected by seed dressing and cultivar in the Sahel and Sudan Savanna of Nigeria. *Archives of Phytopathology and Plant Protection, 43*(4), 404–411.

Atri, A., Bhardwaj, N. R., Roy, A. K., & Kaur, A. (2023). Efficacy of biocontrol agents against *Sclerospora graminicola* causing downy mildew in fodder pearl millet. *Range Management and Agroforestry, 44*(1), 76–83.

Balasubramanian, R., Seshu Reddy, K. V., Govinda, R., & Deviah, M. A. (1975). New record: Armyworm, *Pseudaletia separate* Walker (Lepidoptera: Noctuidae) as a pest of ragi in India. *Journal of the Bombay Natural History Society, 72*, 588–589.

Balikai, R. A. (2010). Insect pest status of pearl millet (*Pennisetum glaucum* (L.) R. Br.) in Karnataka. *International Journal of Plant Protection, 2*(2), 189–190.

Balikai, R. A., & Guggari, A. K. (2006). Evaluation of indigenous technologies for the management of insect pests in pearl millet. *Journal of Entomological Research, 30*(4), 321–324.

Bhatt, R. S. (1946). Studies in the Ustilaginales. 1. The mode of infection of the bajra plant (*Pennisetum typhoides* Stapf. & Hubbard) by the smut, *Tolyposporium penicillariae* Bref. *Journal of the Indian Botanical Society, 25*, 163–186.

Bhavani, B., Sekhar, C. V., Varma, K. P., Lakshmi, B. M., Jamuna, P., & Swapna, B. (2019). Morphological and molecular identification of an invasive insect pest, fall armyworm, *Spodoptera frugiperda* occurring on sugarcane in Andhra Pradesh, India. *Journal of Entomology and Zoology Studies, 7*(4), 12–18.

Bhide, V. P., & Hegde, R. K. (1957). Ergot on bajra [*Pennisetum typhoides* (Burm) Stapf. & Hubbard] in Bombay State. *Current Science, 26*.

Bhosale, G. V., Bharti, D. S., Bhosale, G. B., Jagtap, G. P., & Banne, S. N. (2022). *In vitro* evaluation of biocontrol agents against *Claviceps fusiformis* causing ergot of bajra. *Pharma Innovation Journal, 11*, 410–412.

Biradar, A. (2018). Management of shoot fly in major cereal crops. *International Journal of Pure and Applied Bioscience, 6*(1), 971–975. https://doi.org/10.18782/2320-7051.6120 (accessed on 7th November 2025).

Brefeld, O. (1895). *Untersuchungen. Gesammtgebiete Mykologische, 12*, 99–236.

Chahal, S. S., Rao, V. P., & Thakur, R. P. (1985). Variation in morphology and pathogenicity in *Claviceps fusiformis*, the causal agent of pearl millet ergot. *Transactions of the British Mycological Society, 84*(2), 325–332. https://doi.org/10.1016/s0007-1536(85)80084-x (accessed on 7th November 2025).

Chevalier, A. (1931). Une maladie du penicillaire au Senegal. *Revue de Botanique Appliquée, 11*, 49–50.

Choudhary, S. K., Tandi, B. L., & Singh, S. (2018). Management of white grub, *Holotrichia consanguinea* Blanchard in pearl millet. *Indian Journal of Entomology, 80*, 619–622.

Coutin, R., & Harris, K. M. (1968). The taxonomy, distribution, biology and economic importance of the millet grain midge, *Geromyia penniseti* (Felt.) gen. n. comb. n. (Diptera, Cecidomyiidae). *Bulletin of Entomological Research, 59*, 259–273.

Dashora, K., & Kumar, A. (2008). Management of smut disease of pearl millet through bio-control agents. *Seed Research, 36*, 238–240.

De Carvalho, A., Soares, D., & Carmo, M. (2006). Description of the lifecycle of the pearl millet rust fungus–*Puccinia substriata* var. *penicillariae* with a proposal of reducing var. *indica* to a synonym. *Mycopathologia, 161*, 331–336.

Deepak, S. A., Raj, S. N., Umemura, K., Kono, T., & Shetty, H. S. (2003). Cerebroside as an elicitor for induced resistance against the downy mildew pathogen in pearl millet. *Annals of Applied Biology, 143*(2), 169–173. https://doi.org/10.1111/j.1744-7348.2003.tb00283.x (accessed on 7th November 2025).

Degri, M. M., Mailafiya, D. M., & Mshelia, J. S. (2014). Effect of intercropping pattern on stem borer infestation in pearl millet (*Pennisetum glaucum* L.) grown in the Nigerian Sudan savannah. *Advances in Entomology, 2*(2), 81–86. https://doi.org/10.4236/ae.2014.22014 (accessed on 7th November 2025).

Dhaliwal, T. S., Singh, B. P., & Atwal, A. S. (1974). *Cyaneolytta aceton* (Laporate) (Coleoptera: Meloidae) a new pest of maize and bajra. *Current Science, 43*, 765.

Dieng, A., Ley, G. D., Nair, M. S., Samb, P. I., Gyeye, M., & Mbaye, D. F. (1999). Screening of varieties of millet (*Pennisetum glaucum*) for their resistance to mildew caused by *Sclerospora graminicola*. *Journal of Agriculture and Environment for International Development, 92*(1), 21–32.

Gahukar, R. T. (1985). Some species of *Atherigona* (Diptera: Muscidae) reared from Gramineae in Senegal. *Annals of Applied Biology, 106*, 399–403.

Gahukar, R. T., & Jotwani, M. G. (1980). Present status of field pests of sorghum and millets in India. *Tropical Pest Management, 26*(2), 138–151. https://doi.org/10.1080/09670878009414385 (accessed on 7th November 2025).

Gupta, G. K., & Singh, D. (1996). Studies on the influence of downy mildew infection on yield and yield-contributing plant characters of pearl millet in India. *International Journal of Pest Management, 42*(2), 89–93. https://doi.org/10.1080/09670879609371977 (accessed on 7th November 2025).

Gupta, J. C., & Gupta, D. S. (1970). Note on some new hosts of the painted bug (*Bagrada cruciferarum* Kirk. Pentatomidae: Hemiptera). *Indian Journal of Agricultural Sciences, 4017*, 645–646.

Gwary, D. M., Bdliya, B. S., & Bdliya, J. A. (2009). Appraisal of yield losses in pearl millet due to downy mildew pathogen (*Sclerospora graminicola*) in Nigerian Sudan Savanna. *Archives of Phytopathology and Plant Protection, 42*(11), 1010–1019. https://doi.org/10.1080/03235400701544036 (accessed on 7th November 2025).

Harris, K. M. (1962). Lepidopterous stem borers of cereals in Nigeria. *Bulletin of Entomological Research, 53*, 139–171.

Jotwani, M. G., & Butani, D. K. (1978). Crop and their control: Pearl millet. *Pesticides, 12*, 20–30.

Jotwani, M. G., Dang, K., Saxeena, P. N., & Young, W. R. (1966). Occurrence of *Eublemma silicula* Swinhoe as a pest of bajra in Delhi. *Indian Journal of Entomology, 28*, 407–408.

Jotwani, M. G., Srivastava, B. G., & Dang, K. (1969). Occurrence of *Pyrilla perpusilla* Walker as a serious pest of some lines of bajra (*Pennisetum typhoides*). *Indian Journal of Entomology, 31*, 368–369.

Jouan, B., & Delassus, M. (1971). Principales maladies des mils et sorghos observées au Niger. *L'Agronomie Tropicale, 8*, 830–860.

Kato, H. (2001). Rice blast control. *Pesticide Outlook*, 23–25. In Japan Crop Protection Association (JCPA), Japan.

Khairwal, I. S., & Yadav, O. P. (2005). Pearl millet (*Pennisetum glaucum*) improvement in India–retrospect and prospects. *Indian Journal of Agricultural Sciences, 75*(4), 183–191.

Khairwal, I. S., Rai, K. N., & Diwakar, B. (2007). *Pearl Millet: Crop Management and Seed Production Manual.* International Crops Research Institute for the Semi-Arid Tropics; Patancheru, Andhra Pradesh, India.

Khurana, S. M. P., & Deshmukh, U. (1974). Serious infestation of bajra by the jassid, *Balclutha* sp. *Labdev Journal of Science and Technology, 12*.

King, S. B. (1992). World review of pearl millet diseases: Knowledge and future research needs. In W. A. J. De Milliano, R. A. Frederiksen, & G. D. Bengston (Eds.), *International Crops Research Institute for the Semi-Arid Tropics* (pp. 95–108). Patancheru, Andhra Pradesh, India.

Kishore, P. (1966). Chemical control of stem borers. *Journal of Entomological Research, 20*, 287–297.

Kishore, P. (1996). Changing pest status of earhead caterpillar, *Eublemma silicula* Swinh. on pearl millet, *Pennisetum typhoides* (Burm.). *Journal of Entomological Research, 20*(3), 277–279.

Krishnamurthy, K., & Gnanamanickam, S. S. (1998). Biological control of rice blast by *Pseudomonas fluorescens* strain Pf7-14: Evaluation of a marker gene and formulations. *Biological Control, 3*, 158–165.

Kulkarni, K. S., Zala, H. N., Bosamia, T. C., Shukla, Y. M., Kumar, S., Fougat, R. S., Patel, M. S., Narayanan, S., & Joshi, C. G. (2016). De novo transcriptome sequencing to dissect candidate genes associated with pearl millet-downy mildew (*Sclerospora graminicola* Sacc.) interaction. *Frontiers in Plant Science, 7*.

Kumar, B., & Kumar, J. (2011). Management of blast disease of finger millet (*Eleusine coracana*) through fungicides, bioagents and varietal mixture. *Indian Phytopathology, 64*, 272–274.

Lal, R., Bose, B. N., & Katiyar, R. N. (1976). Incidence of white grubs on bajra around Delhi. *Entomology Newsletter, 6*, 41–42.

Lata, A. M., Rao, K. V., Reddy, T. P., & Reddy, V. D. (2006). Development of transgenic pearl millet (*Pennisetum glaucum*) plants resistant to downy mildew. *Plant Cell Reports, 25*, 927–935.

Latake, S. B., & Kolase, S. V. (2007). Screening of bioagents for control of downy mildew of pearl millet. *International Journal of Agricultural Sciences, 3*(2), 32–35.

Leuck, D. B., Burton, G. W., & Windstrom, N. W. (1977). Insect oviposition and foliage feeding resistance in pearl millet. *Journal of the Georgia Entomological Society, 12*, 138–140.

Leuck, D. B., Taliaferro, C. M., Burton, R. L., Burton, G. W., & Bowman, M. C. (1968). Fall armyworm resistance in pearl millet. *Journal of Economic Entomology, 61*(3), 693–695. https://doi.org/10.1093/jee/61.3.693 (accessed on 7th November 2025).

Loveless, A. R. (1967). *Claviceps fusiformis* sp. nov., the causal agent of an agalactia of sows. *Transactions of the British Mycological Society, 50*, 5015–5018.

Manjunatha, G., Raj, S. N., Shetty, N. P., & Shetty, H. S. (2008). Nitric oxide donor seed priming enhances defense responses and induces resistance against pearl millet downy mildew disease. *Pesticide Biochemistry and Physiology, 91*(1), 1–11. https://doi.org/10.1016/j.pestbp.2007.11.012 (accessed on 7th November 2025).

Mantle, P. G., Shaw, S., & Doling, D. A. (1977). Role of weed grasses in the etiology of ergot disease in wheat. *Annals of Applied Biology, 86*(3), 339–351. https://doi.org/10.1111/j.1744-7348.1977.tb01848.x (accessed on 7th November 2025).

Marley, P. S., Diourt, M., Neya, A., Nutsugah, S. K., Srm, P., Katil, S. O., Hess, D. E., Mbaye, D. F., & Ngoko, Z. (2008). Sorghum and pearl millet diseases in West and Central Africa. In *Sorghum and millets diseases* (pp. 419–425). Iowa State Press: Ames, Iowa, USA.

Math, R., Addangadi, K. C., & Awasthi, D. P. (2022). Current status of bajra/pearl millet diseases and their management. In *Diseases of Field Crops: Diagnosis and Management* (Vol. 2).

Mehta, P. R., Singh, B., & Mathur, S. C. (1953). A new leaf spot disease of bajra (*Pennisetum typhoides* Staph and Hubbard) caused by a species of *Piricularia*. *Indian Phytopathology, 5*, 140–143.

Michelmore, R. W., Pawar, M. N., & Williams, R. J. (1982). Heterothallism in *Sclerospora graminicola*. *Phytopathology, 72*, 1368–1372.

Mishra, R. K., Saxena, R. C., Yadava, C. P. S., & Dadheech, L. N. (1979). Laboratory evaluation of some insecticides against adults *Rhinyptia laeviceps* Arrow. *Pesticides, 13*.

Mittal, V., Rai, G., & Kishore, P. (2006). Monitoring of insect pest problems on pearl millet. *Journal of Entomological Research, 30*(1), 33–34.

N/A. (1970). AICMIP (All India Coordinated Millet Improvement Project). *AICMIP Annual Reports*.

N/A. (2016). *AgriTech Portal*; Coimbatore, Tamil Nadu, India.

N/A. (2017). *Statistics, Statistical Pocketbook*. FAO, Rome, Italy.

N/A. (2019). Punjab Agricultural University Campus, Ludhiana.

N/A. (2020). Directorate of Millets Development. *http://millets.dacfw.nic.in* (accessed on 7th November 2025)

Nandhini, M., Rajini, S. B., Udayashankar, A. C., Niranjana, S. R., Lund, O. S., Shetty, H. S., & Prakash, H. S. (2019). Biofabricated zinc oxide nanoparticles as an eco-friendly alternative for growth promotion and management of downy mildew of pearl millet. *Crop Protection, 121*, 103–112. https://doi.org/10.1016/j.cropro.2019.03.015 (accessed on 7th November 2025).

Natarajan, V. S., Raja, V. D. G., Selvaraj, S., & Anavardham, L. (1973). Extent of damage caused by shoot fly (*Atherigona approximate*) on bajra hybrids. *Madras Agricultural Journal, 60*, 584–585.

Nayaka, C. S., Srivastava, R. K., Udayashankar, A. C., Lavanya, S. N., Prakash, G., Bishnoi, H. R., Kadvani, D. L., Singh, O. V., Niranjana, S. R., Prakash, H. S., & Satyavathi, C. T. (2017). *Magnaporthe* blast of pearl millet in India—Present status and future prospects. All India Coordinated Research Project on Pearl Millet (Indian Council of Agricultural Research).

Nene, Y. L., & Singh, S. D. (1976). Downy mildew and ergot of pearl millet. *PANS, 22*(3), 366–385. https://doi.org/10.1080/09670877609412073 (accessed on 7th November 2025).

Netam, R. S., Tiwari, R., Bahadur, A. N., & Shankar, D. (2014). *In vitro* and *in vivo* efficacy of fungicides against *Pyricularia grisea* causing finger millet blast disease. *International Journal of Plant Protection, 7*, 137–142.

Obeng, E., Cebert, E., Ward, R., Nyochembeng, L. M., Mays, D. A., Singh, H. P., & Singh, B. P. (2015). Insect incidence and damage on pearl millet (*Pennisetum glaucum*) under various nitrogen regimes in Alabama. *Florida Entomologist, 98*(1), 74–79. https://doi.org/10.1653/024.098.0113 (accessed on 7th November 2025).

Pal, S. K., & Sharma, V. P. (1973). Occurrence of *Rhinyptia meridinalis* var. *puncticollis* Arr. (Scarabaeidae: Coleoptera) as pest on bajra in western Rajasthan. *Journal of the Bombay Natural History Society, 70*, 574–575.

Pandey, S. N., & Yadava, C. P. S. (1974). King crow (*Dicrurus macrocerus*), a predator of the adult of *Holotrichia consanguinea* Blanch. *Entomology Newsletter, 4*.

Pandya, R. K., Singh, R., & Tripathi, M. L. (2000). Efficacy of metalaxyl and Aliette on downy mildew of pearl millet. *Crop Research, 20*(1), 134–136.

Patel, R. K., & Dhagat, N. K. (1981). Severe infestation of webworm, *Cryptoblabes gnidiella* (Mill.) on ragi (*Eleusine coracana* G.) in Madhya Pradesh. *Jawaharlal Nehru Krishi Vidhyapith Research Journal, 15*.

Pateliya, V. R., Parmar, G., Kelaiya, M., & Vikani, S. (2019). Ecologically sustainable management of shoot fly and stem borer in pearl millet. *Journal of Entomology and Zoology Studies, 7*(5), 597–599.

Pathak, V. N., Yadav, K. R., Langkilde, N. E., & Mathur, S. B. (1984). Removal of ergot sclerotia from seeds of pearl millet. *Seed Research, 12*(2), 70–81.

Patro, T., Georgia, K. E., Kumar, S. R., Anuradha, N., Rani, Y. S., & Triveni, U. (2020). Management of pearl millet blast through fungicides and biocontrol agents. *International Journal of Chemical Studies, 8*(4), 1357–1359. https://doi.org/10.22271/chemi.2020.v8.i4l.9789 (accessed on 7th November 2025).

Pattanashetti, S. K., Upadhyaya, H. D., Dwivedi, S. L., Vetriventhan, M., & Reddy, K. N. (2016). Pearl millet. In *Genetic and Genomic Resources for Grain Cereals Improvement*. Academic Press.

Peregrine, W. J. H., & Siddiqui, M. A. (1972). A revised and annotated list of plant diseases in Malawi. *CAB Phytopathology Paper, 16*.

Pramesh, D., Maruti, M., Muniraju, K., Mallikarjun, K., Guruprasad, G., Mahantashivayogayya, K., Reddy, B., Gowdar, S., & Chethana, B. (2017). Bio-efficacy of a combination fungicide against blast and sheath blight diseases of paddy. *Journal of Experimental Agriculture International*, *14*(4), 1–8. https://doi.org/10.9734/jeai/2016/28893 (accessed on 7th November 2025).

Pruthi, H. S. (1969). *Textbook of Agricultural Entomology*. Indian Council of Agricultural Research.

Pushpalatha, H. G., Naveen, J., Geetha, N., Hithamani, G., & Shetty, H. S. (2023). Plant growth promotion and biological control of *Sclerospora graminicola* in pearl millet by endophytic *Streptomyces* spp. *Indian Phytopathology*, 1 10.

Pushpavathi, B., Thakur, R. P., & Rao, K. C. (2006). Fertility and mating type frequency in Indian isolates of *Sclerospora graminicola*, the downy mildew pathogen of pearl millet. *Plant Disease*, *90*(2), 211–214. https://doi.org/10.1094/PD-90-0211 (accessed on 7th November 2025).

Rachie, K. O., & Majmudar, J. V. (1981). *Pearl millet*. Pennsylvania State University Press.

Raj, S. H., Shetty, N. P., & Shetty, H. S. (2004). Proline–an inducer of resistance against pearl millet disease caused by *Sclerospora graminicola*. *Phytoparasitica*, *32*(5), 523–527.

Raj, S. N., Chaluvaraju, G., Amruthesh, K. N., Shetty, H. S., Reddy, M. S., & Kloepper, J. W. (2003). Induction of growth promotion and resistance against downy mildew on pearl millet (*Pennisetum glaucum*) by rhizobacteria. *Plant Disease*, *87*(4), 380–384.

Reddy, S. P., Satyavathi, C. T., Khandelwal, V., Patil, H. T., Gupta, P. C., Sharma, L. D., et al. (2021). Performance and stability of pearl millet varieties for grain yield and micronutrients in arid and semi-arid regions of India. *Frontiers in Plant Science*, *12*.

Roonwal, M. L. (1945). Notes on the bionomics of *Hieroglyphus nigrorepletus* Bolivar (Orthoptera: Acrididae) at Banaras, United Province, India. *Indian Entomology Research*, *36*, 339–341.

Rothwell, A. A. (1983). A revised list of plant diseases occurring in Zimbabwe. *Kirkia*, *12*(11), 275.

Sagnia, S. B. (1983). Possible integrated pest management tools for the effective control of cereal stemborers in the Gambia. *Insect Science and Its Application*, *4*, 217–219.

Sandhu, G. S. (1975). Seasonal movement of the painted bug, *Bagrada cruciferarum* Kirklady from cruciferous plants to graminaceous plants and its occurrence as a serious pest of maize, sorghum and pearl millet. *Indian Journal of Entomology*, *37*(2), 215–217.

Santharam, G., Mohanasundaram, M., & Jayaraj, S. (1976). Control of bajra grain midge, *Geromyia penniseti* Felt. with insecticides. *Pesticides*, *10*, 45–46.

Sastry, J. G., Ramakrishna, W., Sivaramakrishnan, S., Thakur, R. P., Gupta, V. S., & Ranjekar, P. K. (1995). DNA fingerprinting detects genetic variability in the pearl millet downy mildew pathogen (*Sclerospora graminicola*). *Züchter Genetic Breeding Research*, *91*(6–7), 856–861. https://doi.org/10.1007/bf00223892 (accessed on 7th November 2025).

Shankara Rao, R., Rao, K. C. S., Singh, S. D., Reddy, M. S., Rao, M. V. B., & Venkateswarlu, B. (1987). Fungicidal control of downy mildew of pearl millet. *Indian Journal of Plant Protection*, *15*(2), 146–151.

Sharma, H. C., & Davies, J. C. (1988). *Insect and Other Animal Pests of Millets*. ICRISAT.

Sharma, H. C., Davies, J. C., & Arora, J. (1981). *Bibliography of Insect and Non-Insect Animal Pests of Millets During 1914–1980*. ICRISAT Sorghum and Millets Information Centre.

Sharma, R., Rao, V. P., Varshney, R. K., Prasanth, V. P., Kannan, S., & Thakur, R. P. (2010). Characterization of pathogenic and molecular diversity in *s*, the causal agent of pearl millet downy mildew. *Archives of Phytopathology and Plant Protection*, *43*(6), 538–551.

Sharma, R., Vishal, L. G., & Madhavan, S. (2018). Evaluation of fungicides for the management of pearl millet (*Pennisetum glaucum* (L.)) blast caused by *Magnaporthe grisea*. *Crop Protection*, *112*, 209–213.

Sharma, S. K., & Singh, B. (1974). Atherigona sp. nr. approximate Malloch (Anthomyiidae: Diptera) as a pest of bajra in Rajasthan. *Indian Journal of Entomology*, *36*, 246–247.

Sharma, Y. P., Singh, R. S., & Tripathi, R. K. (1984). Management of pearl millet ergot by integrating cultural practices and chemical control. *Indian Journal of Mycology and Plant Pathology*, *14*(1), 69–79.

Shetty, H. S. (1987). Biology and epidemiology of downy mildew of pearl millet. In *Proceedings of the International Pearl Millet Workshop ICRISAT Center*; India.

Shishupala, S., Kumar, V. U., Shetty, H. S., & Umesh-Kumar, S. (1996). Screening pearl millet cultivars by ELISA for resistance to downy mildew disease. *Plant Pathology*, *45*(5), 978–983. https://doi.org/10.1111/j.1365-3059.1996.tb02909.x (accessed on 7th November 2025).

Siddaiah, C. N., Prasanth, K. V. H., Satyanarayana, N. R., Mudili, V., Gupta, V. K., Kalagatur, N. K., Satyavati, T., Dai, X. F., Chen, J. Y., Mocan, A., Singh, B. P., & Srivastava, R. K. (2018). Chitosan nanoparticles having higher degree of acetylation induce resistance against pearl millet downy mildew through nitric oxide generation. *Scientific Reports*, *8*(1), 2485. https://doi.org/10.1038/s41598-017-19016-z (accessed on 7th November 2025).

Siddiqui, M. R., & Khan, I. D. (1973). Renaming Claviceps microcephala ergot fungus on Pennisetum typhoides in India as Claviceps fusiformis. *Transactions of the Mycological Society of Japan*, *14*(2), 195–198.

Singh, J. P. (1967). Record of blister beetle. In Coleoptera: Meloidae) on the hybrid bajra in district Budaun; U.P. *Indian Journal of Entomology*, *54*, 251–252.

Singh, M. P., Mishra, P. N., & Bisht, R. S. (2004). Nature and extent of damage of white grub Lachnosterna longipennis (Holotrichia longipennis Blanch.) under various farming situations of Uttaranchal hills. *Indian Journal of Entomology*, *66*(3).

Singh, S. D., & Gopinath, R. (1990). Effect of temperature and light on sporangial germination and zoospore infectivity in *Sclerospora graminicola* on pearl millet. *Canadian Journal of Plant Pathology*, *12*, 25–30.

Singh, S. D., & King, S. B. (1991). Pearl millet rust: Present status and future research needs. *International Journal of Tropical Plant Diseases*, *9*(1), 35–52.

Singh, S. D., & Navi, S. S. (2000). Genetic resistance to pearl millet downy mildew II. Resistance in wilt relatives. *Journal of Mycology and Plant Pathology*, *30*(2), 167–171.

Singh, S. D., Ball, S., & Thakur, D. P. (1986). Problems and strategies in the control of downy mildew. In *Proceedings of the International Pearl Millet Workshop*; India, Patancheru, Andhra Pradesh, pp. 161–172.

Singh, V. S., & Jotwani, M. G. (1973). Efficacy of some systemic insecticides against bajra shoofly. Atherigona approximate Malloch. *Indian Journal of Entomology*, *35*(2), 130–133.

Srivastava, A. S., Srivastava, K. M., Katiyar, S. S., & Bhadauria, A. S. (1971). New records of Chiloloba acuta Wied. (Scarabaeidae: Coleoptera) causing damage to hybrid bajra in Uttar Pradesh. *Indian Journal of Entomology*, *33*.

Subba Rao, K. V., & Thakur, R. P. (1983). *Tolyposporium penicillariae*, the causal agent of pearl millet smut. *Transactions of the British Mycological Society, 81*(3), 597–603. https://doi.org/10.1016/s0007-1536(83)80131-4 (accessed on 7th November 2025).

Sudisha, J., Kumar, S. A., Thakur, R. P., Rao, V. P., & Shetty, H. S. (2009). Molecular characterization of *Sclerospora graminicola*, the incitant of pearl millet downy mildew using ISSR markers: Molecular characterization of *S. graminicola* by ISSR markers. *Journal of Phytopathology, 157*(11–12), 748–755. https://doi.org/10.1111/j.1439-0434.2009.01547.x (accessed on 7th November 2025).

Talati, G. M., & Upadhyay, V. R. (1978). Status of shoot fly, Atherigona approximate (Malloch) as a pest of bajra (*Pennisetum typhoides*) crops Gujarat State. *Gujarat Agricultural University Research Journal, 4*, 30–35.

Tenberge, K. B. (1999). Biology and life strategy of the ergot fungi. In *Ergot* (pp. 40–71). CRC Press.

Thakur, R. P., & King, S. B. (1992). Ergot of pearl millet. In *Ergot of Pearl Millet* (pp. 302–315).

Thakur, R. P., Pushpavathi, B., & Rao, V. P. (1998). Virulence characterization of single-zoospore isolates of *Sclerospora graminicola* from pearl millet. *Plant Disease, 82*(7), 747–751. https://doi.org/10.1094/PDIS.1998.82.7.747 (accessed on 7th November 2025).

Thakur, R. P., Rao, V. P., & Williams, R. J. (1984). The morphology and disease cycle of ergot caused by *Claviceps fusiformis* in pearl millet. *Phytopathology, 74*(2), 201–205.

Thakur, R. P., Sharma, R., & Rao, V. P. (2011). Screening techniques for pearl millet diseases, Information Bulletin No. 89. *International Crops Research Institute for the Semi-Arid Tropics*; Patancheru, Andhra Pradesh, India.

Thakur, R. P., Sharma, R., Rai, K. N., Gupta, S. K., & Rao, V. P. (2009). Screening techniques and resistance sources for foliar blast in pearl millet. *Journal of SAT Agricultural Research, 7*, 1–5.

Thakur, R. P., Subba Rao, K. V., Williams, R. J., Gupta, S. C., Thakur, D. P., Nafade, S. D., Sundaram, N. V., Frowd, J. A., & Guthrie, E. J. (1986). Identification of stable resistance to smut in pearl millet. *Plant Disease, 70*, 38–41.

Thakur, R. P., Williams, R. J., & Rao, V. P. (1982). Development of resistance to ergot in pearl millet. *Phytopathology, 72*(4), 406–408.

Thakur, R. P., Williams, R. J., & Rao, V. P. (1983). Control of ergot in pearl millet through pollen management. *Annals of Applied Biology, 103*(1), 31–36.

Thomas, K. M., Ramakrishnan, T. S., & Srinivasan, K. V. (1945). The natural occurrence of ergot in South India. *Proceedings of the Indian Academy of Sciences, Section B: Biological Sciences, 21*(2), 93–100. https://doi.org/10.1007/bf03049810 (accessed on 7th November 2025).

Upasana, R., Chaudhary, D. P., & Srivastava, M. (2010). Downy mildew and its effect on quality of pearl millet hybrids. *Crop Improvement, 37*(2).

Verma, S. K. (1979). Field control of *Rhinyptia* spp. (Coleoptera: Rutelidae) adults using Fenin compounds. *Annals of Arid Zone, 18*, 274–275.

Verma, S. K. (1980). Field pests of pearl millet in India. *Tropical Pest Management, 26*(1), 13–20. https://doi.org/10.1080/09670878009414277 (accessed on 7th November 2025).

Wilson, J. P., Hanna, W. W., & Gascho, G. J. (1996). Pearl millet grain yield loss from rust infection. *Journal of Production Agriculture, 9*(4), 543–545. https://doi.org/10.2134/jpa1996.0543 (accessed on 7th November 2025).

Yadav, N. K., & Thakur, D. P. (2001). Anatomical defense mechanism in pearl millet leaves against downy mildew infection. *Journal of Mycology and Plant Pathology, 31*(2), 185–187.

Yadava, C. P. S., Pandey, S. N., Eharadwaj, S. C., & Mishra, R. K. (1973). Record of *Rhinyptia laeviceps* Arrow (Coleoptera: Scarabaeidae: Rutelinae) as a pest of bajra. *Indian Journal of Entomology, 35*(3).

Yoshihiro, T., Mitsuo, H., Hayato, H., & Futoshi, K. (2003). Biological control of rice blast disease by *Bacillus subtilis* IK-1080. *Annals of the Phyto Pathological Society of Japan, 69*, 85–93.

CHAPTER 15

Integrated Pest and Disease Management in Finger Millet

K. B. PALANNA,[1,*] PRABHU C. GANIGER,[1] C. NANDINI,[2,*] SUJATA BHAT,[1] T. THARANA PONACH,[1] FAROOQKHAN,[1] SAI BHAVANA,[1] T. V. KRISHNA,[3] and C. NANDINI[4]

[1]*Project Coordinating Unit, ICAR–AICRP on Small Millets, University of Agricultural Sciences, GKVK, Bangalore, Karnataka, India*

[2]*Zonal Agricultural and Horticultural Research Station, Babbur Farm, Hiriyur, KSNUAHS, Karnataka, India*

[3]*ICAR–AICRP on Kharif Pulses, Zonal Agricultural Research Station, University of Agricultural Sciences, GKVK, Bangalore, Karnataka, India*

[4]*Regional Zonal Agricultural and Horticultural Research Station, Babbur Farm, Hiriyur, KSNUAHS, Karnataka, India*

*Corresponding author

15.1 INTRODUCTION

Ragi (*Eleusine coracana*) is a nutrient-rich crop primarily cultivated in marginal and medium agricultural zones of developing countries (Odeny et al., 2020). One subspecies, *Eleusine coracana* (L.) *Gae*rtn. subsp. africana (Kenn.-O'Byrne), is the wild form of finger millet and is believed to have been domesticated in the East African region (Hilu and de Wet, 2025). It is a member of the Poaceae family and genus *Eleusine* within the tribe Eragrostideae (Dida et al., 2007). Currently, finger millet is primarily cultivated in India and Eastern Africa. It is adapted to diverse environmental conditions, including high temperature, moisture deficit, and water stagnation (Lenné et al., 2007; Yogeesh et al., 2016). According to Gupta et al. (2017), ragi is more nutritious in terms of the quality of protein and essential minerals than the staple food crops (rice and wheat).

Many biotic and abiotic factors challenge the production of finger millet. Among the biotic factors, diseases caused by fungal pathogens are major constraints in most of the finger millet-growing regions of the world. They are responsible for many diseases, some of which regularly appear in an intense form and cause significant economic loss, while others appear sporadically and have fewer damaging effects on the crop. Fungal diseases are more prevalent than bacterial and viral diseases. Devastating fungal diseases of finger millet include blast, leaf blight/leaf spot, foot rot, banded sheath blight, smut, rust, and

Millets in Sustainable Agriculture: Genetic Improvement, Nutritional Quality, Cultivation, and Integrated Pest–Disease Management. Roshni Vijayan, A. John Joel, R. Ambika Rajendran, and Adithya P. Balakrishnan (Eds.)

DOI: 10.1201/9781779642332-19

downy mildew. These diseases wreak havoc on the yield and quality of the produce by infecting various plant parts, such as the root, stem, leaves, peduncle, or grain. Similarly, insect pests, such as stem borer, earhead caterpillar, leaf hopper, red hairy caterpillar, and root aphids, presume different levels of significance under different environmental conditions. Under the changing climatic conditions, some of the diseases, like banded sheath blight, *cercospora* leaf spot, and banded blight, are becoming major constraints in specific locations.

15.2 IMPORTANT PEST AND DISEASES

15.2.1 DISEASES OF FINGER MILLET

15.2.1.1 BLAST

Pyricularia grisea (Cke.) Sacc. (Perfect Stage: *Magnaporthe grisea*)

Blast is the most devastating disease and one of the major limiting factors in finger millet production (Palanna et al., 2023a). Finger millet is susceptible to the disease at all growth stages, but the most damaging forms are neck and finger blast (Takan et al., 2012), which can result in a 100% loss in annual average yield and biomass yield (Lenné et al., 2007; Nagaraja et al., 2007). Finger millet blast is more prevalent, followed by blast in pearl millet and foxtail millet, while blast in other millets is less significant. In finger millet, the disease occurs yearly during the rainy season. The disease severity, timing of onset, plant variety, and prevailing climatic conditions all affect yield loss.

1. **Symptoms:** The crop is infected by the pathogen at various growth stages, and elliptical or diamond-shaped lesions are formed on the leaves, fingers, and peduncle. Among them, neck blast causes the most yield loss, followed by finger and leaf blast, because at those respective stages, there is a significant reduction in the number of grains, grain weight, and an increase in spikelet sterility. Blast or blight may affect the young seedlings in the nursery bed or main field. The neck blast is more destructive, followed by the finger and leaf blast. As a consequence of neck infection, there is more loss observed in grain number and grain weight, followed by increased sterility of spikelets. The degree and duration of the infection determine whether the infected ears produce shriveled grains or entirely chaffy grains.

   There are three types of symptoms observed during the crop growth period:

   i. **Leaf Blast:** It is more damaging in the tillering phase. The disease is identified by spindle-shaped spots with gray centers on the leaves surrounded by reddish-brown margins.

   ii. **Finger Blast:** Infection on the fingers causes blackening at the point of infection, and it progresses and results in chaffy grains.

iii. **Neck Blast**: The neck just below the earhead gets damaged during the flowering stage, turning sooty black, and often breaks at this point. Early neck infections cause complete earhead chaffiness and complete absence of grain setting. If grain setting takes place, they are smaller and shriveled (Figure 15.1).

2. **Pathogen:** The mycelium of young hyphae is hyaline and septate and turns brown when it becomes old. In the center of the lesions, a large number of conidiophores and conidia form. Conidiophores have thin walls and are unbranched, emerging alone or in clusters and are pale brown. Conidia are thin-walled, sub-pyriform, hyaline, 2-septate, 3-celled with a prominent hilum. Only end cells form germ tubes upon germination. In culture media, the fungus produces olive brown to dark brown, globose, terminal, or intercalary chlamydospores (Palanna et al., 2023b).
3. **Mode of Spread and Survival:** The pathogen is seed-borne, and other crops and weeds act as sources of inoculum in addition to causing the initial infection through the seed-borne inoculum. Conidia that are carried by the air act as the secondary inoculum.
4. **Favorable Conditions:** A combination of high humidity, cloudy days with sporadic rainfall, and moderate temperatures promotes the rapid spread of the disease.
5. **Yield Loss:** Blast is potentially a serious disease of millets and is caused by the fungi *Pyricularia grisea* in finger, proso, and pearl millet, and on the other hand, by *Pyricularia setariae* in foxtail millet. This is more prominent in finger millet, followed by pearl millet and foxtail millet. The average yield loss ranges from 28–36% and can lead to complete crop failures in endemic areas. Presently, yield loss is increasing in pearl millet and foxtail millet too. Yield loss due to blast disease is reported to be 56% (Mc Rae) and 80% (Venkatarayan) by different authors. In contrast, the plants affected by head blast reported 3–32% test grain weight loss (Sunil and Anilkumar, 2004a). Rao and Hegde reported that when the average disease incidence was 7.69% in the neck and 9.97% in the finger, the loss due to blast was 29.51%. Rao (1990) observed 6.75 to 87.5% yield loss in finger millet. As one% is raised in neck and finger blast infection, it results in an increase of 0.32 and 0.084% in yield losses. Losses were dramatic when the disease appeared within 10 days of ear emergence, and considerable losses were incurred even if infection took place up to 20 days of ear emergence.

15.2.1.2 BROWN SPOT/ LEAF BLIGHT

Helminthosporium nodulosum (Sexual stage: Cochliobolus nodulosum)

Generally brown spot diseases is not common and less damaging unless they are favored by ideal environmental condition with susceptible entry.

i. **Symptoms:** The pathogen affects both seedlings and adult plants. Every portion of the plant gets infected, including the roots, the base of the plants, culms, leaf blades, leaf sheaths, neck of the panicle, and fingers. On young leaves, the disease appears as minute, light brown oval spots. The affected leaves will undergo an early drop, thereby killing the seedlings. In adult plants, spots are oblong and dark brown in color. Several such lesions merge to form large patches in the leaf blade. The infected leaf blades dry prematurely, which leads to seedling death. The spots on the leaf sheath and culms are irregular and are generally found at the junction of the blade and sheath (Figure 15.1).

ii. **Pathogen:** The hypha of the pathogen is light brown colored and septate. Conidiophores are erect, septate, dark brown, often branched, and geniculate. Conidia are sub-cylindrical, obclavate, straight, or ovoid, pale to dark golden brown, and 4–10 septate. It produces sexual fruiting bodies (perithecia) that have long cylindrical beaks and are spherical in shape, and asci are short and straight, containing 1 to 8 ascospores; they are hyaline, filiform, and 11 septate.

iii. **Mode of Spread and Survival:** The pathogen is carried from seed to seed, thereby directly infecting through seed. Also, secondary spread occurs through airborne conidia and stubbles. It survives in soil for several months.

iv. **Favorable Conditions:** Moderate temperature, high humidity with prolonged cloudy weather, and high dew favor disease development. Maintain high relative humidity for 48 hours after inoculation to facilitate infection and disease development.

v. **Yield Loss:** Under favorable conditions, the pathogen may attack the ear head and fingers, causing the rachis to turn brown to dark brown. If grains are affected, they may not develop fully, resulting in shriveled grains and heavy yield loss.

15.2.1.3 SMUT

Melonopsichium eleusinis (Kulk.) Mundk, and Thirum

Grain smut is a common and important disease of finger millet and occurs at the time of grain formation.

i. **Symptoms:** Smut disease appears during Kharif at the grain setting stage. In an ear, only a few grains are damaged. The infected ovaries get transformed into larger greenish gall like bodies than the normal grains. As the disease progresses, the infected grains become swollen and increase in size. They are scattered in an ear. The sori are globose, greenish at first, and later turn dirty black during maturity. The ruptured sorus releases a black mass of spores.

ii. **Pathogen:** It is limited to the spikelets, where it manifests as chlamydospores or hyphae with swollen cells. The globose spores have a spore wall that is pitted, rough, or spiky. Their diameter ranges from 7 to 11 μm, and they easily germinate in water, forming sporidia on septate promycelium. Mature sporidia are ovate to cylindrical, uninucleate, and thin-walled.

iii. **Favorable Conditions:** Low temperature and high relative humidity during grain formation are favorable for disease incidence in severe form.

iv. **Yield Loss:** In India, smuts are found sporadically and are of little importance.

15.2.1.4 FOOT ROT

1. ***Sclerotium rolfsii:*** It is one of the major diseases in finger millet, especially under irrigated conditions and in summer crops in Karnataka and Gujarat states.

   i. **Symptoms:** The pathogen primarily affects the stem portion, followed by the leaf sheath and culm. Consequently, plants become pale, yellowish, and stunted in growth. The infected portion turns brown, and the plant wilts. White mycelial growth is seen between the sheaths and on the basal portion of the stem. Later, mustard-like, small, spherical, dark-colored sclerotial bodies are produced on the surface of the infected plant.

   ii. **Pathogen:** It has a wide host range and is present in most soils. However, it prefers light, well-drained soils. The pathogen produces submerged vegetative mycelium made of single hyphae and aerial mycelium composed of both single hyphae and hyphal strands. On these strands, sclerotial initials are formed. After 4–7 days of mycelial growth, 0.5 to 2.0 mm sized sclerotia begin to grow. White sclerotia initially turn brown.

   iii. **Mode of Spread and Survival:** The fungus is mainly soil-borne, surviving in the soil for several months. Sclerotia act as primary inoculum for disease development as they possess viable hyphae, and secondary spread occurs through irrigation water and other implements.

   iv. **Favorable Conditions:** High soil moisture and high temperature favor disease development. Sandy-loam soil favors disease development. The disease incidence is higher during warm and dry months.

   v. **Yield Loss:** A yield loss of up to 50% has been reported. However, it may vary depending on disease severity and prevailing favorable weather conditions.

15.2.1.5 BANDED LEAF AND SHEATH BLIGHT

1. ***Rhizoctonia solani:*** Banded leaf and sheath blight is one of the major emerging problems in all small millet cultivation. Under hot and humid climates like Vizianagaram (Andhra Pradesh), Jagdalpur (Chhattisgarh), and Ranchi (Jharkhand), banded sheath blight infection is very severe.

   i. **Symptoms:** The pathogen affects all aerial parts of the finger millet, viz., leaves, leaf sheath, stalks, and ears, and manifests as leaf blight. The disease is characterized by oval to irregular-shaped lesions that are light grey to dark brown in color on the lower leaf and leaf sheath. Eventually, the central part of the lesion turns white with narrow reddish-brown margins. Later, the spots are distributed irregularly on the leaf lamina. When the environment is right, lesions grow quickly and coalesce to cover an entire sheath and leaf lamina. At this point, the disease occurs as a series of brown- or copper-colored bands running across the leaves, giving them a distinctly banded appearance. Under extreme cases, irregular to oval-shaped dark brown or purplish to brown necrotic lesions are seen in peduncles, fingers, and glumes as well. Symptoms are observed in every segment of the plant exhibiting a distinctive banded appearance; thereby, it is entitled as banded blight (Dubey, 1995). On and around the lesions, mycelial growth and white to brown sclerotia are seen. Eventually, leaves dry up, and the plants are blighted. The pathogen produces long cells of septate hyphae, which are hyaline when young and later turn brownish. It produces a large number of globose sclerotia. The sclerotia are white initially and later turn brown or purplish brown. De Candolle in 1815 was the first to illustrate the genus concept *Rhizoctonia*. Based on the hyphal fusion of many isolates with varying genotypic and phenotypic characteristics, *R. solani* is a species complex that comprises multiple anastomosis groups (AGs) (Carling et al., 2002). While the genetically distant *R. solani* isolates might not anastomose, the genetically similar isolates are compatible and form a combined hyphal network involving fusion of cell wall, cytoplasm, and nuclei (Carling et al., 2002).

ii. **Mode of Spread and Survival** In dry soil, the pathogen can survive as sclerotia or mycelium for up to 20 months; however, in moist soil, it can survive for only 5 to 8 months. It spreads by irrigation water, movement of contaminated soil, infected plant debris, and other field operations during the cropping season.

iii. **Favorable Conditions:** High relative humidity (>90%), high temperature (28–30°C), closer planting, and heavy doses of nitrogenous fertilizers are favorable for banded blight development.

iv. **Yield loss:** Recent studies (2021–22) on yield loss assessment trials in ICAR-AICRP centers (small millet) revealed that 30–50% yield loss occurred due to banded blight at various stages of disease incidence.

15.2.1.6 DOWNY MILDEW

1. ***Sclerophthora macrospora*:** Downy mildew is another important disease of finger millet. Incidentally, its occurrence is sporadic only. Recently, a severe outbreak of downy mildew was observed in the farmers' fields in Tumkur districts of Karnataka on finger millet's most popular varieties (GPU 28 and PR 202).

i. **Symptoms:** The infection starts at the seedling stage and continues until the death of the plant. The affected plants are stunted with shortened internodes. The characteristic symptoms of downy mildew green ear are more common in pearl millet and foxtail millet; however, they are not observed in finger millet. The leaves are pale green. The disease is usually identified after ear formation. The lemma, palea, and sometimes glumes change into leafy structures. The proliferations take place initially at the basal spikelets, and later, others are involved. Finally, the whole ear presents a brush-like appearance (Thirumalachar and Narasimhan, 1949).

ii. **Pathogen:** Mycelium is coenocytic, multinucleate, and intercellular. Sporangiophores are sympodially branched and haploid in nature. Sporangia are lemon-shaped, born singly at the apices of sporangiophores, and measure 60–100 x 43–64 microns. Sporangia mature quickly, thereby releasing more than 50 motile zoospores from each sporangium. The shape of zoospores is ovoid to pyriform. Under favorable environmental conditions, there would be development of big lemon-shaped sporangia that release 24–48 zoospores; thereby, oospore germination is most likely indirect.

iii. **Mode of Spread and Survival***:* The pathogen is internally seed-borne. Oospores and mycelium can survive on infected leaves and soil. Oospores can survive for a few months in well-drained soils. More sporangia were produced in late spring from oospores that survived in winter.

iv. **Favorable Conditions:** Low soil moisture and cool temperatures favor infection by oospores. High humidity and low temperatures (15–20°C) favor conidial production, infection, and disease development. The frequent use of sprinklers maintains high relative humidity. Spores and infection develop in favor of cool temperatures with high humidity.

v. **Yield Loss:** The disease is sporadic in nature; as infected ears are malformed, the disease could be destructive, leading to complete crop failure. Crop losses may vary depending on the time of infection, cultivars, and weather conditions.

15.2.1.7 CERCOSPORA LEAF SPOT

Cercospora leaf spot is an important foliar disease of finger millet. However, the disease is restricted to the Himalayan region in India and Nepal.

i. **Symptoms:** The early-sown crop is most severely affected by the disease in June. Infection generally begins as spots on the older/lower leaves and spreads to the young leaves. Thus, the severity of the disease decreases from older to younger leaves. They often start as reddish-brown spots with a yellow halo.

Eventually, a number of these specks combine to develop large lesions. Occasionally, the lesions develop to the extent of 15 x 3 mm eye-shaped spots. When the crop reaches maturity, the leaves that are significantly affected become entirely necrotic, shriveled, and dry. The plants are completely blighted at this point. These signs and symptoms are also visible on the fingers, leaf sheath, and stem. (Figure 15.1).

ii. **Pathogen:** Conidiophores are olivaceous brown, with tips dilutely colored, straight to curved, geniculate, with septation at long intervals, not branched, and spore scars prominent; measuring 4–5 x 27–300 μm. Conidia are acicular, hyaline to sub-hyaline, indistinctly multiseptate, straight to curved, with a truncate base, sub-acute tip, and measure 3–4 x 50–260 μm.

iii. **Mode of Spread and Survival**: In plant debris, pathogens survive as pseudo stromata, so conidia persist in debris or seeds. Secondary infection occurs through rain splash and wind, particularly on the lower leaves of finger millet.

iv. **Favorable Conditions:** The disease is generally restricted to cooler regions. The disease is confined to the mid-hills of Nepal, where the average temperature is 20°C with high rainfall.

v. **Yield Loss:** If the disease occurs immediately after heading, a 40% and 21% reduction in yield and 1000 seed weight is noticed (Pradhanang, 1994). However, when the disease incidence hovers around 25%, there would be no loss in yield (Pradhanang and Abington, 1993).

A glance at fungal, bacterial, and viral diseases reported on finger millet is tabulated in Table 15.1.

FIGURE 15.1 Disease of finger millet: a&b) Leaf blast; c&d) Neck blast; e) Finger blast; f&g) Foot rot; h&i) Downy mildew and j) *Cercospora* leaf spot.

TABLE 15.1 Fungal, Bacterial and Viral Diseases Reported From Major Finger Millet Growing Regions of the World

Sl. No.	Disease	Pathogen	Reported by
A) Fungal diseases			
		India	
1	Blast	*Pyricularia grisea*	Mc Rae (1920)
2	Foot Rot	*Sclerotium rolfsil*	Coleman (1920)
3	Leaf spot	*Curvularia lunata*	Shaw (1921)
4	Smut	*Melanopsichium eleusinis*	Kulkarni (1922)
5	Foot rot	*Cercosporella sp.*	Narashimhan (1931)
6	Brown spot	*Helminthosporium nodulsoum*	Mc Rae (1932)
7	Downy mildew	*Sclerophthora macrospora*	Venkatarayan (1946)
8	Blight	*Macrophomina phaseolina*	Anonymous (1950)
9	Narrow leaf spot	*Cercospora eleusinis*	Munjal et.al (1961)
10	Wilt	*Ozonium texanum var parasiticum*	Chakravarti and Ghufran (1964)
11	Leaf blight	*Helminthosporium rostratum*	Lele and Dhanraj (1966)
12	Damping off	*Pythium aphanidermatum*	Raghunathan (1968)
13	Leaf spot	*Helminthosporium tetramera*	Vidhyasekaran (1974)
14	Sheath blight	*Marasmius candidus*	Parambaramani et al. (1975)
15	Rust	*Uromyces eragrostidis*	Dublish and Singh (1976)
16	Brown spot	*Helminthosporium holmi*	Singh and Misra (1978)
17	Banded blight	*Rhizoctonia solani*	Lulu Das and Girija (1989)
18	Root rot	*Fusarium moniliforme*	Roy (1989)
19	Leaf spot	*Curvularia pallescens*	Rao, et al. (1992)
		Uganda	
20	Tar spot	*Phyllochora eleusinis*	Small (1922)
21	Stem rot	*Botryodiplodia theobromae*	Small (1925)
22	Foot rot	*Gibberella saubinetti*	Hansford (1934)
23	Leaf spot	*Cylindrosporium sp.*	Esele (1983)
		Tanganyika	
24	Leaf spot	*Cercospora fusimaculans*	Wallace and Wallace (1947)
		Nyasaland	
25	Blight	*Gloecercospora sp*	Wiehe (1950)
B) Bacterial diseases			
		India	
26	Leaf spot	*Xanthomonas eleusineae*	Rangaswami et al. (1961)
27	Leaf spot	*Xanthomonas coracanae*	Desai et al. (1967)
28	Leaf spot	*Pseudomonas eleusinae*	Billimoria and Hegde (1971)

TABLE 15.1 *(Continued)*

Sl. No.	Disease	Pathogen	Reported by
		Japan	
29	Brown stripe	*Pseudomonas alboprecipitans*	Nishiyama, et al. (1979)
30	Bacterial stripe	*Pseudomonas avenae*	Misawa, et al. (1980)
C) Virus Diseases			
		India	
31	Severe mosaic	Sugarcane mosaic virus	Rao et al. (1965)
32	Mottle streak	Ragi Rhabdo virus	Yaraguntaiah and Keshvamurthy (1969)
33	Streak	Maize streak virus	Nagaraju et al. (1982)
Uganda			
34	Ragi stunt	Rice stunt virus	Adipala (1992)
D) Nematode diseases			
		India	
35	Reniform nematode	*Rotylenchulus reinformis*	Chandrashekharan (1964)
36	Ring nematode	*Criconemoides ornatus*	Narayanaswamy and Govindu (1966)
37	Cyst nematode	*Heterodera grambiensis*	Narayanaswamy and Govindu (1966)
38	Lesion nematode	*Pratylenchus zeae*	Mohanty and Das (1976)
39	Spiral nematode	*Helicotylenchus sp.*	Setty, et al. (1977)
40	Stubby root nematode	*Trichodorus sp.*	Babu, et al. (1996)
E) Disease complexes			
		India	
41	Nematode-fungus complex	*Pellicularia rolfsii, Cyst nematode*	Govindu and Swamy (1966)
42	Virus-fungus complex	*H. nodulsoum + Fusarium sp. + virus*	Govindu and Shivanandappa (1967)

15.3 INSECT PEST OF FINGER MILLET

Small millets offer grains that are nutritionally superior to rice and wheat and provide a cheap source of proteins, minerals, and vitamins to the poor population, who have a greater need for such ingredients. Plant protection deals with insect pests of various kinds and equally varied types of diseases. There are over a dozen pests that, when infesting the crop, damage it to a varying degree depending upon the stage of crop growth (Jagadish and Musthak Ali, 1983; Nagarajan and Channabasavanna, 1977; Puttuswamy and Channabasavanna, 1977). However, only a few of them are severe enough to cause economic losses (Figure 15.2). A brief detail of insect pests found on the finger millet crop is given in Table 15.2.

TABLE 15.2 Insect Pests of Finger Millets and Their Status

Common Name	Scientific Name	Damaging Stage	Plant Part Attacked	Status
Finger millet *(Eleusine coracona*, Gaertn.)				
Pink stem borer	*Sesamia inferens*, Wlk.	Caterpillar	Stem/leaf	High
White stem borer	*Saluria inficita,* Wlk.	Caterpillar	Stem	Low
Sorghum stem borer	*Chilo partellus,* Swin	Caterpillar	Stem	Moderate
Earhead caterpillar	*Cryptoblabes gridiella*, M	Caterpillar	Ear	High
	Heliocoverpa armigera, Hb			High
	Eublemma silicula, Swin			Moderate
	Cacoecia epicyrta, Meur.			High
Phytophagous coccinellid	*Afidentula minima* (Groham)	Grubs	Foliage	Low
Noctuids	*Mythimna separata, M. albistigma, Mocis frugalis*	Larvae	Foliage	Low
Meloid beetle	*Cyaneolytta aceton*	Adults	Ear	Moderate
Red hairy caterpillar	*Amsacta albistriga*	Caterpillar	Leaf	Low
Rusty plum aphid	*Hysteroneura setariae*	Nymphs, adults	Leaf, stem, ear	High
Root aphid	*Tetraneura nigriabdominalis*, Sasaki	Nymph & Adult	Root	Moderate
Leaf hopper	*Cicadulina bipunctella bipunctrella*, Maisumra *Cicodulina chinai*, Ghauri	Nymph & Adult	Leaf	Mod-high
Surface grasshopper	*Chrotogonus* sp.	Nymph & Adult	Leaf	Low

Source: Modified from Murthy and Harinarayana (1989).

15.3.1 RAGI STEM BORER

It is widely distributed in all finger millet-growing countries. In Orissa, Karnataka, Tamil Nadu, and Andhra Pradesh, it occurs very frequently. Two types of borers are reported: Pink borer (*Sesamia inferens*) and Ragi White borer (*Saluria infecita*). Pink borer has been the most common pest of ragi. It has a limited host range: sorghum, wheat, sugarcane, and rice. Krishna Murthy and Osman (1952) made a detailed study on the life history of this pest together with possible control measures, including biological control. Scarification of leaves, followed by pinholes at the openings of the leaves, is observed after larval feeding. The 3rd instar larvae migrate to the base of the plant and bore inside, leading to the production of central shoot starts drying, that is, typical stem borer "dead-heart." The larva bores into the stem of the mature plant and creates tunnels by feeding on the internal tissues.

15.3.2 SUCKING PESTS

Four species from the families Cicadellidae, Pemphigidae, and Aphididae, belonging to the order Homoptera, are the major sucking pests of ragi. We know that species of Cicadellidae are called leafhoppers. They appear during the early crop growth stage. Damage to seedlings in the nursery and main field is caused by both adults and nymphs by sucking sap from the tender leaves. Damaged plants exhibit white dots and specks. In the severe stage, these specks coalesce and give a burnt appearance. They act as vectors for streak and mottle streak virus diseases, increasing vulnerability. In outbreaks of viral diseases, these vectors play a key role in affecting finger millet. The hopper lays eggs inside the plant tissues, and depending on the temperature and humidity, the eggs may hatch. Nymphs feed on the tender parts of the plant after hatching. Besides these two kinds of aphids, viz., shoot and root aphids, appear on the finger millet crop whenever there is a dry spell.

15.3.3 LEAF OR SHOOT APHID

Hysteroneura setariae (Thomas) (Aphididae: Homoptera)

Leaf aphids are also called rusty plum aphids, as the name suggests; they are small brown aphids with dark cornicles. This pest was first reported in India by Raja Singh (1968), who observed it in rice, sorghum, Italian millet, and other crops in Madras. It has been recorded in Madras, Tambaram, Tindivanam, Coimbatore, Madurai, Bangalore, Ananthapur, and Coonoor. It spreads from the base of the spikelet to the entire plant through the feeding of plant sap by adults and nymphs, resulting in reduced vigor of the plant and stunted growth. Generally, infestation starts from the border row and spreads. The infestation is alarming when it appears on crops less than a month old. The occurrence is during the months of August and September. When the crop is caught in a dry spell, infestation continues throughout the crop growth stage. The entire ear heads are fully covered by aphids in cases of severe infestation.

15.3.4 ROOT APHIDS

Tetraneura nigriabdominalis (Sasaki) (Homoptera: Pemphigidae)

According to Fletcher (1914), these aphids cause significant damage to ragi by repeatedly sucking the juice from t2he roots. This causes the crop to wither unless it is constantly irrigated, in which case its yield is significantly decreased. This aphid was considered a minor ragi pest in Mysore, in addition to the Bangalore, Kolar, and Tumkur districts of Karnataka. It was also found feeding on the roots of sugarcane (Krishnamurti, 1930 and 1948). In India, it occurs as an endemic pest in traditional ragi-growing districts of Karnataka. In the year 1970, an outbreak was noticed in Hassan district. The outward signs of this pest's existence are progressively disappearing. In severely afflicted plants, ants that are visiting the roots can be observed; aphid colonies clinging to the roots can be seen by pulling up these plants. They feed on the sap from the roots. Since these aphids are limited to the root zone, their presence is seldom detected until the plants exhibit signs of wilting, excessive tillering, stunted growth, early maturity, and occasionally root drying. Consequently, ill effects on the plant, like yellowing, stunting, and occasionally poor seed setting or producing shriveled grains, are other symptoms. By spreading plant viruses, it can occasionally result in indirect damage. The rusty plum aphid *Hysteroneura setariae* on finger millet (Prabhuraj and Jagadish, 2001) has been recorded. The life cycle of *H. setariae* on finger millet (*Eleusine coracana)* in the laboratory was completed in 7–8 days, and reproduction continued for 13–14 days. The rate of reproduction averaged 2.84–4.55

nymphs/day, and there were two ecdyses at intervals of 2 days. The first nymphs were produced within 7–8 days.

15.3.5 EAR-HEAD PESTS

The earhead pests mainly include caterpillars of three different species of *Lepidoptera* and two species of pentatomid bugs, which are minor in nature.

i. ***Helicoverpa armigera:*** Important among the caterpillars is *H. armigera*, which is a polyphagous pest. It attacks the ear during the dough stage, starts feeding on developing grains, and causes damage. They are severe on local varieties of finger millet, which have closed fingers, giving them a perfect niche for their development and hiding from natural enemies. While the recent high-yielding varieties have semi-compact and open types of earheads, which are not much preferred by these caterpillar pests to invade.

ii. ***Cryptoblabe sangustipennella:*** Hamps. This is found as a pest of ragi in Karnataka. The early instar larva initially gets nourished by damaging the chlorophyll of the lemma, followed by feeding on milky grains. On the other hand, the adult larva damages the earhead by constructing webs and by feeding on flowers and grains. Occasionally, earhead bugs assume pest status, particularly during the dough stage. Both adults and nymphs suck the sap from developing grains and cause partial chaffiness of the earhead.

FIGURE 15.2 Pest of finger millet: a) Ragi stem borer larva b) Ragi stem borer infested plant; c) Leaf or shoot aphid; d) Root aphid; e) Lepidopteran earhead pest (*Cryptoblabe sangustipennella*).

15.4 MANAGEMENT STRATEGIES FOR PESTS AND DISEASES

15.4.1 DISEASE MANAGEMENT STRATEGIES

15.4.1.1 CULTURAL PRACTICES

Numerous agricultural techniques, including timely sowing, maintaining ideal plant populations and spacing, timely weeding, fertilizer balance, crop rotation, deep ploughing in the summer, removing crop residues from the field, cleaning field bunds following the crop season, removing and burning diseased plants from the field, and controlling irrigation water from seeping into other fields, will help lower the risk of disease occurrence. The important cultural practices to be followed for the management of finger millet diseases are given in Table 15.3.

TABLE 15.3 Cultural Practices for the Management of Finger Millet Disease

Sl. No	Cultural Practices	Diseases Managed
1.	Crop Rotation: When the same millet is planted on the same plot of land year after year, disease-causing seed and soil-borne pathogens have a chance to proliferate and often do so to such an extent that the same crop is no longer financially feasible due to crop losses. Crop rotation is a method that can prevent or significantly minimize this accumulation of disease problems.	Footrot/ Wilt, Downy mildew, smut and banded blight
2.	Deep Summer Ploughing: Ploughing during the summer months exposes the resting spores of many fungi and nematodes to solar energy, which are killed due to natural heating and desiccation. It helps in the destruction of crop residues and reduces inoculum of soil-borne diseases.	Downy mildew, Footrot/ Wilt, smut and banded blight
3.	Adjustment of Date of Sowing: The main goal of sowing date selection in relation to crop disease is to minimize the amount of time that the infectious agent comes into contact with the host during its susceptible stage.	Blast and rust
4.	Optimization of Plant Population: The pathogen thrives quickly in the field because of the high plant density, constant high humidity, low temperature, and lack of aeration. Reducing the relative humidity buildup in the field by maintaining an ideal plant population aids in lessening the severity of diseases such as downy mildew, blast, rust, and grain mold in finger millet.	Downy mildew, blast, rust and grain mold
5.	Use of Disease-Free Seeds: For sustainable disease management use of pathogen free seeds is essential, since they are widely spread by infected seeds and seedlings.	Blast, smuts, grain mold and foot rot.
6.	Sanitation: Field sanitation is an essential tool that prevents the spread of diseases besides cultural practices. Plants and their debris are important repositories for any disease organisms. The infected plant acts as inoculum, thereby multiplying in the soil and plants that may eventually cause an epidemic in the respective season.	Banded sheath blight, downy mildew, blast and foot rot
7.	Rouging: Regular removal of diseased plants or plant parts from a population has been found effective in reducing the primary hosts and spread of many diseases. This method is one of the effective recommendations in the control of virus diseases in field crops. The rouging of diseased plants not only checks the spread of disease, but it also reduces the amount of surviving structures.	Grain smut and grain mold
8.	Elimination of Collateral and Alternate Hosts: Plant diseases typically have a broad host range and rely on weeds, wild relatives of the host, or self-sown host plants to actively survive when the primary crop isn't around. Thus, preventing new diseases from entering the field is an important sanitary cultural practice. Diseases including ergot, blast, rust, downy mildew, leaf spots, as well as viral and bacterial infections can all be controlled by eradicating them.	Downy mildew, rust, blast and leaf spots
9.	Fertilizer Management: A plant with adequate nutrition may withstand or endure a pathogen's attack far better than one with either an excess or shortage of nutrients. A plant lacking in nutrients will experience stress and become more vulnerable to disease. Plants that receive excessive fertilizer treatments may also be more prone to disease. The interaction between crops and diseases can be influenced by nutrients in a variety of ways. Controlling the application of nitrogenous fertilizer lowers the prevalence of disease.	Blast and downy mildew

TABLE 15.3 *(Continued)*

Sl. No	Cultural Practices	Diseases Managed
10.	Clean Cultivation: Management of smut diseases requires awareness among farmers. Clean cultivation techniques, such as gathering smutted heads in fabric bags and immersing them in boiling water to destroy the pathogen, minimize the occurrence and lower the inoculum for the following year.	Smuts

15.4.1.2 HOST PLANT RESISTANCE

Utilizing host resistance to manage disease is not only cost-effective but also a sensible requirement in a low-value crop like millet, where the availability of additional financial resources, such as fungicides and other inputs, is restricted. The best way to fight the disease, which is mostly farmed by marginal and resource-poor farmers, is to produce resistant varieties. Disease-resistant varieties identified and released for the various millet-growing regions in India are illustrated in Table 15.4 (www.aicrpsm.res.in).

TABLE 15.4 Identified and Released Cultivars of Finger Millet for Blast Disease Resistance Across India (2000–2018)

Sl.No.	Area/States	Tolerant/ Resistant Cultivar
1.	Karnataka	GPU 26, GPU 28, and GPU 48, KMR 301 KMR 340 and KMR 204
2.	Gujarat, Jharkhand, Karnataka, Madya Pradesh, Maharashtra	GPU 45
3.	Odisha, Madya Pradesh, Gujarat, Andar Pradesh and Tamil Nadu	Chilika (OEB 10)
4.	Uttaranchal	VL 315
5.	Hills of Uttarakhand	PRM 1
6.	Andhra Pradesh	Bharathi (VR 762), Sri Chaitanya, PPR 2700 (Vakula)
7.	Sub-mountain and Ghat zone of Maharashtra	KOPN 235
8.	Odisha, Bihar, Chhattisgarh, Karnataka and Tamil Nadu	OEB 526 and OEB 532
9.	All ragi-growing areas of the country	VL 352
10.	Gujarat	GNN - 6 and GN - 5
11.	Uttarakhand	VL Mandua – 348
12.	Konkan region of Maharashtra	Dapoli – 2 (SCN – 6)
13.	Tamil Nadu	CO 15

15.4.1.3 BIOLOGICAL CONTROL

Biological control provides several benefits to both humans and the ecosystem, acting as the best alternative to synthetic chemical pesticides. It can protect plants from biotic and abiotic stresses, produce high-quality grains, enhance soil fertility, promote sustainability, and ensure environmental safety. Because of their potential to protect crops and promote plant growth in sustainable farming systems, as well as their economic significance, researchers are calling for the development and use of indigenous bioinoculant products. When there is insufficient host resistance to soil-borne diseases of millets, such as foot rot in finger millet, charcoal rot in sorghum, and banded sheath blight in small millets, the application of biocontrol agents can be helpful. For millets with soil- and seed-borne diseases, biocontrol agents, particularly *Trichoderma* and *Pseudomonas* strains, are advantageous.

1. **Blast:** In Ragi, seeds are treated with *P. fluorescens* @ 6g/Kg seed. Under field conditions, *P. fluorescens* formulations are sprayed in the proportion of 2 g/lit of water. The first spray is done as soon as symptoms appear, while the remaining two sprays are at the flowering stage at 15-day intervals (Tharana et al., 2023).

2. **Banded Sheath Blight:** Application of 1 kg bio-agents (*P. fluorescens* + *Bacillus subtilis* + *Trichoderma viride* mixed in 25 Kg FYM or vermicompost) as a talc-based formulation in soil. The above-mentioned proportion was incubated for 15 days and applied over an acre, which showed a profound effect in decreasing the severity of disease in minor millets.

3. **Foot Rot:** A talc-based formulation of *T. viride* and *P. fluorescens* @ 5g/litre was used as a root dip for seedlings, and FYM (50 kg) enriched with *T. viride* and *P. fluorescens* @ 500 g each/acre was applied in soil, which is helpful for reducing the disease incidence.

15.4.1.4 CHEMICAL CONTROL

Due to the high expense of manpower and chemicals, chemical disease management is not usually employed in millet. Nonetheless, there are situations when using it along with a resistant cultivar is required. Most often, fungicides are used as foliar sprays or as seed treatments. Yet better management is achieved when they are combined.

i. **Downy Mildew:** Ridomyl-MZ @ 6g/Kg seed was used for seed dressing, followed by one or two foliar sprays of Ridomyl-MZ @ 3g/L, which showed a profound effect.

ii. **Blast:** Carbamazepine treatment of seeds at a rate of 1 gm/kg of seed. Spray any of the following fungicides: Ediphenphos (0.1%), Propiconazole (0.1%), Tricyclazole (0.1%), Iprobenphos (IBP) (0.1%), Carbendazim (0.2%), or a premixed fungicide (Carbendazim + Mancozeb) (0.1%). Immediately upon observing the symptoms, give the first spray. In finger millet, apply the second and third sprays based on need during the blooming stage, every 15 days, to prevent infections in the neck and fingers.

iii. **Grain Mold:** If there is infrequent rainfall from earhead emergence until the milky stage, to reduce the grain mold incidence, the application of any one of the following fungicides is essential: two to three sprays of 0.2% for Mancozeb + Captan, 0.1% for Propiconazole, and 0.2% for Thiram + Carbendazim.

iv. **Smuts:** To reduce the smut incidence, seed treatment with sulfur @ 4g/Kg seed is recommended.

v. **Banded Sheath Blight:** Seed dressing with Mancozeb (2–3 g/kg). Foliar spray with the fungicides if required, viz., propiconazole @ 1 ml/liter, hexaconazole @ 2 ml/liter, or Validamycin @ 2 ml/liter is highly effective.

vi. **Rust:** Foliar spray of Mancozeb (0.2%), difenconazole (0.1%), hexaconazole (0.1%), and propiconazole (0.1%) to reduce disease prevalence. Two sprays at 15-day intervals immediately after the appearance of symptoms are recommended to control disease incidence.

vii. **Leaf Spot/Leaf Blight:** Carbendazim @ 2g/kg or Mancozeb 0.2% is used for seed treatment and foliar spray if required.

15.4.2 PEST MANAGEMENT STRATEGIES

15.4.2.1 CULTURAL CONTROL

i. **Sanitation:** The persistence of stem borers can be reduced through effective field sanitation, along with uprooting and burning of stubbles. Stalks reserved as animal feed should be used before the onset of the monsoon.

ii. **Tillage:** Hibernating, aestivating, or hidden larvae and pupae of many insect species can be exposed if field tillage is performed before planting and after harvest. The populations of grasshoppers and hairy caterpillars in the field are effectively reduced through deep ploughing.

iii. **Intercropping:** Finger millet with mustard and niger in the ratio of 4:1:5 and 4:2:4 respectively favors the buildup of natural enemies and sustains their population.

iv. **Weed Control:** According to Ndoye et al. (1986), damage caused by *Mythimna separata* can be reduced by ensuring proper and timely weeding of the crop. Maintaining a clean crop is effective, as weeds can provide hiding places and oviposition sites for certain species. However, higher damage has been reported in ploughed fields than in no-tillage systems in certain situations (Shenk and Saunders, 1981). Huis (1981) reported that weeds (*Digitaria* sp. and *E. indica*) tend to harbor larvae of *Spodoptera* sp. and increase crop damage.

15.4.2.2 MECHANICAL METHOD

15.4.2.2.1 Collection and destruction through mechanical means

Collection and destruction of egg masses of hairy caterpillars and other lepidopteran pests. Uprooting and destroying the infested plants can also reduce aphid infestations. You can collect blister beetles by hand and destroy them (Jotwani and Butani,

1978). You can dislodge head bugs and other external feeders from the ear head into a bucket containing water with added kerosene (Nair, 1975).

15.4.2.3 PHYSICAL METHOD

Setting up light traps helps attract and kill the stem borers and other moth pests.

15.4.2.4 HOST PLANT RESISTANCE SOURCES

A good number of genotypes are found to be resistant to stem borers and aphids, which are presented in separate tables 15.1 and 15.2. According to Painter (1951), resistant plants exhibit less damage or infestation by phytophagous insects under similar field conditions. The genotypes tolerant to *Sesamia inferens* are given in Table 15.5, and cultivars resistant to earhead worms, aphids, and shoot fly are presented in Table 15.6.

TABLE 15.5 Finger Millet Genotypes Either Tolerant or Less Susceptible to *Sesamia inferens* Walk

Genotypes	References
IE 932, IE 982 and IE 1037	Lingappa (1979)
PES 9, PES 144, PES 224, KM 1, KM 14, HR 228, JNR 1008 and T 36-B	Kishore and Jotwani (1980)
KM 1, RAU 1, RAU 3, Indaf ;7, Indaf 8, HR 154, HR 374, HR 1523, PES 110, PES 400, WR 9, VL 110	Murthi and Harinarayana (1986)
VL 109, VR 530, PR 202, HR 374	Dhamdhere (1988)

TABLE 15.6 Promising Finger Millet Cultivars Resistant to Earhead Worms, Aphids and Shoot Fly Identified in Coordinated Multi-Location Screening

Earhead Worms		Aphids		Shoot Fly
1980–86	1990–96	1980–86	1990–96	1992
Indaf 7	KM 229	PES 176	PES 400	HR 374
Indaf 8	KM 230	RAU 1	VL 149	PES 400
PR 202	PR 202	HR 374	GPU 36	VL 162
PR 177	PPR 2618		KM 225	VL 171
HR 374	HR 374			VL 231
HR 1523	PES 400			VR 584
PES 110	PES 110			VMEC 35
PES 1877	GPM 36			RAU 8
TNAU 1877	TNAU 551			MR 5
TNAU 294	VMEC 35			MR 6
VL 110	VL 149			L 15-1
	VR 530			
	VR 696			
	VR 708			

15.4.2.5 CHEMICAL CONTROL

- Dusting the crop with Malathion 5% D or Methyl parathion powder for management of red-headed hairy caterpillar, grass hoppers, weevils and also reduce the leaf-eating caterpillars.
- Quinalphos or Methyl parathion 2ml mixed with one liter of water then sprayed for the management of leaf-eating caterpillars.
- Use of poison bait: Use 10 kg of rice bran or wheat bran mixed with one kg of jaggery and keep it overnight. During the evening hour, insecticide 100 ml quinalphos was mixed with a fermented mixture of rice bran or wheat bran. Then sprinkle throughout the field to manage *Spodoptera* sp.
- Use of systemic insecticide that is dimethoate 1.7ml/lit of water for the management of leaf aphid.

15.5 CONCLUSION

Compared to other cereals, finger millet provides essential nutritional benefits. Almost all millet crops have built-in defenses against biotic and environmental stresses like heat, drought, lodging, and waterlogging, but over climate change these stresses still pose a threat to millets' production. There is major concern in developing countries, as they are more prone for loss in food production due to climate change. Developing and improving the nutritional quality of local variety are highly adaptable and survive under biotic and abiotic condition.

It is very difficult to develop a staple food that satisfies all primary criteria, however there is wide range of diversity in millets, so climate -resilient crop can be selected from them. the enormous range and diversity of local food crops (such as minor millets) gives us a number of climate-resilient crops to choose from. Crop yield is being increased while also reducing the effects of biotic and abiotic pressures through the application of techniques like CRISPR/Cas9 and PGPRs. The CRISPR/Cas9 system can be used to modify specific candidate genes in order to enhance finger millet growth and production. Already, there is genome sequence information in sorghum and foxtail millet, while in finger millet and pearl millet genomic study is in progress and research on biotic and abiotic stress tolerance helps in crop improvement program. High throughput sequencing technology helps to figure out the regulation of stress tolerance at transcriptional, post-transcriptional, and epigenetic levels, additionally helps to overcome the complexity finger millet genome. In future developing a super grain is possible by incorporating important agronomic traits into single millet genotype. Thus, improvement in finger millet would be outstanding, if molecular breeding and genetic engineering combined with modern Omics technologies is used. The alternative approaches for managing biotic and abiotic stresses will be supplemented by research in the disciplines of genomics, transcriptomics, metabolomics, proteomics, and other related areas. Geneticists, biotechnologists, breeders, germplasm conservationists, and others should focus more on improving finger millet in order to increase global food security, as the production of staple crops is being negatively impacted by climate change.

KEYWORDS

- **banded leaf and sheath blight**
- **blast and leaf spots**
- **downy mildew**
- **finger millet disease management**
- **genetic engineering**
- **leaf aphids**
- **pentatomid bugs**
- **stem borer**

REFERENCES

Annual progress report of Indian Council of Agricultural Research (ICAR)–All India Co-ordinated Research Project (AICRP) on small millets. (n.d.). *2019–2020*.

Annual progress report of Indian Council of Agricultural Research (ICAR)–All India Co-ordinated Research Project (AICRP) on small millets. (n.d.). *2020–2021*.

Carling, D. E., Gitaitis, R. D., Brainard, K. A., & Kuninaga, S. (2002). Characterization of AG-13, a newly reported anastomosis group of *Rhizoctonia solani*. *Phytopathology*, *92*, 893–899.

Das, I. K., & Padmaja, P. G. (2016). *Biotic stress resistance in millets*. Academic Press: USA.

Das, I. K., Nagaraja, A., & Tonapi, V. A. (2016). *Diseases of millets: A ready reckoner* (Vol. 500030). Indian Institute of Millets Research, Rajendranagar, Hyderabad.

Dida, M. M., Srinivasachary, Ramakrishnan, S., Bennetzen, J. L., Gale, M. D., & Devos, K. M. (2007). The genetic map of finger millet, *Eleusine coracana*. *Züchter Genet. Breed. Res.*, *114*(2), 321–332. https://doi.org/10.1007/s00122-006-0435-7

Dubey, S. C. (1995). Banded blight of finger millet caused by *Thanatephorus cucumeris*. *Indian Journal of Mycology and Plant Pathology*, *25*, 315–316.

Fletcher, T. (1920). *Life histories of Indian insects (Microlepidoptera)*. *Memoirs of the Department of Agriculture, India (Entomology Series)*, *6*.

Gupta, D. K., Jain, A. K., & Kumar, A. (2016). Evaluation of finger millet genotypes for partial resistance to leaf blast caused by *Pyricularia grisea*. *Electronic Journal of Plant Breeding*, *7*(4), 939. https://doi.org/10.5958/0975-928X.2016.00128.9

Kishore, P., & Jotwani, M. G. (1980). Screening of some improved genotypes of the finger millet (*Eleusine coracana* Gaertn.) for resistance to *Sesamia inferens* Walker and *Myllocerus maculosus* Desbrochers. *Journal of Entomological Research*, *4*(2), 221–223.

Krishnamurti, B. (1952). The ragi stem borer. *Sesamia inferens* Wlk. *Department of Agriculture, Mysore, Entomology Series Bulletin*.

Lenné, J. M., Takan, J. P., Mgonja, M. A., Manyasa, E. O., Kaloki, P., Wanyera, N., & Tamale, M. S. (2007). Finger millet blast management: A key entry point for fighting malnutrition and poverty in East Africa. *Outlook on Agriculture*, *36*, 101–108.

Leslie, J. F. (2003). *Sorghum and millets diseases*. John Wiley & Sons.

Lingappa, S. (1979). Development of artificial diet for mass rearing *Sesamia inferens* Walker (Lepidoptera: Noctuidae) and screening for resistance in finger millet germplasm. *Mysore Journal of Agricultural Sciences*, *8*(3).

Murthi, T. K., Harinarayana, G., Seetharam, A., Riley, K. W., & Harinarayana, G. (1986). Insect pests of small millets and their management in India. In *Small millets in global agriculture: Proceedings of the First International Small Millets Workshop* (pp. 255–270). Pvt Ltd: Bangalore, India; New Delhi.

Nagaraja, A., Kumar, J., Jain, A. K., Narasimhudu, Y., Raghuchander, T., Kumar, B., & Gowda, B. H. (2007). *Compendium of small millets diseases*. AICSMIP.

Nageshchandra, B. K. (1981). Insects in ragi production. Paper presented at All India Coordinated Millets Workshop held at College of Agriculture, 26–28.

Odeny, D. A., Niazi, A., Tesfaye, K., Lule, D., Wanyonyi, S., & Kunguni, J. (2020). Genomic designing for climate-smart finger millet. In *[Book title not provided]* (pp. 287–307).

Palanna, K. B., Vinaykumar, H. D., Koti, P. S., Jeevan, B., Rajashekara, H., Raveendra, H. R., Patro, T. S. S. K., Ramesh, G. V., Mudalagiriyappa, Rawat, L., Netam, P., Ekka, S., Rajesha, G., Das, I. K., Chandranayaka, S., Nagaraja, T. E., & Satyavathi, C. T. (2024). Geographic distribution, host preference and phylogenetic relationships among *Pyricularia* species inciting millet and rice blast disease in India. *Plant Pathology*, *73*(3), 692–705. https://doi.org/10.1111/ppa.13830

Palanna, K. B., Vinaykumar, H. D., Prasanna, S. K., Rajashekara, H., Devanna, B. N., Anilkumar, C., Jeevan, B., Raveendra, H. R., Khan, F., Bhavana, C. H. S., Upadhyay, V., Patro, T. S. S. K., Rawat, L., Rajesh, M., Saravanan, P. T., Netam, P., Rajesha, G., Das, I. K., Patil, H. E., Jain, A. K., Saralamma, S., Nayaka, S. C., Prakash, G., & Nagaraja, T. E. (2023). Exploring the diversity of virulence genes in the *Magnaporthe* population infecting millets and rice in India. *Frontiers in Plant Science*, *14*, 1131315. https://doi.org/10.3389/fpls.2023.1131315

Parambaramani, C., Subramanian, C. L., & Kandaswamy, T. K. (1975). Sheath blight and wilt of ragi (*Eleusine coracana*) caused by *Marasmius candidus* Bolt. *Current Science*, *44*.

Prabhuraj, A., & Jagadish, P. S. (2001). Aphididae: Homoptera on finger millet. *Hysteroneura setariae* (Thomas), *14*, 470–473.

Pradhanang, P. M. (1994). *Quantification of the relationship between Cercospora leaf spot disease (Cercospora eleusine) and yield loss of finger millet.*

Pradhanang, P. M., Abington, J., Riley, K. W., Gupta, S. C., Seetharam, A., & Mushonga, J. N. (1993). *Cercospora* leaf spot disease (*Cercospora eleusine*) of finger millet in Nepal. In *Advances in small millets*. Oxford IBH Pub. Co: New Delhi.

Pradhanang, P., Pradhanang, P. M., Abington, J. B., Riley, K. W., Gupta, S. C., Seetharam, A., & Mushonga, J. N. (1993). Quantification of the relationship between *Cercospora* leaf spot disease (*Cercospora eleusine*) and yield loss of finger millet. Tech. Paper – Lumle Agricultural Centre No. In *Advances in small millets* (Vol. 94, pp. 12–17). Oxford IBH Pub. Co: New Delhi.

Puttarudriah, M., & Channabasavanna, G. P. (1950). Host range of *Cacoecia epicyrta*. *Indian Journal of Entomology*, *12*, 267–268.

Raghunathan, V. (1968). Damping off of greengram, cauliflower, diancha, ragi and clusterbean. *Indian Phytopathology*, *21*, 456–457.

Rao, A., & Hegde, R. K. (1987). Effect of disease incidence on the grain yield of ragi due to blast. *Current Research*, *16*, 26–27.

Rao, V. (1930). Annual report of work done in the Mycological Section during the year 1928–29. *Annual Report, Agriculture Department, Mysore for the Year 1928–29. Part I*, 18–21.

Report of the Imperial Mycologist. (1922). *Scientific Report, Agricultural Research Institute, Pusa*. 1921–22.

Shaw, F. J. F. (1921). *Report of the Imperial Mycologist*. Science Reports. Agricultural Research Institute, Pusa, 34–40.

Sneh, B., Burpee, L., & Ogoshi, A. (1991). *Identification of Rhizoctonia species*. The American Phytopathological Society: St. Paul.

Subramanian, C. V. (1953). Fungi Imperfecti from Madras V. *Proceedings of the Indian Academy of Sciences*, *38*, 27–39.

Sunil, M. B., & Anulkumar, T. B. (2003). Effect of head blast on seed quality and weight in finger millet. *International Sorghum and Millet Newsletter*.

Takan, J. P., Chipili, J., Muthumeenakshi, S., Talbot, N. J., Manyasa, E. O., Bandyopadhyay, R., & Sreenivasaprasad, S. (2012). *Magnaporthe oryzae* populations adapted to finger millet and rice exhibit distinctive patterns of genetic diversity, sexuality and host interaction. *Molecular Biotechnology*, *50*, 145–158. https://doi.org/10.1007/s12033-011-9429-z

Tharana Poonacha, T., Sai Bhavana, C. H., Farooqkhan, Ramesh, G. V., Gavayi, N., Koti, P. S., Palanna, K. B., Rajashekara, H., Rajesh, G., & Das, I. K. (2024). Blast disease of millets: Present status and future perspectives. In *Millets – Rediscover ancient grains*. Intech Open.

Thirumalachar, M. J., & Narasimhan, M. (1949). Downy mildew of *Eleusine coracana* and *Iscilema laxum* in Mysore. *Indian Phytopathology*, *2*, 46–51.

Venkatarayan, S. V. (1946). Disease of ragi (*Eleusine coracana*). *Mysore Agricultural Journal*, *24*, 50–51.

Venkatarayan, S. V. (1947). Disease of ragi (*Eleusine coracana*). *Mysore Agricultural Journal*, *24*, 50–52, 57.

Yogeesh, L. N., Naryanareddy, A. B., Nanjareddy, Y. A., & Gowda, M. (2016). High temperature tolerant genotypes of finger millet (*Eleusine coracana* L.). *Nature Environment and Pollution Technology*, *15*, 1293–1296.

CHAPTER 16

Integrated Pest and Disease Management in Small Millets

R. R. RACHANA,[1,*] U. AMALA,[1] and SENTHILRAJA CHINNAIAH[2]

[1]ICAR–National Bureau of Agricultural Insect Resources (ICAR–NBAIR), Bangalore, Karnataka, India

[2]Texas A&M AgriLife Research, Amarillo, Texas, USA

*Corresponding author

16.1 INTRODUCTION

The small millets, excluding ragi, are barnyard millet (*Echinochloa frumentaceae*), foxtail millet (*Setaria italica*), kodo millet (*Paspalum scrobiculatum*), little millet (*Panicum sumatrense*), and proso millet (*Panicum miliaceum*). They are cultivated in less productive soils with the lowest management practices. They are high-energy, nutritious foods equivalent to other cereals, and a few of them are even better with respect to protein and mineral content. They are mainly low in phytic acid and abundant in dietary fiber, iron, calcium, and B vitamins. As they are consumed by the poor, they guard them from food and nutritional insecurity imposed by several agronomic, socio-economic, and political factors. They can thus act as a shield against nutritional deficiency syndromes and provide nutritional security. Supplementation of cereal-based products with small millet has become increasingly popular because of nutritional and economic advantages. These crops relatively require fewer inputs and are known for their soil rejuvenation capacity, forming an integral component of dryland farming. Though these crops are hardy in nature, insect pests and diseases cause considerable yield loss under favorable climatic conditions. Owing to the fact that most of the pests and diseases of small millets are common across the crops, the available information on pests and diseases has been presented in the following sections.

16.2 INSECT PESTS

16.2.1 SHOOT FLY (ATHERIGONA SPP.)

They injure all small millets and cause grain yield as well as dry fodder loss. This is a seedling pest generally occurring in the 1st to 6th week after germination. Maggots feed on the

Millets in Sustainable Agriculture: Genetic Improvement, Nutritional Quality, Cultivation, and Integrated Pest–Disease Management. Roshni Vijayan, A. John Joel, R. Ambika Rajendran, and Adithya P. Balakrishnan (Eds.)

DOI: 10.1201/9781779642332-20

growing point, causing leaf wilting and later drying of the central leaf, referred to generally as the dead heart. If the infestation happens a little later, damaged plants develop side tillers, which are also infested. The adult looks like a housefly, dark grey, and about 3–4 mm long. The female flies lay singly whitish cigar-shaped eggs on the lower surface of the leaves. The maggot is yellowish, and pupation occurs at the base of the plant (Nigus and Damte, 2018) (Figure 16.1). The pest population usually is very low in April to June, tends to increase in July, and reaches its peak in August. The population slowly declines from September onwards and stays at a moderate level until March (Kalaisekar and Padmaja, 2017).

A variety of shoot fly species infest small millets (Nageshchandra and Musthak Ali, 1983). *Atherigona destructor* alone causes a 36% yield reduction in proso millet (Natarajan et al., 1974) and 39% in little millet (Selvaraj et al., 1974). Yield reduction to the extent of 44.90% in barnyard millet, 90.90% in proso millet, 78.50% in little millet, 35.0% in kodo millet, and 1.80% in foxtail millet was reported (Nageshchandra and Musthak Ali, 1983). *Atherigona atripalpis* was recorded in India, Sri Lanka, Myanmar, Nepal, and China. *Atherigona miliaceae* was first reported as the China fly (Ramchandra Rao, 1924). This species was reported from India on Panicum spp. *Atherigona falcata* is distributed throughout the Oriental region. The species is comparatively larger than other species.

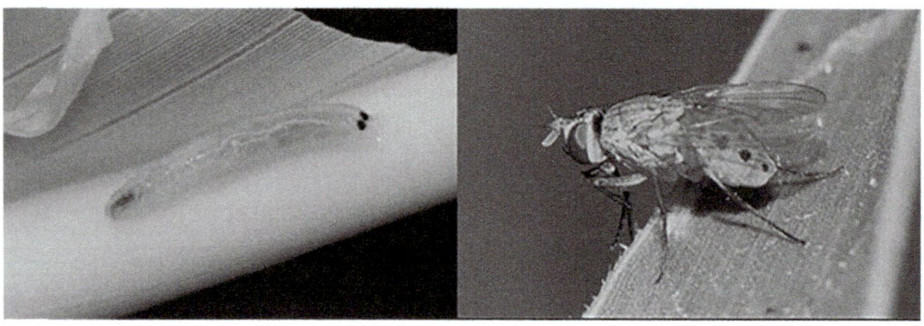

FIGURE 16.1 Maggot and adult of *Atherigona* sp.

Source: Reprinted with permission ICAR-NBAIR database on agriculturally important insects © ICAR-NBAIR.

16.2.2 GALL MIDGE (ORSEOLIA SP.) (CECIDOMYIIDAE: DIPTERA)

It infests little and kodo millets. The infestation on spikelets ranged from 3 to 50% in little millet. They were found to be hypertrophied due to the feeding of the maggots on the ovary. This pest is also a common occurrence in kodo-growing areas of Madhya Pradesh (Kalaisekar and Padmaja, 2017).

16.2.3 STEM BORERS

16.2.3.1 SPOTTED STEM BORER CHILO PARTELLUS SWINHOE (PYRALIDAE: LEPIDOPTERA)

This pest infests foxtail millet and proso millet. Infestation starts from the 2nd week after sowing until the maturity of the crop. Primarily, the caterpillars feed on the upper surface of leaves, keeping the lower surface intact as clear windows. Later, holes and scrapes on the epidermis appear conspicuously. Occasionally, young plants develop dead heart symptoms due to early attacks. Subsequently, the caterpillars bore into the stem, causing extensive tunneling. Peduncle tunneling results in breakage of the whole or partial chaffy panicle. The adult is a medium-sized and straw-colored moth (Figure 16.2). The female lays approximately 500 eggs in 10–80 masses on the lower surface of the leaf near the midrib. The egg hatches in 4–5 days; the larval period lasts 19–27 days. Pupation occurs inside the stem, and the adult develops in 7–10 days. Larval diapause happens during the dry season and survives in harvested stalks as well as leftover stubbles in the field. The diapause breaks down with the commencement of the rainy season and goes for pupation (Kalaisekar and Padmaja, 2017; Shyam Prasad et al., 2015).

FIGURE 16.2 Larva and adult of *Chilo partellus*.

Source: Reprinted with permission ICAR-NBAIR database on agriculturally important insects © ICAR-NBAIR.

16.2.3.2 RAGI STEM BORER SESAMIA INFERENS WALKER (NOCTUIDAE: LEPIDOPTERA)

It injures all small millet except a little millet. The larvae are pink in color, gregarious, gather inside the leaf whorls, and devour the central leaves, causing pinhole symptoms (Figure 16.3). The central shoot turns brown and dries out, though the lower leaves stay green and healthy, known as dead heart. Bored holes are filled with excreta. White ear head symptoms form during the panicle initiation stage with straight and whitish empty panicles. The adult is a pale pink-colored moth with dark brown bands on the forewings and white hindwings. The female lays around 150 creamy-white and hemispherical eggs, arranged in two or three rows amid the leaf sheath and stem of the host. The egg period remains about 7 days. The fully grown larva is about 25 mm, pale yellow with a purple-pink shade and a reddish-brown head. Pupation occurs in the stem tunnel, and the adult emerges in 12 days (Shyam Prasad et al., 2015).

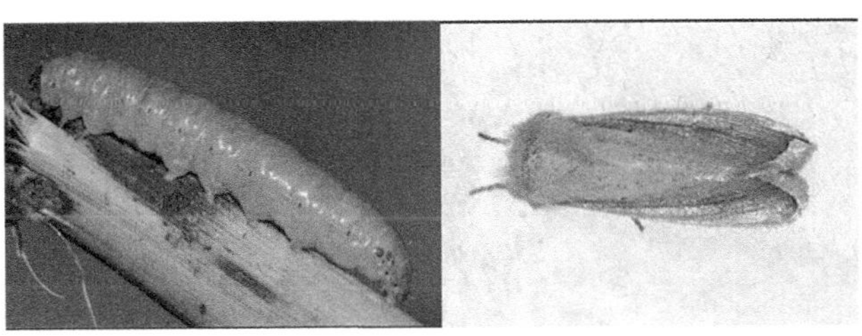

FIGURE 16.3 Larva and adult of *Sesamia inferens*

Source: Reprinted with permission ICAR-NBAIR database on agriculturally important insects © ICAR-NBAIR.

16.2.4 FOLIAGE PESTS

16.2.4.1 ARMY WORMS MYTHIMNA SEPERATA WALKER, SPODOPTERA LITURA (FAB) (NOCTUIDAE: LEPIDOPTERA)

They infest foxtail millet, kodo millet, and proso millet. Caterpillars feed on foliage, causing defoliation. They scrape the green leaf tissue, giving leaves a skeletonized appearance. The young armyworm feeds on plants and later cuts off foliage and panicles. They emerge at night to feed on the roots and shoots and hide inside the soil during morning hours. The field appears as if grazed by cattle. The larvae migrate even during the day. They cut tender stems of young and growing plants. Larvae hide during the day in the soil and become active at dusk. In severe cases, the entire leaf is eaten. The adults of *M. separata* are brownish with black spots in the center of the forewing (Shyam Prasad et al., 2015) (Figures 16.4 and 16.5).

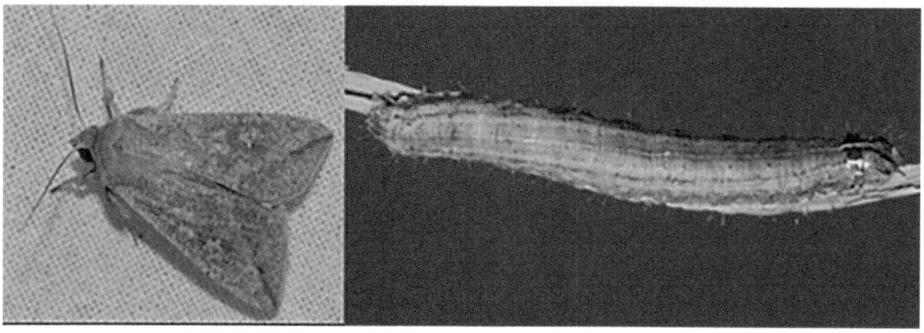

FIGURE 16.4 Larva and adult of *Mythimna separata*.

Source: Reprinted with permission ICAR-NBAIR database on agriculturally important insects © ICAR-NBAIR.

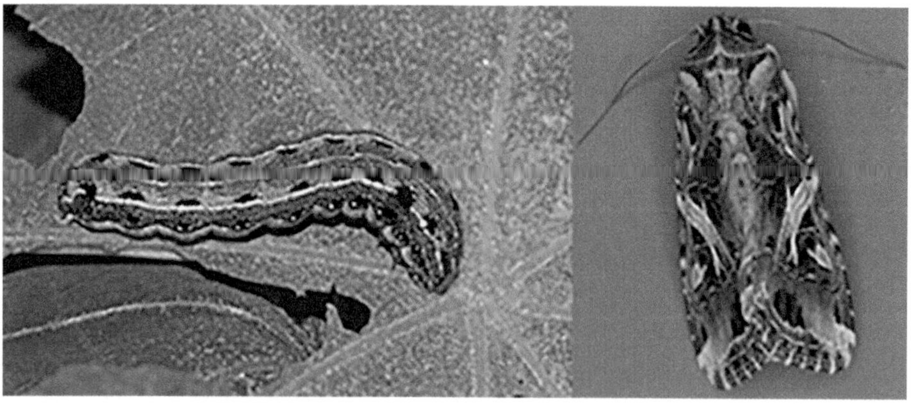

FIGURE 16.5 Larva and adult of *Spodoptera litura*.

Source: Reprinted with permission ICAR-NBAIR database on agriculturally important insects © ICAR-NBAIR.

16.2.4.2 GRASSHOPPERS HIEROGLYPHUS NIGROREPLETUS (BOLIVAR) H. BANIAN (FAB), COLEMANIA SPHENEROIDES (BOLIVAR), CHROTOGONUS SPP. (ACRIDIDAE: ORTHOPTERA)

Both nymphs and adults feed on leaves. They mark marginal notching's or holes on the foliage. Severe infestation results in defoliation, leaving the field with a grazed appearance. Adult grasshoppers lay eggs in groups inside the soil. Eggs are white, yellowish-green, tan, or brown. Newly hatched immatures are white. Later, they adopt the distinctive colors and markings of adults. Nymphal duration is 35 to 50 days. Adults are stout, greenish-brown with a yellow stripe laterally (Kalaisekar and Padmaja, 2017; Shyam Prasad et al., 2015) (Figure 16.6).

FIGURE 16.6 a) ***Hieroglyphus nigrorepletus*** b) ***Hieroglyphus banian*** c) ***Colemania spheneroides***

Source: Reprinted with permission ICAR-NBAIR database on agriculturally important insects © ICAR-NBAIR.

16.2.4.3 LEAF FOLDER/ROLLER CNAPHALOCROCIS MEDINALIS GUENEE, (PYRALIDAE: LEPIDOPTERA)

Infests foxtail millet, kodo millet, and proso millet. The larva scrapes the leaf tissues, folds a leaf blade, and glues it with silken strands. The caterpillar feeds inside a folded leaf, making longitudinal whitish patches on the leaf lamina. Each female lays around 300 oblong, whitish-yellow eggs on either side of the leaf blade, which is oblong and whitish-yellow. The egg period lasts for 4–6 days. The larva is yellowish-green with a black head (Figure 16.7). One or two pairs of subdorsal spots are present on the pronotum. The larval period lasts for 15–20 days. The larvae pupate in the leaf itself, and the pupal period lasts 6–10 days. The adult moth is whitish or golden yellow with three black oblique lines on the forewing and a broad anal area on the hindwings (Shyam Prasad et al., 2015).

FIGURE 16.7 Larva and adult of *Cnaphalocrocis medinalis*.

Source: Reprinted with permission ICAR-NBAIR database on agriculturally important insects © ICAR-NBAIR.

16.2.4.4 FLEA BEETLE CHAETOCNEMA BASALIS MAULIK (COLEOPTERA)

It attacks foxtail millet and little millet. Adult makes holes in the leaves of young plants, reducing their vigor. Adults are oblong, dark blue beetles with distended hind femur (Figure 16.8) (Kalaisekar and Padmaja, 2017).

FIGURE 16.8 *Chaetocnema* sp.

Source: Reprinted with permission ICAR-NBAIR database on agriculturally important insects © ICAR-NBAIR.

A list of insect pests attacking small millets along with a brief detailing is given in Table 16.1.

TABLE 16.1 Insect Pests of Small Millets

Crop	Insect Pest		Damaging Stage	Plant Part Damaged
	Common Name	**Scientific Name**		
Foxtail millet	Shoot fly	Atherigona atripalpis Malloch	Maggot	Growing point
	Flea beetle	*Chaetocnema basalis* Baly *Madurasia* sp.	Adult	Leaf
	Armyworm	*Mythimna separata* (Walker) *Spodoptera frugiperda* (J.E.Smith)	Caterpillar	Leaf
	Leaf roller	*Cnaphalocrocis medinalis* *Cnaphalocrosis trapezolis* *Marasmia trapezolis* Wlk.	Caterpillar	Leaf
	Stem borer	*Chilo partellus Sesamia inferens Ostrinia furnacalis*	Caterpillar	Stem
	Surface grasshopper	*Chrotogonus* sp.	Nymph and Adult	Leaf
	Ant	*Sima* sp.	Adult	Ear
	Leaf beetle	*Oulema melanopus*	Adult	Leaf
Kodo millet	Shoot fly	*Atherigona simplex*	Maggot	Growing point
	Gall midge	*Orseolia* sp.	Maggot	Spikelet
	Stem borer	*Sesamia inferens*	Caterpillar	Stem
	Leaf roller	*Cnaphalocrosis patnalis Marasmia trapezolis* Wlk.	Caterpillar	Leaf
	Jassid	*Hecalus* sp.	Nymph and Adult	Leaf
	Grain bug	*Leptocorisa acuta*	Nymph and Adult	Ear
	Army worm	*Mythimna separata* (Walker)	Caterpillar	Leaf
	Grasshopper	*Acrida exalta*	Nymph and Adult	Leaf
	leaf miner	*Hydrellia philippina*	Nymph and Adult	Leaf
	Mealy bug	*Brevennia rehi*	Nymph and Adult	Leaf
	Skipper	*Pelopidas mathias*	Caterpillar	Leaf
	Thrips	*Stenchaetothrips biformis*	Larva and Adult	Leaf
Barnyard millet	Shoot fly	*Atherigona falcata*	Maggot	Growing point
	Stemborer	*Sesamia inferens*	Caterpillar	Stem
	Leafhopper	*Nephotettix cincticeps*	Nymph and Adult	Leaf
	Aphid	*Hysteroneura setariae* (Thomas)	Nymph and Adult	Leaf and shoot
	Grass hopper	*Acrida exaltata* Wlk.	Nymph and Adult	Leaf
	Leaf caterpillar	*Euproctis* sp.	Caterpillar	Leaf
	White grub	*Anomala dimidiata* Burm; *Holotrichia seticollis* Mos.	Grub	Root
Proso millet	Shoot fly	*Atherigona pulla*	Maggot	Growing point
	Stemborers	*Chilo partellus, C. suppressalis, C. orichalcociliellus, S. inferens* and *Ostrinia furnacalis*	Caterpillar	Stem
	Webworm	*M. trapezalis*	Caterpillar	Leaf and Ear

TABLE 16.1 *(Continued)*

Crop	Insect Pest		Damaging Stage	Plant Part Damaged
	Common Name	**Scientific Name**		
Foxtail millet	Shoot fly	Atherigona atripalpis Malloch	Maggot	Growing point
	Armyworms	*M. separate, M. unipuncta, Spodoptera exempta* and *S. frugiperda*	Caterpillar	Leaf
	Hairy caterpillar	*Spilarctia obliqua*	Caterpillar	Leaf
	Leaf folder	*Cnaphalocrocis medinalis*	Caterpillar	Leaf
	Thrips	*Haplothrips aculeatus*	Larva and adult	Leaf and shoot
	Grasshoppers	*Hieroglyphus banian* and *Oxya chinensis*	Nymph and Adult	Leaf
	Aphids	*Sipha flava*	Nymph and Adult	Leaf and shoot
	Field cricket	*Brachytrupes* sp.	Nymph and Adult	Leaf
Little millet	Shoot fly	*Atherigona miliaceae*	Maggot	Growing point
	Gall midge	*Orseolia* sp.	Maggot	Spikelet
	Stink bug	*Nezara viridula*	Nymph and Adult	Ear
	Black pentatomid bug	*Dolycoris indicus*	Nymph and Adult	Ear
	Jassid	*Kolla mimica*	Nymph and Adult	Leaf
	Grasshopper	*Acrida exalta*	Nymph and Adult	Leaf
	Flea beetle	*Chaetocnema* sp.	Adult	Leaf

Source: Adapted from Das (2023); Kalaisekar and Padmaja (2017).

16.3 INTEGRATED PEST MANAGEMENT

Pesticides have been used comprehensively as an instant remedy, mainly on high-yielding hybrids and varieties. Studies need to be intensified on varietal resistance, mass rearing techniques of beneficial insects, and their population manipulation by cultural practices. This leads to the possibility of utilizing all available methods on an integrated basis for the pest management of small millets. The different management measures are detailed below (Base data from Das, 2023; Mahendra et al., 2023; Shyam Prasad et al., 2023).

16.3.1 CULTURAL METHODS

- Collect and burn stubbles and infested earheads.
- Deep ploughing one month before planting.
- Sowing with no nitrogen to reduce the incidence of shoot fly.
- Adopt synchronous as well as timely or early sowings of cultivars with similar maturity to reduce the infestation of shoot fly, midge, and grain bugs.
- Rotate small millets with cotton, groundnut, or sunflower to reduce the damage caused by shoot fly, midge, and head bugs.
- Adopt a high seed rate (1.5 times more) and delay thinning to minimize the damage caused by shoot fly.

16.3.2 MECHANICAL METHODS

- Set up light traps till midnight to monitor, attract and kill adults of midge and other moth pests.
- Set up fishmeal traps impregnated with Arpocarb insecticides @ 12/ha till the crop is 30 days old to monitor the shoot fly population.

16.3.3 BIORATIONAL METHODS

- Balanced application of fertilizers to reduce shoot fly damage.

16.3.4 CHEMICAL METHODS

- Seed treatment with Thiamethoxam 70 WS @ 3 g/1 kg seed to improve plant stand, seedling vigor, and reduce the damage by shoot fly and sucking pests.
- **Shoot Fly:** Furrow application of Carbofuran 3G @ 20 kg/ha before sowing to protect the seedlings from shoot fly.
- When shoot fly damage reaches 5%–10% of plants, spray Cypermethrin 10% EC @ 1 mL/L.
- **Stem Borers:** Whorl application of Carbofuran 3G @ 8–12 kg a.i./ha, or spray the entire field with Metasystox @ 2 mL/L.
- **Midge:** Spray Cypermethrin 25 EC @ 0.5 mL/L at the 50% flowering stage.
- Dust Carbaryl 10G @ 20–25 kg/ha at the 50% flowering and grain formation stage.
- **Cutworms, Red Hairy Caterpillars, and Semi-Loopers:** Apply poison baits comprising 10 kg rice bran + 1 kg jaggery + 1 L Quinolphos (25% EC). Prepare small balls and broadcast them in the fields in the evening.
- **Sucking Pests:** Apply Dimethoate 30% EC @ 1.5 mL/L.
- Looking at the pest scenario in the changing climate, an integrated approach to the management of these pests is very important.

16.4 DISEASES

16.4.1 BLAST

Casual organism - *Pyricularia grisea* (little millet and barnyard millet), *Pyricularia seteriae* (foxtail millet)

16.4.1.1 SYMPTOMS

The disease infects any stage of the crop, from seedling to grain formation stage. Initial symptoms appear as small circular brown-colored spots that develop into elongated spindle-shaped spots; later, several spots coalesce, and the infected leaf dries and withers. The disease infects the neck region of the panicle, resulting in a black-colored spot with partial or complete inhibition of seed set, and the panicle breaks at the neck point. Severely infected crops develop a burnt appearance with stunted growth, resulting in lodging of the crop. Infected panicles yield chaffy and shriveled grains, resulting in yield loss (Das et al., 2016; Pordel et al., 2018). Yield loss differs based on the onset of the disease, severity, variety of crop, and existing weather conditions. An average loss of 30 to 40% is reported in foxtail millet. Moderate temperatures, high humidity, and cloudy days with intermittent rains encourage the disease to spread (Jain and Ashish Kumar, 2023). The blast pathogen is very sensitive to temperature changes and other environmental factors. The fungal life cycle in nature is very complicated. It can parasitize several grasses. The primary inoculum originates from weeds or collateral hosts. The fungus spreads mostly by airborne conidia. A temperature range of 25–30°C and an RH >90% with sporadic rainfall are favorable for the rapid spread of the disease. The incidence of the disease increases with rising N levels (Kumar and Rashmi, 2012) (Figure 16.9).

FIGURE 16.9 Blast symptoms on small millets.

Source: Reprinted with permission Das et al. (2016); © ICAR-IIMR.

16.4.2 RUST

Causal organism - *Uromyces setariae italica* (Foxtailmillet), *Puccinia substriata* (Kodo millet), *Uromyces linearis* (Little millet)

Among small millets, rust is observed on foxtail, kodo, and barnyard millets. Two different pathogens cause rust on them. Rust on millets is not a common disease; it occurs sporadically, causing minor damage to the crop.

16.4.2.1 SYMPTOMS

Many times, the incidence occurs sporadically towards the end of the cropping season and causes minor damage to the crop. If rust appears before ear formation, it becomes severe and destroys the crop. The rust symptoms appear as minute reddish to dark brown broken pustules on both surfaces of the lower leaves. The rust is more severe in the upper half of the leaves compared to the lower half. The rust pustules may also occur in the leaf sheath, culms, and stems. Under severe conditions, the leaves wither off prematurely, and poor grain set may occur. As the disease progresses, the infection spreads to the younger leaves.

The disease may become severe under cool, humid weather (Jain and Ashish Kumar, 2023). The pathogen can persist on grasses. Brinjal is identified as an alternate host in which aecia are formed (de Carvalho et al., 2006) (Figure 16.10).

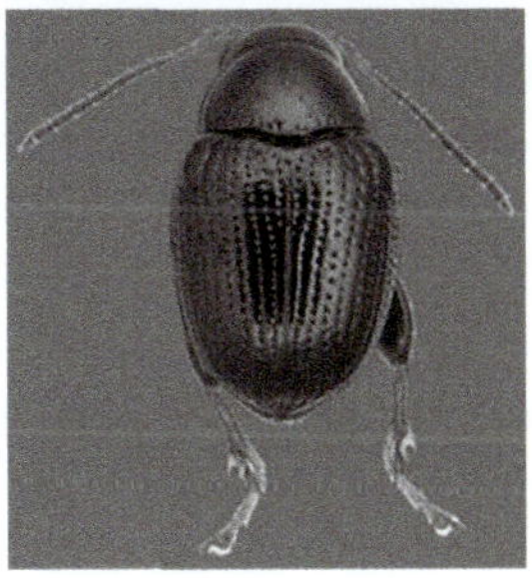

FIGURE 16.10 Rust of foxtail millet.

Source: Reprinted with permission Das et al. (2016); © ICAR-IIMR.

16.4.3 UDBATTA

Causal organism – *Euphelis* sp.

This is a common panicle disease infecting foxtail millet and kodo millet.

16.4.3.1 SYMPTOMS

The infected panicles get transformed into a compact agarbatti-like shape, hence named "Udbatta." The infected plants remain stunted with white mycelia and conidia on the flag leaves along the veins before panicle emergence. The disease is noticed on numerous host plants such as grasses: *Cynodon dactylon, Pennisetum sp.,* and *Eragrostis tenufolia.* It is an irregular disease, and incidence goes up to 19.6% (Nagaraja et al., 2010) (Figure 16.11).

FIGURE 16.11 Udbatta symptoms on small millets.

Source: Reprinted with permission Das et al. (2016); © ICAR-IIMR.

16.4.4 SMUT

Casual organism - *Ustilago crameri* (Foxtail millet), *Ustilago crusgalli* and *U. paradoxa* (Barnyard millet)

Among the smuts, grain smut is common on barnyard and foxtail millet, while head smuts are common in barnyard, kodo, and proso millet. Normally, they are minor diseases.

16.4.4.1 SYMPTOMS

Smut disease converts the entire or part of an earhead into smut sorus. Head smut is soil-borne, whereas grain smut is mainly carried on the surface of the seed. Spores get attached to the seed and germinate in the soil along with the seed. The disease infects the crop at the grain setting stage, with smutted grains randomly in the earhead. The affected grains get converted into gall-like structures, bigger compared to normal healthy grains. The infected grains later turn pink in color, then dark black, and dry off, causing direct yield loss. In barnyard millet, the head smut disease infects the crop at the grain formation stage, producing large gall-like swellings of the stems, the nodes of shoots, and the axils of the older leaves.

In kodo millet, head smut is important and causes substantial yield loss in Andhra Pradesh, Bihar, Karnataka, Madhya Pradesh, and Tamil Nadu. The disease is mostly seed-borne. The spores adhere to the grain surface and infect the seedlings. Two types of smuts are reported on barnyard millet, namely head smut and grain smut. Head smut is also reported from Uttarakhand (Jain and Ashish Kumar, 2023; Kumar et al., 2008) (Figure 16.12).

FIGURE 16.12 Smut symptoms on small millets.

Source: Reprinted with permission Das et al. (2016); © ICAR-IIMR.

16.4.5 *DOWNY MILDEW*

Causal organism - *Sclerospora graminicola*

Downy mildew incidence on small millets is not severe.

16.4.5.1 SYMPTOMS

The infection begins at the seedling stage and continues until plant maturity. The infected plant appears stunted in growth, with shortened internodes and closely formed leaves, giving it a bunchy/bushy appearance. The infected earhead bears abnormally shaped spikelets. The infected plant has an earhead covered with green, narrow, sterile, leaf-like growth, creating a crowded appearance referred to as the "green ear" symptom. A cool environment and high humidity favor the production of spores and the infection. The disease is generally observed in peninsular India, particularly in Karnataka and Tamil Nadu. The fungus is both internally and externally seed-borne. It has an extensive host range, including maize, wheat, oats, *Eragrostis pectinacea*, and *Digitaria marginata* (Jain and Ashish Kumar, 2023) (Figure 16.13).

FIGURE 16.13 Downy mildew symptoms on small millets.

Source: Reprinted with permission Das et al. (2016); © ICAR-IIMR.

16.4.6 BROWN SPOT/SEEDLING BLIGHT/LEAF BLIGHT

Causal organism - *Bipolaris panici milicei* (Proso millet), *Alternaria* sp. (Little millet), *Alternaria tenuissima* (Kodo millet)

16.4.6.1 SYMPTOMS

The disease infects crops grown in prolonged drought or nutrient deficiency conditions. The infected leaves show dark brown to black-colored spots on their lamina, with a whitish mycelial growth in the center of the spot. As the infection spreads further under favorable humid climatic conditions, the neck of the infected plants breaks at the point (Figure 16.14).

FIGURE 16.14 Leaf blight symptoms on small millets.

Source: Reprinted with permission Das et al. (2016); © ICAR-IIMR.

16.4.7 *BACTERIAL LEAF BLIGHT*

Causal organism – *Pseudomonas avenae.*

16.4.7.1 *SYMPTOMS*

The symptom appears as brown colored stripes that spread upwards along the leaf margin of the lamina covering the entire leaf sheath. Table 16.2 for a consolidated list of diseases affecting small millet crops.

TABLE 16.2 Diseases of Small Millets

Crop	Disease	Pathogen
Foxtail millet	Blast	*Pyricularia setariae*
	Rust	*Uromyces setariae italicae*
	Smut	*Ustilago crameri*
	Downy mildew	*Sclerospora graminicola*
	Udbatta	*Ephelis* sp.
	Bacterial leaf blight	*Pseudomonas avenae*
Kodo millet	Head smut	*Soropsorium paspali*
	Rust	*Puccinia substriata*
	Udbatta	*Ephelis* sp.
Barnyard millet	Head smut	*Ustilago crusgalli*
	Kernel smut	*Ustilago paradoxa*
	Bacterial leaf blight	*Pseudomonas avenae*
Proso millet	Head smut	*Sphacelotheca destructans*
	Bacterial leaf blight	*Pseudomonas avenae*
Little millet	Rust	*Uromyces linearis*

Source: Adapted from Das (2017); Kumar and Singh (2010).

16.5 INTEGRATED DISEASE MANAGEMENT

16.5.1 *CULTURAL METHODS*

- Clean cultivation by the destruction of crop residues will prevent the spread of soil-borne diseases by eradicating the inoculum.
- Use disease-free seeds.
- Use resistant cultivars.
- Rogue out infected plants.
- Removal of alternate hosts like weeds and wild crop species.
- Early sowing.
- Hot water seed treatment.
- Regulating nitrogen fertilizers.

- Crop rotation with pulses and oilseeds can minimize the primary inoculum of the downy mildew pathogen.
- Mechanical destruction of infected ear heads, particularly the head smuts, will prevent the spread of fungal spores to infect the healthy ear heads.
- Shallow sowing and application of nitrogenous and phosphatic fertilizers can minimize the smut incidence.

16.5.2 *BIOLOGICAL METHODS*

- Seed treatment with antifungal antagonists like *Trichoderma viride* helps in the management of soil-borne diseases like foot rot, charcoal rot, etc.
- Seed treatment with plant growth-promoting rhizobacteria, *Pseudomonas fluorescens*, helps in imparting resistance against the infection by a few fungal pathogens.
- **Blast** - Three foliar sprays of *Pseudomonas fluorescens* @ 2 g/l, first spray at the immediate notice of the symptoms, second and third sprays at the flowering stage at 15-day intervals.

16.5.3 *CHEMICAL METHODS*

16.5.3.1 BLAST

- Seed treatment with Tricyclazole @ 8 g/kg seed followed by one foliar spray of *Prosophis juliflora* (10%) plant extract.
- Two sprays of Carbendazim (0.05%) or first spray of Chlorothalonil (0.2%) and second spray of Mancozeb (0.2%).
- Seed treatment with *Trichoderma harzianum* and one spray of *Pseudomonas fluorescens* @ 0.3% at the time of flowering followed by a second spray of *P. fluorescens* 10 days later.
- Spray Carbendazim 50 WP @ 500 g or Iprobenphos (IBP) @ 500 ml/ha. First spray at the immediate notice of the symptoms, second and third sprays at the flowering stage at 15-day intervals to control neck and finger infection.

16.5.3.2 RUST

Foliar spray of Chlorothalonil (0.1%) or Mancozeb (0.2%) or Hexaconazole (0.1%) or Difenoconazole (0.1%) or Propiconazole (0.1%) just after the initiation of symptoms. Two sprays at 15-day intervals after the appearance of symptoms are recommended.

16.5.3.3 SMUT

- Seed treatment with *Trichoderma viride*, *T. harzianum* @ 4–6 g seed/kg or seed treatment with Carboxin, Carbendazim, a mixture of Carboxin + Thiram, Carbendazim + Mancozeb @ 2 g seed/kg.

- Seed treatment and foliar spray of Triozol (Triadimenol) @ 0.2% at 5–7-day intervals.
- Seed treatment and foliar spray of Carbendazim or Tebuconazole @ (1%) is recommended for grain smut of barnyard millet.
- Seed treatment with *T. viride* + 1 foliar spray of Carbendazim is very effective against grain smut of barnyard millet.
- Seed treatment with Carbendazim, Carboxin, Raxil, Chlorothalonil @ 2 g per kg seed is effective against kodo millet head smut.

16.5.3.4 DOWNY MILDEW AND BROWN SPOT

- Seed treatment with Metalaxyl @ 2 g a.i./kg seed.
- Seed treatment with a single foliar spray of Metalaxyl or Mancozeb (0.2%).
- Seed dressing with Apron Star 42 WS.
- Spray Mancozeb @ 2 g/l.

(Base data from Das, 2017; Das et al., 2023; Jain and Ashish Kumar, 2023; Kumar and Singh, 2010).

16.6 CONCLUSION

The available information suggests small millets are relatively less subject to pests and diseases compared with other cereal crops. The information on the pests of small millets concerning data on economic injury levels, yield loss, effectiveness of beneficial insects, and their use in existing ecosystems is scanty. These crops' pest and disease monitoring systems are in their initial stage and permit particular emphasis under the changing climate scenario.

KEYWORDS

- **antifungal antagonists**
- **armyworm**
- **caterpillar**
- **diseases of minor millet**
- **kodo millet**
- **leafhopper**
- **pests in minor millet**
- **plant growth-promoting rhizobacteria**
- **proso millet**
- **udbatta disease**

REFERENCES

Das, I. K. (2017). Millet diseases: Current status and their management. In *Millets and sorghum* (pp. 291–322). John Wiley & Sons, Ltd: Chichester, UK.

Das, I. K., Nagaraja, A., & Tonapi, V. (2016). *Diseases of Millets–A Ready Reckoner* (p. 71).

Das, I. K., Palanna, K. B., & Rajesh, G. (2023). Emerging pests of millets and their management. *Indian Farming, 73*(01), 67–70.

Das, S. B. (2023). *Insect Pests of Millets & Their Management* (pp. 19–30).

De Carvalho, A., Soares, D., & Carmo, M. (2006). Description of the lifecycle of the pearl millet rust fungus–*Puccinia substriata* var. *penicillariae* with a proposal of reducing var. *indica* to a synonym. *Micropathological, 161*, 331–336.

Jain, A. K., & Kumar, A. (2023). Salvaging millets productivity through disease management options (pp. 31–38).

Kalaisekar, A., & Padmaja, P. G. (2017). Insect pests of millets and their host plant relations. In *Millets and Sorghum* (pp. 267–290). John Wiley & Sons, Ltd: Chichester, UK.

Kumar, B., & Rashmi, Y. (2012). Influence of nitrogen fertilizer dose on blast disease of finger millet caused by *Pyricularia grisea. Indian Phytopathology, 65*(1), 52–55.

Kumar, B., & Singh, K. P. (2010). Important small millet diseases in India and their management. In *Microbial Diversity and Plant Disease Management*. Muller GmbH and Co. KG: Germany/USA/U.K.

Kumar, B., Kumar, J., & Srinivas, P. (2008). First record of head smut of barnyard millet from mid hills of Uttarakhand. *Journal of Mycology and Plant Pathology, 38*(1).

Mahendra, K. R., Archana, A., & Swagata, T. (2023). A brief overview of nutri-millet's insect pests and their management. *Indian Farming, 73*(03), 17–20.

Nagaraja, A., Anjaneya Reddy, B., & Govindappa, M. R. (2010). Occurrence of udabatta disease on kodo millet. In *A New Report from South India; Paspalum scrobiculatum L., 48*, 163–164.

Nageshchandra, B. K., & Ali, M. (1983). Losses due to shoot fly on minor millets. *MILWAI Newsletter, 2*.

Natarajan, U. S., Selvaraj, S., & Regupathy, A. (1974). Anthomyiidae: Diptera) in certain varieties of panivaragu, *Panicum miliaceum. Atherigona Destructor* M. *Current Science, 40*(12), 502–504.

Nigam, C., & Damte, T. (2018). Identification of the tef shoot fly species from tef, *Eragrostis tef* (Zucc.), Trotter growing areas of Ethiopia. *African Journal of Insects, 5*, 181–184.

Pordel, A., Tharreau, D., Cros-Arteil, S., Shams, E., Moumeni, A., Mirzadi Gohari, A., & Javan-Nikkhah, M. (2018). *Pyricularia oryzae* causing blast on foxtail millet in Iran. *Plant Disease, 102*(9), 1853. doi:10.1094/PDIS-01-18-0091-PDN (accessed on 7th November 2025).

Prasad, S., Bhagwat, G., & Srinivasa Rao, V. R. (2015). Insect pests of millets. *Indian Farming, 65*(4), 46–49.

Prasad, S., Stanley, G., Babu, J., Kalisekar, K. S., Padmaja, A., & Subbarayudu, P. G. (2023). Emerging pests of millets and their management. *Indian Farming, 73*(01), 64–66.

Rao, R. (1923). The genitalia of certain anthomyiid flies (*Atherigona* spp.). In *Report of the Proceedings of the 5th Entomology Meeting*, Pusa, Bihar, India (pp. 330–335).

Selvaraj, S., Natarajan, U. S., & Ragupathy, A. (1974). Note on the occurrence of shootfly, and its damage on some varieties of little millet. *Indian Journal of Agricultural Sciences, 44*(8), 556–557.

Index

C

D

E

H

M

N

Q

R

T

Z

For Product Safety Concerns and Information please contact our EU representative GPSR@taylorandfrancis.com Taylor & Francis Verlag GmbH, Kaufingerstraße 24, 80331 München, Germany